BIOFUELS AND BIOREFINING

BIOFUELS AND BIOREFINING

Volume 2: Intensification Processes and Biorefineries

Edited by

CLAUDIA GUTIÉRREZ-ANTONIO

FERNANDO ISRAEL GÓMEZ CASTRO

ELSEVIER

Elsevier
Radarweg 29, PO Box 211, 1000 AE Amsterdam, Netherlands
The Boulevard, Langford Lane, Kidlington, Oxford OX5 1GB, United Kingdom
50 Hampshire Street, 5th Floor, Cambridge, MA 02139, United States

Copyright © 2022 Elsevier Inc. All rights reserved.

No part of this publication may be reproduced or transmitted in any form or by any means, electronic or mechanical, including photocopying, recording, or any information storage and retrieval system, without permission in writing from the publisher. Details on how to seek permission, further information about the Publisher's permissions policies and our arrangements with organizations such as the Copyright Clearance Center and the Copyright Licensing Agency, can be found at our website: www.elsevier.com/permissions.

This book and the individual contributions contained in it are protected under copyright by the Publisher (other than as may be noted herein).

Notices
Knowledge and best practice in this field are constantly changing. As new research and experience broaden our understanding, changes in research methods, professional practices, or medical treatment may become necessary.

Practitioners and researchers must always rely on their own experience and knowledge in evaluating and using any information, methods, compounds, or experiments described herein. In using such information or methods they should be mindful of their own safety and the safety of others, including parties for whom they have a professional responsibility.

To the fullest extent of the law, neither the Publisher nor the authors, contributors, or editors, assume any liability for any injury and/or damage to persons or property as a matter of products liability, negligence or otherwise, or from any use or operation of any methods, products, instructions, or ideas contained in the material herein.

ISBN: 978-0-12-824117-2

For information on all Elsevier publications
visit our website at https://www.elsevier.com/books-and-journals

Publisher: Charlotte Cockle
Acquisitions Editor: Peter Adamson
Editorial Project Manager: Jai Marie Jose
Production Project Manager: Sojan P. Pazhayattil
Cover Designer: Greg Harris

Typeset by STRAIVE, India

Contents

Contributors   ix

1. Process intensification in biofuels production   1
Salvador Hernández

    1.1 Introduction   1
    1.2 Basic concepts on process intensification   5
    1.3 Conventional processes for the production of biofuels   13
    1.4 Intensified processes for the production of biofuels   22
    1.5 Conclusions   30
    References   30

2. Sustainable bioalcohol production: Pretreatment, separation, and control strategies leading to sustainable processes   41
Thiago Edwiges, Maria Cinta Roda-Serrat, Juan Gabriel Segovia-Hernández, Eduardo Sánchez-Ramírez, Stefania Tronci, and Massimiliano Errico

    2.1 Introduction   41
    2.2 Lignocellulosic biomass   43
    2.3 Pretreatment technologies for lignocellulosic biomass   44
    2.4 Distillation-based bioalcohols separation processes   55
    2.5 Membrane-based processes for separation of bioalcohols   61
    2.6 Process control   68
    2.7 Bioethanol separation by membrane-assisted reactive distillation: A case study   72
    2.8 Conclusions   78
    Acknowledgments   79
    References   79

3. A review of intensification technologies for biodiesel production   87
Lai Fatt Chuah, Jiří Jaromír Klemeš, Awais Bokhari, Saira Asif, Yoke Wang Cheng, Chi Cheng Chong, and Pau Loke Show

    3.1 Introduction   87
    3.2 Intensification technologies for biodiesel production   88
    3.3 Conclusion   111
    Acknowledgments   112
    References   112

4. Intensified technologies for the production of renewable aviation fuel — 117
Araceli Guadalupe Romero-Izquierdo and Salvador Hernández

4.1 Introduction — 117
4.2 Processes intensification strategies in the production of biojet fuel — 121
4.3 Intensified processes to produce biojet fuel — 130
4.4 Implementation to industrial scale — 132
4.5 Case of study — 134
4.6 Conclusions — 149
Acknowledgments — 149
References — 149

5. Opportunities in the intensification of the production of biofuels for the generation of electrical and thermal energy — 157
Noemí Hernández-Neri, Julio Armando de Lira-Flores, Araceli Guadalupe Romero-Izquierdo, Juan Fernando García-Trejo, and Claudia Gutiérrez-Antonio

5.1 Introduction — 157
5.2 Conventional production processes — 160
5.3 Proposals for the intensification of the production processes — 170
5.4 Intensification of the production of fuel pellets — 174
5.5 Future trends — 190
5.6 Concluding remarks — 190
Acknowledgments — 191
References — 191

6. Intensified and hybrid distillation technologies for production of high value-added products from lignocellulosic biomass — 197
Le Cao Nhien, Junaid Haider, Nguyen Van Duc Long, and Moonyong Lee

6.1 Introduction — 197
6.2 Application of intensified and hybrid distillation technologies for high value-added processes from biomass — 207
6.3 Conclusions — 224
References — 224

7. Intensification of biodiesel production through computational fluid dynamics — 231
Harrson S. Santana, Marcos R.P. de Sousa, and João L. Silva Júnior

7.1 Introduction — 231
7.2 Brief introduction to CFD concepts — 235

7.3 Literature review	247
7.4 Case study	254
7.5 Conclusions and future perspectives	267
References	268

8. Control properties of intensified distillation processes: Biobutanol purification — 273

Ernesto Flores-Cordero, Eduardo Sánchez-Ramírez, Gabriel Contreras-Zarazúa, César Ramírez-Márquez, and Juan Gabriel Segovia-Hernández

8.1 Introduction	273
8.2 Separation process for biobutanol	276
8.3 Closed-loop analysis	278
8.4 Conclusions	290
References	291

9. Assessment of modular biorefineries with economic, environmental, and safety considerations — 293

Alexandra Barron, Natasha Chrisandina, Antioco López-Molina, Debalina Sengupta, Claire Shi, and Mahmoud M. El-Halwagi

9.1 Introduction to modular biorefineries	293
9.2 Approach	295
9.3 Case study	297
9.4 Conclusions	301
References	302

10. Production of biofuels and biobased chemicals in biorefineries and potential use of intensified technologies — 305

Alvaro Orjuela and Andrea del Pilar Orjuela

10.1 Introduction	305
10.2 Evolution and implementation of industrial biorefineries	309
10.3 Biorefineries and biobased chemicals market	316
10.4 Key biobased fuels and chemicals for current industrial biorefineries	319
10.5 Potential for process intensification at the biorefineries	322
10.6 Implemented intensification technologies for bioderived chemicals	328
10.7 Process intensification in action—Succinic acid production	337
10.8 Some challenges and future directions	341
10.9 Concluding remarks	349
Acknowledgments	350
References	350

11. Modeling and optimization of supply chains: Applications to conventional and intensified biorefineries — 361
Fernando Israel Gómez-Castro, Yulissa Mercedes Espinoza-Vázquez, and José María Ponce-Ortega

11.1 Introduction	362
11.2 Objective functions for supply chain optimization	364
11.3 Elements of the supply chain for a biorefinery	366
11.4 Process intensification and supply chain	367
11.5 Modeling and optimization of the supply chain	368
11.6 Case study: Optimization of the supply chain for the production of bioethanol and high value-added products in Mexico	373
11.7 Conclusion	386
References	386

12. Life cycle approach for the sustainability assessment of intensified biorefineries — 389
M. Collotta, P. Champagne, G. Tomasoni, and W. Mabee

12.1 Introduction	389
12.2 LCA methodology	390
12.3 LCA tools	391
12.4 State of the art of LCA in the biorefineries sector	394
12.5 Social LCA methodology	396
12.6 LCA at the early stage of development—Case study	398
12.7 Conclusions	401
References	401

13. Social impact assessment in designing supply chains for biorefineries — 405
Sergio Iván Martínez-Guido, Juan Fernando García-Trejo, and José María Ponce-Ortega

13.1 Introduction	405
13.2 Current perspective of biorefinery systems: Conventional fuel replacement by bioenergy sources	406
13.3 Biorefineries evaluation using optimization tools	408
13.4 Supply chain optimization	409
13.5 Problem statement	410
13.6 Mathematical model	410
13.7 Mathematical modeling solution	415
13.8 Case study	416
13.9 Conclusions	422
Acknowledgments	422
References	424

Index — 427

Contributors

Saira Asif
Sustainable Process Integration Laboratory, SPIL, NETME Centre, Faculty of Mechanical Engineering, Brno University of Technology, VUT Brno, Brno, Czech Republic; Faculty of Sciences, Department of Botany, PMAS Arid Agriculture University, Rawalpindi, Punjab, Pakistan

Alexandra Barron
Department of Chemical Engineering, Texas A&M University; Gas and Fuels Research Center, Texas A&M Engineering Experiment Station, College Station, TX, United States

Awais Bokhari
Sustainable Process Integration Laboratory, SPIL, NETME Centre, Faculty of Mechanical Engineering, Brno University of Technology, VUT Brno, Brno, Czech Republic; Chemical Engineering Department, COMSATS University Islamabad (CUI), Punjab, Lahore, Pakistan

P. Champagne
Institut national de la recherche scientifique, Québec City, QC, Canada

Yoke Wang Cheng
Department of Chemical Engineering, School of Engineering and Computing, Manipal International University, Negeri Sembilan, Malaysia

Chi Cheng Chong
Department of Chemical Engineering, School of Engineering and Computing, Manipal International University, Negeri Sembilan, Malaysia

Natasha Chrisandina
Department of Chemical Engineering, Texas A&M University; Gas and Fuels Research Center, Texas A&M Engineering Experiment Station, College Station, TX, United States

Lai Fatt Chuah
Faculty of Maritime Studies, Universiti Malaysia Terengganu, Kuala Terengganu, Terengganu, Malaysia

M. Collotta
DIMI, Department of Mechanical and Industrial Engineering, University of Brescia, Brescia, Italy

Gabriel Contreras-Zarazúa
Chemical Engineering Department, University of Guanajuato, Guanajuato, Mexico

Julio Armando de Lira-Flores
Facultad de Química, Universidad Autónoma de Querétaro, Centro Universitario, Querétaro, Mexico

Marcos R.P. de Sousa
University of Campinas, School of Chemical Engineering, Campinas, SP, Brazil

Thiago Edwiges
Department of Biological and Environmental Sciences, Federal University of Technology, Medianeira, Parana, Brazil

Mahmoud M. El-Halwagi
Department of Chemical Engineering, Texas A&M University; Gas and Fuels Research Center, Texas A&M Engineering Experiment Station, College Station, TX, United States

Massimiliano Errico
Faculty of Engineering, Department of Green Technology, University of Southern Denmark, Odense, Denmark

Yulissa Mercedes Espinoza-Vázquez
Departamento de Ingeniería Química, División de Ciencias Naturales y Exactas, Universidad de Guanajuato, Guanajuato, Guanajuato, Mexico

Ernesto Flores-Cordero
Biotechnology Engineering Department, University of Guanajuato, Campus Celaya-Salvatierra, Guanajuato, Gto., Mexico

Juan Fernando García-Trejo
Facultad de Ingeniería, Universidad Autónoma de Querétaro, Amazcala, Querétaro, Mexico

Fernando Israel Gómez-Castro
Departamento de Ingeniería Química, División de Ciencias Naturales y Exactas, Universidad de Guanajuato, Guanajuato, Guanajuato, Mexico

Claudia Gutiérrez-Antonio
Facultad de Ingeniería, Universidad Autónoma de Querétaro, Amazcala, Querétaro, Mexico

Junaid Haider
Sustainable Process Analysis, Design, and Engineering Laboratory, Energy and Chemical Engineering Department, Ulsan National Institute of Science and Technology (UNIST), Ulsan, South Korea

Salvador Hernández
Departamento de Ingeniería Química, División de Ciencias Naturales y Exactas, Universidad de Guanajuato, Guanajuato, Mexico

Noemí Hernández-Neri
Facultad de Ingeniería, Universidad Autónoma de Querétaro, Amazcala, Querétaro, Mexico

Jiří Jaromír Klemeš
Sustainable Process Integration Laboratory, SPIL, NETME Centre, Faculty of Mechanical Engineering, Brno University of Technology, VUT Brno, Brno, Czech Republic

Moonyong Lee
School of Chemical Engineering, Yeungnam University, Gyeongsan, South Korea

Nguyen Van Duc Long
School of Engineering, University of Warwick, Coventry, United Kingdom; School of Chemical Engineering and Advanced Materials, University of Adelaide, Adelaide, SA, Australia

Antioco López-Molina
Universidad Juárez Autónoma de Tabasco, Jalpa de Méndez, Mexico

W. Mabee
Queen's University, Department of Geography and Planning, Mackintosh-Corry Hall, Kingston, ON, Canada

Sergio Iván Martínez-Guido
Facultad de Ingeniería, Universidad Autónoma de Querétaro, Amazcala, Querétaro, Mexico

Le Cao Nhien
School of Chemical Engineering, Yeungnam University, Gyeongsan, South Korea

Alvaro Orjuela
Department of Chemical and Environmental Engineering, Universidad Nacional de Colombia, Bogotá D.C., Colombia

Andrea del Pilar Orjuela
Process Solutions and Equipment SAS, Engineering Division, Bogotá D.C., Colombia

José María Ponce-Ortega
Facultad de Ingeniería Química, División de Estudios de Posgrado, Universidad Michoacana de San Nicolás de Hidalgo, Morelia, Michoacán, Mexico

César Ramírez-Márquez
Chemical Engineering Department, University of Guanajuato, Guanajuato, Mexico

Maria Cinta Roda-Serrat
Faculty of Engineering, Department of Green Technology, University of Southern Denmark, Odense, Denmark

Araceli Guadalupe Romero-Izquierdo
Facultad de Ingeniería, Universidad Autónoma de Querétaro, Amazcala, Querétaro, Mexico

Eduardo Sánchez-Ramírez
Chemical Engineering Department, University of Guanajuato, Guanajuato, Mexico

Harrson S. Santana
University of Campinas, School of Chemical Engineering, Campinas, SP, Brazil

Juan Gabriel Segovia-Hernández
Chemical Engineering Department, University of Guanajuato, Guanajuato, Mexico

Debalina Sengupta
Gas and Fuels Research Center, Texas A&M Engineering Experiment Station, College Station, TX, United States

Claire Shi
Department of Chemistry, Rice University, Houston, TX, United States

Pau Loke Show
Department of Chemical and Environmental Engineering, University of Nottingham—Malaysia Campus, Semenyih, Malaysia

João L. Silva Júnior
Federal University of ABC, CECS—Center for Engineering, Modeling and Applied Social Sciences, Alameda da Universidade, São Bernardo do Campo, SP, Brazil

G. Tomasoni
DIMI, Department of Mechanical and Industrial Engineering, University of Brescia, Brescia, Italy

Stefania Tronci
Dipartimento di Ingegneria Meccanica, Chimica e dei Materiali, Universitá degli Studi di Cagliari, Cagliari, Italy

CHAPTER 1

Process intensification in biofuels production

Salvador Hernández
Departamento de Ingeniería Química, División de Ciencias Naturales y Exactas, Universidad de Guanajuato, Guanajuato, Mexico

1.1 Introduction

Through history, the society has evolved due to the research and technological advances in all the knowledge areas. These advances have allowed better quality life of the society, through the medical advances, transportation means, home comforts, education, as well as recreational activities. Even in the last decades, the globalization has made possible the contact between people located in different parts of the world, which can interchange experiences, culture, goods, news, and even real-time events with just a click on a computer with internet access. All these improvements and benefits to the society has one common factor: energy.

In 2019, the worldwide energy consumption was 14,406 Mtoe (IEA, 2020a); this amount of energy proceeds from oil (31%), coal (26%), natural gas (23%), renewables (14%), and nuclear (6%) (IEA, 2020a, 2020b). The forecasts indicated that this amount of energy would have increased in 10% for 2028 (IEA, 2020c). However, the forecasts changed due to the appearance and spread of the Sars-CoV-2, who has modified the known world's dynamics. Potential new practices and social forms being facilitated by the pandemics are having impacts on energy demand and consumption, which has, in general, declined (Jiang, Van Fan, & Klemeš, 2021). In spite of two vaccines have been successfully developed at an unprecedented speed (Huang, Zeng, & Yan, 2021), its large-scale production is now the bottleneck. Nowadays, not all the world population has received the vaccine; therefore there are still many economic sectors that are detained or with low activity, such as the aviation sector which recovery process seems much slower than anticipated (Dube, Nhamo, & Chikodzi, 2021). It is important to mention that the pandemic situation, as well as its effects, is a constantly changing situation.

In this context, the International Energy Agency has proposed, in collaboration with the International Monetary Fund, a Sustainable Recovery Plan (IEA, 2020d). This plan is focused on boosting the economic growth, creating jobs, and building more resilient and cleaner energy systems; it is important to mention that this plan is intended to be implemented in the period 2021–23. The plan includes policies, investments, and measures to accelerate the deployment of six key areas (IEA, 2020d), which are shown in Fig. 1.1.

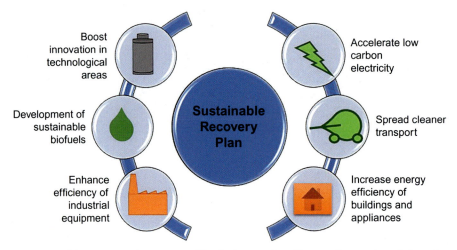

Fig. 1.1 Sustainable recovery plan proposed by the International Energy Agency in collaboration with the International Monetary Fund.

Fig. 1.2 Classification of biomass based on its chemical nature.

From Fig. 1.1, it can be seen that the development of biofuels plays a key role in this Sustainable Recovery Plan.

Biofuels are defined as those fuels that are generated from the conversion of biomass, which is a complex natural renewable material with enormous chemical variability (Bonechi et al., 2017). The biomass is defined as all the organic matter originating from living plants, organisms, as well as some types of residues from agricultural, agroindustrial, food, domestic, and other sectors (Pang, 2016; Soria-Ornelas, Gutiérrez-Antonio, & Rodríguez, 2016). Biomass is considered a suitable source for renewable energy and biobased products due to its organic nature, carbon stability, and abundant supply (Gent, Twedt, Gerometta, & Almberg, 2017). The biomass can be classified according to several criteria, such as arable, edible, residual, among others. An interesting classification considering the chemical nature of the biomass is as follows: triglyceride, lignocellulosic, sugar, and starch (Maity, 2015) (Fig. 1.2).

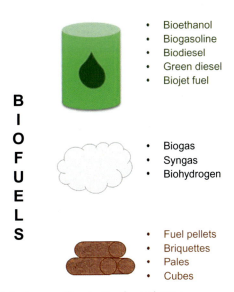

Fig. 1.3 Classification of biofuels according to its physical state.

The triglyceride biomass contains fatty acids (palmitic acid, linoleic acid, ricinoleic acid) and triglycerides (palmitin, linolein, ricinolein); this type of biomass includes oils from soybean, castor bean or microalgae, as well as fats from poultry, fish, beef. On the other hand, the lignocellulosic biomass contains lignin, cellulose, and hemicellulose as main components; this type of biomass includes all the agricultural residues, leaves, wood, grass, as well as energetic crops. Finally, as the name suggest, the sugar and starch biomass contain pentose, hexose, glucose, amylose, and amylopectin; this type of biomass includes sugarcane, potatoes, apples, as well as other edible crops. This classification allows to group all the biofuel conversion processes based on the chemical nature of the biomass, in spite of it is edible, nonedible, or residual.

The biomass can be converted into biofuels in liquid, gaseous, or solid state (Fig. 1.3). Among liquid biofuels, those destined mainly for transport sector are found, such as biogasoline, green diesel, and biojet fuel; however, they also can be used to generate electricity or heat. Regard the solid biofuels, the most popular are the fuel pellets, which can be used to produce electrical or thermal energy, similar to the gaseous biofuels, among which biogas is the most popular.

The biofuels are obtained from the conversion of biomass through chemical, biochemical, thermochemical, and biological processes (Fig. 1.4). In the chemical processes, the biomass (or a fraction of it) is converted through a set of chemical reactions, which would require additional reactants, solvents, catalysts, and moderate to high temperature and pressure; examples of this type of process are transesterification, hydrodeoxygenation, hydrocracking, and oligomerization, among others. On the other hand, in the

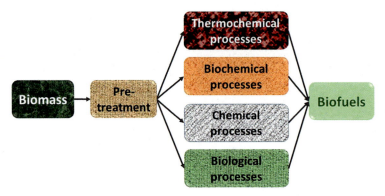

Fig. 1.4 Conversion processes for the production of biofuels.

biochemical processes, the large molecules that constitute the biomass (or a fraction of it) are converted in smaller ones through the action of microorganisms; this type of process usually requires water and low temperature and pressure. The fermentation, hydrolysis, as well as digestion, are biochemical processes.

On the other hand, in the thermochemical processes, the main objective is converting the biomass (or a fraction of it) into gases, liquids, or even solid compounds, releasing the energy contained in them as heat. The heat can be used directly or employed to produce electricity; at the same time, the compounds generated can be transformed in other valuable products. In thermochemical process is required air or inert atmospheres as well as high temperature and pressure. The pyrolysis, gasification, and combustion are examples of this type of processes. Finally, the biomass (or a fraction of it) can also be converted through biological processes; in this case, the biomass is used as feed for the organisms, which generated new compounds as results of the digestion or new organisms which can be further processed to generate other valuable products. In this category, we can mention the culture of black soldier fly as well as worms.

It is important to mention that all biofuels are renewables, since they are generated from biomass; however, they do not necessarily are sustainable, since this depends on the kind of biomass and processing route. Each one of these processes have different energetic efficiencies, obtained products, yields as well as operation and investment costs. To have biofuels that are renewable and sustainable, it must be ensured that the whole supply chain has reduced carbon footprint, with special emphasis on the conversion processes.

In the last years, researchers have focused their efforts on the development of production processes for the production of biofuels. In the literature, there are studies for the production of bioethanol (Ayodele, Alsaffar, & Mustapa, 2020; Greetham, Zaky, Makanjuola, & Du, 2018; Mohd Azhar et al., 2017; Sharma, Larroche, & Dussap, 2020), biobutanol (Huzir et al., 2018; Ibrahim, Kim, & Abd-Aziz, 2018; Wang et al., 2017; Yeong et al., 2018), biogasoline (Hassan, Sani, Abdul Aziz, Sulaiman, & Daud,

2015; Mascal & Dutta, 2020; Shamsul, Kamarudin, & Rahman, 2017), green diesel (Ameen, Azizan, Yusup, Ramli, & Yasir, 2017; Amin, 2019; Arun, Sharma, & Dalai, 2015; Kordulis, Bourikas, Gousi, Kordouli, & Lycourghiotis, 2016), biojet fuel (Galadima & Muraza, 2015; Gutiérrez-Antonio, Gómez-Castro, de Lira-Flores, & Hernández, 2017; Kandaramath Hari, Yaakob, & Binitha, 2015; Vásquez, Silva, & Castillo, 2017), biogas (Pramanik, Suja, Zain, & Pramanik, 2019; Alavi-Borazjani, Capela, & Tarelho, 2020; Kovačić et al., 2021; Liu, Ren, Yang, Liu, & Sun, 2021; Liu, Wei, & Leng, 2021), syngas (Aziz, Setiabudi, Teh, Annuar, & Jalil, 2019; Leonzio, 2018; Ren, Cao, Zhao, Yang, & Wei, 2019; Yeo, Ashok, & Kawi, 2019), biohydrogen (Fagbohungbe, Komolafe, & Okere, 2019; Chen, Wei, & Ni, 2021; Dahiya, Chatterjee, Sarkar, & Mohan, 2021; Fajín & Cordeiro, 2021), fuel pellets (Bajwa, Peterson, Sharma, Shojaeiarani, & Bajwa, 2018; He et al., 2018; Mamvura & Danha, 2020; Pradhan, Mahajani, & Arora, 2018), and briquettes (Bajwa et al., 2018; Kaliyan & Vance Morey, 2009; Zhang, Sun, & Xu, 2018). In these studies, several biomasses are analyzed as well as different conversion pathways, with the main objective of obtaining feasible processes with high yields. Nevertheless, biofuels must also be competitive from the economic point of view with its fossil counterparts; this implies that the production costs of biofuels must be as small as possible. In this context, process intensification plays a key role, since it could help to have compact process, with reduced energy consumption, safer, and environmentally friendly.

Therefore, in this chapter the potential advantages on using process intensification in the biofuel production processes will be explored. For this, it will be presented the generalities of the conventional production processes for biofuels, as well as the concept of process intensification. Based on these concepts, the necessity of applying process intensification in the production processes for biofuels will be exposed. Finally, the current state of the intensified biofuel production processes will be described.

1.2 Basic concepts on process intensification

The production processes can be defined as a succession of unit operations, where the raw materials are adequated, transformed, and purified to obtain the product of interest; this type of production processes are usually known as conventional ones. Most of the conventional equipment has as main troublesome the presence of dead zones, shortcuts in the processing, or limited heat and mass transfers; as consequence, they are usually oversized, which directly impacts its investment and operation costs. In this context, process intensification arises in order to overcome these limitations.

Process intensification is defined as any chemical engineering development that leads to a substantially smaller, cleaner, and more energy-efficient technology (Stankiewicz & Moulijn, 2000). The intensification of a process considers two main strategies: the use of highly efficient equipment or the combination of two, or more, unit operations. In the first case, new equipment has been proposed, whose main characteristic is the high rate of heat and/or mass transfer; as consequence, its size is small in comparison with its conventional counterpart. In the second case, the thermodynamic synergy is used in order to carry out two, or more, unit operations in the same vessel, which are usually called as hybrid equipment; as consequence, the investment and operation costs are reduced.

The application of process intensification strategy has many advantages. The first one is that this type of process are inherently safer; since the equipment are smaller or hybrid, the amount of reactants, solvents, as well as energy are lower. Thus in case of an accident, the potential consequences can be managed more easily. The second one is that the plants are smaller, respect to the conventional ones, since higher rates of heat and/or mass transfer are observed; in consequence, minor areas for the construction of the plant are required, and in some cases, less pipes and additional equipment (like pumps). The third one is that the intensified processes are more competitive from the economic point of view, due to the efficient use of energy and raw materials. Finally, the intensified process has a reduced carbon footprint without lose productivity. On the other hand, process intensification has two main disadvantages. The main disadvantage is that not all process intensification alternatives have a better performance in all type of process, in comparison with the conventional one; thus the alternatives must be evaluated. The second disadvantage is that process intensification implies the replacement of the equipment, for a smaller one or for a hybrid technology.

Until now, several advances have been reported in the literature in relation to the proposal of intensified equipment for reaction, separation, and conditioning tasks. These alternatives will be presented next.

1.2.1 Reaction equipment

A reactor is a vessel where a chemical reaction is carried out. The reactors can be classified based on the number of phases interacting on it. The homogeneous reactors are those where the reactants, products, and catalysts are in the same phase; on the other hand, the heterogeneous reactors are those where the reactants, products, and catalysts are in at least two different phases. The reactors can also be classified based on its form: tank with agitation and tubular (Fig. 1.5).

In spite of its form or the number of phases present, these types of conventional reactors present dead zones, shortcuts, and it is necessary to introduce the reactants in excess in order to obtain high conversion rates. In order to overcome these limitations, new reactor equipment has been proposed, among which can mention multifunctional reactors, microreactors, microchannel reactors, spinning disk reactor, heat exchanger reactor, oscillatory flow reactor (Fig. 1.6).

The multifunctional reactors allow to carry on several reactions in the same vessel, which favors the generation of some products of interest. In this type of reactor, the

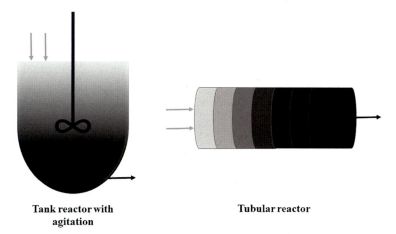

Fig. 1.5 Conventional reactors.

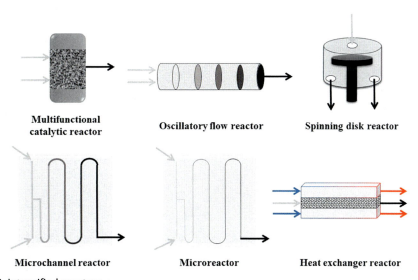

Fig. 1.6 Intensified reactors.

intensification allows performing several reactions in the same vessel; however, the type of reactor is conventional. For instance, the hydroprocessing of one step for the conversion of oil into hydrocarbons is an example of this type of multifunctional reactor (Gutiérrez-Antonio et al., 2017).

On the other hand, the other equipment showed in Fig. 1.6 are new designs, where the heat and mass transfer are intensified in such a way that the size of the equipment is small.

Microreactors are devices consisting of single or multiple small-diameter channels, typically between 10 and 1000 μm (Moulijn & Stankiewicz, 2017). Some of the reactions that have been study in microreactors include hydrogen production and ethylene partial oxidation (Keiski, Ojala, Huuhtanen, Kolli, & Leiviskä, 2011). The main difference between microreactors and microchannel reactors is the size of the channels. Microchannel reactors can be defined as reactors consisting of channels with a hydraulic diameter (Dh) more than 3 mm, while mini-channels can be considered as reactors with channels with a hydraulic diameter in the range of 200 μm–3 mm and microchannels with a hydraulic diameter in the range of 10–200 μm (Kiani, Makarem, Farsi, & Rahimpour, 2020). Microchannel reactors have been used for the conversion of syngas to alcohols (Reay, Ramshaw, & Harvey, 2013a). The oscillatory flow reactors superimpose an oscillatory flow to the net movement through a flow reactor, with the aim to effectively have a plug flow (Bianchi, Williams, & Kappe, 2020); this type of reactors has been used for the conversion of jatropha oil to biodiesel (Ghazi, Resul, Yunus, & Yaw, 2008). The heat exchanger reactor allows the combination of the reaction and heat exchange, which increases the selectivity and prevent runaway reactions (Hesselgreaves, Law, & Reay, 2017a, 2017b); this type of equipment has been used for the steam reforming of LPG, ethanol, and methanol and catalytic combustion of LPG and methanol (Kolb et al., 2007). Finally, the spinning disk reactor has high fluid dynamic intensity, which favors the rapid transmission of heat, mass, and momentum (Reay, Ramshaw, & Harvey, 2013b), thereby making it an ideal vehicle for performing fast endothermic reactions such as catalytic isomerization of α-pinene oxide to campholenic aldehyde (Reay et al., 2013a, 2013b).

The intensified reactors offer higher transfer rates of heat and mass, which reduce dead zones and, as consequence, the size of the equipment decreases. An important fact is that in this type of equipment the conversion is higher, and the runaway reactions are better controlled.

1.2.2 Separation equipment

Once the raw materials have been transformed into products, the purification of them must be made. Usually, the purification implies the separation of unreacted compounds as well as byproducts generated. There are several unit operations that can be used to separate the products of interest, such as liquid-liquid extraction, adsorption, absorption, membranes, evaporation, and distillation; nevertheless, distillation is the most common method for the separation of fluid mixtures. Thus in this section, the focus will be given on this unit operation.

Distillation is a unit operation that allows the purification of homogeneous mixtures of fluids, through the transfer of mass between liquid and vapor streams; these last ones are created with a reboiler and a condenser, located in the bottom and top of the distillation

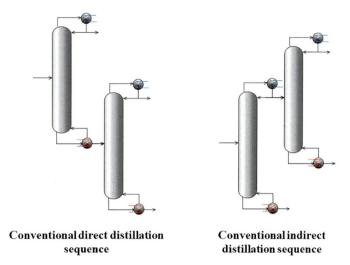

Conventional direct distillation sequence **Conventional indirect distillation sequence**

Fig. 1.7 Conventional distillation sequences.

columns, respectively. A conventional distillation column can separate one product at the time, by top or bottom; thus usually a train of distillation columns is required. These distillation trains are considered as conventional ones, and the two more known sequences are the direct (where the most of the products are obtained at the top of the column) and the indirect (where the most of the products are obtained at the bottom of the column); the conventional distillation trains are presented in Fig. 1.7 for the purification of a ternary mixture.

The distillation columns are very flexible in their design for the separation of mixtures of different characteristics; their main disadvantage is their low thermodynamic efficiency, due to the remixing effect. To overcome this weakness, new distillation configurations have been proposed, and they are known as thermally coupled distillation sequences (Fig. 1.8).

The thermally coupled distillation schemes consist of distillation columns linked between them through liquid and vapor interconnection flows. Since the supply of liquid and vapor requirements is satisfied with the interconnection flows, it is possible to eliminate a condenser and/or a reboiler; this helps to decrease the investment cost for the separation of the mixture. Moreover, in type of schemes, it is possible to reduce the energy requirements since the interconnection flows are located to avoid the remixed effect. According to the literature, thermally coupled distillation sequences can reduce the energy requirements between 30% and 50%, in comparison with conventional distillation sequences (Caballero, 2009; Dejanović, Matijašević, & Olujić, 2010; Gómez-Castro et al., 2016; Yildirim, Kiss, & Kenig, 2011). There are several types of thermally coupled distillation sequences such as the direct and indirect ones, as well as the Petlyuk

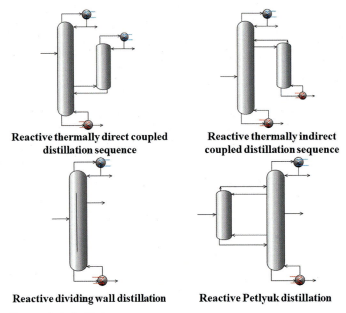

Fig. 1.8 Thermally coupled distillation sequences.

distillation column and the dividing wall distillation column. In particular, the dividing wall distillation column is a very promising technology allowing a significant energy requirement reduction (Yildirim et al., 2011), as well as reduction in space requirements and piping and installation costs.

Moreover, thermally coupled distillation columns have been used for the separation of multicomponent mixtures (Avendaño, Pinzón, & Orjuela, 2020; Kiss, Ignat, Flores Landaeta, & de Haan, 2013; Rong, 2011; Vazquez-Castillo et al., 2009), azeotropic (Waltermann, Münchrath, & Skiborowski, 2017; Yang et al., 2019; Zhang et al., 2020), extractive mixtures (Aniya, De, Singh, & Satyavathi, 2018; Murrieta-Dueñas, Gutiérrez-Guerra, Segovia-Hernández, & Hernández, 2011; Staak & Grützner, 2017; Yang et al., 2020) as well as reactive processes (Murrieta-Dueñas et al., 2011; Weinfeld, Owens, & Eldridge, 2018).

1.2.3 Conditioning equipment

In all conversion process, reaction and purification zones, as well as the conditioning equipment, are essentials. Usually, the conditioning equipment is used to modify the temperature, pressure, or size of the raw materials; thus in this category of equipment, heat exchangers, pumps, compressors, turbines, grinders, and crushers can be found (Fig. 1.9). Among these equipment, the design of heat exchangers has been studied

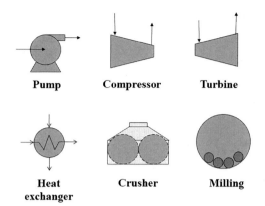

Fig. 1.9 Conventional conditioning equipment.

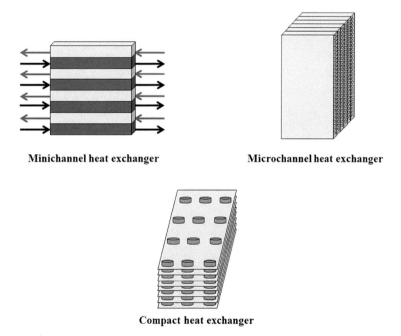

Fig. 1.10 Intensified conditioning equipment.

and improved through different strategies (Awais & Bhuiyan, 2018; Dixit & Ghosh, 2015; Klemeš et al., 2020).

The traditional heat exchangers employ conventional tubes (≥ 6 mm) with various cross-sections and orientations; even in those with enhanced surface textures, this technology is nearing its limits (Khan & Fartaj, 2011). In contrast, intensified equipment for heat exchanger is of smaller size respect to conventional technology (Fig. 1.10).

Microchannel heat exchangers (≤1 mm) represent an improved alternative due to their higher heat transfer and reduced both weight and space requirements (Khan & Fartaj, 2011). There are also minichannel heat exchangers with diameters greater than 200 μm but minors or equal than 3 mm (Dhar, 2017). Other type of intensified equipment is the compact heat exchanger, which is formed of layers of plates or finned channels of fixed length and width (Hesselgreaves et al., 2017a, 2017b); among this kind of heat exchangers, the following are included: the plate heat exchangers, the printed circuit heat exchanger, the Chart-flow unit of chart heat exchangers, and the polymer film heat exchanger (Reay, Ramshaw, & Harvey, 2008).

1.2.4 Design methodologies for intensified equipment

The design of intensified equipment can be addressed with three different approaches: shortcut methodologies, optimization strategies, and the use of computational fluid dynamics. Shortcut methodologies are defined as design procedures that employ mainly material balances and simple equations for the modeling of the thermodynamic behavior of the components involved. For instance, in the design of ideal conventional reactors, mass balance along with the kinetic model is employed (Smith, 1981), while in conventional distillation sequences, the Fensk-Underwood-Gilliland equations are used (Henley & Seader, 1981). On the other hand, optimization procedures can be used as a design tool; usually the objectives considered are the yield, energy consumption, or volume. The optimization procedures can include mathematical programming as well as metaheuristic strategies, and even can be linked to process simulators (Modak, Lobos, Merigó, Gabrys, & Lee, 2020; Pistikopoulos et al., 2021). Finally, computational fluid dynamics is an efficient computerized method of studying fluid mechanics based on numerical analysis (Junka, Daly, & Yu, 2013). Thus computational fluid dynamics is a powerful tool for the simulation and design of intensified equipment, especially those which are miniaturized.

Regard the design of intensified equipment for chemical reactions, optimization procedures as well as computational fluid dynamics are the most common tools. According to the literature, optimization procedures have been applied to obtain the optimal design of tailor-made reactors (Freund, Maußner, Kaiser, & Xie, 2019), a multitubular reactor for ethylene production (Ovchinnikova, Banzaraktsaeva, & Chumachenko, 2019), and capillary-based LSC-photomicroreactors (Zhao et al., 2020). On the other hand, computational fluid dynamics methodologies have been used to design a microchannel reactor heat exchanger (Engelbrecht, Everson, Bessarabov, & Kolb, 2020), gas-solid vortex reactor for oxidative coupling of methane (Vandewalle, Marin, & Van Geem, 2021), photocatalytic reactors in the gas phase (Oliveira de Brito Lira, Riella, Padoin, & Soares, 2021), microwave heating in heterogeneous catalysis (Yan, Stankiewicz, Eghbal Sarabi, & Nigar, 2021), as well as pneumatically agitated slurry reactors (Geng, Mao, Huang, & Yang, 2021).

The design of intensified distillation schemes can be made with shortcut methods as well as optimization procedures. Among the shortcut methodologies, those can be cited are the ones proposed for the design of thermally coupled distillation (Hernández & Jiménez, 1996), the synthesis of intensified nonsharp distillation systems (Rong, 2014), and simple column configurations for multicomponent distillation (Rong & Errico, 2012). On the other hand, the generation of optimal designs through optimization strategies is reported through the use of multicomponent intensified distillation systems (Errico, Pirellas, Torres-Ortega, Rong, & Segovia-Hernandez, 2014), intensified nonsharp distillation configurations (Torres-Ortega, Strieker, Errico, & Rong, 2015), intensified thermally coupled distillation sequences (Caballero & Reyes-Labarta, 2016), synthesis of intensified sequences for multicomponent zeotropic mixtures (Li, Demirel, & Hasan, 2019), dividing wall columns for extractive separations (Li et al., 2021), separating normal alkanes via multi-objective optimization (Liu, Ren, et al., 2021; Liu, Wei, & Leng, 2021). For the design of reactive distillation, there are shortcut methodologies as well as optimization strategies. Regard the shortcut methodologies, the works of Barbosa and Doherty (1988), Dragomir and Jobson (2004), Carrera-Rodríguez, Segovia-Hernández, and Bonilla-Petriciolet (2011), Flores-Estrella and Iglesias-Silva (2016) can be cited. In addition, optimization strategies have been developed for the optimal design of reactive distillation, as it can be seen in the works of Ciric and Gumus (2009), Tsatse, Oudenhoven, ten Kate, and Sorensen (2021), and Tian, Pappas, Burnak, Katz, and Pistikopoulos (2021).

Regard the intensified heat exchangers, most of the reported works are experimental; the studies on the design of intensified heat exchangers are scarce, and they employed computational fluid dynamics as design tool (Alimoradi, Olfati, & Maghareh, 2017; Bahiraei, Mazaheri, & Hanooni, 2021; Jamshidmofid, Abbassi, & Bahiraei, 2021; Piriyarungrod et al., 2018).

1.3 Conventional processes for the production of biofuels

As it was mentioned before, the biomass can be converted into biofuels in liquid, gaseous or solid state. In this section, the definition, main uses, as well as the conventional process for its production will be presented.

1.3.1 Liquid biofuels

Among liquid biofuels, we can mention bioethanol, biobutanol, biogasoline, biodiesel, green diesel, and renewable aviation fuel.

Bioethanol, as well as biobutanol, is an alcohol, whose chemical composition is exactly the same of its fossil counterpart; the main difference is that bioethanol and biobutanol are not generated from petroleum (Alam & Tanveer, 2020). Bioethanol can be used as an additive in internal combustion engines that operate with gasoline or as a fuel in engines specifically designed for it. In addition, bioethanol can be used in the beverage

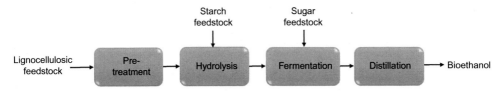

Fig. 1.11 Main steps in the conventional process for the production of bioethanol.

industry as well as raw material for the generation of ethylene and derivative compounds (Ferreira, Agnihotri, & Taherzadeh, 2019). On the other hand, biobutanol has captured the attention for its potential use as gasoline replacement (Roberts & Patterson, 2014). Biobutanol can be used as an additive in internal combustion engines that operates with gasoline, until 12.5% volume according to the standard ASTM D7862-19 (ASTM, 2019). In addition, biobutanol is used as a solvent in cosmetics, hydraulic fluids, detergent formulations, drugs, antibiotics, hormones, and vitamins, as an intermediate in chemical synthesis, and also as an extractant in the manufacturing of pharmaceuticals (Isomäki, Pitkäaho, Niemistö, & Keiski, 2017).

Bioethanol and biobutanol can be produced through fermentation of sugars, which can be obtained for cultivated or residual biomass. Fig. 1.11 shows the main process in the production of bioethanol.

As it can be seen in Fig. 1.11, bioethanol can be produced from lignocellulosic, sugar, and starch feedstock. In all cases, it is necessary to grind the feedstock in order to increase the available area for the release of the sugars contained in the raw material. In the case of sugar feedstock, it is usually required to perform a solid–liquid extraction, while hydrolysis is used for starch feedstock; in the case of lignocellulosic feedstock, pretreatments are carried on to separate cellulose and hemicellulose from biomass followed by hydrolysis and saccharification. Later, the released sugars are fermented usually with *Saccharomyses cerevisiae* in order to generate bioethanol (Ciani, Comitini, & Mannazzu, 2008). Typical yields to bioethanol are in the order of 5%–12%, since at major concentrations the microorganisms are inactivated (Dimian, Bildea, & Kiss, 2014). Due to this, the separation is necessary in order to obtain anhydrous bioethanol. Typically, distillation has been used for the purification of bioethanol, for which the azeotropic point must be overpass; this usually requires the use of a solvent, such as ethylene glycol (Pacheco-Basulto et al., 2012). In the conventional process for the production of bioethanol, its purification is the operation with higher-energy consumption.

On the other hand, the main steps involved in the production process of biobutanol are shown in Fig. 1.12.

From Fig. 1.12, it is important to mention that the pretreatment and hydrolysis steps are exactly the same previously described for bioethanol production. The main difference relies on the fermentation and distillation steps. The released sugars are converted

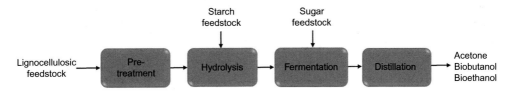

Fig. 1.12 Main steps in the conventional process for biobutanol production.

through acetone-butanol-ethanol (ABE) fermentation, whose name indicates that these products are generated. The concentrations of butanol range from 2.75 to 12 g/L, obtained with *Clostridium acetobutylicum* ATCC 824 (Sindhu et al., 2019). Due to the low yields, the separation is necessary in order to obtain biobutanol; according to the literature, distillation and liquid-liquid extraction have been used for the purification of the products of ABE fermentation (Kaymak, 2018; Sánchez-Ramírez, Quiroz-Ramírez, Segovia-Hernández, Hernández, & Ponce-Ortega, 2016). In the conventional process for the production of biobutanol, its purification is the operation with higher-energy consumption.

On the other hand, renewable gasoline or biogasoline consists of hydrocarbons in the range of C4–C12, generated from biomass; in a strict sense, biogasoline would exclude the alcohols since alcohols are typically oxygenated, in contrast with oil-derived fuels (Pagliuso, 2010). Thus the composition of biogasoline and gasoline are the same, as well as its properties (ASTM, 2021). Thus the biogasoline can be used in mixtures with fossil gasoline or at 100% in internal combustion engines that operate with fossil gasoline. In addition, biogasoline can be used to produce steam in boilers, or to generate electricity in small power plants. Fig. 1.13 shows the routes to produce biogasoline.

As it can be observed from Fig. 1.13, biogasoline can be produced from triglyceride, lignocellulosic, sugar, and starch feedstock to different processing pathways. If triglyceride feedstock is used, it must be hydroprocessed in order to generate biogasoline, as well as light gases and diesel fuel; in addition, biooil derived from the pyrolysis of lignocellulosic feedstock can also be hydroprocessed to generate renewable gasoline. On the other hand, it is possible to produce biogasoline from sugar and starch feedstock, through its conversion to alcohol and later oligomerized. In spite of the type of raw material used, distillation is included for the purification of the hydrocarbons generated. The reported selectivities for renewable gasoline are 90% or higher (Duan et al., 2020; Ohayon Dahan, Porgador, Landau, & Herskowitz, 2020). In these conversion pathways, the pyrolysis, hydrodeoxygenation, as well as distillation are the most energy-intensive processes.

Bioethanol and biobutanol have been proposed as replacements or additives for gasoline, while biodiesel and green diesel have been developed for fossil diesel. In particular, biodiesel is a mixture of long chain fatty acid esters that are generated through a transesterification reaction (Du, Kamal, & Zhao, 2019); biodiesel differs from fossil diesel in composition, since the last one consists of hydrocarbons in the range of C17–C28.

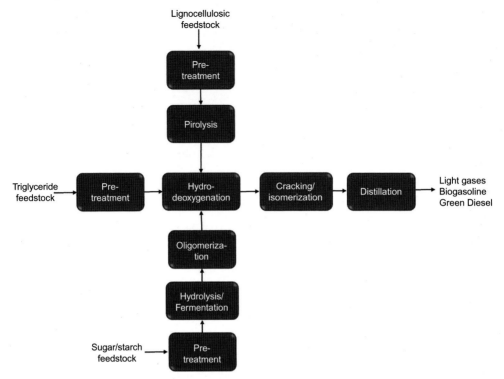

Fig. 1.13 Main steps in the conventional process for biogasoline production.

Fig. 1.14 Main steps in the conventional process for biodiesel production.

Biodiesel can be used in mixtures in internal combustion engines that operate with fossil diesel or at 100% in engines specifically designed for this biofuel (ASTM, 2020a). In addition, it can be used as fuel in boilers to generate steam or in small-generation power plants. Fig. 1.14 shows the main steps in the conventional process for the production of biodiesel.

Biodiesel can be produced from triglyceride feedstock, which may require a pretreatment especially if the raw material is residual. The triglycerides are converted into fatty acid esters and glycerol, through transesterification and esterification reactions; these products must be separated by decantation. Finally, the biodiesel is purified to eliminate the impurities and reach the purity required for its use as fuel. The most used process involves the homogeneous catalysis, which can be acid or basic; the conversions are 98% or greater (Brito Cruz, Souza, & Barbosa Cortez, 2014). In the conventional process

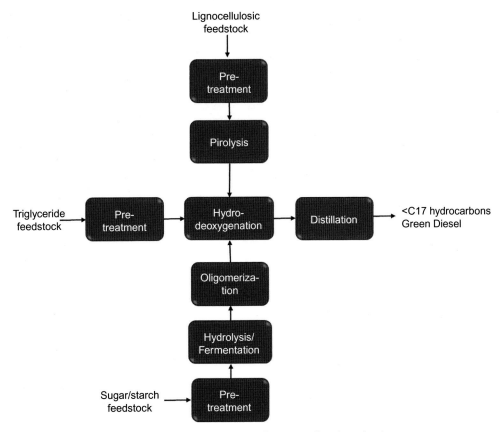

Fig. 1.15 Main steps in the conventional processes for green diesel production.

for the production of biodiesel, its purification is the operation with higher-energy consumption and also with the greater hydric footprint.

On the other hand, green diesel consists of hydrocarbons in the range of C17–C28, which correspond to the same composition of fossil diesel. Green diesel can be used in mixtures in internal combustion engines that operate with fossil diesel or at 100% in engines designed to operate with fossil diesel (ASTM, 2020a). Moreover, green diesel can be used as fuel in boilers to generate steam or in small generation power plants. Fig. 1.15 shows the main steps in the conventional process for the production of green diesel. Similar to the production of biogasoline, green diesel can be generated from sugar, starch, triglyceride, or lignocellulosic feedstock for different conversion pathways; in spite of the processing pathway, the purification of the generated hydrocarbons is carried out through distillation. The reported selectivities for the production of green diesel range from 80% to 94% (Ameen et al., 2020; Papanikolaou et al., 2020). In these

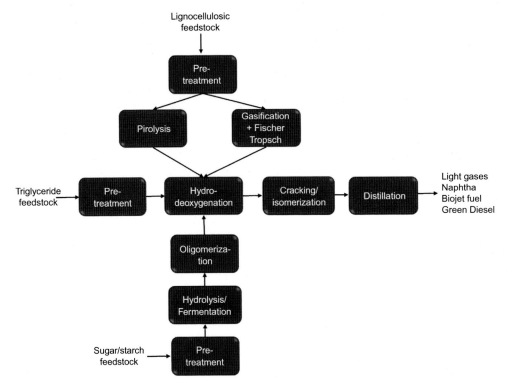

Fig. 1.16 Main steps in the conventional processes for renewable aviation fuel production.

conversion pathways, the pyrolysis, hydrodeoxygenation as well as distillation are the most energy-intensive processes.

Finally, renewable aviation biofuel consists of hydrocarbons from C8 to C16, and this biofuel must be drop-in; this means that the composition and properties of biojet fuel are the same of its fossil counterpart. Renewable aviation fuel can be used in mixtures until 50% in volume with fossil jet fuel (ASTM, 2020b) in commercial flights for the transport of passengers and goods. Renewable aviation fuel can be produced from triglyceride, lignocellulosic, sugar, and starch feedstock through several conversion processes (Fig. 1.16). Sugar and starch biomasses can be transformed to alcohols, which later can be oligomerizated and hydroprocesssed in order to produce biojet fuel. Moreover, triglyceride feedstock as well as biooil, derived from the pyrolysis of lignocellulosic biomass, can be hydroprocessed to produce renewable aviation fuel. In all cases, it is necessary a purification stage where usually distillation is employed. The selectivities reported for biojet fuel range from 15% to 80%, depending on the type of raw material used (Gutiérrez-Antonio, Romero-Izquierdo, Gómez-Castro, & Hernández, 2021). In these conversion pathways, the pyrolysis, gasification, hydrodeoxygenation as well as distillation are the most energy-intensive processes.

It is important to mention that in the production of renewable aviation fuel, as well as green diesel and biogasoline, in the hydroprocessing an important reactant is hydrogen; this reactant is usually generated from the reforming of natural gas, a nonrenewable fuel. Thus it is necessary to develop processes for its production from renewable sources, which will be described in the next section.

1.3.2 Gaseous biofuels

Among gaseous biofuels, we can mention biogas, syngas as well as renewable hydrogen; a summary of the most important aspects of these biofuels will provided next.

Biogas is a mixture of gases mainly represented by methane and carbon dioxide with small quantities of hydrogen sulfide and ammonia and saturated with water vapor (Aparicio et al., 2020); in this mixture of compounds, methane is the one that most contributes with its calorific power. Biogas is similar in composition to natural gas, which is composed of methane, ethane, propane, butanes, and pentanes, and it can also include carbon dioxide, helium, hydrogen sulfide, and nitrogen (Viswanathan, 2017); however, methane is also the main component of natural gas. Biogas can be used to produce electricity or as fuel for transport sector (replacing natural gas) (Kougias & Angelidaki, 2018). Fig. 1.17 shows the conventional process for the production of biogas.

Based on Fig. 1.17, it can be observed that biogas can be produced from lignocellulosic, sugar, and starch biomasses, being the last two the most used. For sugar and starch biomasses, usually grinding is required before the hydrolysis, while in the case of lignocellulosic biomass, grinding is necessary along with other pretreatments such as steam explosion (Cotana, Cavalaglio, Petrozzi, & Coccia, 2015). For both type of feedstock, hydrolysis and anaerobic digestion is carried out in order to generate biogas.

Depending on the raw material and operating conditions, methane represents between 50% and 75% of biogas, followed by carbon dioxide, 25%–50%, and nitrogen, maximum 10%, (Harun et al., 2019). After biogas is obtained, methane can be purified

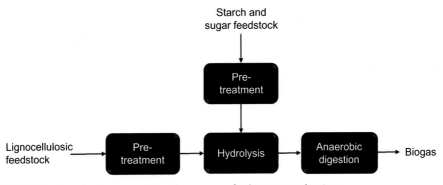

Fig. 1.17 Main steps in the conventional processes for biogas production.

Fig. 1.18 Main steps in the conventional process for syngas production.

through absorption with water. In the conventional production processes of biogas, the major energy consumptions are observed in the pretreatments as well in the anaerobic digestion.

On the other hand, syngas, or synthetic gas, is a gas mixture consisting primarily of hydrogen (30%–45%), carbon monoxide (20%–30%), methane (8%–12%), and some carbon dioxide (15%–25%) (Yang & Ge, 2016). Syngas can be used in fuel cells and also in turbine fuels (Sikarwar & Zhao, 2017); moreover, it can be used as raw material to produce other fuels such as biogasoline, green diesel, and renewable aviation fuel (Gutiérrez-Antonio et al., 2021). In addition, it can be employed to produce ammonia and methanol, and also it is a source of hydrogen (Yang & Ge, 2016). Fig. 1.18 shows the main step in the production of biogas from biomass.

The main type of biomass used for the production of syngas is the lignocellulosic one. This biomass usually is grinded and dehydrated, in order to improve the yields during the gasification. After that, gasification takes place, usually at 750–1150°C, reaching conversions until 95% (Wei et al., 2020). Based on the gasification agents used, biomass gasification processes can be divided into air gasification, oxygen gasification, steam gasification, carbon dioxide gasification, and supercritical water gasification (Zhang et al., 2019). In this process, gasification is the step with major energy requirements, followed by dehydration and grinding.

Finally, renewable hydrogen is another biofuel that has captured the attention of the scientific community due to its use not only as biofuel but also as an energy carrier. Renewable hydrogen has the same composition of its fossil counterpart, which is produced from the high-temperature reaction of natural gas or naphtha with steam (Simeons, 1980). Hydrogen can be used as fuel for transport sector (ASTM, 2011) or power generation, and it can be also employed to store energy, mainly the renewable one which is intermittent. The conventional processes for the production of renewable hydrogen are presented in Fig. 1.19 (Martinez-Burgos et al., 2021).

According to Martinez-Burgos et al. (2021), renewable hydrogen can be produced from lignocellulosic, sugar, and starch feedstock to several processing routes. The renewable hydrogen can be produced from lignocellulosic biomass through thermochemical processes, such as pyrolysis, gasification, combustion, or liquefaction; later, the purification of hydrogen must be performed. Moreover, this type of biomass can be hydrolyzed and digested in order to produce biomethane, which later can be reformed to generate hydrogen. On the other hand, the sugar and starch feedstock can be fermented to produce renewable hydrogen. For the lignocellulosic feedstock, the thermochemical

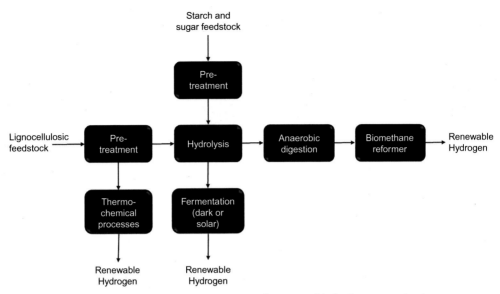

Fig. 1.19 Main steps in the conventional processes for renewable hydrogen production.

processes are the ones with major energy consumption, while for sugar and starch feedstock, the pretreatments contribute the most with the energy consumption.

It is important to mention that in the case of thermochemical processes a solid residue is generated that can be used as solid biofuel, whose production processes are discussed next.

1.3.3 Solid biofuels

Among solid biofuels, it can be found firewood, rod, fuel pellets, and briquettes; in this section, information about the last two will be provided, since they are the most popular solid biofuels.

Briquettes are defined as a densified fuel produced from lignocellulosic biomass; the briquettes can be produced in form of cylinder, hexahedron or parallelepiped (ISO, 2021). Briquettes can be used as a renewable replacement for coal, which is a nonrenewable resource. According to standard ISO 17225-3 (ISO, 2021), the minimum net calorific power of briquettes ranges from 14.9 to 15.5 MJ/kg, depending on the type of raw material; in comparison, the calorific power of mineral coal is 29.3 MJ/kg (IEA, 2020). Briquettes can be utilized for domestic and industrial applications to generate heat and also to produce electricity in power plants. The conventional process to produce briquettes can be observed in Fig. 1.20.

This type of solid biofuel is produced from lignocellulosic feedstock, especially those with high content of lignin. The lignocellulosic feedstock is grinded and the conditioning regard its moisture content. After that, the biomass is densified, and the final form of the

Fig. 1.20 Main steps in the conventional processes for briquette production.

Fig. 1.21 Main steps in the conventional processes for fuel pellet production.

briquette depends on the matrix used in the densification machine. The grinding and densification stages are the one with major energy consumptions.

On the other hand, fuel pellets are the most popular solid biofuels. The fuel pellets are a densified biofuel made from milled biomass, with or without additives, usually with a cylindrical form, with typical length between 5 and 40 mm (ISO, 2014). Similar to briquettes, the fuel pellets can be used for the generation of thermal energy for industrial and domestic applications, as well as electricity in power generation plants. According to standard ISO 17225-6 (ISO, 2014), the minimum net calorific power of fuel pellets is 16.5 MJ/kg. Fig. 1.21 shows the main steps in the conventional processes for the production of fuel pellets.

Fuel pellets are produced from lignocellulosic feedstock, especially residues from agriculture, forest, or agroindustrial sectors. The production process of fuel pellets is similar to the one for briquettes, with the main difference that the size of this solid biofuel is smaller. Also, fuel pellets are produced only in cylindrical form. For the production of pellets and briquettes, it is desirable that the biomass contains high amounts of lignin, since this polymer melts during the densification and helps to increase the durability of the solid biofuels. Similar to briquette production process, the grinding and densification stages are the one with major energy consumptions.

1.4 Intensified processes for the production of biofuels

From Section 1.3, it can be concluded that there are several conversion pathways for the production of liquid, gaseous, and solid biofuels; this means that the production of these biofuels is feasible from the technical point of view. However, it is also necessary that the price of these biofuels is competitive with its fossil counterparts, and also the production processes must be of reduced environmental impact. In this way, the biofuels generated will be sustainable. In the following subsections, the main advances regard the intensification processes for the production of liquid, gaseous, and solid biofuels in the last 6 years will be presented.

1.4.1 Liquid biofuels

The intensification of the processes for the production of bioethanol has been focus on pretreatment, reactive, and separation zones.

Regard the intensification of the conversion processes, several proposals have been presented. In 2015, Bharadwaja, Shuchi, and Vijayanand (2015) proposed a simplified process for the production of bioethanol, which comprises pretreatment, hydrolysis assisted with ultrasound, and fermentation. Their results show that is possible to obtain 0.26 g of ethanol for each gram of biomass (*Parthenium hysterophorus*). In the same year, Subhedar and Gogate (2015) also proposed the use of ultrasound-assisted alkaline treatment to eliminate the lignin in newspaper, which was the raw material selected for the production of bioethanol; moreover, the ultrasound was also implemented during the fermentation. Their results showed that their approach increased the productivity of ethanol from 7.8 to 14.1 g/L compared with the nonsonicated control fermentation. On the other hand, Terán Hilares et al. (2018) proposed the use of hydrodynamic cavitation assisted to remove lignin from sugarcane bagasse; later the hydrolysis was performed in a packed bed flow-through column reactor, and finally, the fermentation was carried out in a bubble column reactor. They reported that the concentration of ethanol was 31.5 g/L, while the productivity was 0.49 g of bioethanol per gram of raw material. Pal, Kumar, and Ghosh (2018) proposed a novel multistaged membrane process for bioethanol from sugarcane juice using *S. cerevisiae*; in addition, the final dehydration of bioethanol was carried out on a solar-driven direct contact membrane distillation configuration. Their results indicated that it is possible to produce 14.6 g/L of bioethanol. Recently, Konopacka, Rakoczy, and Konopacki (2019) presented the use of rotating magnetic field on the production of bioethanol using yeast strain modified with ferromagnetic nanoparticles; as control, the fermentation was carried out without exposure of the magnetic field and with yeast without nanoparticles. The results show that it is possible to increase in 50% of the production of bioethanol when a rotating magnetic field is used in combination with modified yeasts.

Respect to the purification of bioethanol, Torres-Ortega and Rong (2016) proposed the intensification of the purification of bioethanol, considering as feed of the fermentation stream, which contained several components. Their results showed that the use of a dividing wall column generates a reduction of 21.42% in the total annual costs, in comparison with a conventional separation train. Later, Singh, da Cunha, and Rangaiah (2019) presented the use of distillation-pressure swing adsorption in combination with heat-pump-assisted distillation; they compared this proposal with distillation-pressure swing adsorption with double-effect distillation. The results show that the first option reduces in 38% of the energy consumption and 19% of the total annual cost. Khalid et al. (2019) and Khalid et al. (2019) compared the use of distillation, membranes, and hybrid processes (distillation with membranes) for the purification of bioethanol.

Their results showed that distillation with pervaporation allows a reduction of 49% in the production cost of bioethanol, respect the conventional process. In 2020, Errico et al. (2020) presented a study to analyze the use of reactive distillation and membrane-assisted reactive distillation. Their results show that, in comparison with extractive distillation, an ordinary distillation column along with a reactive distillation column is the best alternative. Recently, Biasi et al. (2021) proposed the use of parastillation and metastillation (distillation columns with phase division) applied to the purification of bioethanol. In comparison with conventional distillation columns, the new schemes allow a reduction in 35% in total annual costs and also 42% less carbon dioxide emissions.

On the other hand, the intensification of the processes for the production of biobutanol has been focus on reactive and separation zones.

Regard the reactive zones, the main advances in the intensification have been reported in the last years. An interesting approach was the use of microemulsion-based butanol extraction (pentane, hexane, and heptane) from fermentation broth (Gedam, Raut, & Dhamole, 2019). The best results were obtained with hexane, allowing a recuperation of 62% of biobutanol in a single step; in comparison with the traditional two-step process. Recently, Seifollahi and Amiri (2020) analyzed the simultaneous cosaccharification and fermentation of cellulose for the production of biobutanol. As result, 24 g/L of ABE production was obtained; however, some products of the fermentation inhibit the cellulase. Finally, Rochón et al. (2020) proposed the use of IBE fermentation with *Clostridium beijerinckii* DSM 6423 coupled to an in situ gas stripping pervaporation. This proposal allows us to obtain a butanol concentration of 559 g/L, and 15% less energy is required respect to ABE fermentation.

Respect to the intensification of the purification of biobutanol, it can be found several works. Okoli and Adams (2015) presented the use of a dividing wall column for the purification of biobutanol. Respect to the conventional distillation sequence, the intensified scheme requires 31% less energy along with a decreasing in capital costs of 15%. Later, Rom, Miltner, Wukovits, and Friedl (2016) studied three pervaporation membranes for the separation of biobutanol. Their results show that the combination of a pervaporation membrane with the distillation process allows a reduction of 50% in energy consumption, regard the state of the art of distillation. Sánchez-Ramírez, Quiroz-Ramírez, Hernández, Segovia-Hernández, and Kiss (2017) presented the use of liquid-liquid extraction along with a dividing wall column for the purification of biobutanol from an ABE mixture. This scheme exhibited the minimal total annual cost as well as good values for eco indicator 99 and condition number (control properties). Respect to the conventional separation scheme, a reduction of 24.5% in the total annual cost was observed and also a reduction of 11.8% respect to the environmental indicator (Errico, Sanchez-Ramirez, Quiroz-Ramirez, Segovia-Hernandez, & Rong, 2016). In 2018, van Wyk, van der Ham, and Kersten (2018) analyzed the use of a pervaporation membrane as pretreatment of the ABE fermentation outlet stream with an alternative distillation column for the purification of biobutanol. The results of the new proposal were compared with a conventional distillation process. The intensified scheme decreases energy consumption in 53% respect to the conventional separation scheme.

The production of bioalcohols has been intensified mainly in the simultaneous saccharification and fermentation, as well as in the use of thermally coupled distillation and membranes for its purification. At the moment, the use of intensified reactors, such as microreactors or microchannel reactors is still missing in the literature, as well as the use of microdistillation. The intensification of the processes for the production of biogasoline has not been reported in the literature.

On the other hand, the intensification of the production of biodiesel has been widely studied. Unlike the review of the mainly advances presented for the intensification of the production of bioethanol and biobutanol, in the case of biodiesel, it will be presented the most relevant review articles on intensification technologies.

Skorupskaite, Makareviciene, and Gumbyte (2016) presented a revision of opportunities to perform the microalgae oil extraction simultaneously with the transesterification for the production of biodiesel. The authors conclude that this intensification strategy is promissory, especially if the microalgae have high oil content; however, the main limitation relies in the drying costs, which represent nearly 70% of the total costs. This limitation was addressed by the scientific community in order to convert microalgae via wet in situ transesterification (Kim et al., 2019). According to these authors, the water content of the microalgae is the main variable that affect the process; moreover, when the reaction is carry out between 101°C and 200°C, there is a reduction in the consumption of reactants, catalyst, and cosolvents. On the other hand, Sivaramakrishnan and Incharoensakdi (2018) presented a revision on the in situ ultrasound-assisted oil extraction with transesterification; the raw material was microalgae. The authors conclude that the reaction time decreases considerably with the use of ultrasound, and a reduction in the production costs is also observed. With the same objective of making a revision on the in situ ultrasound-assisted oil extraction with transesterification, Tan, Lim, Ong, and Pang (2019) made a revision but expanding the revision to nonedible vegetable oils. Moreover, the authors include a recompilation of the existing pilot reactors on the ultrasound-assisted transesterification. In the same line of the use of microalgae, Ortiz-Martínez et al. (2019) presented a revision on the production of biodiesel under supercritical conditions; the revision included wet and dry microalgae biomass. Authors conclude that several factors affect the properties of biodiesel, such as microalgae type as well as operation variables. Regard the use of microreactors, Tiwari, Rajesh, and Yadav (2018) presented a revision of the effects of design and operation parameters in the production of biodiesel, including its continuous production and inline purification. The authors conclude that the liquid-liquid flow patterns are mainly affected by the reaction temperature and the inlet mixer type. In the same topic, Santana, Silva, and Taranto (2019) made a recompilation of the experimental and simulation studies for the production of biodiesel in microreactors.

The authors conclude that the use of computational fluid dynamics is a powerful tool to design efficient microreactors, which overcome the mixing problems observed in the experimental studies; moreover, it is necessary to perform more studies related to the scale-up of this intensified equipment. The feasibility of using microchannel reactors for the production of biodiesel was reviewed by Natarajan et al. (2019). The authors focused on reactor design, fabrication, different types of mixers and mixing inside the channel. The conclusions indicated that a homogeneous catalyst is the most used in this intensified equipment, followed by heterogeneous catalysts, while the enzymatic transesterification is still an opportunity area. Also in 2019, Wong, Ng, Chong, Lam, and Chong (2019) published a review focus on the intensification of biodiesel production through catalytic enhancement and emerging reactor designs. According to the authors, the advances are converging to modular systems and improved sustainability, keeping the economic feasibility of the process. Recently, a complete review on the use of ionic liquid in catalytic systems for the production of biodiesel from agricultural products and microalgae was published (Ong et al., 2021). The authors conclude that ionic liquids with brønsted acidity have low environmental impact and allows to obtain high yields of biodiesel without the need of pretreatment. Moreover, Mohiddin et al. (2021) presented the new technologies for the sustainable production of biodiesel; among these technologies, the use of biochar as a catalyst and a magnetic catalyst for transesterification of biodiesel highlights.

Respect to the production of green diesel, the advances in the intensification has been focus on reactive zones.

Regard the reaction, Boffito, Galli, Pirola, and Patience (2017) studied the simultaneous transesterification and cracking of vegetable oil to produce isopropyl esters and green diesel. The process is carried out at 450°C, and the maximum yield to hydrocarbons in the range C5–C19 was 42%. However, it is important to recall that green diesel consists of hydrocarbons from C17 to C28. Zheng et al. (2020) studied the catalytic pyrolysis vapor upgrading of oleic acid. They found that it is possible to reach 85.68% of hydrocarbons generated (C8–C17), where 26.77% correspond to C17. An important aspect of this work is that there is no need to use hydrogen as reactant.

The intensification of the processes for the production of biojet fuel has been focused on reactive and separation zones, as well as the proposal of hybrid process.

Regard the intensification of the reactive process, Verma, Kumar, Rana, and Sinha (2011) developed a multifunctional catalyst (Ni-W catalyst supported on an acidic zeolitic ZSM-5 support with hierarchical structure) that allows the performance of all hydrotreating reactions in the same vessel; this is also called one-step hydrotreating. This catalyst was tested with microalgae oil, and it was reported a yield of 77% to hydrocarbons in the boiling point range of biojet fuel. Also, with a one-step hydroprocessing, Hanafi et al. (2016) studied the conversion of waste chicken fat into hydrocarbons using the DHC-8 commercial hydrocracking catalyst. Their results show that is possible to generate 40% of hydrocarbons in the boiling point range of aviation fuel, but the authors did not report the use of hydrogen. On the other hand, Lee, Lee, Kim, Shin, and Choi (2019) reported the hydroprocessing of palm oil to produce renewable aviation fuel CO-tolerant PtRe catalyst supported on USY. The maximum yield to aviation fuel reported was 41%, and all the required fuel specifications are fulfilled.

Regard the purification of biojet fuel, some works have been developed mainly with the use of thermally coupled distillation. In 2015, Gutiérrez-Antonio, Gómez-Castro, Hernández, and Briones-Ramírez (2015) proposed the intensification of the separation zone of the hydrotreating process. The results show that is possible to reduce in 21% the energy consumption in the intensified configuration; moreover, they incorporated a turbine that allows the production of 75 W. Later, the simultaneous energy integration and intensification of the hydroprocessing of *Jatropha curcas* oil was presented (Gutiérrez-Antonio, Romero-Izquierdo, Gómez-Castro, Hernández, & Briones-Ramírez, 2016). Their results show that is possible to reduce the energy requirements due to the energy integration; however, the capital costs increased. Moreover, a reduction of 87% in carbon dioxide emissions was obtained. Finally, considering the hydroprocessing of two stages, the use of reactive distillation for the production of biojet fuel was proposed (Gutiérrez-Antonio, Gómez-De la Cruz, Romero-Izquierdo, Gómez-Castro, & Hernández, 2018; Gutiérrez-Antonio, Soria Ornelas, Gómez-Castro, & Hernández, 2018). The reactive distillation considers the cracking and isomerization reactor. The results indicated that is possible decrease considerably the operation pressure, from 80 bar in the conventional process to 10 bar in the intensified one. In addition, it was reported an increase of 1% in the selectivity to biojet fuel and a decreasing of 25% in carbon dioxide emissions.

Considering the experimental works reported by Verma et al. (2011) and Hanafi et al. (2016), the intensification of the separation zone in the one-step hydroprocessing was studied. Gutiérrez-Antonio, Gómez-De la Cruz, et al. (2018) and Gutiérrez-Antonio, Soria Ornelas, et al. (2018) analyzed the intensification of the one-step hydroprocessing considering microalgae oil for the production of renewable aviation fuel. The carbon dioxide emissions were reduced in 34%, while the price of biojet fuel was decreased in 78%, respect to its fossil counterpart. Based on the work of Hanafi et al. (2016), Moreno-Gómez, Gutiérrez-Antonio, Gómez-Castro, and Hernández (2021) presented the intensification of the one-step hydroprocessing of chicken fat to produce biojet fuel. The results show that the use of thermally coupled distillation allows a decreasing in the 33.8% operation costs and also 22% in carbon dioxide emissions.

All the previous works have been focused on the intensification of the hydroprocessing. Then, only one work where the intensification of the alcohol to jet process is performed was presented by Romero-Izquierdo, Gómez-Castro, Gutiérrez-Antonio, Hernández, and Errico (2021). The intensification was applied to separation zones through the use of thermally coupled distillation. The results show that it is possible to obtain a reduction of 5.31% in the energy consumption, while the total annual cost and carbon dioxide emissions diminish in 4.83% and 4.99%, respectively.

1.4.2 Gaseous biofuels

The intensification of the processes for the production of biogas has been focus on pretreatment, reactive, and separation zones.

The application of nanoferrosonication pretreatment to sludge was studied in order to improve the solubilization and anaerobic digestion for the production of biogas (Córdova Lizama, Figueiras, Gaviria, Pedreguera, & Ruiz Espinoza, 2019). The authors reported that the biogas yield increased from 106 to 308 mL/g. On the other hand, the use of hydrodynamic cavitation as pretreatment was presented by Saxena, Saharan, and George (2019). This pretreatment was applied to tannery waste effluent, and the results indicated that biogas yield increase by 11.8-fold, respect to the process where this pretreatment is not applied. Other pretreatment studied was the vortex-based-cavitation, which was applied to distillery vinasses (Nagarajan & Ranade, 2020). According to the authors, the use of this pretreatment allows an increase in 22% of the biogas yield.

Regard the reactive zone, a solar-assisted bioreactor for the production of biogas was proposed (Khalid, Aslam, et al., 2019; Khalid, Siddique, et al., 2019). The raw material was cattle manure codigested with palm oil mill effluent. The maximum amount of biogas produced was 1567 mL, with a content of 64.13% of methane.

Regard the intensification in the production of syngas, some studies have been realized in the reactive zones mainly.

Ebrahimi and Rahmani (2017) presented an experimental study of the use of a microreactor for syngas production through the chemical looping reforming. The results show that a 100% increase of CO_2:CH_4 ratio is possible due to the reduction in 38% of the H_2:CO ratio. Later, Fedotov, Antonov, Uvarov, and Tsodikov (2018) proposed a hybrid membrane catalytic reactor for the coproduction of syngas and hydrogen. The raw materials were biomethane and bioethanol as well as dimethyl ether. The authors report 83% of hydrogen recovery at 5 atm and 700°C.

Ngoenthong et al. (2019) presented an experimental study where a microchannel reactor is used for the production of syngas from the H_2O/CO_2 cosplitting. The reaction was feasible with an optimal temperature of 700°C and yields of 2266 μmol/g of H_2, 705 μmol/g of CO, and 67% of solid conversion. In the same year, a solar biomass gasification reactor coupled with a parabolic dish solar concentrator was presented by Leiva Butti, Núñez Mc Leod, and Rivera (2020). In this reactor, it was performed the gasification of grape marc for the production of syngas. The results show that is possible to produce 3180 Nm^3 (36.68 GJ) for the total syngas yield. Liu, Zhang, Hong, and Jin (2020) reported an experimental study where syngas is produced through a honeycomb reaction via chemical looping cycle. According to the authors, the conversion of methane is around 95% and the concentration of syngas is 90%. In this intensified technology, the methane conversion is increased in more than 20%.

Recently, Gao, Quiroz-Arita, Diaz, and Lister (2021) presented the simultaneous capture and electrolysis of carbon dioxide for the production of syngas. The results show that it was reached 70% of conversion of the carbon dioxide at a current density up to $0.20 A/cm^2$ during 14 h of operation. An interesting proposal was presented by Chuayboon, Abanades, and Rodat (2018) for the continuous steam gasification of wood biomass in a high-temperature particle-fed solar reactor. The reported energy conversion efficiencies were above 20%, and one important fact is that the large particles were used (3–5 mm). Also in 2021, Chan et al. presented a revision of the production of syngas through the gasification of several wastes feedstock. In this review, an interesting fact is the inclusion of microwave-driven CO_2 gasification, which has a conversion of 99% of the carbon dioxide via Bourdon reaction (Lahijani, Zainal, Mohamed, & Mohammadi, 2014). On the other hand, the reforming of methane for the production of syngas through a structured catalytic wall-plate microreactor was studied by Hamzah, Fukuda, Ookawara, Yoshikawa, and Matsumoto (2021). The authors realized experimental and CFD simulation studies. The results show that at 650°C, the methane conversion is 80%, slightly minor than the conversion reported for a tubular packed bed reactor.

A novel process for the simultaneous production of hydrogen and syngas without purification was presented by Hosseini, Khosravi-Nikou, and Shariati (2019). The process consists of the steam reforming of methane in a microreactor along with CeO_2-Fe_2O_3 oxides as oxygen carriers. The authors report that the optimal conditions are 850°C with a relation Fe/Ce of one.

Finally, the intensified production of renewable hydrogen has been addressed in the literature, mainly for the reactive zones.

In 2016, Tsodikov et al. (2016) studied the hydrogen production using dry reforming of fermentation products through porous ceramic membrane catalytic converters. According to the authors, specific syngas productivity of Ni-Co_3O_4 (50%–50%) membrane reaches 85,000 L/h dm^3. Later, Durán, Sanz-Martínez, Soler, Menéndez, and Herguido (2019) presented a two zone fluidized bed coupled for methane dry reforming along with hydrogen permselective membranes. In this equipment, the reaction of biogas and separation of hydrogen is achieved. Nearly 70%–85% of hydrogen is obtained purely through the membrane. On the other hand, Parente, Soria, and Madeira (2020) proposed the production of hydrogen through combined dry and steam reforming of biogas in a membrane reactor. According to the authors, it is possible to obtain 73 mol of H_2 by 100 mol of biogas.

1.4.3 Solid biofuels

Regard the production of fuel pellets and briquettes, to the author's knowledge, there are not intensification processes reported in the literature.

1.5 Conclusions

Biofuels represent the most promissory alternative for the energetic transition without changes in the actual infrastructure. Moreover, biofuels play a key role in the sustainable recovery of the different economic sectors after the pandemic. According to the literature, there are different processes for the production of liquid, gaseous, and solid biofuels; however, the fact the production of these biofuels is feasible from the technical point of view is not enough. Nowadays, it is necessary to produce sustainable biofuels; this means that biofuels meet the technical specifications with a competitive price and also minimum environmental impact. In this context, the application of process intensification strategies to the production processes of biofuels is a promissory route. In spite of some efforts have been made to intensify the production of biofuels, there are still many opportunities areas such as the development of better catalysts, hybrid operations, as well as photocatalysis technologies.

References

Alam, M. S., & Tanveer, M. S. (2020). Conversion of biomass into biofuel: A cutting-edge technology. In L. Singh, A. Yousuf, & D. M. Mahapatra (Eds.), *Bioreactors* (pp. 55–74). Elsevier.

Alavi-Borazjani, S. A., Capela, I., & Tarelho, L. A. C. (2020). Over-acidification control strategies for enhanced biogas production from anaerobic digestion: A review. *Biomass and Bioenergy, 143*, 105833.

Alimoradi, A., Olfati, M., & Maghareh, M. (2017). Numerical investigation of heat transfer intensification in shell and helically coiled finned tube heat exchangers and design optimization. *Chemical Engineering and Processing Process Intensification, 121*, 125–143.

Ameen, M., Azizan, M. T., Yusup, S., Ramli, A., Shahbaz, M., & Aqsha, A. (2020). Process optimization of green diesel selectivity and understanding of reaction intermediates. *Renewable Energy, 149*, 1092–1106.

Ameen, M., Azizan, M. T., Yusup, S., Ramli, A., & Yasir, M. (2017). Catalytic hydrodeoxygenation of triglycerides: An approach to clean diesel fuel production. *Renewable and Sustainable Energy Reviews, 80*, 1072–1088.

Amin, A. (2019). Review of diesel production from renewable resources: Catalysis, process kinetics and technologies. *Ain Shams Engineering Journal, 10*, 821–839.

Aniya, V., De, D., Singh, A., & Satyavathi, B. (2018). Design and operation of extractive distillation systems using different class of entrainers for the production of fuel grade tert-butyl alcohol: A techno-economic assessment. *Energy, 144*, 1013–1025.

Aparicio, E., Rodríguez-Jasso, R. M., Lara, A., Loredo-Treviño, A., Aguilar, C. N., Kostas, E. T., et al. (2020). Biofuels production of third generation biorefinery from macroalgal biomass in the Mexican context: An overview. In M. D. Torres, S. Kraan, & H. Dominguez (Eds.), *Advances in green and sustainable chemistry. Sustainable seaweed technologies* (pp. 393–446). Elsevier.

Arun, N., Sharma, R. V., & Dalai, A. K. (2015). Green diesel synthesis by hydrodeoxygenation of bio-based feedstocks: Strategies for catalyst design and development. *Renewable and Sustainable Energy Reviews, 48*, 240–255.

ASTM. (2011). *WK34574 new test methods for determination of organic halides, Total non-methane hydrocarbons and formaldehyde in hydrogen fuel by gas chromatography (GC) and mass spectrometry (MS)*. ASTM Int.

ASTM. (2019). *ASTM D7862-19, standard specification for butanol for blending with gasoline for use as automotive spark-ignition engine fuel*. ASTM Int.

ASTM. (2020a). *ASTM D6751-20a standard specification for biodiesel fuel blend stock (B100) for middle distillate fuels*. ASTM Int.

ASTM. (2020b). *ASTM D7566-20c, standard specification for aviation turbine fuel containing synthesized hydrocarbons*. ASTM Int.

ASTM. (2021). *ASTM D4814-21a standard specification for automotive spark-ignition engine fuel*. ASTM Int.

Avendaño, S. J., Pinzón, J. S., & Orjuela, A. (2020). Comparative assessment of different intensified distillation schemes for the downstream separation in the oxidative coupling of methane (OCM) process. *Chemical Engineering and Processing Process Intensification, 158*, 108172.

Awais, M., & Bhuiyan, A. A. (2018). Heat and mass transfer for compact heat exchanger (CHXs) design: A state-of-the-art review. *International Journal of Heat and Mass Transfer, 127*, 359–380.

Ayodele, B. V., Alsaffar, M. A., & Mustapa, S. I. (2020). An overview of integration opportunities for sustainable bioethanol production from first- and second-generation sugar-based feedstocks. *Journal of Cleaner Production, 245*, 118857.

Aziz, M. A. A., Setiabudi, H. D., Teh, L. P., Annuar, N. H. R., & Jalil, A. A. (2019). A review of heterogeneous catalysts for syngas production via dry reforming. *Journal of the Taiwan Institute of Chemical Engineers, 101*, 139–158.

Bahiraei, M., Mazaheri, N., & Hanooni, M. (2021). Performance enhancement of a triple-tube heat exchanger through heat transfer intensification using novel crimped-spiral ribs and nanofluid: A two-phase analysis. *Chemical Engineering and Processing Process Intensification, 160*, 108289.

Bajwa, D. S., Peterson, T., Sharma, N., Shojaeiarani, J., & Bajwa, S. G. (2018). A review of densified solid biomass for energy production. *Renewable and Sustainable Energy Reviews, 96*, 296–305.

Barbosa, D., & Doherty, M. F. (1988). Design and minimum-reflux calculations for single-feed multicomponent reactive distillation columns. *Chemical Engineering Science, 43*, 1523–1537.

Bharadwaja, S. T. P., Shuchi, S., & Vijayanand, S. M. (2015). Design and optimization of a sono-hybrid process for bioethanol production from *Parthenium hysterophorus*. *Journal of the Taiwan Institute of Chemical Engineers, 51*, 71–78.

Bianchi, P., Williams, J. D., & Kappe, C. O. (2020). Oscillatory flow reactors for synthetic chemistry applications. *Journal of Flow Chemistry, 10*, 475–490.

Biasi, L. C. K., Batista, F. R. M., Zemp, R. J., Romano, A. L. R., Heinkenschloss, M., & Meirelles, A. J. A. (2021). Parastillation and metastillation applied to bioethanol and neutral alcohol purification with energy savings. *Chemical Engineering and Processing Process Intensification, 162*, 108334.

Boffito, D. C., Galli, F., Pirola, C., & Patience, G. S. (2017). CaO and isopropanol transesterify and crack triglycerides to isopropyl esters and green diesel. *Energy Conversion and Management, 139*, 71–78.

Bonechi, C., Consumi, M., Donati, A., Leone, G., Magnani, A., Tamasi, G., et al. (2017). Biomass: An overview. In F. Dalena, A. Basile, & C. Rossi (Eds.), *Bioenergy systems for the future* (pp. 3–42). Woodhead Publishing.

Brito Cruz, C. H., Souza, G. M., & Barbosa Cortez, L. A. (2014). Biofuels for transport. In T. M. Letcher (Ed.), *Future energy* (2nd ed., pp. 215–244). Boston: Elsevier.

Caballero, J. A. (2009). Thermally coupled distillation. In R. M. de Brito Alves, C. A. O. do Nascimento, & E. C. Biscaia (Eds.), *27. Computer aided chemical engineering* (pp. 59–64). Elsevier.

Caballero, J. A., & Reyes-Labarta, J. A. (2016). Mathematical programming approach for the design of intensified thermally coupled distillation sequences. In Z. Kravanja, & M. Bogataj (Eds.), *38. Computer aided chemical engineering* (pp. 355–360). Elsevier.

Carrera-Rodríguez, M., Segovia-Hernández, J. G., & Bonilla-Petriciolet, A. (2011). Short-cut method for the design of reactive distillation columns. *Industrial and Engineering Chemistry Research, 50*, 10730–10743.

Chen, Z., Wei, W., & Ni, B.-J. (2021). Cost-effective catalysts for renewable hydrogen production via electrochemical water splitting: Recent advances. *Current Opinion in Green and Sustainable Chemistry, 27*, 100398.

Chuayboon, S., Abanades, S., & Rodat, S. (2018). Experimental analysis of continuous steam gasification of wood biomass for syngas production in a high-temperature particle-fed solar reactor. *Chemical Engineering and Processing Process Intensification, 125*, 253–265.

Ciani, M., Comitini, F., & Mannazzu, I. (2008). Fermentation. In S. E. Jørgensen, & B. D. Fath (Eds.), *Encyclopedia of ecology* (pp. 1548–1557). Oxford: Academic Press.

Ciric, A., & Gumus, Z. (2009). MINLP: Reactive distillation column synthesis MINLP: Reactive distillation column synthesis. In C. A. Floudas, & P. M. Pardalos (Eds.), *Encyclopedia of optimization* (pp. 2183–2190). Boston, MA: Springer US.

Cotana, F., Cavalaglio, G., Petrozzi, A., & Coccia, V. (2015). Lignocellulosic biomass feeding in biogas pathway: State of the art and plant layouts. *Energy Procedia, 81*, 1231–1237.

Dahiya, S., Chatterjee, S., Sarkar, O., & Mohan, S. V. (2021). Renewable hydrogen production by dark-fermentation: Current status, challenges and perspectives. *Bioresource Technology, 321*, 124354.

Dejanović, I., Matijašević, L., & Olujić, Ž. (2010). Dividing wall column—A breakthrough towards sustainable distilling. *Chemical Engineering and Processing Process Intensification, 49*, 559–580.

Dhar, P. L. (2017). Modeling of thermal equipment. In P. L. Dhar (Ed.), *Thermal system design and simulation* (pp. 147–296). Academic Press.

Dimian, A. C., Bildea, C. S., & Kiss, A. A. (2014). Process intensification. In A. C. Dimian, C. S. Bildea, & A. A. Kiss (Eds.), *35. Computer aided chemical engineering* (pp. 397–448). Elsevier.

Dixit, T., & Ghosh, I. (2015). Review of micro- and mini-channel heat sinks and heat exchangers for single phase fluids. *Renewable and Sustainable Energy Reviews, 41*, 1298–1311.

Dragomir, R. M., & Jobson, M. (2004). Conceptual design of reactive distillation columns with non-reactive sections. In A. Barbosa-Póvoa, & H. Matos (Eds.), *18. Computer aided chemical engineering* (pp. 385–390). Elsevier.

Du, W., Kamal, R., & Zhao, Z. K. (2019). 3.06—Biodiesel. In M. Moo-Young (Ed.), *Comprehensive biotechnology* (3rd ed., pp. 66–78). Oxford: Pergamon.

Duan, D., Zhang, Y., Wang, Y., Lei, H., Wang, Q., & Ruan, R. (2020). Production of renewable jet fuel and gasoline range hydrocarbons from catalytic pyrolysis of soapstock over corn cob-derived activated carbons. *Energy, 209*, 118454.

Dube, K., Nhamo, G., & Chikodzi, D. (2021). COVID-19 pandemic and prospects for recovery of the global aviation industry. *Journal of Air Transport Management, 92*, 102022.

Durán, P., Sanz-Martínez, A., Soler, J., Menéndez, M., & Herguido, J. (2019). Pure hydrogen from biogas: Intensified methane dry reforming in a two-zone fluidized bed reactor using permselective membranes. *Chemical Engineering Journal, 370*, 772–781.

Ebrahimi, H., & Rahmani, M. (2017). A novel intensified microreactor for syngas production by coupling reduction-oxidation reactions in chemical looping reforming process. *Journal of Cleaner Production, 167*, 376–394.

Engelbrecht, N., Everson, R. C., Bessarabov, D., & Kolb, G. (2020). Microchannel reactor heat-exchangers: A review of design strategies for the effective thermal coupling of gas phase reactions. *Chemical Engineering and Processing Process Intensification, 157*, 108164.

Errico, M., Madeddu, C., Flemming Bindseil, M., Dall Madsen, S., Braekevelt, S., & Camilleri-Rumbau, M. S. (2020). Membrane assisted reactive distillation for bioethanol purification. *Chemical Engineering and Processing Process Intensification, 157*, 108110.

Errico, M., Pirellas, P., Torres-Ortega, C. E., Rong, B.-G., & Segovia-Hernandez, J. G. (2014). A combined method for the design and optimization of intensified distillation systems. *Chemical Engineering and Processing Process Intensification, 85*, 69–76.

Errico, M., Sanchez-Ramirez, E., Quiroz-Ramìrez, J. J., Segovia-Hernandez, J. G., & Rong, B.-G. (2016). Synthesis and design of new hybrid configurations for biobutanol purification. *Computers and Chemical Engineering, 84*, 482–492.

Fagbohungbe, M. O., Komolafe, A. O., & Okere, U. V. (2019). Renewable hydrogen anaerobic fermentation technology: Problems and potentials. *Renewable and Sustainable Energy Reviews, 114*, 109340.

Fajín, J. L. C., & Cordeiro, M. N. D. S. (2021). Light alcohols reforming towards renewable hydrogen production on multicomponent catalysts. *Renewable and Sustainable Energy Reviews, 138*, 110523.

Fedotov, A. S., Antonov, D. O., Uvarov, V. I., & Tsodikov, M. V. (2018). Original hybrid membrane-catalytic reactor for the co-production of syngas and ultrapure hydrogen in the processes of dry and steam reforming of methane, ethanol and DME. *International Journal of Hydrogen Energy, 43*, 7046–7054.

Ferreira, J. A., Agnihotri, S., & Taherzadeh, M. J. (2019). Waste biorefinery. In M. J. Taherzadeh, K. Bolton, J. Wong, & A. Pandey (Eds.), *Sustainable resource recovery and zero waste approaches* (pp. 35–52). Elsevier.

Flores-Estrella, R. A., & Iglesias-Silva, G. A. (2016). Reactive McCabe-Thiele: Short cut method including reactive vapor-liquid efficiency. *Revista Mexicana de Ingeniería Química*, *15*(1), 193–207.

Freund, H., Maußner, J., Kaiser, M., & Xie, M. (2019). Process intensification by model-based design of tailor-made reactors. *Current Opinion in Chemical Engineering*, *26*, 46–57.

Galadima, A., & Muraza, O. (2015). Catalytic upgrading of vegetable oils into jet fuels range hydrocarbons using heterogeneous catalysts: A review. *Journal of Industrial and Engineering Chemistry*, *29*, 12–23.

Gao, N., Quiroz-Arita, C., Diaz, L. A., & Lister, T. E. (2021). Intensified co-electrolysis process for syngas production from captured CO_2. *Journal of CO2 Utilization*, *43*, 101365.

Gedam, P. S., Raut, A. N., & Dhamole, P. B. (2019). Microemulsion extraction of biobutanol from surfactant based-extractive fermentation broth. *Chemical Engineering and Processing Process Intensification*, *146*, 107691.

Geng, S., Mao, Z.-S., Huang, Q., & Yang, C. (2021). Process intensification in pneumatically agitated slurry reactors. *Engineering*, *7*, 304–325.

Gent, S., Twedt, M., Gerometta, C., Almberg, E., & Gent, S. (2017). Introduction to feedstocks. In M. Twedt, C. Gerometta, & E. Almberg (Eds.), *Theoretical and applied aspects of biomass torrefaction* (pp. 17–39). Butterworth-Heinemann.

Ghazi, A. T. I. M., Resul, M. F. M. G., Yunus, R., & Yaw, T. C. S. (2008). Preliminary design of oscillatory flow biodiesel reactor for continuous biodiesel production from jatropha triglycerides. *Journal of Engineering Science and Technology*, *3*(2), 138–145.

Gómez-Castro, F. I., Ramírez-Vallejo, N. E., Segovia-Hernández, J. G., Gutiérrez-Antonio, C., Errico, M., Briones-Ramírez, A., et al. (2016). Energy consumption maps for quaternary distillation sequences. In Z. Kravanja, & M. Bogataj (Eds.), *38. Computer aided chemical engineering* (pp. 121–126). Elsevier.

Greetham, D., Zaky, A., Makanjuola, O., & Du, C. (2018). A brief review on bioethanol production using marine biomass, marine microorganism and seawater. *Current Opinion in Green and Sustainable Chemistry*, *14*, 53–59.

Gutiérrez-Antonio, C., Gómez-Castro, F. I., de Lira-Flores, J. A., & Hernández, S. (2017). A review on the production processes of renewable jet fuel. *Renewable and Sustainable Energy Reviews*, *79*, 709–729.

Gutiérrez-Antonio, C., Gómez-Castro, F. I., Hernández, S., & Briones-Ramírez, A. (2015). Intensification of a hydrotreating process to produce biojet fuel using thermally coupled distillation. *Chemical Engineering and Processing Process Intensification*, *88*, 29–36.

Gutiérrez-Antonio, C., Gómez-De la Cruz, A., Romero-Izquierdo, A. G., Gómez-Castro, F. I., & Hernández, S. (2018). Modeling, simulation and intensification of hydroprocessing of micro-algae oil to produce renewable aviation fuel. *Clean Technologies and Environmental Policy*, *20*, 1589–1598.

Gutiérrez-Antonio, C., Romero-Izquierdo, A. G., Gómez-Castro, F. I., & Hernández, S. (2021). *Production processes of renewable aviation fuel*. Elsevier.

Gutiérrez-Antonio, C., Romero-Izquierdo, A. G., Gómez-Castro, F. I., Hernández, S., & Briones-Ramírez, A. (2016). Simultaneous energy integration and intensification of the hydrotreating process to produce biojet fuel from *Jatropha curcas*. *Chemical Engineering and Processing Process Intensification*, *110*, 134–145.

Gutiérrez-Antonio, C., Soria Ornelas, M. L., Gómez-Castro, F. I., & Hernández, S. (2018). Intensification of the hydrotreating process to produce renewable aviation fuel through reactive distillation. *Chemical Engineering and Processing Process Intensification*, *124*, 122–130.

Hamzah, A. B., Fukuda, T., Ookawara, S., Yoshikawa, S., & Matsumoto, H. (2021). Process intensification of dry reforming of methane by structured catalytic wall-plate microreactor. *Chemical Engineering Journal*, *412*, 128636.

Hanafi, S. A., Elmelawy, M. S., Shalaby, N. H., El-Syed, H. A., Eshaq, G., & Mostafa, M. S. (2016). Hydrocracking of waste chicken fat as a cost effective feedstock for renewable fuel production: A kinetic study. *Egyptian Journal of Petroleum*, *25*, 531–537.

Harun, N., Othman, N. A., Zaki, N. A., Mat Rasul, N. A., Samah, R. A., & Hashim, H. (2019). Simulation of anaerobic digestion for biogas production from food waste using SuperPro designer. *Materials Today: Proceedings*, *19*, 1315–1320.

Hassan, S. N., Sani, Y. M., Abdul Aziz, A. R., Sulaiman, N. M. N., & Daud, W. M. A. W. (2015). Biogasoline: An out-of-the-box solution to the food-for-fuel and land-use competitions. *Energy Conversion and Management*, *89*, 349–367.

He, C., Tang, C., Li, C., Yuan, J., Tran, K.-Q., Bach, Q.-V., et al. (2018). Wet torrefaction of biomass for high quality solid fuel production: A review. *Renewable and Sustainable Energy Reviews*, *91*, 259–271.

Henley, E. J., & Seader, J. D. (1981). *Equilibrium-stage separation operations in chemical engineering*. John Wiley & Sons.

Hernández, S., & Jiménez, A. (1996). Design of optimal thermally-coupled distillation systems using a dynamic model. *Chemical Engineering Research and Design*, *74*, 357–362.

Hesselgreaves, J. E., Law, R., & Reay, D. A. (2017a). Chapter 1—Introduction. In J. E. Hesselgreaves, R. Law, & D. A. Reay (Eds.), *Compact heat exchangers* (2nd ed., pp. 1–33). Butterworth-Heinemann.

Hesselgreaves, J. E., Law, R., & Reay, D. A. (2017b). Chapter 7—Thermal design. In J. E. Hesselgreaves, R. Law, & D. A. Reay (Eds.), *Compact heat exchangers* (2nd ed., pp. 275–360). Butterworth-Heinemann.

Hosseini, S. Y., Khosravi-Nikou, M. R., & Shariati, A. (2019). Production of hydrogen and syngas using chemical looping technology via cerium-iron mixed oxides. *Chemical Engineering and Processing Process Intensification*, *139*, 23–33.

Huang, Q., Zeng, J., & Yan, J. (2021). COVID-19 mRNA vaccines. *Journal of Genetics and Genomics*, *48*(2), 107–114.

Huzir, N. M., Aziz, M. M. A., Ismail, S. B., Abdullah, B., Mahmood, N. A. N., Umor, N. A., et al. (2018). Agro-industrial waste to biobutanol production: Eco-friendly biofuels for next generation. *Renewable and Sustainable Energy Reviews*, *94*, 476–485.

Ibrahim, M. F., Kim, S. W., & Abd-Aziz, S. (2018). Advanced bioprocessing strategies for biobutanol production from biomass. *Renewable and Sustainable Energy Reviews*, *91*, 1192–1204.

IEA. (2020). *Coal information—Database documentation*. Int. Energy Agency.

IEA. (2020a). *World energy outlook 2020—Analysis—IEA*.

IEA. (2020b). *World total final consumption by source, 1973–2018—Charts—Data & Statistics—IEA*.

IEA. (2020c). *Global primary energy demand growth by scenario, 2019–2030—Charts—Data & Statistics—IEA*.

IEA. (2020d). *Sustainable recovery—Analysis—IEA*.

ISO. (2014). *ISO—ISO 17225-6:2014—Solid biofuels—Fuel specifications and classes—Part 6: Graded non-woody pellets*. Int. Stand. Organ.

ISO. (2021). *ISO - ISO 17225-3:2014—Solid biofuels—Fuel specifications and classes—Part 3: Graded wood briquettes*. Int. Stand. Organ.

Isomäki, R., Pitkäaho, S., Niemistö, J., & Keiski, R. L. (2017). Biobutanol production technologies. In M. A. Abraham (Ed.), *Encyclopedia of sustainable technologies* (pp. 285–291). Oxford: Elsevier.

Jamshidmofid, M., Abbassi, A., & Bahiraei, M. (2021). Efficacy of a novel graphene quantum dots nanofluid in a microchannel heat exchanger. *Applied Thermal Engineering*, *189*, 116673.

Jiang, P., Van Fan, Y., & Klemeš, J. J. (2021). Impacts of COVID-19 on energy demand and consumption: Challenges, lessons and emerging opportunities. *Applied Energy*, *285*, 116441.

Junka, R. A., Daly, L. E., & Yu, X. (2013). Bioreactors for evaluating cell infiltration and tissue formation in biomaterials. In M. Jaffe, W. Hammond, P. Tolias, & T. Arinzeh (Eds.), *Woodhead publishing series in biomaterials. Characterization of biomaterials* (pp. 138–181). Woodhead Publishing.

Kaliyan, N., & Vance Morey, R. (2009). Factors affecting strength and durability of densified biomass products. *Biomass and Bioenergy*, *33*, 337–359.

Kandaramath Hari, T., Yaakob, Z., & Binitha, N. N. (2015). Aviation biofuel from renewable resources: Routes, opportunities and challenges. *Renewable and Sustainable Energy Reviews*, *42*, 1234–1244.

Kaymak, D. B. (2018). A novel process design for biobutanol purification from ABE fermentation. *Chemical Engineering Transactions*, *69*, 445–450.

Keiski, R. L., Ojala, S., Huuhtanen, M., Kolli, T., & Leiviskä, K. (2011). Partial oxidation (POX) processes and technology for clean fuel and chemical production. In M. R. Khan (Ed.), *Woodhead publishing series in energy. Advances in clean hydrocarbon fuel processing* (pp. 262–286). Woodhead Publishing.

Khalid, A., Aslam, M., Qyyum, M. A., Faisal, A., Khan, A. L., Ahmed, F., et al. (2019). Membrane separation processes for dehydration of bioethanol from fermentation broths: Recent developments, challenges, and prospects. *Renewable and Sustainable Energy Reviews*, *105*, 427–443.

Khalid, Z. B., Siddique, M. N. I., Nasrullah, M., Singh, L., Wahid, Z. B. A., & Ahmad, M. F. (2019). Application of solar assisted bioreactor for biogas production from palm oil mill effluent co-digested with cattle manure. *Environmental Technology and Innovation*, *16*, 100446.

Khan, M. G., & Fartaj, A. (2011). A review on microchannel heat exchangers and potential applications. *International Journal of Energy Research*.

Kiani, M. R., Makarem, M. A., Farsi, M., & Rahimpour, M. R. (2020). Novel gas-liquid contactors for CO_2 capture: Mini- and micro-channels, and rotating packed beds. In M. R. Rahimpour, M. Farsi, & M. A. Makarem (Eds.), *Advances in carbon capture* (pp. 151–170). Woodhead Publishing.

Kim, B., Heo, H. Y., Son, J., Yang, J., Chang, Y.-K., Lee, J. H., et al. (2019). Simplifying biodiesel production from microalgae via wet in situ transesterification: A review in current research and future prospects. *Algal Research*, *41*, 101557.

Kiss, A. A., Ignat, R. M., Flores Landaeta, S. J., & de Haan, A. B. (2013). Intensified process for aromatics separation powered by Kaibel and dividing-wall columns. *Chemical Engineering and Processing Process Intensification*, *67*, 39–48.

Klemeš, J. J., Wang, Q.-W., Varbanov, P. S., Zeng, M., Chin, H. H., Lal, N. S., et al. (2020). Heat transfer enhancement, intensification and optimisation in heat exchanger network retrofit and operation. *Renewable and Sustainable Energy Reviews*, *120*, 109644.

Kolb, G., Schürer, J., Tiemann, D., Wichert, M., Zapf, R., Hessel, V., et al. (2007). Fuel processing in integrated micro-structured heat-exchanger reactors. *Journal of Power Sources*, *171*, 198–204.

Konopacka, A., Rakoczy, R., & Konopacki, M. (2019). The effect of rotating magnetic field on bioethanol production by yeast strain modified by ferrimagnetic nanoparticles. *Journal of Magnetism and Magnetic Materials*, *473*, 176–183.

Kordulis, C., Bourikas, K., Gousi, M., Kordouli, E., & Lycourghiotis, A. (2016). Development of nickel based catalysts for the transformation of natural triglycerides and related compounds into green diesel: A critical review. *Applied Catalysis B: Environmental*, *181*, 156–196.

Kougias, P. G., & Angelidaki, I. (2018). Biogas and its opportunities—A review. *Frontiers of Environmental Science & Engineering*, *12*, 14.

Kovačić, Đ., Rupčić, S., Kralik, D., Jovičić, D., Spajić, R., & Tišma, M. (2021). Pulsed electric field: An emerging pretreatment technology in a biogas production. *Waste Management*, *120*, 467–483.

Lahijani, P., Zainal, Z. A., Mohamed, A. R., & Mohammadi, M. (2014). Microwave-enhanced CO2 gasification of oil palm shell char. *Bioresource Technology*, *158*, 193–200.

Lee, K., Lee, M.-E., Kim, J.-K., Shin, B., & Choi, M. (2019). Single-step hydroconversion of triglycerides into biojet fuel using CO-tolerant PtRe catalyst supported on USY. *Journal of Catalysis*, *379*, 180–190.

Leiva Butti, J. M., Núñez Mc Leod, J. E., & Rivera, S. S. (2020). Solar gasification of grape marc for syngas production: Solar-dish-coupled reactor basic engineering and optimization. *Chemical Engineering and Processing Process Intensification*, *156*, 108050.

Leonzio, G. (2018). State of art and perspectives about the production of methanol, dimethyl ether and syngas by carbon dioxide hydrogenation. *Journal of CO2 Utilization*, *27*, 326–354.

Li, J., Demirel, S. E., & Hasan, M. M. F. (2019). Systematic process intensification involving zeotropic distillation. In S. G. Muñoz, C. D. Laird, & M. J. Realff (Eds.), *47. Computer aided chemical engineering* (pp. 421–426). Elsevier.

Li, M., Cui, Y., Shi, X., Zhang, Z., Zhao, X., Zhu, X., & Gao, J. (2021). Simulated annealing-based optimal design of energy efficient ternary extractive dividing wall distillation process for separating benzene-isopropanol-water mixtures. *Chinese Journal of Chemical Engineering*, *33*, 203–210.

Liu, J., Ren, J., Yang, Y., Liu, X., & Sun, L. (2021). Effective semicontinuous distillation design for separating normal alkanes via multi-objective optimization and control. *Chemical Engineering Research and Design*, *168*, 340–356.

Liu, M., Wei, Y., & Leng, X. (2021). Improving biogas production using additives in anaerobic digestion: A review. *Journal of Cleaner Production*, *297*, 126666.

Liu, X., Zhang, H., Hong, H., & Jin, H. (2020). Experimental study on honeycomb reactor using methane via chemical looping cycle for solar syngas. *Applied Energy*, *268*, 114995.

Lizama, A. C., Figueiras, C. C., Gaviria, L. A., Pedreguera, A. Z., & Ruiz Espinoza, J. E. (2019). Nano-ferrosonication: A novel strategy for intensifying the methanogenic process in sewage sludge. *Bioresource Technology*, *276*, 318–324.

Maity, S. K. (2015). Opportunities, recent trends and challenges of integrated biorefinery: Part I. *Renewable and Sustainable Energy Reviews*, *43*, 1427–1445.

Mamvura, T. A., & Danha, G. (2020). Biomass torrefaction as an emerging technology to aid in energy production. *Heliyon*, *6*(3), e03531.

Martinez-Burgos, W. J., de Souza Candeo, E., Pedroni Medeiros, A. B., Cesar de Carvalho, J., Oliveira de Andrade Tanobe, V., Soccol, C. R., et al. (2021). Hydrogen: Current advances and patented technologies of its renewable production. *Journal of Cleaner Production*, *286*, 124970.

Mascal, M., & Dutta, S. (2020). Synthesis of highly-branched alkanes for renewable gasoline. *Fuel Processing Technology*, *197*, 106192.

Modak, N. M., Lobos, V., Merigó, J. M., Gabrys, B., & Lee, J. H. (2020). Forty years of computers & chemical engineering: A bibliometric analysis. *Computers and Chemical Engineering*, *141*, 106978.

Mohd Azhar, S. H., Abdulla, R., Jambo, S. A., Marbawi, H., Gansau, J. A., Mohd Faik, A. A., et al. (2017). Yeasts in sustainable bioethanol production: A review. *Biochemistry and Biophysics Reports*, *10*, 52–61.

Mohiddin, M. N. B., Tan, Y. H., Seow, Y. X., Kansedo, J., Mubarak, N. M., Abdullah, M. O., et al. (2021). Evaluation on feedstock, technologies, catalyst and reactor for sustainable biodiesel production: A review. *Journal of Industrial and Engineering Chemistry*, *98*, 60–81.

Moreno-Gómez, A. L., Gutiérrez-Antonio, C., Gómez-Castro, F. I., & Hernández, S. (2021). Modelling, simulation and intensification of the hydroprocessing of chicken fat to produce renewable aviation fuel. *Chemical Engineering and Processing Process Intensification*, *159*, 108250.

Moulijn, J. A., & Stankiewicz, A. (2017). Process intensification☆. In M. A. Abraham (Ed.), *Encyclopedia of sustainable technologies* (pp. 509–518). Oxford: Elsevier.

Murrieta-Dueñas, R., Gutiérrez-Guerra, R., Segovia-Hernández, J. G., & Hernández, S. (2011). Analysis of control properties of intensified distillation sequences: Reactive and extractive cases. *Chemical Engineering Research and Design*, *89*, 2215–2227.

Nagarajan, S., & Ranade, V. V. (2020). Pre-treatment of distillery spent wash (vinasse) with vortex based cavitation and its influence on biogas generation. *Bioresource Technology Reports*, *11*, 100480.

Natarajan, Y., Nabera, A., Salike, S., Dhanalakshmi Tamilkkuricil, V., Pandian, S., Karuppan, M., et al. (2019). An overview on the process intensification of microchannel reactors for biodiesel production. *Chemical Engineering and Processing Process Intensification*, *136*, 163–176.

Ngoenthong, N., Tongnan, V., Sornchamni, T., Siri-nguan, N., Laosiripojana, N., & Hartley, U. W. (2019). Application of a micro-channel reactor for process intensification in high purity syngas production via H_2O/CO_2 co-splitting. *International Journal of Hydrogen Energy*, *46*(48), 24581–24590.

Ohayon Dahan, H., Porgador, B., Landau, M. V., & Herskowitz, M. (2020). Conversion of hydrous bioethanol on $Zn_xZr_yO_z$ catalyst to renewable liquid chemicals and additives to gasoline. *Fuel Processing Technology*, *198*, 106246.

Okoli, C. O., & Adams, T. A. (2015). Design of dividing wall columns for butanol recovery in a thermochemical biomass to butanol process. *Chemical Engineering and Processing Process Intensification*, *95*, 302–316.

Oliveira de Brito Lira, J., Riella, H. G., Padoin, N., & Soares, C. (2021). An overview of photoreactors and computational modeling for the intensification of photocatalytic processes in the gas-phase: State-of-art. *Journal of Environmental Chemical Engineering*, *9*, 105068.

Ong, H. C., Tiong, Y. W., Goh, B. H. H., Gan, Y. Y., Mofijur, M., Fattah, I. M. R., et al. (2021). Recent advances in biodiesel production from agricultural products and microalgae using ionic liquids: Opportunities and challenges. *Energy Conversion and Management*, *228*, 113647.

Ortiz-Martínez, V. M., Andreo-Martínez, P., García-Martínez, N., Pérez de los Ríos, A., Hernández-Fernández, F. J., & Quesada-Medina, J. (2019). Approach to biodiesel production from microalgae under supercritical conditions by the PRISMA method. *Fuel Processing Technology*, *191*, 211–222.

Ovchinnikova, E. V., Banzaraktsaeva, S. P., & Chumachenko, V. A. (2019). Optimal design of ring-shaped alumina catalyst: A way to intensify bioethanol-to-ethylene production in multi-tubular reactor. *Chemical Engineering Research and Design*, *145*, 1–11.

Pacheco-Basulto, J.Á., Hernández-McConville, D., Barroso-Muñoz, F. O., Hernández, S., Segovia-Hernández, J. G., Castro-Montoya, A. J., et al. (2012). Purification of bioethanol using extractive batch distillation: Simulation and experimental studies. *Chemical Engineering and Processing Process Intensification*, *61*, 30–35.

Pagliuso, J. D. (2010). Biofuels for spark-ignition engines. In H. Zhao (Ed.), *Advanced direct injection combustion engine technologies and development* (pp. 229–259). Woodhead Publishing.

Pal, P., Kumar, R., & Ghosh, A. K. (2018). Analysis of process intensification and performance assessment for fermentative continuous production of bioethanol in a multi-staged membrane-integrated bioreactor system. *Energy Conversion and Management, 171*, 371–383.

Pang, S. (2016). Fuel flexible gas production: Biomass, coal and bio-solid wastes. In J. Oakey (Ed.), *Fuel flexible energy generation* (pp. 241–269). Boston: Woodhead Publishing.

Papanikolaou, G., Lanzafame, P., Giorgianni, G., Abate, S., Perathoner, S., & Centi, G. (2020). Highly selective bifunctional Ni zeo-type catalysts for hydroprocessing of methyl palmitate to green diesel. *Catalysis Today, 345*, 14–21.

Parente, M., Soria, M. A., & Madeira, L. M. (2020). Hydrogen and/or syngas production through combined dry and steam reforming of biogas in a membrane reactor: A thermodynamic study. *Renewable Energy, 157*, 1254–1264.

Piriyarungrod, N., Kumar, M., Thianpong, C., Pimsarn, M., Chuwattanakul, V., & Eiamsa-ard, S. (2018). Intensification of thermo-hydraulic performance in heat exchanger tube inserted with multiple twisted-tapes. *Applied Thermal Engineering, 136*, 516–530.

Pistikopoulos, E. N., Barbosa-Povoa, A., Lee, J. H., Misener, R., Mitsos, A., Reklaitis, G. V., et al. (2021). Process systems engineering – The generation next? *Computers and Chemical Engineering, 147*, 107252.

Pradhan, P., Mahajani, S. M., & Arora, A. (2018). Production and utilization of fuel pellets from biomass: A review. *Fuel Processing Technology, 181*, 215–232.

Pramanik, S. K., Suja, F. B., Zain, S. M., & Pramanik, B. K. (2019). The anaerobic digestion process of biogas production from food waste: Prospects and constraints. *Bioresource Technology Reports, 8*, 100310.

Reay, D., Ramshaw, C., & Harvey, A. (2008). Compact and micro-heat exchangers. In D. Reay, C. Ramshaw, & A. Harvey (Eds.), *Process intensification: Engineering for efficiency, sustainability and flexibility* (pp. 77–101). Oxford: Butterworth-Heinemann.

Reay, D., Ramshaw, C., & Harvey, A. (2013a). Chapter 8—Application areas—Petrochemicals and fine chemicals. In D. Reay, C. Ramshaw, & A. Harvey (Eds.), *Process intensification (second edition), isotopes in organic chemistry* (pp. 259–321). Oxford: Butterworth-Heinemann.

Reay, D., Ramshaw, C., & Harvey, A. (2013b). Chapter 5—reactors. In D. Reay, C. Ramshaw, & A. Harvey (Eds.), *Process intensification (second edition), isotopes in organic chemistry* (pp. 121–204). Oxford: Butterworth-Heinemann.

Ren, J., Cao, J.-P., Zhao, X.-Y., Yang, F.-L., & Wei, X.-Y. (2019). Recent advances in syngas production from biomass catalytic gasification: A critical review on reactors, catalysts, catalytic mechanisms and mathematical models. *Renewable and Sustainable Energy Reviews, 116*, 109426.

Roberts, L. G., & Patterson, T. J. (2014). Biofuels. In P. Wexler (Ed.), *Encyclopedia of toxicology* (3rd ed., pp. 469–475). Oxford: Academic Press.

Rochón, E., Cortizo, G., Cabot, M. I., García Cubero, M. T., Coca, M., Ferrari, M. D., et al. (2020). Bioprocess intensification for isopropanol, butanol and ethanol (IBE) production by fermentation from sugarcane and sweet sorghum juices through a gas stripping-pervaporation recovery process. *Fuel, 281*, 118593.

Rom, A., Miltner, A., Wukovits, W., & Friedl, A. (2016). Energy saving potential of hybrid membrane and distillation process in butanol purification: Experiments, modelling and simulation. *Chemical Engineering and Processing Process Intensification, 104*, 201–211.

Romero-Izquierdo, A. G., Gómez-Castro, F. I., Gutiérrez-Antonio, C., Hernández, S., & Errico, M. (2021). Intensification of the alcohol-to-jet process to produce renewable aviation fuel. *Chemical Engineering and Processing Process Intensification, 160*, 108270.

Rong, B.-G. (2011). Synthesis of dividing-wall columns (DWC) for multicomponent distillations—A systematic approach. *Chemical Engineering Research and Design, 89*, 1281–1294.

Rong, B.-G. (2014). A systematic procedire for synthesis of intensified nonsharp distillation systems with fewer columns. *Chemical Engineering Research and Design, 92*(10), 1955–1968.

Rong, B.-G., & Errico, M. (2012). Synthesis of intensified simple column configurations for multicomponent distillations. *Chemical Engineering and Processing Process Intensification, 62*, 1–17.

Sánchez-Ramírez, E., Quiroz-Ramírez, J. J., Hernández, S., Segovia-Hernández, J. G., & Kiss, A. A. (2017). Optimal hybrid separations for intensified downstream processing of biobutanol. *Separation and Purification Technology, 185*, 149–159.

Sánchez-Ramírez, E., Quiroz-Ramírez, J. J., Segovia-Hernández, J. G., Hernández, S., & Ponce-Ortega, J. M. (2016). Economic and environmental optimization of the biobutanol purification process. *Clean Technologies and Environmental Policy, 18*, 395–411.

Santana, H. S., Silva, J. L., & Taranto, O. P. (2019). Development of microreactors applied on biodiesel synthesis: From experimental investigation to numerical approaches. *Journal of Industrial and Engineering Chemistry, 69*, 1–12.

Saxena, S., Saharan, V. K., & George, S. (2019). Modeling & simulation studies on batch anaerobic digestion of hydrodynamically cavitated tannery waste effluent for higher biogas yield. *Ultrasonics Sonochemistry, 58*, 104692.

Seifollahi, M., & Amiri, H. (2020). Enhanced production of cellulosic butanol by simultaneous co-saccharification and fermentation of water-soluble cellulose oligomers obtained by chemical hydrolysis. *Fuel, 263*, 116759.

Shamsul, N. S., Kamarudin, S. K., & Rahman, N. A. (2017). Conversion of bio-oil to bio gasoline via pyrolysis and hydrothermal: A review. *Renewable and Sustainable Energy Reviews, 80*, 538–549.

Sharma, B., Larroche, C., & Dussap, C.-G. (2020). Comprehensive assessment of 2G bioethanol production. *Bioresource Technology, 313*, 123630.

Sikarwar, V. S., & Zhao, M. (2017). Biomass gasification. In M. A. Abraham (Ed.), *Encyclopedia of sustainable technologies* (pp. 205–216). Oxford: Elsevier.

Simeons, C. (1980). Hydrogen. In C. Simeons (Ed.), *Hydro-power* (pp. 164–174). Pergamon.

Sindhu, R., Binod, P., Pandey, A., Ankaram, S., Duan, Y., & Awasthi, M. K. (2019). Biofuel production from biomass: Toward sustainable development. In S. Kumar, R. Kumar, & A. Pandey (Eds.), *Current developments in biotechnology and bioengineering* (pp. 79–92). Elsevier.

Singh, A., da Cunha, S., & Rangaiah, G. P. (2019). Heat-pump assisted distillation versus double-effect distillation for bioethanol recovery followed by pressure swing adsorption for bioethanol dehydration. *Separation and Purification Technology, 210*, 574–586.

Sivaramakrishnan, R., & Incharoensakdi, A. (2018). Microalgae as feedstock for biodiesel production under ultrasound treatment—A review. *Bioresource Technology, 250*, 877–887.

Skorupskaite, V., Makareviciene, V., & Gumbyte, M. (2016). Opportunities for simultaneous oil extraction and transesterification during biodiesel fuel production from microalgae: A review. *Fuel Processing Technology, 150*, 78–87.

Smith, J. M. (1981). *Chemical engineering kinetics. 18* (pp. 1–612). McGraw-Hill.

Soria-Ornelas, M. L., Gutiérrez-Antonio, C., & Rodríguez, J. M. (2016). Biocombustibles de cara al futuro: un panorama actual. *Digital Ciencia@UAQRO* (pp. 1–16). Universidad Autónoma de Querétaro.

Staak, D., & Grützner, T. (2017). Process integration by application of an extractive dividing-wall column: An industrial case study. *Chemical Engineering Research and Design, 123*, 120–129.

Stankiewicz, A. I., & Moulijn, J. A. (2000). Process intensification: Transforming chemical engineering. *Chemical Engineering Progress*, 22–34.

Subhedar, P. B., & Gogate, P. R. (2015). Ultrasound-assisted bioethanol production from waste newspaper. *Ultrasonics Sonochemistry, 27*, 37–45.

Tan, S. X., Lim, S., Ong, H. C., & Pang, Y. L. (2019). State of the art review on development of ultrasound-assisted catalytic transesterification process for biodiesel production. *Fuel, 235*, 886–907.

Terán Hilares, R., Kamoei, D. V., Ahmed, M. A., da Silva, S. S., Han, J.-I., & dos Santos, J. C. (2018). A new approach for bioethanol production from sugarcane bagasse using hydrodynamic cavitation assisted-pretreatment and column reactors. *Ultrasonics Sonochemistry, 43*, 219–226.

Tian, Y., Pappas, I., Burnak, B., Katz, J., & Pistikopoulos, E. N. (2021). Simultaneous design & control of a reactive distillation system – A parametric optimization & control approach. *Chemical Engineering Science, 230*, 116232.

Tiwari, A., Rajesh, V. M., & Yadav, S. (2018). Biodiesel production in micro-reactors: A review. *Energy for Sustainable Development, 43*, 143–161.

Torres-Ortega, C.-E., & Rong, B.-G. (2016). Intensified separation processes for the recovery and dehydration of bioethanol from an actual lignocellulosic fermentation broth. In Z. Kravanja, & M. Bogataj (Eds.), *26th European symposium on computer aided process engineering, computer aided chemical engineering* (pp. 727–732). Elsevier.

Torres-Ortega, C. E., Strieker, K., Errico, M., & Rong, B.-G. (2015). Design and optimization of intensified non-sharp distillation configurations. In K. V. Gernaey, J. K. Huusom, & R. Gani (Eds.), *37. Computer aided chemical engineering* (pp. 1055–1060). Elsevier.

Tsatse, A., Oudenhoven, S. R. G., ten Kate, A. J. B., & Sorensen, E. (2021). Optimal design and operation of reactive distillation systems based on a superstructure methodology. *Chemical Engineering Research and Design, 170*, 107–133.

Tsodikov, M. V., Fedotov, A. S., Antonov, D. O., Uvarov, V. I., Bychkov, V. Y., & Luck, F. C. (2016). Hydrogen and syngas production by dry reforming of fermentation products on porous ceramic membrane-catalytic converters. *International Journal of Hydrogen Energy, 41*, 2424–2431.

van Wyk, S., van der Ham, A. G. J., & Kersten, S. R. A. (2018). Pervaporative separation and intensification of downstream recovery of acetone-butanol-ethanol (ABE). *Chemical Engineering and Processing Process Intensification, 130*, 148–159.

Vandewalle, L. A., Marin, G. B., & Van Geem, K. M. (2021). CFD-based assessment of steady-state multiplicity in a gas-solid vortex reactor for oxidative coupling of methane. *Chemical Engineering and Processing Process Intensification, 165*, 108434.

Vásquez, M. C., Silva, E. E., & Castillo, E. F. (2017). Hydrotreatment of vegetable oils: A review of the technologies and its developments for jet biofuel production. *Biomass and Bioenergy, 105*, 197–206.

Vazquez-Castillo, J. A., Venegas-Sánchez, J. A., Segovia-Hernández, J. G., Hernández-Escoto, H., Hernández, S., Gutiérrez-Antonio, C., et al. (2009). Design and optimization, using genetic algorithms, of intensified distillation systems for a class of quaternary mixtures. *Computers and Chemical Engineering, 33*, 1841–1850.

Verma, D., Kumar, R., Rana, B. S., & Sinha, A. K. (2011). Aviation fuel production from lipids by a single-step route using hierarchical mesoporous zeolites. *Energy & Environmental Science, 4*, 1667–1671.

Viswanathan, B. (2017). Natural gas. In B. Viswanathan (Ed.), *Energy sources* (pp. 59–79). Amsterdam: Elsevier.

Waltermann, T., Münchrath, D., & Skiborowski, M. (2017). Efficient optimization-based design of energy-intensified azeotropic distillation processes. In A. Espuña, M. Graells, & L. Puigjaner (Eds.), *40. Computer aided chemical engineering* (pp. 1045–1050). Elsevier.

Wang, Y., Ho, S.-H., Yen, H.-W., Nagarajan, D., Ren, N.-Q., Li, S., et al. (2017). Current advances on fermentative biobutanol production using third generation feedstock. *Biotechnology Advances, 35*, 1049–1059.

Wei, R., Li, H., Chen, Y., Hu, Y., Long, H., Li, J., et al. (2020). Environmental issues related to bioenergy. In *Reference module in earth systems and environmental sciences* Elsevier.

Weinfeld, J. A., Owens, S. A., & Eldridge, R. B. (2018). Reactive dividing wall columns: A comprehensive review. *Chemical Engineering and Processing Process Intensification, 123*, 20–33.

Wong, K. Y., Ng, J.-H., Chong, C. T., Lam, S. S., & Chong, W. T. (2019). Biodiesel process intensification through catalytic enhancement and emerging reactor designs: A critical review. *Renewable and Sustainable Energy Reviews, 116*, 109399.

Yan, P., Stankiewicz, A. I., Eghbal Sarabi, F., & Nigar, H. (2021). Microwave heating in heterogeneous catalysis: Modelling and design of rectangular traveling-wave microwave reactor. *Chemical Engineering Science, 232*, 116383.

Yang, A., Chun, W., Sun, S., Shi, T., Ren, J., & Shen, W. (2020). Dynamic study in enhancing the controllability of an energy-efficient double side-stream ternary extractive distillation of acetonitrile/methanol/benzene with three azeotropes. *Separation and Purification Technology, 242*, 116830.

Yang, L., & Ge, X. (2016). Biogas and syngas upgrading. In Y. Li, & X. Ge (Eds.), *Advances in bioenergy* (pp. 125–188). Elsevier.

Yang, A., Jin, S., Shen, W., Cui, P., Chien, I.-L., & Ren, J. (2019). Investigation of energy-saving azeotropic dividing wall column to achieve cleaner production via heat exchanger network and heat pump technique. *Journal of Cleaner Production, 234*, 410–422.

Yeo, T. Y., Ashok, J., & Kawi, S. (2019). Recent developments in sulphur-resilient catalytic systems for syngas production. *Renewable and Sustainable Energy Reviews, 100*, 52–70.

Yeong, T. K., Jiao, K., Zeng, X., Lin, L., Pan, S., & Danquah, M. K. (2018). Microalgae for biobutanol production—Technology evaluation and value proposition. *Algal Research, 31*, 367–376.

Yildirim, Ö., Kiss, A. A., & Kenig, E. Y. (2011). Dividing wall columns in chemical process industry: A review on current activities. *Separation and Purification Technology*, *80*, 403–417.

Zhang, Y., Cui, Y., Chen, P., Liu, S., Zhou, N., Ding, K., et al. (2019). Gasification technologies and their energy potentials. In M. J. Taherzadeh, K. Bolton, J. Wong, & A. Pandey (Eds.), *Sustainable resource recovery and zero waste approaches* (pp. 193–206). Elsevier.

Zhang, G., Sun, Y., & Xu, Y. (2018). Review of briquette binders and briquetting mechanism. *Renewable and Sustainable Energy Reviews*, *82*, 477–487.

Zhang, Q., Yang, S., Shi, P., Hou, W., Zeng, A., Ma, Y., et al. (2020). Economically and thermodynamically efficient heat pump-assisted side-stream pressure-swing distillation arrangement for separating a maximum-boiling azeotrope. *Applied Thermal Engineering*, *173*, 115228.

Zhao, F., Chen, Z., Fan, W., Dou, J., Li, L., & Guo, X. (2020). Reactor optimization and process intensification of photocatalysis for capillary-based PMMA LSC-photomicroreactors. *Chemical Engineering Journal*, *389*, 124409.

Zheng, Y., Wang, J., Liu, C., Lu, Y., Lin, X., Li, W., et al. (2020). Efficient and stable Ni-Cu catalysts for ex situ catalytic pyrolysis vapor upgrading of oleic acid into hydrocarbon: Effect of catalyst support, process parameters and Ni-to-Cu mixed ratio. *Renewable Energy*, *154*, 797–812.

CHAPTER 2

Sustainable bioalcohol production: Pretreatment, separation, and control strategies leading to sustainable processes

Thiago Edwiges[a], Maria Cinta Roda-Serrat[b], Juan Gabriel Segovia-Hernández[c], Eduardo Sánchez-Ramírez[c], Stefania Tronci[d], and Massimiliano Errico[b]

[a]Department of Biological and Environmental Sciences, Federal University of Technology, Medianeira, Parana, Brazil
[b]Faculty of Engineering, Department of Green Technology, University of Southern Denmark, Odense, Denmark
[c]Chemical Engineering Department, University of Guanajuato, Guanajuato, Mexico
[d]Dipartimento di Ingegneria Meccanica, Chimica e dei Materiali, Università degli Studi di Cagliari, Cagliari, Italy

2.1 Introduction

Biofuel is a term that became part of the common language. According to the definition provided by Radionova et al. (2017), *"biofuels are referred to the energy-enriched chemicals generated through the biological processes or derived from the biomass of living organisms, such as microalgae, plants and bacteria."* This general definition, also reported in different forms by other authors (Demirbas, 2008; Luque & Clark, 2010; Soetaert & Vandamme, 2009), includes two fundamental concepts that are indissolubly connected to biofuels: biological processes and biomass. The former identifies the type of process used for their production, while the latter highlights the nature of the feedstock.

Once the framework for identification of biofuels is given, their classification is not so straightforward. A basic classification can be done based on their use either as an unprocessed or a processed form. In this case, biofuels are named primary and secondary, respectively. The primary ones are those used directly in their natural form like firewood, animal waste, or crop residues. Secondary biofuels are produced from plants and microorganisms like bioethanol and biobutanol from fermentation of starch. Another classification can be done based on the feedstock source. In this case, biofuels are divided according to different generations (Alalwan, Alminshid, & Aljaafari, 2019). First generation biofuels are produced from edible biomass and nowadays represent the most applied technology at an industrial level (Naik, Goud, Rout, & Dalai, 2010). However, this option is not under consideration for further expansion anymore, mainly due to ethical reasons related to the food-versus-fuel debate (Filip, Janda, Kristoufek, & Zilberman, 2019). Lignocellulosic biomass and waste animal oils paved the way to the second-generation biofuels from nonedible sources. This generation is still facing some technological barriers mainly due to the full exploitation of the feedstock recalcitrant structure (Sheldon, 2017). Microbial organisms, including microalgae, are the feedstock for the third generation and they are considered a possibility to overcome the limitations of the first two generations (Chowdhury & Loganathan, 2019). In the fourth generation, synthetic biology applied to design genetically modified microorganisms is used to improve the algae-to-biofuel third generation (Moravvej, Makarem, & Reza Rahimpour, 2019).

Biofuels can be also classified based on their physical state as solid, liquid, or gaseous. In particular, liquid biofuels are further divided into vegetable oils and biodiesels, bioalcohols, biocrude, and biosynthetic oils (Demirbas, Balat, & Balat, 2011). Bioalcohols are the focus of this chapter and they are considered as a green alternative to conventional oil-derived fuels. What attracted the attention of research, industry, and society around this biofuel category was their application in the transport sector. In 2017, in the IEA countries, passenger cars alone used more energy than the whole residential sector and, together with freight road vehicles, they accounted for almost a third of final energy-related CO_2 emissions (IEA, 2020). According to the International Energy Agency, in 2017, the transport sector as a whole accounted for 36% of final energy consumption, ranking first among the different contributions. In this context, it is understandable why so much research effort has been dedicated in bringing bioalcohol production, or in general biofuels, to be economically profitable or in any case competitive with respect to the nonrenewable counterpart. This challenge is still open, and research efforts are still necessary to improve the bioalcohol/biofuel sustainability. Challenges related to the feedstock selection still remain the main concern for the overall process economy, and as highlighted by Yan et al. (2020), physical properties of bulk biomass have a great impact on biomass processing and conversion. Different residual biomasses have been recently reviewed by Papadaki (2020).

The pretreatment processes, fundamental to make the sugars available for the fermentation step, are also a topic in continuous evolution. Diluted acid, liquid hot water (LHW), steam explosion, ammonia fiber explosion, and organosolv pretreatments are the most studied methods (Rodrigues Gurgel da Silva, Errico, & Rong, 2018a; Rodrigues Gurgel da Silva, Giuliano, Errico, Rong, & Barletta, 2019), but green-oriented techniques are also emerging (Haldar & Rurkait, 2021).

Another section of particular importance for the overall process economy is the separation and purification of the products. In the specific case of bioalcohols, the low concentration of the components and the presence of azeotropes limit the possibility of using well-developed unit operations like ordinary distillation. Different methods have been proposed in literature as new enhanced distillation systems or as a hybrid combination of unit operations (Errico, 2017; Errico, Rong, Tola, & Spano, 2013a, 2013b; Errico, Sanchez-Ramirez, Quiroz-Ramirez, Rong, & Segovia-Hernandez, 2017a; Errico,

Sanchez-Ramirez, Quiroz-Ramirez, Segovia-Hernandez, & Rong, 2016; Singh & Rangaiah, 2017). Separation by membranes is also earning a spot among promising technologies for bioalcohol concentration since they are not limited by azeotropes (Abdellatif, Babin, Arnal-Herault, David, & Jonquieres, 2018; Tin, Lin, Ong, & Chung, 2011). As a natural consequence of introducing more complex separation units, process control was also called to contribute to the topic, thus proving the controllability and reliability of the different options proposed (Ramirez-Marquez, Segovia-Hernandez, Hernandez, Errico, & Rong, 2013; Sanchez-Ramirez et al., 2017; Segovia-Hernandez et al., 2014).

As stated by Alizadeh, Lund, and Soltanisehat (2020), reducing the costs of technologies is the focal challenge for the development of a biofuel economy. It is also clear that research efforts are still required to fully realize a biofuel transition without depending on national subsidies.

In the present chapter, four main aspects of bioalcohol production are explored:
1. the main pretreatment technologies to enhance fermentation of sugars for bio-alcohol production
2. separation processes obtained as intensified enhanced distillation configurations
3. membrane-based processes and their performance in bio-alcohol separation
4. control strategies applied in the biorefining context.

Feedstock selection was deliberately excluded since it has a strong geographical dependence. However, a brief discussion is given about lignocellulosic structure in order to properly introduce the pretreatment methods.

A case study for the separation of bioethanol by membrane-assisted-reactive distillation completes the chapter.

2.2 Lignocellulosic biomass

A wide variety of biomass sources can be used for bioalcohol production. The main feedstock sources include wood and wood wastes, agricultural crops, and their waste byproducts, municipal solid waste, animal manure, food waste, fruit and vegetable waste, aquatic plants, and algae. The main advantage of using biowastes is their availability in large quantities and that they do not compete with food supplies, like in the case of first-generation biofuels from starch or sugar crops. Lignocellulosic biomass is thus a promising feedstock for the production of biofuels since the carbohydrate components (cellulose and hemicellulose) are fermentable after hydrolysis and can be converted to bioalcohols (bioethanol, biobutanol, biopropanol) or other fermentation products like biomethane and biohydrogen. The conversion of lignin-rich substrates to fermentable sugars usually has technical and economic challenges due to the recalcitrance of the lignocellulose. Improving the biodegradability of the lignocellulose is a crucial step to overcome the operational drawbacks in the production of biofuels.

Lignocellulose is composed of cellulose, hemicellulose, and lignin, and it is a heterogeneous and rigid structure that is naturally resistant to biological degradation.

Cellulose is one of the main components of the plant cell wall matrix (approximately 30% of the biomass of the plant), and it is a polysaccharide which is composed of a chain of glucose strongly linked by linear β (1–4) glycosidic bonds. The linear, fibrous, and moist structure of cellulose contains about 10,000 glucose units linked together by hydrogen bonds and van der Waals forces resulting in microfibrils with high tensile strength. Cellulose is characterized by two different orientations throughout its structure: an amorphous region (low crystallinity) and a crystalline region (high crystallinity).

Hemicellulose is a short heterogeneous polymer composed of various pentoses (xylose and arabinose), hexoses (galactose, mannose, and glucose) and some organic acids (glucuronic, methyl glucuronic, and galacturonic acids). The main differences between cellulose and hemicellulose include the size of the chain (approximately 50–300 sugar units), the presence of branching in the main chain molecules, and the amorphous structure, making hemicellulose less resistant to biological, thermal, and chemical hydrolysis.

As opposed to cellulose and hemicellulose, lignin is not composed of sugar units. The amorphous structure of lignin is characterized by phenyl propane units. The dominant building blocks of the three-dimensional lignin structure include p-coumaril, coniferyl, and sinapyl alcohols. Softwood lignin is rich in coniferyl alcohol. While hardwood lignin contains both coniferyl and sinapyl alcohols, grass lignin contains all three alcohols. The concentration of lignin in the cell wall of plants depends on the species, age, and part of the plant, ranging from 15% to 45%. The function of lignin is to provide rigidity and cohesion to the cell wall, creating a hygroscopic surface that is also very resistant to microbial attack.

Cellulose and hemicellulose can be converted into simple sugars like glucose, xylose, arabinose, galactose, and/or mannose. Hexoses (glucose, galactose, and mannose) are easily converted to alcohols during fermentation by naturally occurring organisms, but a few native strains can ferment pentoses (xylose and arabinose). Thus converting the carbohydrate fractions of biowastes like lignocellulosic biomass into bioalcohols or biofuels in general is a key factor to increase process economic feasibility.

2.3 Pretreatment technologies for lignocellulosic biomass

Pretreatment technologies are applied to lignocellulosic biomass to remove or redistribute plant cell wall components, decrease recalcitrance, increase cellulose accessibility, and as final result, enhance the cost competitiveness of biofuel plants. Considering the ethanol production from lignocellulosic substrate as an example, pretreatment is one of the crucial steps, followed by enzymatic hydrolysis, separation of pentose-rich hydrolysates, fermentation of hexose and pentose hydrolysates, and product separation. This process sequence

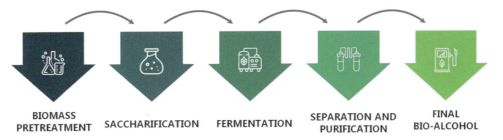

Fig. 2.1 Block flow diagram for bioalcohol production.

can be generalized to the production of different bioalcohols, as reported in the block flow diagram of Fig. 2.1.

In general, the choice of the optimal pretreatment strategy depends on the following factors (Kumar, Prasad, Giri, & Singh, 2019; Kumar, Ravikumar, Thenmozhi, Kumar, & Shankar, 2019):

1. nature of the feedstock
2. heterogeneity of lignin polymer
3. generation of toxic inhibition compounds
4. energy requirement
5. possibility of recycling the chemicals used
6. waste management

Several pretreatment strategies have been applied to lignocellulosic biomass prior to the enzymatic hydrolysis, and they can be generally classified in physical, chemical, biological, and combined process.

2.3.1 Physical pretreatment

The use of physical pretreatments increases the surface area of substrates either by reducing their particle size, through electromagnetic waves, using high temperatures, or by combination of different methods. The most used physical pretreatments are as follows: mechanical, LHW, steam explosion, ultrasound, and microwaves.

2.3.1.1 Mechanical

Mechanical pretreatment can modify the physical structure of substrates by decreasing the particle size and the degree of polymerization of cellulose, thus increasing the specific surface area and disrupting the cellulose crystallinity. The smaller particles of the cellulosic biomass after mechanical pretreatment are more amenable to subsequent enzymatic hydrolysis. Grinding and milling (hammer- and ball-milling) machines are normally used, but extrusion machines using heat to create compression and shear forces can also be applied. These pretreatments can normally increase hydrolysis yields by 5%–25% (Hendriks & Zeeman, 2009).

Mechanical pretreatments are very useful in biogas plants since smaller particle sizes can result in higher hydrolysis rate and biogas yields. However, it is not very common as a single pretreatment method applied to bioethanol/bioalcohols at commercial-scale level. This is because it can be energy intensive and it does not result in lignin removal, as it is desired to improve cellulose accessibility. Thus mechanical pretreatments are usually applied in combination with other methods.

2.3.1.2 Liquid hot water (LHW)

LHW pretreatment solubilizes hemicellulose and part of lignin when applied at temperatures of 140–200°C at neutral pH, making cellulose available to enzymatic hydrolysis. At higher temperatures, the solubility of monosaccharides that form hemicellulose (xylose, arabinose, glucose, mannose, and galactose) increases, allowing higher bioalcohols/bioethanol yields. One of the major benefits of using LHW compared to chemical pretreatments is that the sugar-enriched prehydrolysates generated after this neutral pretreatment contain little or no inhibitor and can be directly fermented. Since the pH of the hydrolysate is close to neutral, structural problems due to corrosion of pipes and equipment are less frequent, making this an attractive method for industrial-scale applications.

Pretreatments with LHW are conducted with water at 140–200°C added to the biomass for 1–4 h. The pressure effect allows water to penetrate biomass, hydrating cellulose, and solubilizing part of hemicellulose and lignin. Keeping pH at neutral range is mandatory to avoid inhibitory compounds, since solubilized hemicellulose provides the formation of aldehydes like furfural ($C_5H_4O_2$) and hydroxymethylfurfural ($C_6H_6O_3$). Moreover, lignin solubilization during pretreatments with temperatures above 180°C can lead to the formation phenolic compounds like vanillin ($C_8H_8O_3$), which if not quickly removed can inhibit the activity of anaerobic microorganisms depending on the concentration inside the reactors. Thus determining the severity of the pretreatment by balancing the formation of simple sugars and inhibitory compounds from the solubilization of lignocellulose is a key factor for a successful thermal pretreatment.

The combination of LHW for 10 min with disk milled (DM) (mechanical pretreatment) applied to sugarcane bagasse was investigated by Wang, Dien, Rausch, Tumbleson, and Singh (2018), and the results are presented in Table 2.1.

Pretreatment temperature was reported to positively affect sugar yields from bagasse. According to Table 2.1, glucose and xylose yields were 273% and 180% higher when temperature increased from 140°C to 200°C. The addition of disk milling as a mechanical refining pretreatment after LHW resulted in an improvement of glucose yields by 166%, 177%, 41%, and 10% at 140°C, 160°C, 180°C, and 200°C, respectively, indicating that higher severity factors decreased pretreatment efficiency. Moreover, inhibitory compounds were also formed at higher severity factors, with furfural concentration 7.6 times higher at 200°C than at 180°C. The combination of low-severity LHW

Table 2.1 Sugar yields and inhibitor concentrations of LHW combined with DM pretreatment in sugarcane bagasse.

Pretreatment condition	SF	Glucose yield (g/g)	Xylose yield (g/g)	Furfural (g/L)
Control (untreated)	Not applied	0.095	0.035	Not applied
140°C LHW	2.18	0.105	0.047	ND
140°C LHW + DM		0.279	0.124	Not applied
160°C LHW	2.77	0.109	0.050	ND
160°C LHW + DM		0.302	0.141	Not applied
180°C LHW	3.36	0.268	0.094	0.187
180°C LHW + DM		0.278	0.154	Not applied
200°C LHW	3.94	0.392	0.132	1.42
200°C LHW + DM		0.432	0.125	Not applied

ND, not detected by HPLC; *SF*, severity factor based on pretreatment time and temperature.

pretreatment with mechanical refining pretreatment is then indicative of a promising strategy for industrial-scale bioethanol production.

2.3.1.3 Steam explosion (autohydrolysis)

Steam explosion is a cost-effective and one of the most common pretreatment technologies applied to lignocellulosic biomass. The substrate is exposed to steam at high temperature (160–260°C) with pressure in the range of 0.6–4.8 MPa for 1–30 min using an autoclave or a jacketed reactor, where the temperature and the pressure are abruptly reduced, causing an "explosion" in the biomass structure. Organic acids like acetic acid and other acids formed from acetyl or other functional groups released from biomass hydrolyze the hemicellulose. At high temperatures, water also has acidic properties, which catalyze the hydrolysis of hemicellulose. Important parameters for the optimization of the steam-explosion pretreatment are the following: the treatment time, the temperature, the particle size, and the moisture.

The steam explosion was also considered by Boluda-Aguilar and López-Gómez (2013) for lemon (*Citrus limon* L.) peel wastes in wet condition (14% total solids) using steam at 160°C and 6 bar (0.6 MPa) for 5 min. The fermentable and total sugar yields after hydrolysis of untreated lemon peel wastes were in the range of 19–22 and 44–49 g sugars per 100 g peel dry matter. The enzymatic hydrolysis after steam explosion resulted in fermentable and total sugar yields increased up to 27–37 and 47–62 g sugar per 100 g peel dry matter.

Different pressure and time conditions during steam-explosion pretreatment were investigated by Wang et al. (2012). Rice straw (60% total solids) was pretreated during 0.5–8 min at 2.0 MPa and during 2 min at 1.5, 2.5, and 3.0 MPa to evaluate time and pressure effects on lignocellulose degradation and sugar yields for bioethanol production. The summary of the results obtained is presented in Table 2.2.

Table 2.2 Steam explosion effect on sugar degradation and sugar yields from rice straw.

Pretreatment condition		Degradation		Sugar yields	
Pressure (MPa)	Time (min)	Cellulose (%)	Hemicellulose (%)	Glucose (g/g)	Xylose (g/g)
Control (untreated)		–	–	0.019	0.018
2.0	0.5	2.1	50.2	0.027	0.050
2.0	1.0	5.0	67.4	0.029	0.060
2.0	**2.0**	**6.6**	**76.3**	**0.030**	**0.055**
2.0	4.0	8.4	79.6	0.027	0.057
2.0	8.0	9.2	82.1	0.026	0.045
1.5	2.0	4.5	67.0	0.027	0.038
2.5	2.0	7.4	76.9	0.022	0.040
3.0	2.0	8.0	78.1	0.026	0.031

Cellulose and hemicellulose degradation increased from 2% to 9% and from 50% to 82%, respectively when the time of pretreatment was increased from 0.5 to 8 min. In addition, cellulose and hemicellulose degradations were also increased from 4% to 8% and from 67% to 78%, respectively, when the pressure was increased from 1.5 to 3.0 MPa. The best operational conditions regarding enzymatic hydrolysis were obtained at 2.0 MPa for 2 min, when glucose and xylose were 0.030 and 0.055 g/g, when compared to only 0.019 and 0.018 from the untreated rice straw.

2.3.1.4 Ultrasound

Ultrasound is an acoustic energy in the form of waves with a frequency above the range of human hearing. The physical effect of ultrasound treatment is the collapse of cavitation bubbles generated during the process, changing the chemical nature of the biomass and resulting in the destruction of cell walls. Ultrasound pretreatment increases the surface area available and reduces the degree of polymerization of cellulose. The ultrasound field (usually >20 kHz) can be applied to different types of biomass, but its moisture content can affect the pretreatment efficiency. Lignocellulosic substrates usually have a better response to pretreatment by ultrasound when mixed with water because the transmission of sound at the solid–liquid interface is more effective than at the solid-air interface. Results from different biomass pretreated with ultrasound at different sonication conditions were summarized by Subhedar and Gogate (2013) and are presented in Table 2.3.

Nikolic, Mojovic, Rakin, Pejin, and Pejin (2010) investigated the possibility of improving glucose yield and ethanol productivity by applying an ultrasound pretreatment in the bioethanol production by simultaneous saccharification and fermentation of corn meal. They tested different sonication times and temperatures observing that the maximum increase in glucose concentration (3%) was achieved after 5 min of ultrasound

Table 2.3 Summary of ultrasound pretreatment characteristics applied to different biomass.

Biomass	Sonication condition	Benefit	Energy density (J/mL)
Cassava chips	2.2 kW/20 kHz	Fermentation time reduced 40% and ethanol yield increased 29%	192
Corn meal	0.6 kW/40 kHz	Ethanol yield increased 11%	360
Sugarcane bagasse	0.4 kW/24 kHz	Glucose yield improved by 91%	1.350

pretreatment. Longer durations (>5 min) were reported to increase glucose concentration, but at the end of the liquefaction step, the final glucose concentration decreased due to enzyme inhibition. Since ultrasound treatment consumes a large amount of energy, keeping sonication time as low as possible is mandatory from the economic point of view. The combination of 40 kHz at 60°C for 5 min was reported as the best operational conditions, with a glucose concentration and ethanol yield increased by 7% and 11.5%, respectively. The use of ultrasound-assisted pretreatment methods offers a possibility in defining intensified enzymatic hydrolysis processes due to the increase in the mass transfer rate enhanced by cavitation (Subhedar & Gogate, 2013).

2.3.1.5 Microwave

Microwaves are short waves of electromagnetic energy between 0.3 and 300 GHz. As an example, domestic and industrial microwaves usually operate at a frequency of 2.45 GHz. These waves increase the kinetic energy of the water present in the biomass (or added during pretreatment) leading to the boiling point. The rapid generation of heat and the increase of pressure contribute to the hydrolysis of the biomass, forcing the components out of the cell walls. During microwave pretreatment, the energy generated from an electromagnetic field is applied to biomass, promoting a fast heating in its structure (115–300°C) with reduced thermal gradient, leading to changes in the structure of cellulose and degradation of hemicellulose and lignin. This can be an alternative to traditional thermal pretreatments because it can heat a larger amount of biomass in less time (1–60 min), reducing the energy consumption during the process. The process economy for ethanol production from microwave-pretreated biomass is strongly dependent on the energy consumption.

Microwave-assisted hydrothermal pretreatment was applied to brewer's spent grains for biobutanol production by López-Linares, García-Cubero, Lucas, González-Benito, and Coca (2019). The hydrothermal parameter was combined with microwave to replace acid or alkali solutions normally applied to microwave pretreatments, resulting in a greener pretreatment strategy. The optimized conditions were reported to be 192.7°C for 5.4 min, resulting in 64% hemicellulosic sugar recovery and 70% glucose recovery in enzymatic hydrolysate. The combination of microwave with diluted sulfuric acid (0.2 M) at different power (300, 600, and 1200 W), pressure (54, 93, and 152 PSI), and time (10, 15, and 20 min) conditions were applied by Mikulski, Kłosowski, Menka, and Koim-Puchowska (2019) to evaluate the cellulosic ethanol production from maize distillery stillage. The best operational condition (300 W, 54 PSI, and 15 min) resulted in high glucose concentration (104 mg/g dry weight) and in the highest yield of enzymatic cellulose hydrolysis (76%).

2.3.2 Chemical pretreatment

Pretreatments using chemical solutions are one of the most investigated and applied strategies. Chemical pretreatments are used to alter the physical and chemical composition of lignocellulosic biomass and to facilitate contact between microorganisms and the biodegradable fraction of the substrates. For this type of pretreatment, alkaline reagents, acids, or solvents are used, in order to facilitate the breakdown of the biomass structural bonds and remove lignin and hemicellulose.

2.3.2.1 Acid

Acid pretreatment can be applied using concentrated (10%–70%) or diluted (0.1%–10%) solutions of sulfuric acid (H_2SO_4), nitric acid (HNO_3), hydrochloric acid (HCl), phosphoric acid (H_3PO_4), or acetic acid (CH_3COOH). Concentrated acids are very effective for the hydrolysis of cellulose, but they are also toxic, corrosive, and as hazardous materials, they require adequate storage and final treatment. In addition, the formation of inhibitory compounds like furfural, 5-hydroximethilfurfural, and levulinic acid is strictly conditioned by process parameters such as temperature, concentration of acid, and the type of biomass. Recovery of concentrated acids can improve the economic feasibility of this type of pretreatment. Alternatively, diluted acid solutions solubilize hemicellulose and preserve cellulose and lignin structure, improving cellulose degradability. Diluted acidic pretreatment under elevated temperature is considered a relatively cheap and economically profitable method. Moreover, economic feasibility is more attractive when diluted acid pretreatments are applied, and this is the reason why diluted acid pretreatments have been more explored when compared to the concentrated counterpart (Tomás-Pejó, Alvira, Ballesteros, & Nego, 2011).

The efficiency of dilute sulfuric acid pretreatment in the production of cellulosic ethanol was investigated by Mikulski and Kłosowski (2018). Different conditions in terms of temperature (121 or 131°C), exposure time (30 or 60 min), and acid concentration (0.1 or 0.2 M) were applied to distillery stillage. Cellulose, hemicellulose, and lignin concentration ranged from 17% to 32%, 21% to 34%, and 3% to 16%, respectively. Inhibitory compounds like furfural, 5-hydroxymethilfurfural or levulinic acid were not detected, and this was attributed to the relatively low hydrolysis temperature and limited concentration of sulfuric acid. Sugar concentrations increased with the increasing temperature, prolonged pretreatment time, and increasing sulfuric acid concentration (0.2 M H_2SO_4, 131°C and 60 min).

2.3.2.2 Alkaline

Alkaline pretreatment is one of the major technologies applied to increase bioethanol production yields from lignocellulosic feedstocks. In alkaline pretreatments, chemical reactions occur through alkaline catalysts that cause saponification of intermolecular ester bonds between hemicellulose and other lignocellulosic components, causing swelling and increasing porosity of the biomass. In addition, alkaline pretreatment promotes disruption of lignin structure and separation of structural linkages between lignin and carbohydrates. This effect decreases the degree of polymerization and crystallinity of the lignin and facilitates the contact of exoenzymes with cellulose and hemicellulose. Sodium hydroxide (NaOH) is one of the most investigated solutions, but potassium hydroxide (KOH) and calcium hydroxide ($Ca(OH)_2$) can also be used. One of the most attractive benefits of alkaline pretreatments in comparison with acid pretreatments is the possibility of being applied at lower or ambient temperatures and pressures with satisfactory results.

The alkaline pretreatment using NaOH at 8% (w/v) was applied to pine wood using different temperature conditions by Bay, Karimi, Esfahany, and Kumar (2020), and the main results are presented in Table 2.4.

Alkali pretreatment resulted in lower concentrations of hemicellulosic carbohydrates (xylan, galactan, and mannan) when compared to the untreated biomass, and the increased temperature also resulted in lower content of hemicellulose and total lignin due to their solubilization during pretreatment. In the same study, it was also reported that alkali pretreatment with NaOH at 93°C for 2 h increased glucose yield by 56% and total ethanol production by 315% when compared to the untreated biomass.

2.3.2.3 Organosolv

Pretreatments with organic or aqueous organic solvents (organosolv) are applied to break the internal lignin and hemicellulose bonds to obtain free lignin of relatively high purity that can be converted into biofuels. Organic solvents with low molecular weight like ethanol and methanol are preferred due to their low cost and easy recover at the end

Table 2.4 Composition of untreated and alkali-pretreated pine wood.

Pretreatment condition	Glucan (%)	Xylan (%)	Galactan (%)	Mannan (%)	Total lignin (%)	Lignin and hemicellulose removal (%)
Untreated	47.4	10.1	6.4	10.6	25.2	0.0
NaOH/0°C	50.3	8.9	4.2	8.1	24.0	6.9
NaOH/25°C	48.9	7.2	5.3	7.4	24.7	7.6
NaOH/93°C	54.6	5.4	4.3	7.1	23.8	11.7

of pretreatment. Alcohols with high molecular weight and high boiling point, like ethylene glycol (EG) and glycerol, demand high energy consumption for the solvent recovery. Acid or alkaline solutions that serve as catalysts in the process can be used to increase solubilization of hemicellulose and accelerate delignification. The usual temperature range of the organosolv pretreatment is 150–200°C. The high cost of additional solutions, flammability, and volatility of solvents can make this method less advantageous when compared to other pretreatment strategies. The recovery and recycling of the organic solvents used are relatively simple, making the process interesting from the technoeconomic point of view.

The organosolv pretreatment applied to spruce, a softwood tree from the Pinacea family composed of 45% glucan, 26% xylan, 28% lignin, 3% acetate, and 2% ash (dry matter), was investigated by Rodrigues Gurgel da Silva, Errico, and Rong (2017, 2018b). The pretreatment consisted of preheating at 130°C followed by ethanol mixing at 50% (w/w) in a solvent-dry biomass ratio of 5:1. Sulfuric acid (1.75 w/w) was added as a catalyst in this study. Ethanol productivity was reported to be 25,084.5 kg/h with a concentration of 11.1%. These results also indicated better performance of the organosolv pretreatment compared to diluted acid pretreatment, whose ethanol productivity was 16,202.5 kg/h with a concentration of only 5.4%.

2.3.3 Biological pretreatment

Wood degrading microorganisms that solubilize recalcitrant components of biomass during metabolic reactions are used in biological pretreatments. The main advantages of biological pretreatments are related to avoiding the use of chemicals, the low energy input, and mild environmental conditions, making this type of pretreatment an environmentally friendly alternative.

2.3.3.1 Fungi

Some species of lignolytic fungi (white, brown, and soft-rot fungi) are used to produce extracellular enzymes capable of partially degrading lignin and phenolic compounds, resulting in an easily degradable biomass for bioethanol production. Generally, brown and soft-rot fungi mainly degrade cellulose, while white-rot fungi are more effective in lignin degradation. Some disadvantages of the biological pretreatment are usually reported. Cellulose or hemicellulose can be lost during the pretreatment due to the consumption of these components in metabolic reactions, thus decreasing the yield of bioethanol. Other disadvantages include the necessity of careful control of growth conditions, the large area requirement, and the prolonged time, which can be as long as 4–8 weeks. Some fungal strains with preference of lignin degradation like *Pleurotus ostreatus*, *Ceriporiopsis subvermispora*, *Cyathus stercoreus*, and genetically modified *Phanerochaete chrysosporium* are investigated to produce less cellulase activity.

Since the choice of biological pretreatment is strongly dependent on the biomass source, García-Torreiro, López-Abelairas, Lu-Chau, and Lema (2016) evaluated the biological pretreatment efficiency in four different agricultural residues (corn stover, barley straw, corncob, and wheat straw) with moisture in the range of 7.3%–8.5% using white-rot fungus *Irpex lacteus* at 30°C for 21 days. The main results are presented in Table 2.5.

In general, glucan reduction and increase of xylan content were reported. The glucan fraction was significantly reduced only with wheat straw (from 33.7 ± 1.30 to 27.6 ± 1.0), while the increase of xylan content was significant for corn stover, barley straw, and corncob. Regarding lignin content, it was significantly reduced only with wheat straw and corn stover. As mentioned before, these results agree with the fact that fungal pretreatment has different effects depending on the feedstock. The total weight loss (TWL) achieved with wheat straw and corn stover was 30% and 21%, respectively, while the TWL for barley straw and corncob was only 2%. Moreover, the lignin removal was 42% and 46% for the wheat straw and corn stover, respectively, while corncob resulted in the lower value of 17%. Overall, the best results are expected when the lignin reduction is the highest together with the lowest sugar consumption in the shortest time.

2.3.3.2 Microorganism consortia

The use of microbial consortia appears as a way to replicate the same chemical reactions that occur in nature. The synergy between different microorganism communities that naturally degrade recalcitrant organic matter (e.g., leaves and trunks) can be applied as a strategy to degrade the lignocellulosic biomass before bioethanol production. As an advantage, this method has shorter contact time, better pH control, more efficient capability of selective lignin degradation, and microorganisms can easily adapt to different biomass (Lin, Zheng, & Dong, 2020).

2.3.4 Combined pretreatment

The combination of two or more pretreatment strategies is an interesting alternative to increase biodegradability of lignocellulosic biomass and bioethanol yields. Combined pretreatments can include the combination of mechanical, thermal, chemical, and biological strategies. A single pretreatment technique may not result in the best efficiency due to the specific function of each type of pretreatment. Therefore, combining different strategies can meet the specific need of each type of biomass, considering cost-effectiveness benefits.

The addition of chemical solutions as a catalyst to irradiation pretreatments has been performed by combining an acid or alkaline solution together with microwave or ultrasound pretreatments (Garcia, Alriols, Llano-Ponte, & Labidi, 2011). Thermochemical pretreatments combining the effect of high temperature with a catalytic effect promoted by the acid or alkaline chemical reaction was reported to improve hemicellulose solubilization and decrease optimal temperature range, when compared to the isolate thermal

Table 2.5 Composition (% dry basis) of the lignocellulosic biomass before and after biological pretreatment.

Compound	Wheat straw		Corn stover		Barley straw		Corncob	
	Raw	Pretreated	Raw	Pretreated	Raw	Pretreated	Raw	Pretreated
Glucan	33.7	27.6	35.3	33.9	34.4	28.5	30.5	24.7
Xylan	29.9	31.8	29.3	38.9	25.7	35.4	33.9	40.7
Lignin	23.4	19.5	23.2	16.4	21.1	20.2	23.6	22.1
Ash	4.2	7.6	4.6	5.4	5.9	8.1	2.1	3.2
Others	8.9	13.5	7.6	5.4	12.9	9.8	9.8	9.4

pretreatment (LHW) (Monlau, Barakat, Steyer, & Carrere, 2012). Diluted acid solutions are usually applied to minimize the formation of inhibitory compounds. Catalyzed steam-explosion combining the effect of heat and pressure with the addition of acidic chemicals as a catalyst to impregnate the biomass prior to traditional steam explosion was reported to increase hemicellulose removal and enzymatic digestibility of feedstock with less generation of inhibitory compounds (Hendriks & Zeeman, 2009).

2.4 Distillation-based bioalcohols separation processes

The separation of bioalcohols is the step that addresses the purification of the fermentation products according to the market requests. The production of bioalcohols is associated with common problems like the dilution of the stream and the presence of azeotropes between the alcohols and water. The first issue is related to the biomass used and the toxicity of the alcohol toward the microorganisms used in the fermentation that limits its maximum concentration. The presence of azeotropes is related to the thermodynamic properties of the mixture and it represents a limit in the maximum purity achievable when simple unit operations based on liquid-vapor equilibria are used. The most common unit operation at an industrial level is distillation, and it is the first obvious choice to be applied in the production of concentrated bio-alcohols. Since ordinary distillation cannot overcome the azeotropic composition, different enhanced distillation alternatives have been introduced, and they are here discussed separately for the bioethanol and biobutanol case.

2.4.1 Distillation processes for bioethanol separation

Bioethanol produced by fermentation is a diluted stream with a content of ethanol up to 10 wt%.

Extractive distillation is one of the most studied alternatives for its separation. It is based on the use of an entrainer to alter the liquid phase activity coefficient and allow the separation of pure (bio)ethanol (Meirelles, Weiss, & Herfurth, 1992; Pacheco-Basulto et al., 2012). The classic configuration is composed by three columns as shown in Fig. 2.2. According to the figure, the first column is used to approach the azeotropic composition, the second one is the extractive column where pure bioethanol is obtained as distillate, and the third is the solvent recovery column. Even if not reported in the figure, the solvent is cooled before being recycled to the extractive column.

Optimization studies on extractive distillation applied to bioethanol production were reported by different authors (Garcia-Herreros, Gomez, Gil, & Rodriguez, 2011; Gerbaud et al., 2019; Li & Bai, 2012). However, in some cases, the results suffered from the simplification of not considering the first preconcentration column. Different combinations of recycling streams and configurations with partial or total condensers were analyzed by Errico et al. (2013a), resulting in the configurations shown in Fig. 2.3 as best options.

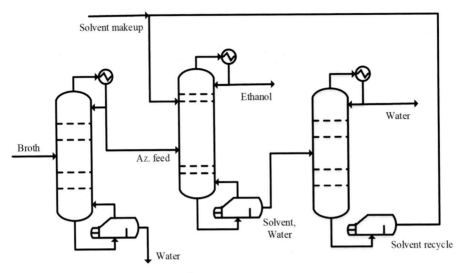

Fig. 2.2 Classic extractive distillation configuration.

Compared to the classic extractive distillation configuration reported in Fig. 2.2, the alternatives of Fig. 2.3 achieved a higher ethanol recovery by recycling the distillate of the solvent recovery column, producing only one high-purity water stream.

These configurations were then used as starting point to generate more complex configurations, including thermally coupled structures and extractive divided wall columns (DWCs) (Errico et al., 2013b). The synthesis procedure for the alternative configuration reported in Fig. 2.4 followed the principle of column section recombination explained elsewhere (Errico & Rong, 2012; Errico, Rong, Tola, & Turunen, 2009; Rong & Errico, 2012).

The configuration reported in Fig. 2.4 is classified as an intensified alternative since the number of columns was reduced from 3 to 2. Moreover, the ethanol separation and the solvent recovery are performed in a single column using an extractive DWC, where the partition is located in the stripping section of the column. The distillate obtained from the partition could be in liquid or vapor phase according to the type of condenser used, and it is recycled to the first column to avoid ethanol losses. Compared to the sequences reported in Fig. 2.3, the intensified one offered the benefit of a 23% reduction of the capital costs with the same energy consumption (Errico et al., 2013b).

Other intensified alternatives using DWC as an intensified separation method were reported by Kiss and Ignat (2012) and Kiss and Suszwalak (2012). In particular, the extractive dividing-wall reported in Fig. 2.5 integrates the separation tasks of the classical extractive distillation sequence of Fig. 2.2 in a single column (Kiss & Ignat, 2012). In this configuration, the column section above the wall acts as the extractive column, the section on the left of the wall as the preconcentrator, and the section of the right is dedicated

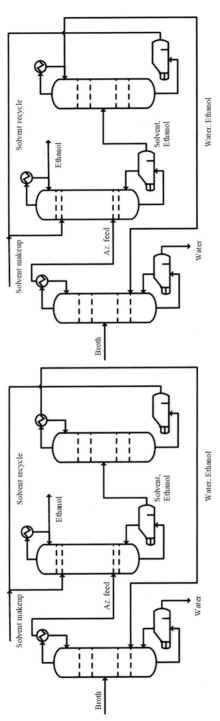

Fig. 2.3 Modified extractive distillation configurations using simple columns. Third column equipped with partial condenser (left) or with total condenser (right).

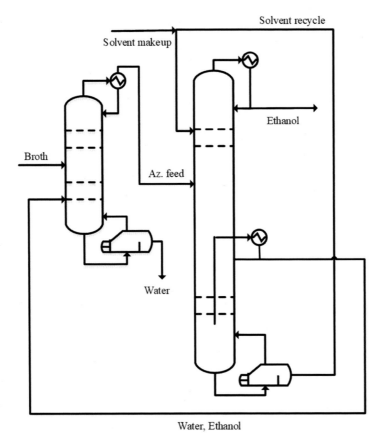

Fig. 2.4 Intensified extractive divided wall configuration.

to the solvent recovery. The water stream is removed as liquid side stream and a reboiler is used to provide the necessary vapor boil-up to the section above the withdraw. The liquid reflux is assured by the liquid feed. Overall, this configuration was able to achieve a saving of 17% in the total annual costs (TACs) when compared to the classic extractive distillation sequence.

Another possible way to overcome the azeotropic composition is by using heterogeneous azeotropic distillation. In this case, an entrainer that forms a binary or ternary heterogeneous azeotrope and then two immiscible liquid phases is used. Fig. 2.6 presents the conventional azeotropic distillation scheme and the intensified alternative proposed by Kiss and Suszwalak (2012).

According to the conventional scheme of Fig. 2.6, the feed with a composition close to the azeotropic point is sent to the first column, where pure ethanol is recovered as bottom stream, the azeotrope water-ethanol-solvent is collected as distillate, and using a decanter two phases are obtained. The organic phase rich in the solvent is recycled

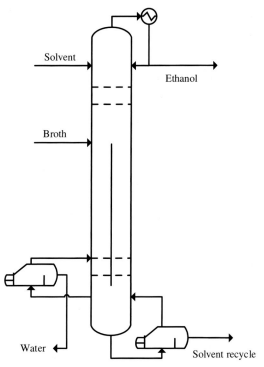

Fig. 2.5 Intensified single-column extractive divided wall configuration.

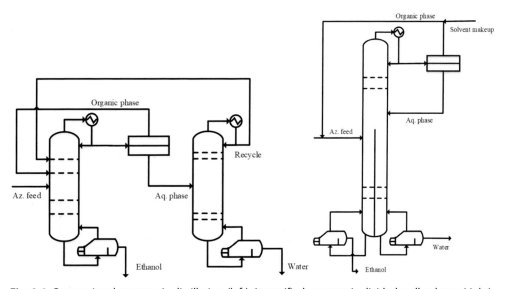

Fig. 2.6 Conventional azeotropic distillation (left), intensified azeotropic divided wall column (right).

to the first column, while the aqueous phase is sent to the second column to recover high-purity water as bottom stream. The distillate of the second column is recycled to the first one. The corresponding intensified alternative is also reported in Fig. 2.6. The separation is performed in a single column with a divided bottom section and a common overhead section. The column is equipped with two reboilers and a single condenser. Ethanol and water are the two bottom products, while the azeotropic top stream is condensed and two liquid phases are obtained through a decanter. The organic phase is recycled with the azeotropic feed, and the aqueous phase is sent to the opposite side of the dividing wall. The intensified option achieved a remarkable 20% energy saving compared to the two-column traditional scheme when n-pentane is used as an entrainer.

2.4.2 Distillation processes for biobutanol separation

Biobutanol is mainly produced by anaerobic fermentation through the so-called ABE process where together with butanol, ethanol and acetone are also produced. The typical mass proportion between acetone, butanol, and ethanol is 3:6:1 (Jiang et al., 2019). Analogous to the case of bioethanol, when lignocellulosic feedstocks are considered, pretreatments are necessary and as revised in Section 3 they still represent a considerable part of the global economy of the production process. Research efforts are still required in genetic engineering and fermentation technology in order to increase the biobutanol yield and make the process competitive (Zheng, Tashiro, Wang, & Sonomoto, 2015). In particular, the increase of the ABE fermentation yield has a positive impact on the economy of the separation section. Different distillation alternatives were proposed for this separation task. An example is reported in Fig. 2.7A and B, where the

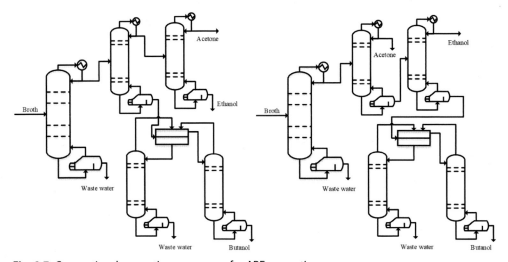

Fig. 2.7 Conventional separation sequences for ABE separation.

configurations proposed by Marlatt and Datta (1986) and Roffler, Blank, and Wilke (1987) have been reported, respectively.

These configurations have similar features, and they differ in the order of separation of acetone and ethanol. In both cases, it is possible to notice that ethanol can be recovered as an aqueous solution and different waste streams are also obtained. Butanol is recovered by stripping of the heterogeneous azeotrope phase splitting in the decanter. van der Merwe, Cheng, Gorgens, and Knoetze (2013) compared the options of Fig. 2.7 with a hybrid configuration composed by liquid-liquid extraction followed by distillation proving how this approach is the only one able to compete with the petrochemical production pathway. Following this principle Errico et al. (2017a); Errico, Sanchez-Ramirez, Quiroz-Ramirez, Rong, and Segovia-Hernandez (2017b) defined the optimal configuration for liquid-liquid assisted simple column distillation and used this configuration to generate a set of intensified alternatives composed by liquid-liquid extraction and conventional and nonconventional DWCs. The reference configuration and the corresponding best intensified configuration are reported in Fig. 2.8.

The configurations were simulated using hexyl-acetate as a solvent and the optimization was performed minimizing a multiobjective function defined to take into account the economy, the environmental impact, and the controllability. In particular, the authors used the total annualized cost (TAC) as an indication of the process economy, the eco-indicator 99 (EI99) to quantify the environmental load of the configurations over the live cycle, and the condition number (CN) as controllability index. The intensified alternative resulted in a 22% reduction of the TAC when compared to the conventional liquid-liquid assisted distillation. Moreover, a 18% reduction of the EI99 was also observed together with a better CN.

2.5 Membrane-based processes for separation of bioalcohols

Membrane technology is defined as the family of separation processes in which a semipermeable membrane allows the selective permeation of components in a mixture

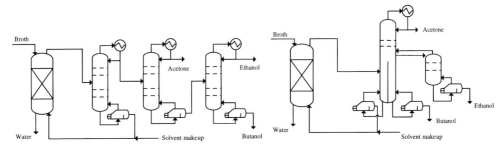

Fig. 2.8 Conventional liquid-liquid assisted distillation (left), liquid-liquid-assisted divided wall intensified configuration.

(Baker, 2004). The main advantages of membrane processes are high selectivity, low energy consumption, no need for additives, and easy scalability due to their modular design (Lipnizki, Hausmanns, Laufenberg, Field, & Kunz, 2000). These features combined with the research efforts on development of novel materials and configurations make them ideal candidates for process intensification and sustainable industrial production of bioalcohols. Two basic approaches are encountered for this application: (1) continuous bioalcohol removal during fermentation and (2) bioalcohol enrichment in postfermentation broth.

As previously mentioned, the accumulation of bioalcohols in the fermentation media inhibits of the fermentation reaction (Stanley, Bandara, Fraser, Chambers, & Stanley, 2010) and thus relatively diluted solutions and high volumes are necessary to achieve high conversion. This results in increased size and cost of fermentation and downstream equipment, and high energy demand (Jain & Chaurasia, 2014). Membrane bioreactors (MBRs) emerged as hybrid units that combine fermentation with membrane separation, where continuous removal of bioalcohols aims at keeping the concentration low enough to minimize inhibition. The membrane can be implemented either as a side-stream unit or submerged in the fermentor. There are different advantages and disadvantages to these two operation modes. On the one hand, side-stream units allow for high membrane areas and easy module replacement, while high shear stress due to pumping can be detrimental to the cells. On the other hand, submerged units are simpler to operate and do not require external circulation, but they have limited membrane area and can suffer from enhanced concentration polarization and fouling (Lipnizki et al., 2000). The selection of operation mode depends highly on the characteristics of each feedstock. For instance, lignocellulosic feedstocks with high content of suspended solids like wheat straw hydrolysate can benefit from MBRs with submerged units providing more robust filtration performance (Mahboubi et al., 2017).

Continuous closed-circulating fermentation (CCCF) systems have been shown to maintain a steady-state low ethanol concentration in the fermentor with high glucose utilization rate, high ethanol yield, and low final discharge as compared to the equivalent batch process (Ding, Wu, Tang, Yuan, & Xiao, 2011). The configuration for a CCCF with side-stream pervaporation (PV) is reported in Fig. 2.9.

The membrane-based processes used for bioalcohol purification are described in the following subsections. While the main focus is on PV, other membrane-based technologies like reverse osmosis (RO), microfiltration, ultrafiltration, membrane distillation (MD), and forward osmosis (FO) are also reported.

2.5.1 Pervaporation

PV is defined as the membrane process in which the feed and retentate are both in the liquid form and the permeate leaves the membrane as a vapor (Koros, Ma, & Shimidzu,

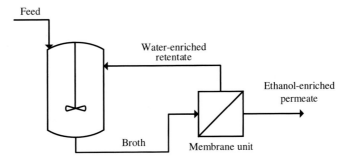

Fig. 2.9 Continuous closed-circulating fermentation configuration with side-stream membrane unit.

1996). The feed is in contact with one side of the membrane, while typically a vacuum or sweep gas is applied to the other side. The driving force for the mass transfer is the chemical potential gradient that results from the difference of partial pressure across the membrane, while selective separation takes place according to the different diffusion rates of the components through the membrane material.

In bioalcohol separation, hydrophobic PV membranes are used to selectively permeate the alcohols while the feed is enriched in water, whereas hydrophilic PV results in water permeation and dehydration of the feed (Vane, 2005). Due to the high membrane selectivity, the PV permeates can have a composition very different from the vapor phase obtained after a free vapor–liquid thermodynamic equilibrium process. In fact, the main historical application of PV is the dehydration of organic solvents that form azeotropic mixtures with water (Jonquières et al., 2002). A very representative example is the production of anhydrous ethanol, where cross-contamination by the third component used in conventional azeotropic distillation is avoided (Aptel, Challard, Cuny, & Neel, 1976).

Even though hydrophobic PV can be very selective, the permeates still contain water, and a secondary hydrophilic PV unit can be used for production of anhydrous alcohol (Gaykawad et al., 2013). The benchmark PV membranes for purification of bioethanol and biobutanol are based on dense polymethylsiloxane (PDMS) due to its excellent selectivity, stability, and low price (Liu, Wei, & Jin, 2014; Peng, Shi, & Lan, 2010).

Ceramic PV membranes and composites based on molecular sieves or zeolites have shown superior performance in terms of flux and separation efficiency (Huang, Tamaswamy, Tschirner, & Ramarao, 2008; Vane, 2005). Furthermore, they are stable throughout a much higher temperature and pressure range. However, they can be difficult to manufacture as defect free and can be 10–50 times more expensive (Caro, Noack, & Kölsch, 2005).

Even though most of R&D efforts in PV are focused on synthesis of novel and superior membrane materials, the integration of PV as a complementary unit with reliable performance has the potential to become the major breakthrough for this technology (Van der Bruggen & Luis, 2014).

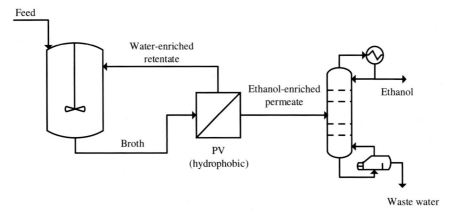

Fig. 2.10 Continuous fermentation, hydrophobic PV, and distillation.

Different approaches to cascading and integration are found across literature. Continuous fermentation and purification have been implemented via a fermentor coupled to a hydrophobic PV unit, where the permeate (42 wt% ethanol) was sent to a distillation column for final purification (95 wt% ethanol) as shown in Fig. 2.10 (O'Brien, Roth, & McAloon, 2000).

Hydrophilic PV can also be implemented as a dehydration step after distillation, as shown in Fig. 2.11A (Cardona & Sánchez, 2007). The azeotropic mixture that leaves the distillation column (95 wt% ethanol) is upgraded to anhydrous ethanol (>99.5% ethanol). This configuration was further developed by Gaykawad et al. (2013) who added a hydrophobic PV step prior to distillation in order to reduce the energy costs associated with distillation, as shown in Fig. 2.11B. In this process, the original broths contained ~3 wt% ethanol and were sent to a hydrophobic PV unit with PDMS membranes. The resulting permeates had compositions in the range of 11–13 wt% ethanol for different feedstocks. After that, the ethanol-enriched stream was further sent to distillation and hydrophilic PV for dehydration.

As shown in Fig. 2.12, a purely membrane-based system combines the fermentor with a microfiltration unit for cell recycle and to prevent fouling on the PV unit. The microfiltration permeate is directed to a hydrophobic PV, where the ethanol-rich permeate is sent to a hydrophilic PV for final dehydration, while the water-rich retentate is split between the fermentor and the bleed (Huang et al., 2008).

Huang, Baker, and Vane (2010) reported a distillation-PV hybrid system where the compressed overhead vapor from the distillation column is sent to a hydrophilic PV unit. As shown in Fig. 2.13, the water-enriched permeate is returned to the column while ethanol-enriched retentate is split between the product and the liquid reflux. This configuration can achieve product purities >99 wt% ethanol while consuming at least

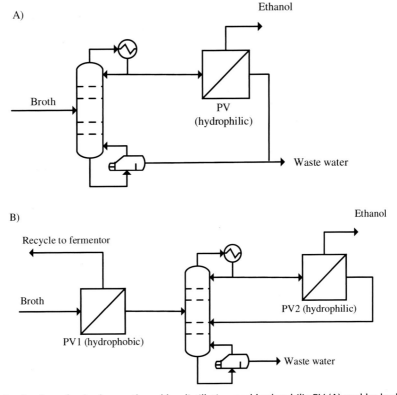

Fig. 2.11 Production of anhydrous ethanol by distillation and hydrophilic PV (A) and by hydrophobic PV, distillation, and hydrophilic PV (B).

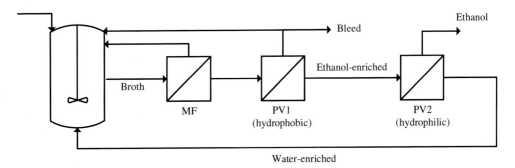

Fig. 2.12 Fermentation hydrophobic-hydrophilic pervaporation system.

50% less energy than conventional distillation equipped with a molecular sieve dryer unit. One of the possible limitations of this approach is that there are few membrane materials that can be stable in hot ethanol/water mixtures at the temperatures required by this process.

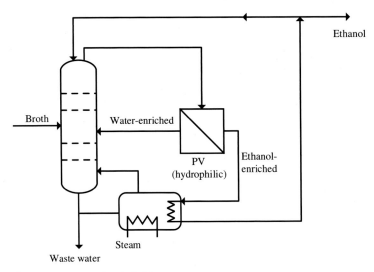

Fig. 2.13 Distillation-pervaporation hybrid system.

2.5.2 Membrane distillation

In MD a microporous hydrophobic membrane is used as a contactor between feed and permeate. The driving force for the mass transfer is the vapor pressure difference across the membrane. Vapor molecules and volatiles are transported through the membrane by evaporation on the feed side and condensation on the permeate side. As opposed to PV, selectivity in MD is governed by vapor-liquid equilibrium and not by the membrane material (Gostoli & Sarti, 1989). Vacuum membrane distillation (VMD) using polypropylene membranes has been reported for separation of volatiles including ethanol and acetic acid during continuous fermentation (Gryta & Barancewicz, 2011). Li et al. (2018) reported a VMD bioreactor with polytetrafluoroethylene membranes where a fraction of the permeate was cooled and condensed in a vacuum condenser, and the remaining vapor was sent to a mechanical vapor compression heat pump. They identified that in this process 90% of the energy was consumed in the evaporation of the feed for VMD. Thus the recovery of the latent heat from the permeate vapor by the heat pump could provide an advantage to reduce energy consumption of this process. Using this technology, fermentation broth containing 30 g/L ethanol was upgraded to 315 g/L in the final condensate.

VMD has also been reported in the pretreatment of hydrolysates for removal of inhibitors prior to fermentation. Furfural and acetic acid in lignocellulosic hydrolysates were partially removed by VMD resulting in overall 17.8% increased bioethanol production as compared to the original hydrolysate (Chen et al., 2013).

2.5.3 Forward osmosis

In FO, the difference in osmotic pressure between two sides of a semipermeable membrane acts as driving force for water transport (Cath, Childress, & Elimelech, 2006). Water molecules are thus transferred from a diluted feed toward a highly concentrated draw solution, while solutes and ions are rejected. NaCl is perhaps the most common draw agent; however, different compounds from organic or inorganic origin have also been evaluated (Achilli, Cath, & Childress, 2010; Bowden, Achilli, & Childress, 2012). As it is implied by the process definition, draw solutions need to be regenerated in order to maintain their high concentration and osmotic power.

One of the possible applications of FO in bioalcohol production is the concentration of diluted byproducts from biomass pretreatment. Zhang et al. (2018) employed a NaClO-treated polyamide-polysulfone FO membrane and saturated $MgSO_4$ as draw to concentrate the liquid fraction resulting from hydrothermal treatment of rice straw. During this process, sugars (xylose) were concentrated while inhibitors (furfural and acetic acid) permeated through the membrane and showed no concentration change. The resulting concentrate was fermented and produced higher ethanol yields as opposed to the nontreated solution. The same application was further investigated by Shibuya et al. (2017), who combined dia-nanofiltration for improved removal of inhibitors, with FO for concentration of sugars.

Dewatering of fermentation broth via FO has also been reported. Zhang, Ning, Wang, and Diniz da Costa (2013) presented a FO process for dehydration of water-ethanol mixtures using a cellulose triacetate membrane and saturated NaCl as draw solution. The process was shown to be feasible for partial dewatering, being the main challenges the permeation of product at high ethanol concentrations, and the deterioration of membrane after long contact times with ethanol. Ambrosi et al. (2017) reported ethanol separation from aqueous mixtures using an FO TCA membrane with different draw solutions and concluded that organic draw agents produced higher overall fluxes and separation factors as opposed to inorganic draw agents. However, the separation factors could not match those observed in hydrophobic PV. Investigation of novel membrane materials, draw agents, and process configurations makes FO a plausible unit to be integrated in hybrid bioalcohol production processes.

2.5.4 Pressure-driven membrane processes

RO is a pressure-driven operation in which a semipermeable membrane allows for separation of water from dissolved solids. The operating pressure needs to be high enough to overcome the osmotic pressure difference between the feed and the permeate (water). When this condition is fulfilled, solvent molecules are forced to permeate through the membrane via solution-diffusion, while most of the solutes are rejected. Some studies have reported RO for processing of alcohol-water mixtures. Diltz, Marolla, Henley, and Li (2007) performed RO on model solutions of alcohols and volatiles and post fermentation broth using a polyamide thin-film composite membrane. The study was focused on water reclamation and obtained rejections of volatile and alcohols of 94% on average, with individual components ranging from 90.3% to 100%.

Preconcentration of ethanol has also been reported via RO. In this configuration, a train of RO units is installed prior to the distillation-PV hybrid system, as shown in Fig. 2.14 (Kanchanalai, Lively, Realff, & Kawajiri, 2013). For dilute feeds, the RO pretreatment allows to reduce the energy consumption and the reboiler and operating costs of distillation. However, for processes aiming at a high recovery, the capital costs of RO can overrule the advantage gained in energy savings. Further improvement on RO costs is needed before this alternative is feasible also for higher feed concentrations.

Other pressure-driven membrane operations like microfiltration (MF) and ultrafiltration (UF) have been suggested in production of bio-alcohols, mostly in combination with PV. The role of these units is mainly to remove remains of cells, macromolecules, and/or nonvolatiles that could cause fouling at the PV membrane. The most common option is to combine fermentation, microfiltration, and PV in a hybrid process, as shown in Fig. 2.12. The main benefit of this solution is that both the feed to PV and the bleed are free of cells, and the possibilities of fouling at the PV membrane are minimized (Lipnizki, 2010).

2.6 Process control

The development of competitive production of bioalcohols starting from residual biomass requires adequate control strategies. The main difficulties encountered when designing the control of a plant for the production of biofuels are the following: high

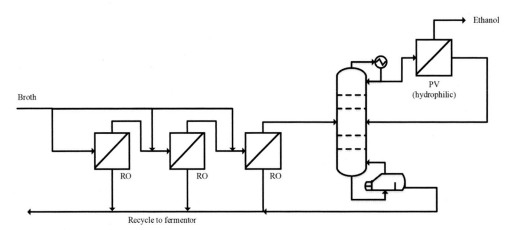

Fig. 2.14 Reverse osmosis treatment prior to the hybrid distillation pervaporation system.

variability of the feedstock composition, nonlinear characteristics of system dynamics, sluggish dynamic responses, the presence of living organisms in the fermentation step, the lack of suitable and robust on-line sensors for measurement of biomass or product concentrations, and model uncertainties. One of the critical units in a biorefinery is the bioreactor where the fermentation of the sugars occurs. Bioreactor yield and productivity are affected by the pretreatment process and the type of feedstock. Additionally, the composition of the reactor products affects the successive separation process; therefore it is important to guarantee the required target.

2.6.1 State estimation for the fermentation process

Real-time monitoring of bioreactors is regarded as an essential part of effective bioprocess control that can lead to improved efficiency, productivity, quality control, thus optimizing overall costs. The main difficulty for controlling the fermenter is the lack of suitable sensors to have real-time information about concentration of products, substrate, and biomass. Only few variables are commonly measured on-line in bioreactors, namely, temperature, pressure, dissolved oxygen, pH, stirring speed, and gas and liquid flow rates, which are generally referred to as process variables or engineering data (Lourenço, Lopes, Almeida, Sarraguça, & Pinheiro, 2012). For fermentation processes, specifically, additional probes are used for variables such as dissolved carbon dioxide, oxidation-reduction potential, and optical density or turbidity. In recent years, several researchers proposed efficient bioprocess-monitoring strategies, including the development of optical sensors, chemosensors, and biosensors (Holzberg et al., 2018). Despite the improved performance and reliability of the on-line sensors developed for bioprocess monitoring, industrial applications are rare because most tools are still expensive, require frequent maintenance, and are normally limited to single-property analysis (Lourenço et al., 2012).

A possible solution to solve the on-line monitoring problem in a bioreactor is to develop software sensors, which use available measurement devices and a mathematical model, in order to obtain continuous time estimates of nonmeasured variables on-line (Ödman, Johansen, Olsson, Gernaey, & Lantz, 2009). A review on monitoring key process variables for ethanol production from lignocellulosic biomass was proposed by Lopez et al. (2020), where different soft sensor strategies applicable to bioreactors have been discussed. Some examples for ethanol production are reported in the following, to show possible solutions when addressing the on-line monitoring problem of a fermentation process.

A monitoring tool based on continuous-discrete extended Kalman filter was proposed for the production of lignocellulosic ethanol by Iglesias, Gernaey, and Huusom (2015). Biomass, furfural, and acetic acid were estimated by measuring glucose, xylose, ethanol, and pH. In this application off-line measurements were used. For this reason, the estimation algorithm was based on different sampling rates. It is worth noting that the extended Kalman filter led to reasonably good results even in conditions of simulated contamination by lactic acid bacteria.

The use of off-line laboratory analysis could be not applicable in the industrial case, where the interval between two successive samples can be quite large (one day). In this situation, the mismatch between estimation and actual value can increase so much that the estimated outputs cannot be used in a control loop because of convergence issues. An alternative solution is to use on-line not delayed measurements to develop the monitoring tool, using a procedure that helps the designer find the best estimation structure for the available set of measurements (Lisci, Grosso, & Tronci, 2020a). On the basis of robust exponential estimability arguments, Lisci, Grosso, and Tronci (2020b) found that it is possible to distinguish all the six unmeasured states in a bioreactor for ethanol production, if temperature and dissolved oxygen concentration measurements are combined with substrate concentration. The proposed estimator structure was validated through numerical simulations considering two different measurement processor algorithms: the geometric observer and the extended Kalman filter.

Hybrid approaches that combine data-driven and first-principles models can also be used to obtain information on the fermentation process. In Lopez et al. (2020), a data-driven model used spectral data to estimate the concentration of glucose, that is then used in a kinetic model to obtain the other states (xylose, ethanol, furfural, and acetic acid concentrations) during cellulose fermentation for the obtainment of ethanol. In this case, a state estimator that updated the estimated concentration was not developed, but the authors suggested it as a possible solution to improve state prediction when measurements were not reliable enough.

2.6.2 Control strategies for the fermenter

The possibility to develop adequate control strategies for the fermentation bioreactor is related to the availability of on-line measurements of the key variables or to guarantee the respect of the targets by using the control of secondary variables. Temperature control is a possible choice in the latter case because it affects the substrate conversion and also the gas mass transfer. As the fermentation process exhibits nonlinear behavior, temperature control is not an easy task, as evidenced by the following papers on such topic.

Nagy (2007) developed a detailed model of a bioreactor for the obtainment of ethanol by fermentation and developed a neural network-based control of the reactor temperature. In the results, the authors evidenced the inadequacy of traditional feedback controllers, such as PID. Using the same model for the process simulation, several authors designed different control strategies for the control of the bioreactor temperature, using advanced algorithms or traditional PID algorithms. For example, Imtiaz, Assadzadeh, Jamuar, and Sahu (2013) proposed a temperature controller using inverse neural networks. A robust model-based predictive control with integral action was presented in

Bakosova, Oravec, Vasickaninova, Meszaros, and Artzova (2019), showing that the developed method can ensure high product yield and minimize energy consumption. Recently, Pachauri, Rani, and Singh (2017) compared a conventional PID with a modified factional order IMC-PID, while Kumar, Prasad, et al. (2019) and Kumar, Ravikumar, et al. (2019) applied an IMC-PID controller. A MIMO control system was presented in Imtiaz, Jamuar, Sahu, and Ganesan (2014), where a nonlinear auto regressive moving average controller was applied to control reactor temperature, and a two degree of freedom PID was used to control pH and dissolved oxygen concentration.

Using a simpler model where only the dynamics of three states were described, Skupin and Metzger (2018) addressed the problem of stabilizing ethanol concentration in continuous fermentation processes, due to the delayed response of microorganism to inhibitory effects of ethanol concentration. They assumed that ethanol is measured on-line, which is not common practice in industrial plants. Mathematical modeling was employed to assess the dynamic behavior of the flash fermentation process for the production of butanol in Mariano et al. (2010). Based on the study of the dynamics of the process, suitable feedback control strategies were proposed to deal with disturbances related to the process. The aim of the control consisted of keeping sugar and/or butanol concentrations in the fermenter constant in the face of disturbances in the feed substrate concentration. The performances of an advanced controller, the dynamic matrix control, and the classical proportional-integral controller were evaluated.

2.6.3 Control of the separation processes

Enhanced distillation methods are valid alternatives for the purification of bioalcohols from fermentation broth. In particular, thermally coupled structures and extractive DWCs can be used to improve separation efficiency while saving energy. The synthesis of alternative sequences is usually carried out by considering performance indicator, such as specific energy used per amount of bioalcohols produced and capital cost saving with respect to the reference case (Hernandez & Jimenez, 1999). Enhanced distillation sequences exhibit a complex structure, with recycle streams, that may affect their controllability properties. The analysis of control properties is strongly recommended in order to avoid that their industrial implementation may show potential control problems or it cannot be fairly operated in practice. Indeed, integration leads to changes in the operating and controllability of the integrated system. Jimenez, Hernandez, Montoy, and Zavala-García (2001) demonstrated that controllability properties of intensified distillation structures can be carried out applying the singular value decomposition (SVD) technique. This procedure has been successfully used to address control properties in highly intensified distillation systems for the purification of bio-alcohols from fermentation broth. Sanchez-Ramirez et al. (2017) performed a controllability analysis for the

ABE fermentation, using the SVD technique and a closed-loop dynamic analysis on several hybrid distillation processes including conventional, thermally coupled, thermodynamically equivalent, and intensified designs. Torres-Ortega et al. (2018) presented the analysis of process features and control properties of the intensified systems for lignocellulosic bioethanol separation and dehydration through dividing wall columns. The control properties were based on SVD and dynamic performances under mild disturbances and changes of set point in Aspen Dynamics.

Kaymak (2019) presented the design and control for a two-column process of bioethanol purification from fermentation broth. The enhanced configuration involved a preconcentrator column that was followed by a reactive distillation column. Alternative control structures were designed for the optimum steady-state configuration, and their robustness was tested using several disturbances.

An overview on the classical control strategies for enhanced bioethanol dehydration by extractive and azeotropic distillation in DWCs of different control structures for dividing wall columns was reported in Kiss and Bildea (2011). Varying from the classic three-point control structure and PID controllers in a multiloop framework to model predictive control, they also considered advanced control strategies (such as linear quadratic gaussian, loop shaping design procedure, H∞, and μ-synthesis).

2.7 Bioethanol separation by membrane-assisted reactive distillation: A case study

Reactive distillation (RD) can be considered as the highest example of process intensification, where the reaction and the separation tasks are combined in a single equipment (Harmsen, 2007; Segovia-Hernandez, Hernandez, & Bonilla-Petriciolet, 2015). This technology was proved to be a valuable option for production of biodiesel (Cossio-Vargas, Hernandez, Segovia-Hernandez, & Cano-Rodriguez, 2011; Miranda-Galindo et al., 2009; Poddar, Jagannath, & Almansoori, 2017) as well as to bypass azeotropes (Guo, Chin, & Lee, 2004; Lee, Hauan, & Westerberg, 2000) making it an attractive alternative for the separation of bioalcohols. Considering bioethanol as reference bioalcohol, different authors proposed a reactive distillation arrangement based on the irreversible hydration reaction of ethylene oxide (EO) to EG (An, Lin, Chen, & Zhu, 2014; Kaymak, 2019; Kumar & Daoutidis, 1999; Tavan & Hosseini, 2013). This possibility was recently reconsidered by Errico et al. (2020), who developed an alternative composed by membrane, reactive distillation, and ordinary distillation, as reported in Fig. 2.15.

The configuration is composed of a preliminary membrane separation used to reduce the duty of the following preconcentration column used to approach the azeotropic composition. In the reactive distillation, EO is used to remove the water and allow the separation of pure ethanol as distillate. A further distillation column is used to purify the byproduct and recover the ethanol and water through a recycle to the first column.

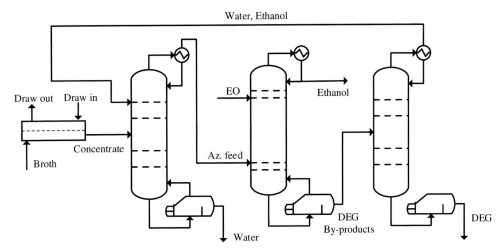

Fig. 2.15 Membrane-assisted reactive distillation configuration.

This configuration was selected after examining different possibilities in combining the membrane separation in different ways within the configuration.

Aquaporin-based biomimetic FO membranes were used. The membranes were fed with a hypothetical fermentation broth containing 5 wt% of ethanol, the draw solution was a solution of 4.17 M of NaCl, and the trans-membrane pressure was kept at 0.2 bar. More details of the membrane performances were reported by Errico et al. (2020).

In the RD, the following reaction scheme has been considered:

$$C_2H_4O\ (EO) + H_2O \rightarrow C_2H_6O_2\ (EG)$$

$$C_2H_4O\ (EO) + C_2H_6O_2\ (EG) \rightarrow C_4H_{10}O_3\ (DEG)$$

In the design and optimization procedure, the definition of the pressure is one of the most important parameters. Different authors based its definition on the reactant conversion and the EG selectivity (An et al., 2014; Kaymak, 2019). However, this approach does not offer a complete description of the system since the choice of the pressure defines also the pressure in the reboiler that is limited by the decomposition temperature of the reaction products. The optimization was performed considering the EO/water ratio, the distillate flow rate, the reflux ratio, the feed locations, and the liquid hold-up. The optimization procedure started in defining the pressure value. This value was used to define the optimal EO/water ratio. Once the optimal value was defined, the pressure optimization was repeated. With this approach, the optimized value is used to check its influence on the previous variable. The number of reactive, stripping, and rectifying stages were optimized last.

Considering as a reference the extractive distillation configuration reported in Fig. 2.3 (left), this achieved a TAC of 332 k\$ yr^{-1}, while the membrane-assisted RD has a value of 284 k\$ yr^{-1}. Nevertheless, a comparison based on the TAC alone does not give a complete description of the two alternatives. It is possible to introduce an energy index (EI) based on the ratio of the total reboiler duty and flowrate of ethanol produced. The EI for the extractive distillation configuration was evaluated in 6.86 and 6.16 MJ kg^{-1} for the membrane-assisted RD. Even if there is a clear preference in using the intensified alternative, this comparison appeared to be improved since the two configurations used different chemicals. For this reason, a further comparison was performed including the controllability, the carbon dioxide emissions, the inherent risk, the mass intensity (MI), and effective mass yield. These performance indicators may not appear to have a real connection. However, in a framework of green chemistry evaluation, aiming at the promotion of sustainable processes that support the transition toward a circular economy, they all contribute by generating and selecting sustainable processes and products. Jiménez-González, Constable, and Ponder (2012) published an interesting review of metrics to evaluate the "greens" of some process. In this review, several metrics were highlighted, such as MI, greenhouse gas emissions, real-time analysis (controllability analysis), or inherent safety associated with the process. By evaluating these metrics, an assessment of progress toward the broad goal of environmental sustainability can be generated. There are a few basic things that need to be weighed along with this type of metric in addition to these green metrics. Using energy for output is a crucial factor. The prevention of industrial waste products is another key factor, thus preventing a greater environmental effect than that used for the industry's operating activities. The general principles of these indexes are described as follows.

2.7.1 Greenhouse gas emissions

CO_2 emissions are related to the amount of fuel burned to generate a certain amount of energy. It is possible to calculate these emissions according to Eq. 2.1.

$$[CO_2]_{Emisions} = \left(\frac{Q_{fuel}}{NHV}\right)\left(\frac{\%C}{100}\right)\alpha \tag{2.1}$$

where Q_{fuel} (kW) is the amount of fuel burned, $\alpha = 3.67$ is the ratio of CO_2 and C molar masses, and NHV (kJ/kg) represents the net heating value of fuel with a carbon content of $C\%$ (Gadalla, Olujic, Jansens, Jobson, & Smith, 2005).

2.7.2 Mass intensity

The MI, calculated according Eq. 2.2, measures the amount of material necessary to synthesize the desired product. Ideally, a value close to 1 would be expected (Jiménez-González et al., 2012).

$$MI = \frac{\text{Total mass used in a process or process step (kg)}}{\text{Mass of product (kg producto)}} \tag{2.2}$$

2.7.3 Effective mass yield

Effective mass yield (EMY) is defined as the percentage of the mass of desired product relative to the mass of all nonbenign materials used in its synthesis. EMY is reported in Eq. (2.3).

$$\text{EMY (\%)} = \frac{\text{Mass of products} \times 100}{\text{Mass of nonbeningn reagents}} \qquad (2.3)$$

2.7.4 Inherent safety

The assessment of the inherent safety of a process allows the identification of the risk associated with the use of certain substances in a chemical process. A quantitative risk analysis (QRA) assesses the inherent safety and represents it as an individual risk analysis (IR). The QRA estimates the predicted risk of such accidents, as well as the adverse effects of deaths or injuries (Freeman, 1990). Subsequently, the frequencies of the catastrophic events associated with and form of mass release are evaluated via an event tree. The event tree considers BLEVE, UVCE, Flash Fire, and Toxic Release for instant release of matter for the distillation columns and Jet fire, Flash Fire, and toxic release for continuous release of matter (Freeman, 1990). IR is independent of the number of people and is mathematically defined by Eq. (2.4):

$$IR = \sum f_i P_{x,y} \qquad (2.4)$$

where f_i is the frequency of occurrence of an incident i, while $P_{x,y}$ is the probability of injury or death caused by incident i.

2.7.5 Controllability analysis

Processes that are not properly controlled may generate more waste and consume more raw material and energy per product unit. In certain situations, failure to keep the process under control means noncompliance with product requirements. For an open-loop control test, the CN is a qualitative measure of a process's complex behavior. A large CN suggests an impractical control system that hardly meets the entire collection of control goals (regardless of the control strategy used). Therefore a poorly conditioned system appears to amplify modeling errors, nonlinearities, disruptions, and therefore systems with small or close to unity CN are favored. The CN is obtained dividing the maximum singular value (σ^*) by the minimum singular value (σ_*) according to Eq. 2.5. The singular values are previously obtained by decomposing the singular values of the relative gain matrix that in an open-loop policy represents the dynamic behavior of the method.

$$\gamma = \frac{\sigma^*}{\sigma_*} \qquad (2.5)$$

With respect to closed-loop analysis, a composition control test was carried out inside the Aspen Dynamics simulator as follows: a negative strep change was induced at the set point for each product composition under single-input and single-output feedback control at each output flow rate. Structures based on energy balance considerations were used with regard to the manipulated variables of control in the distillate column, bottom, and side stream output composition. The so-called LV control structure yields this structure (Häggblom & Waller, 1992). An initial value of proportional gain was set to tune-up each controller, and a range of integral reset time was checked with a fixed value until a local optimum in the integral absolute error (IAE) value was obtained.

2.7.6 Results

Table 2.6 summarizes the values of the indicators previously explained evaluated for the extractive distillation configuration used as reference and the RD scheme.

Regarding the controllability study, Figs. 2.16 and 2.17 and Table 2.7 show the main results for the open-loop and closed-loop analysis.

According to Table 2.6, neither of the two schemes analyzed has the best value in all performance indexes. In other words, there is a direct commitment between several of the performance indexes. Considering only the economic indicators, the reactive system has a clear advantage. Even when the energy used per kilogram of product is weighted, the reactive system is also superior. However, when considering the emission of greenhouse gases by total energy consumption, the extraction process is slightly better than the reactive system.

On the other hand, evaluating the MI, the reactive system presents better efficiency considering 1 as the base utopian number. Mass intensity takes the yield, stoichiometry, solvent, and reagent used in the mixture of the reaction into account and expresses this rather than a percentage on a weight/weight basis. With the exception of water, total mass contains anything used in a process or process phase, that is, reactants, reagents, solvents, catalysts, and so on. This indicator is consistent with EMY. This performance index shows that the relationship between products and reagents is better for the reagent scheme. That is, more product of interest is generated compared to the number of nonproducts.

When evaluating process safety, several factors come into play, such as reboiler duty, operating pressure, and obviously the nature of the chemicals in the process. The design

Table 2.6 Performance indexes for both extractive and reactive distillation.

	TAC (k\$ yr^{-1})	EI (MJ kg^{-1})	CO_2 (ton h^{-1})	MI (kg kg^{-1})	EMY (%)	IR (yr^{-1})
Extractive Distillation	332	6.86	15.48	21.23	50.76	2.65 exp. -4
RD	284	6.16	15.96	20.56	77.29	8.01 exp. -6

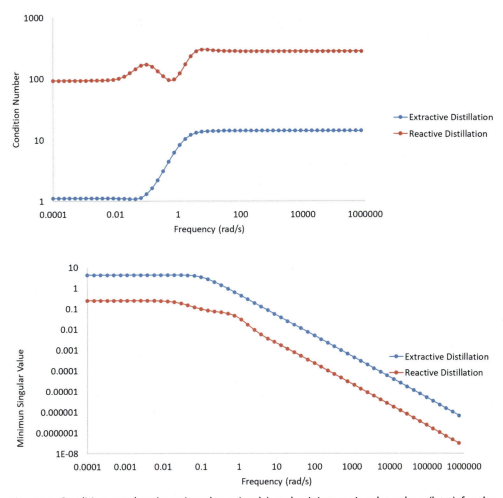

Fig. 2.16 Condition number (mentioned previously) and minimum singular values (later) for the configurations considered.

parameters for both schemes can be retrieved from Errico et al. (2020). In this sense, the determining factor for the safety of the extractive scheme, as opposed to the reactive scheme, was the presence of EO in the reactive column. Due to the nature and physicochemical properties of this compound, the probability of a catastrophic event increases considerably.

Finally, in the controllability analysis of both closed and open loop, the extractive scheme presents better characteristics. Through the open-loop study, the extractive scheme showed lower CNs and a higher minimum singular value. These results indicate that the scheme is better conditioned to possible disturbances and modeling errors. Additionally, it is expected that using some closed-loop control strategy, the extractive scheme

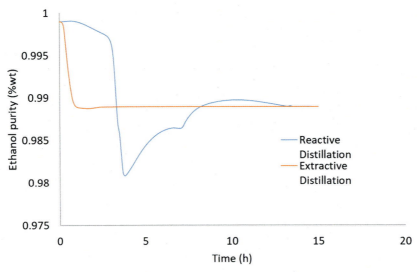

Fig. 2.17 Dynamic response—closed-loop test for the configurations considered.

Table 2.7 Tunning values for both extractive and reactive distillation.

	K_c	τ_i	IAE
Extractive distillation	250	20	0.00505243
Reactive distillation	40	80	0.04999065

will respond better. These results are totally conclusive when the closed-loop control analysis was performed at a setpoint change. The reactive scheme showed higher values of IAE and higher settling times.

2.8 Conclusions

Production of bioalcohols was discussed based on the main challenges related to the pretreatment step and the separation of the final product. It can be concluded that:

1. The use of pretreatment strategies can improve the feasibility of bioalcohol production from cellulosic feedstocks. Each type of pretreatment has peculiar characteristics and requires adequate analysis in terms of demand for area, time of exposure, energy spend during pretreatment, demand for chemical reagents, and their disposal costs. A technical and economic feasibility study considering the specificities of the biomass to be treated, such as the generation characteristics, the input acquisition costs, and the handling of the residual material, is an essential step for the successful application of the pretreatment.

2. Separation of diluted and complex mixtures is a challenging step, and intensified and hybrid separation schemes can contribute in improving the overall process economy. In particular, membrane processes offer a toolbox with several opportunities for industrial bioalcohol production. There are two main aspects where membranes can play an important role: removal of inhibitors during continuous fermentation and overcoming the azeotropic nature of the alcohol-water mixture. Whereas PV appears as the most effective and energy saving technique, the choice of optimal configuration will depend on the feedstock characteristics and throughput. Thus the integration of different membrane unit operations into hybrid systems has a large potential to combine the best qualities of each process to boost productivity and reduce energy consumption in the cellulosic and lignocellulosic biorefineries of the future.
3. Automatic control of bioalcohols plant is still a challenge because of the lack of measurements for key variables and the complexity of the process. State estimation techniques in conjunction with advanced control strategies proved to be valuable tools for competitive and sustainable production of bioalcohols.
4. The evaluation of intensified schemes composed by membrane and reactive distillations appears as a promising solution. Although the reactive scheme resulted in a lower cost, it did not demonstrate a better performance in other indexes, such as controllability and safety compared the extractive distillation alternative chosen as reference. It is important to evaluate intensified schemes beyond their economic performance in order to include relevant aspects like sustainability, controllability, and the process safety. In future works, it may be relevant to include some of these indexes in the objective function used for the design optimization.

Acknowledgments

M. Errico acknowledges the funding received from the European Union's Horizon 2020 research and innovation programme under Marie Skłodowska-Curie grant agreement No 778168.

References

Abdellatif, F. H. H., Babin, J., Arnal-Herault, C., David, L., & Jonquieres, A. (2018). Grafting cellulose acetate with ionic liquids for biofuel purification membranes: Influence of the anion. *Carbohydrate Polymers*, *196*, 176–186.

Achilli, A., Cath, T. Y., & Childress, A. E. (2010). Selection of inorganic-based draw solutions for forward osmosis applications. *Journal of Membrane Science*, *364*, 233–241.

Alalwan, H. A., Alminshid, A. H., & Aljaafari, H. A. S. (2019). Promising evolution of biofuel generations. Subject review. *Renewable Energy Focus*, *28*, 127–139.

Alizadeh, R., Lund, P. D., & Soltanisehat, L. (2020). Outlook on biofuels in future studies: A systematic literature review. *Renewable and Sustainable Energy Reviews*, *134*, 110326.

Ambrosi, A., Correâ, G. L., de Vargas, N. S., Gabe, L. M., Cardozo, N. S. M., & Tessaro, I. C. (2017). Impact of osmotic agent on the transport of components using forward osmosis to separate ethanol from aqueous solutions. *AICHE Journal*, *63*(10), 4499–4507.

An, W., Lin, Z., Chen, J., & Zhu, J. (2014). Simulation and analysis of a reactive distillation column for removal of water from ethanol-water mixtures. *Industrial & Engineering Chemistry Research, 53*, 6056–6064.

Aptel, P., Challard, N., Cuny, J., & Neel, J. (1976). Application of the pervaporation process to separate azeotropic mixtures. *Journal of Membrane Science, 1*, 271–287.

Baker, R. W. (2004). *Membrane technology and applications*. Chichester, England: John Wiley & Sons, Ltd.

Bakosova, M., Oravec, J., Vasickaninova, A., Meszaros, A., & Artzova, P. (2019). Advanced control of a biochemical reactor for yeast fermentation. *Chemical Engineering Transactions, 76*, 769–774.

Bay, M. S., Karimi, K., Esfahany, M. N., & Kumar, R. (2020). Structural modification of pine and poplar wood by alkali pretreatment to improve ethanol production. *Industrial Crops and Products, 152*, 112506.

Boluda-Aguilar, M., & López-Gómez, A. (2013). Production of bioethanol by fermentation of lemon (*Citrus limon* L.) peel wastes pretreated with steam explosion. *Industrial Crops and Products, 41*, 188–197.

Bowden, K. S., Achilli, A., & Childress, A. E. (2012). Organic ionic salt draw solutions for osmotic membrane bioreactors. *Bioresource Technology, 122*, 207–216.

Cardona, C. A., & Sánchez, Ó. J. (2007). Fuel ethanol production: Process design trends and integration opportunities. *Bioresource Technology, 98*(12), 2415–2457.

Caro, J., Noack, M., & Kölsch, P. (2005). Zeolite membranes: From the laboratory scale to technical applications. *Adsorption, 11*, 215–227.

Cath, T. Y., Childress, A. E., & Elimelech, M. (2006). Forward osmosis: Principles, applications, and recent developments. *Journal of Membrane Science, 281*, 70–87.

Chen, J., Zhang, Y., Wang, Y., Xiaosheng, J., Zhang, L., Mi, X., et al. (2013). Removal of inhibitors from lignocellulosic hydrolyzates by vacuum membrane distillation. *Bioresource Technology, 144*, 80–683.

Chowdhury, H., & Loganathan, B. (2019). Third-generation biofuels from microalgae: A review. *Current Opinion in Green and Sustainable Chemistry, 20*, 39–44.

Cossio-Vargas, E., Hernandez, S., Segovia-Hernandez, J. G., & Cano-Rodriguez, M. I. (2011). Simulation study of the production of biodiesel using feedstock mixtures of fatty acids in complex reactive distillation columns. *Energy, 36*, 6289–6297.

Demirbas, A. (2008). Biofuels sources, biofuel policy, biofuel economy and global biofuel projections. *Energy Conversion and Management, 49*, 2106–2116.

Demirbas, M. F., Balat, M., & Balat, H. (2011). Biowastes-to-biofuels. *Energy Conversion and Management, 52*, 1815–1828.

Diltz, R. A., Marolla, T. V., Henley, M. V., & Li, L. (2007). Reverse osmosis processing of organic model compounds and fermentation broths. *Bioresource Technology, 98*(3), 686–695.

Ding, W. W., Wu, Y. T., Tang, X. Y., Yuan, L., & Xiao, Z. Y. (2011). Continuous ethanol fermentation in a closed-circulating system using an immobilized cell coupled with PDMS membrane pervaporation. *Journal of Chemical Technology & Biotechnology, 86*(1), 82–87.

Errico, M. (2017). *Process synthesis and intensification of hybrid separations*. Berlin, Germany: De Gruyter (chapter 5).

Errico, M., Madeddu, C., Flemming Bindseil, M., Madsen, S. D., Braekevelt, S., & Camilleri-Rumbau, M. S. (2020). Membrane assisted reactive distillation for bioethanol purification. *Chemical Engineering and Processing: Process Intensification, 157*, 108110.

Errico, M., & Rong, B.-G. (2012). Synthesis of new separation processes for bioethanol production by extractive distillation. *Separation and Purification Technology, 96*, 58–67.

Errico, M., Rong, B.-G., Tola, G., & Spano, M. (2013a). Optimal synthesis of distillation systems for bioethanol separation. Part 1: Extractive distillation with simple columns. *Industrial & Engineering Chemistry Research, 53*, 1612–1619.

Errico, M., Rong, B.-G., Tola, G., & Spano, M. (2013b). Optimal synthesis of distillation systems for bioethanol separation. Part 2: Extractive distillation with complex columns. *Industrial & Engineering Chemistry Research, 53*, 1620–1626.

Errico, M., Rong, B.-G., Tola, G., & Turunen, I. (2009). A method for systematic synthesis of multicomponent distillation systems with less than N-1 columns. *Chemical Engineering and Processing: Process Intensification, 48*, 907–920.

Errico, M., Sanchez-Ramirez, E., Quiroz-Ramirez, J. J., Rong, B.-G., & Segovia-Hernandez, J. G. (2017a). Multiobjective optimal acetone-butanol-ethanol separation systems using liquid-liquid extraction-assisted divided wall columns. *Industrial & Engineering Chemistry Research, 56*, 11575–11583.

Errico, M., Sanchez-Ramirez, E., Quiroz-Ramirez, J. J., Rong, B.-G., & Segovia-Hernandez, J. G. (2017b). Biobutanol purification by hybrid extraction-divided wall column configurations. *Computer Aided Chemical Engineering*, *40*, 1027–1032.

Errico, M., Sanchez-Ramirez, E., Quiroz-Ramirez, J. J., Segovia-Hernandez, J. G., & Rong, B.-G. (2016). Synthesis and design of new hybrid configurations for biobutanol purification. *Computers and Chemical Engineering*, *84*, 482–492.

Filip, O., Janda, K., Kristoufek, L., & Zilberman, D. (2019). Food versus fuel: An updated expanded evidence. *Energy Economics*, *82*, 152–166.

Freeman, R. A. (1990). CCPS guidelines for chemical process quantitative risk analysis. *Process Safety Progress*, *9*(4), 231–235.

Gadalla, M. A., Olujic, Z., Jansens, P. J., Jobson, M., & Smith, R. (2005). Reducing CO_2 emissions and energy consumption of heat-integrated distillation systems. *Environmental Science & Technology*, *39*, 6860–6870.

Garcia, A., Alriols, M. G., Llano-Ponte, R., & Labidi, J. (2011). Ultrasound-assisted fractionation of the lignocellulosic material. *Bioresource Technology*, *102*, 6326–6330.

Garcia-Herreros, P., Gomez, J. M., Gil, I. D., & Rodriguez, G. (2011). Optimization of the design and operation of an extractive distillation system for the production of fuel grade ethanol using glycerol as entrainer. *Industrial & Engineering Chemistry Research*, *20*, 3977–3985.

García-Torreiro, M., López-Abelairas, M., Lu-Chau, T. A., & Lema, J. M. (2016). Fungal pretreatment of agricultural residues for bioethanol production. *Industrial Crops and Products*, *89*, 486–492.

Gaykawad, S. S., Zha, Y., Punt, P. J., van Groenestijn, J. W., van der Wielen, L. A., & Straathof, A. J. (2013). Pervaporation of ethanol from lignocellulosic fermentation broth. *Bioresource Technology*, *129*, 469–476.

Gerbaud, V., Rodriguez-Donis, I., Hegely, L., Lang, P., Denes, F., & You, X. Q. (2019). Review of extractive distillation. Process design, operation, optimization and control. *Chemical Engineering Research and Design*, *141*, 229–271.

Gostoli, C., & Sarti, G. (1989). Separation of liquid mixtures by membrane distillation. *Journal of Membrane Science*, *41*, 211–224.

Gryta, M., & Barancewicz, M. (2011). Separation of volatile compounds from fermentation broth by membrane distillation. *Polish Journal of Chemical Technology*, *13*(3), 56–60.

Guo, Z., Chin, J., & Lee, J. W. (2004). Feasibility of continuous reactive distillation with azeotropic mixtures. *Industrial & Engineering Chemistry Research*, *43*, 3758–3769.

Häggblom, W. K., & Waller, K. V. (1992). *Control structures, consistency, and transformations*. NY: Springer.

Haldar, D., & Rurkait, M. K. (2021). A review on the environment-friendly emerging techniques for pretreatment of lignocellulosic biomass: Mechanistic insight and advancements. *Chemosphere*, *264*, 128523.

Harmsen, G. J. (2007). Reactive distillation: The front-runner of industrial process intensification: A full review of commercial applications, research, scale-up, design and operation. *Chemical Engineering and Processing: Process Intensification*, *46*, 774–780.

Hendriks, A. T. W. M., & Zeeman, G. (2009). Pretreatments to enhance the digestibility of lignocellulosic biomass. *Bioresource Technology*, 10–18.

Hernandez, S., & Jimenez, A. (1999). Controllability analysis of thermally coupled distillation systems. *Industrial & Engineering Chemistry Research*, *38*(10), 3957–3963.

Holzberg, T. R., Watson, V., Brown, S., Andar, A., Ge, X., Kostov, Y., et al. (2018). Sensors for biomanufacturing process development: Facilitating the shift from batch to continuous manufacturing. *Current Opinion in Chemical Engineering*, *22*, 115–127.

Huang, Y., Baker, R. W., & Vane, L. M. (2010). Low-energy distillation-membrane separation process. *Industrial & Engineering Chemistry Research*, *49*(8), 3760–3768.

Huang, H.-J., Tamaswamy, S., Tschirner, U. W., & Ramarao, B. (2008). A review of separation technologies in current and future biorefineries. *Separation and Purification Technology*, *62*(1), 1–21.

IEA. (2020). *Energy efficiency indicators 2020*. Paris: IEA. https://www.iea.org/reports/energy-efficiency-indicators-2020. Accessed 01.12.2020.

Iglesias, M. M., Gernaey, K. V., & Huusom, J. K. (2015). State estimation in fermentation of lignocellulosic ethanol. Focus on the use of pH measurements. *Computer Aided Chemical Engineering*, *37*, 1769–1774.

Imtiaz, U., Assadzadeh, A., Jamuar, S. S., & Sahu, J. N. (2013). Bioreactor temperature profile controller using inverse neural network (INN) for production of ethanol. *Journal of Process Control*, *23*(5), 731–742.

Imtiaz, U., Jamuar, S. S., Sahu, J. N., & Ganesan, P. B. (2014). Bioreactor profile control by a nonlinear auto regressive moving average neuro and two degree of freedom PID controllers. *Journal of Process Control*, *24*(11), 1761–1777.

Jain, A., & Chaurasia, S. P. (2014). Bioethanol production in membrane bioreactor (MBR) system: A review. *International Journal of Environmental Research and Development*, *4*(4), 387–394.

Jiang, Y., Lv, Y., Wu, R., Sui, Y., Chen, C., Xin, F., et al. (2019). Current status and perspectives on biobutanol production using lignocellulosic feedstocks. *Bioresource Technology Reports*, *7*, 100245.

Jimenez, A., Hernandez, S., Montoy, F. A., & Zavala-García, M. (2001). Analysis of control properties of conventional and nonconventional distillation sequences. *Industrial & Engineering Chemistry Research*, *40*(17), 3757–3761.

Jiménez-González, C., Constable, D. J. C., & Ponder, C. S. (2012). Evaluating the "greenness" of chemical processes and products in the pharmaceutical industry—A green metrics primer. *Chemical Society Reviews*, *41*, 1485–1498.

Jonquières, A., Clément, R., Lochon, P., Néel, J., Dresch, M., & Chrétien, B. (2002). Industrial state-of-the-art of pervaporation and vapour permeation in the western countries. *Journal of Membrane Science*, *206*, 87–117.

Kanchanalai, P., Lively, R. P., Realff, M. J., & Kawajiri, Y. (2013). Cost and energy savings using an optimal design of reverse osmosis membrane pretreatment for dilute bioethanol purification. *Industrial & Engineering Chemistry Research*, *52*(32), 11132–11141.

Kaymak, D. B. (2019). Design and control of an alternative bioethanol purification process via reactive distillation from fermentation broth. *Industrial & Engineering Chemistry Research*, *58*, 1675–1685.

Kiss, A. A., & Bildea, C. S. (2011). A control perspective on process intensification in dividing-wall columns. *Chemical Engineering and Processing: Process Intensification*, *50*(3), 281–292.

Kiss, A. A., & Ignat, R. M. (2012). Innovative single step bioethanol dehydration in an extractive dividing-wall column. *Separation and Purification Technology*, *98*, 290–297.

Kiss, A. A., & Suszwalak, D. J.-P. C. (2012). Enhanced bioethanol dehydration by extractive and azeotropic distillation in dividing-wall columns. *Separation and Purification Technology*, *86*, 70–78.

Koros, W., Ma, Y., & Shimidzu, T. (1996). Terminology for membranes and membrane processes (IUPAC recommendations 1996). *Pure and Applied Chemistry*, *68*(7), 1479–1489.

Kumar, A., & Daoutidis, P. (1999). Modeling, analysis and control of ethylene glycol reactive distillation column. *AICHE Journal*, *45*, 51–68.

Kumar, M., Prasad, D., Giri, B. S., & Singh, R. S. (2019). Temperature control of fermentation bioreactor for ethanol production using IMC-PID controller. *Biotechnology Reports*, *22*, e00319.

Kumar, M. N., Ravikumar, R., Thenmozhi, S., Kumar, M. R., & Shankar, M. K. (2019). Choice of pretreatment technology for sustainable production of bioethanol from lignocellulosic biomass: Bottle necks and recommendations. *Waste and Biomass Valorization*, *10*, 1693–1709.

Lee, J. W., Hauan, S., & Westerberg, A. W. (2000). Circumventing and azeotrope in reactive distillation. *Industrial & Engineering Chemistry Research*, *39*, 1061–1063.

Li, G., & Bai, P. (2012). New operation strategy for separation of ethanol-water by extractive distillation. *Industrial & Engineering Chemistry Research*, *51*, 2723–2729.

Li, J., Zhou, W., Fan, S., Liu, Y., Liu, J., Qiu, B., et al. (2018). Bioethanol production in vacuum membrane distillation bioreactor by permeate fractional condensation and mechanical vapor compression with polytetrafluoroethylene (PTFE) membrane. *Bioresource Technology*, *268*, 708–714.

Lin, Y., Zheng, H., & Dong, L. (2020). Enhanced ethanol production from tree trimmings via microbial consortium pretreatment with selective degradation of lignin. *Biomass and Bioenergy*, *142*, 105787.

Lipnizki, F. (2010). Membrane process opportunities and challenges in the bioethanol industry. *Desalination*, *250*(3), 1067–1069.

Lipnizki, F., Hausmanns, S., Laufenberg, G., Field, R., & Kunz, B. (2000). Use of pervaporation-bioreactor hybrid processes in biotechnology. *Chemical Engineering & Technology: Industrial Chemistry-Plant Equipment-Process Engineering-Biotechnology*, *23*(7), 569–577.

Lisci, S., Grosso, M., & Tronci, S. (2020a). A geometric observer-assisted approach to tailor state estimation in a bioreactor for ethanol production. *PRO*, *8*(4), 480.A.

Lisci, S., Grosso, M., & Tronci, S. (2020b). A robust nonlinear estimator for a yeast fermentation biochemical reactor. *Computer Aided Chemical Engineering, 48*, 1303–1308.

Liu, G., Wei, W., & Jin, W. (2014). Pervaporation membranes for biobutanol production. *ACS Sustainable Chemistry & Engineering, 2*(4), 546–560.

Lopez, P. C., Udugama, I. A., Thomsen, S. T., Roslander, C., Junicke, H., Mauricio-Iglesias, M., et al. (2020). Towards a digital twin: a hybrid data-driven and mechanistic digital shadow to forecast the evolution of lignocellulosic fermentation. *Biofuels, Bioproducts and Biorefining, 14*(5), 1056–1060.

López-Linares, J. C., García-Cubero, M. T., Lucas, S., González-Benito, G., & Coca, M. (2019). Microwave assisted hydrothermal as greener pretreatment of brewer's spent grains for biobutanol production. *Chemical Engineering Journal, 368*, 1045–1055.

Lourenço, N. D., Lopes, J. A., Almeida, C. F., Sarraguça, M. C., & Pinheiro, H. M. (2012). Bioreactor monitoring with spectroscopy and chemometrics: A review. *Analytical and Bioanalytical Chemistry, 404*(4), 1211–1237.

Luque, R., & Clark, J. (2010). *Handbook of biofuels production processes and technologies*. Cambridge, UK: Woodhead Publishing.

Mahboubi, A., Ylitervo, P., Doyen, W., De Wever, H., Molenberghs, B., & Taherzadeh, M. J. (2017). Continuous bioethanol fermentation from wheat straw hydrolysate with high suspended solid content using an immersed flat sheet membrane bioreactor. *Bioresource Technology, 241*, 296–308.

Mariano, A. P., Costa, C. B. B., Maciel, M. R. W., Maugeri Filho, F., Atala, D. I. P., de Angelis, D. D. F., et al. (2010). Dynamics and control strategies for a butanol fermentation process. *Applied Biochemistry and Biotechnology, 160*(8), 2424–2448.

Marlatt, J. A., & Datta, R. (1986). Acetone-butanol fermentation process development and economic evaluation. *Biotechnology Progress, 2*, 23–28.

Meirelles, A., Weiss, S., & Herfurth, H. (1992). Ethanol dehydration by extractive distillation. *Journal of Chemical Technology and Biotechnology, 53*, 181–188.

Mikulski, D., & Kłosowski, G. (2018). Efficiency of dilute sulfuric acid pretreatment of distillery stillage in the production of cellulosic ethanol. *Bioresource Technology, 268*, 424–433.

Mikulski, D., Kłosowski, G., Menka, A., & Koim-Puchowska, B. (2019). Microwave-assisted pretreatment of maize distillery stillage with the use of dilute sulfuric acid in the production of cellulosic ethanol. *Bioresource Technology, 278*, 318–338.

Miranda-Galindo, E. Y., Segovia Hernandez, J. G., Hernandez, S., de la Rosa Alvarez, G., Gutierrez-Antonio, C., & Briones-Ramirez, A. (2009). Design of reactive distillation with thermal coupling for the synthesis of biodiesel using genetic algorithms. *Computer Aided Chemical Engineering, 26*, 549–554.

Monlau, F., Barakat, A., Steyer, J. P., & Carrere, H. (2012). Comparison of seven types of thermo-chemical pretreatments on the structural features and anaerobic digestion of sunflower stalks. *Bioresource Technology, 120*, 241–247.

Moravvej, Z., Makarem, M. A., & Reza Rahimpour, M. (2019). *The fourth generation of biofuels*. Oxford, UK: Elsevier (chapter 20).

Nagy, Z. K. (2007). Model based control of a yeast fermentation bioreactor using optimally designed artificial neural networks. *Chemical Engineering Journal, 127*, 95–109.

Naik, S. N., Goud, V. V., Rout, P. K., & Dalai, A. K. (2010). Production of first and second generation biofuels: A comprehensive review. *Renewable and Sustainable Energy Reviews, 14*(2), 578–597.

Nikolic, S., Mojovic, L., Rakin, M., Pejin, D., & Pejin, J. (2010). Ultrasound-assisted production of bioethanol by simultaneous saccharification and fermentation of corn meal. *Food Chemistry, 122*, 216–222.

O'Brien, D. J., Roth, L. H., & McAloon, A. J. (2000). Ethanol production by continuous fermentation–pervaporation: A preliminary economic analysis. *Journal of Membrane Science, 166*(1), 105–111.

Ödman, P., Johansen, C. L., Olsson, L., Gernaey, K. V., & Lantz, A. E. (2009). On-line estimation of biomass, glucose and ethanol in *Saccharomyces cerevisiae* cultivations using in-situ multi-wavelength fluorescence and software sensors. *Journal of Biotechnology, 144*(2), 102–112.

Pachauri, N., Rani, A., & Singh, V. (2017). Bioreactor temperature control using modified fractional order IMC-PID for ethanol production. *Chemical Engineering Research and Design, 122*, 97–112.

Pacheco-Basulto, J. A., Hernandez-McConville, D., Barroso-Munoz, F. O., Hernandez, S., Segovia-Hernandez, J. G., Castro-Montoya, A. J., et al. (2012). Purification of bioethanol using extractive batch distillation: Simulation and experimental studies. *Chemical Engineering and Processing: Process Intensification*, *61*, 30–35.

Papadaki, M. (2020). *Waste biomass suitable as feedstock for biofuels production*. UK: John Wiley & Sons Ltd. (chapter 2).

Peng, P., Shi, B., & Lan, Y. (2010). A review of membrane materials for ethanol recovery by pervaporation. *Separation Science and Technology*, *46*(2), 234–246.

Poddar, T., Jagannath, A., & Almansoori, A. (2017). Use of reactive distillation in biodiesel production: A simulation-based comparison of energy requirements and profitability indicators. *Applied Energy*, *185*, 985–997.

Radionova, M. V., Poudyal, R. S., Tiwari, I., Voloshin, R. A., Zharmukhamedov, S. K., Nam, H. G., et al. (2017). Biodiesel production: Challenges and opportunities. *International Journal of Hydrogen Energy*, *42*, 8450–8461.

Ramirez-Marquez, C., Segovia-Hernandez, J. G., Hernandez, S., Errico, M., & Rong, B.-G. (2013). Dynamic behavior of alternative separation processes for ethanol dehydration by extractive distillation. *Industrial & Engineering Chemistry Research*, *52*, 17554–17561.

Rodrigues Gurgel da Silva, A., Errico, M., & Rong, B.-G. (2017). Techno-economic analysis of organosolv pretreatment process from lignocellulosic biomass. *Clean Technologies & Environmental Policy*, *20*, 1401–1412.

Rodrigues Gurgel da Silva, A., Errico, M., & Rong, B.-G. (2018a). Systematic procedure and framework for synthesis and evaluation of bioethanol production processes from lignocellulosic biomass. *Bioresource Technology Reports*, *4*, 29–39.

Rodrigues Gurgel da Silva, A., Errico, M., & Rong, B.-G. (2018b). Evaluation of organosolv pretreatment for bioethanol production from lignocellulosic biomass: Solvent recycle and process integration. *Biomass Conversion and Biorefinery*, *8*, 397–411.

Rodrigues Gurgel da Silva, A., Giuliano, A., Errico, M., Rong, B.-G., & Barletta, D. (2019). Economic value and environmental impact analysis of lignocellulosic ethanol production: Assessment of different pretreatment processes. *Clean Technologies and Environmental Policy*, *21*, 637–654.

Roffler, S., Blank, H. W., & Wilke, C. R. (1987). Extractive fermentation of acetone and butanol: Process design and economic evaluation. *Biotechnology Progress*, *3*, 131–140.

Rong, B.-G., & Errico, M. (2012). Synthesis of intensified simple column configurations for multicomponent distillations. *Chemical Engineering and Processing: Process Intensification*, *62*, 1–17.

Sanchez-Ramirez, E., Alcocer-Garcia, H., Quiroz-Ramirez, J. J., Ramirez-Marquez, C., Segovia-Hernandez, J. G., Hernandez, S., et al. (2017). Control properties of hybrid distillation processes for the separation of biobutanol. *Journal of Chemical Technology and Biotechnology*, *92*, 959–970.

Segovia-Hernandez, J. G., Hernandez, S., & Bonilla-Petriciolet, A. (2015). Reactive distillation: A review of optimal design using deterministic and stochastic techniques. *Chemical Engineering and Processing: Process Intensification*, *97*, 134–143.

Segovia-Hernandez, J. G., Vazquez-Ojeda, M., Gomez-Castro, F. I., Ramirez-Marquez, C., Errico, M., Tronci, S., et al. (2014). Process control analysis for intensified bioethanol separation systems. *Chemical Engineering and Processing: Process Intensification*, *75*, 119–125.

Sheldon, R. A. (2017). *Enzymatic conversion of first- and second-generation sugars*. Switzerland: Springer (chapter 7).

Shibuya, M., Sasaki, K., Tanaka, Y., Yasukawa, M., Takahashi, T., Kondo, A., et al. (2017). Development of combined nanofiltration and forward osmosis process for production of ethanol from pretreated rice straw. *Bioresource Technology*, *235*, 405–410.

Singh, A., & Rangaiah, G. P. (2017). Review of technological advanced in bioethanol recovery and dehydration. *Industrial & Engineering Chemistry Research*, *56*, 5147–5163.

Skupin, P., & Metzger, M. (2018). PI control for a continuous fermentation process with a delayed product inhibition. *Journal of Process Control*, *72*, 30–38.

Soetaert, W., & Vandamme, E. J. (2009). *Biofuels*. John Wiley & Sons, Ltd.

Stanley, D., Bandara, A., Fraser, X., Chambers, P., & Stanley, G. A. (2010). The ethanol stress response and ethanol tolerance of *Saccharomyces cerevisiae*. *Journal of Applied Microbiology*, *109*(1), 13–24.

Subhedar, P. B., & Gogate, P. R. (2013). Intensification of enzymatic hydrolysis of lignocellulose using ultrasound for efficient bioethanol production: A review. *Industrial and Engineering Chemistry Research, 52*, 11816–11828.

Tavan, Y., & Hosseini, S. H. (2013). A novel integrated process to break the ethanol/water azeotrope using reactive distillation—Part 1: Parametric study. *Separation and Purification Technology, 118*, 455–462.

Tin, P. S., Lin, H. Y., Ong, R. C., & Chung, T.-S. (2011). Carbon molecular sieve membranes for biofuel separation. *Carbon, 49*, 369–375.

Tomás-Pejó, E., Alvira, P., Ballesteros, M., & Nego, M. J. (2011). *Pretreatment technologies for lignocellulose-to-bioethanol conversion*. Oxford, UK: Academic Press (chapter 7).

Torres-Ortega, C. E., Ramírez-Márquez, C., Sánchez-Ramírez, E., Quiroz-Ramírez, J. J., Segovia-Hernandez, J. G., & Rong, B.-G. (2018). Effects of intensification on process features and control properties of lignocellulosic bioethanol separation and dehydration systems. *Chemical Engineering and Processing-Process Intensification, 128*, 188–198.

Van der Bruggen, B., & Luis, P. (2014). Pervaporation as a tool in chemical engineering: A new era? *Current Opinion in Chemical Engineering, 4*, 47–53.

van der Merwe, A. B., Cheng, H., Gorgens, J. F., & Knoetze, J. H. (2013). Comparison of energy efficiency and economics of process designs for biobutanol production from sugarcane molasses. *Fuel, 105*, 451–548.

Vane, L. M. (2005). A review of pervaporation for product recovery from biomass fermentation processes. *Journal of Chemical Technology & Biotechnology: International Research in Process, Environmental & Clean Technology, 80*(6), 603–629.

Wang, Z., Dien, S. B., Rausch, D. D., Tumbleson, M. E., & Singh, V. (2018). Fermentation of undetoxified sugarcane bagasse hydrolyzates using a two-stage hydrothermal and mechanical refining pretreatment. *Bioresource Technology, 261*, 313–321.

Wang, K., Xiong, X., Chen, J., Chen, L., Su, X., & Liu, Y. (2012). Comparison of gamma irradiation and steam explosion pretreatment for ethanol production from agricultural residues. *Biomass and Bioenergy, 46*, 301–308.

Yan, J., Oyedeji, O., Leal, J. H., Donohoe, B. S., Semelsberger, T. A., Li, C., et al. (2020). Characterizing variability in lignocellulosic biomass: A review. *ACS Sustainable Chemistry & Engineering, 8*(22), 8059–8085.

Zhang, Y., Nakagawa, K., Shibuya, M., Sasaki, K., Takahashi, T., Shintani, T., et al. (2018). Improved permselectivity of forward osmosis membranes for efficient concentration of pretreated rice straw and bioethanol production. *Journal of Membrane Science, 566*, 15–24.

Zhang, X., Ning, Z., Wang, D. K., & Diniz da Costa, J. C. (2013). A novel ethanol dehydration process by forward osmosis. *Chemical Engineering Journal, 232*, 397–404.

Zheng, J., Tashiro, Y., Wang, Q., & Sonomoto, K. (2015). Recent advances to improve fermentative butanol production: Genetic engineering and fermentation technology. *Journal of Bioscience and Bioengineering, 119*, 1–9.

CHAPTER 3

A review of intensification technologies for biodiesel production

Lai Fatt Chuah[a], Jiří Jaromír Klemeš[b], Awais Bokhari[b,c], Saira Asif[b,d], Yoke Wang Cheng[e], Chi Cheng Chong[e], and Pau Loke Show[f]

[a]Faculty of Maritime Studies, Universiti Malaysia Terengganu, Kuala Terengganu, Terengganu, Malaysia
[b]Sustainable Process Integration Laboratory, SPIL, NETME Centre, Faculty of Mechanical Engineering, Brno University of Technology, VUT Brno, Brno, Czech Republic
[c]Chemical Engineering Department, COMSATS University Islamabad (CUI), Punjab, Lahore, Pakistan
[d]Faculty of Sciences, Department of Botany, PMAS Arid Agriculture University, Rawalpindi, Punjab, Pakistan
[e]Department of Chemical Engineering, School of Engineering and Computing, Manipal International University, Negeri Sembilan, Malaysia
[f]Department of Chemical and Environmental Engineering, University of Nottingham—Malaysia Campus, Semenyih, Malaysia

3.1 Introduction

Sustainable development in industrialization, motorization, and shipping (Chuah et al., 2021) has significantly increased the utilization of fossil fuel (Chuah, Yusup, Abd Aziz, Klemeš, et al., 2016). This phenomenon results in high greenhouse gas (GHG) emissions to the environment (Fózer et al., 2020). Environmental concerns are the key factor to inspire researchers to search for renewable and sustainable fuel (Chuah, Klemeš, Yusup, Bokhari, & Akbar, 2017a). Biodiesel is an alternative fuel to provide a solution to replace petroleum diesel. Biodiesel can be directly utilized in a diesel engine without any modifications due to its similarity properties to diesel fuel, such as heating value and viscosity (Chuah, Bokhari, Yusup, & Saminathan, 2017). Biodiesel contributes to environmental protection due to biodegradable, renewable, nontoxic, reduces carbon footprint, sulfur oxide emissions and GHGs (Chuah, Bokhari, et al., 2016). The sulfur content of biodiesel is 0.0018 wt%, which is about 28-fold lower than diesel fuel of 0.0500 wt% (Chuah, Abd Aziz, Yusup, Klemeš, & Bokhari, 2016). Chuah, Abd Aziz, et al. (2015) reported that by employing B50 compared to diesel fuel, the brake-specific fuel consumption and exhaust gas temperature were higher about 9.0% and 6.8%, while brake power, torque, and brake thermal efficiency were lower 6.7%, 5.2%, and 18.4%, respectively. In terms of engine emissions, the release particulates, unburned hydrocarbons, sulfur oxide, and carbon monoxide were lowered by 40%, 68%, 100%, and 26.3%, but higher carbon dioxide (38.5%) and nitrogen oxide (19.0%) using biodiesel compared to diesel fuel. Although higher carbon dioxide amounts were emitted, the use of biodiesel greatly reduced the life cycle circulation of carbon dioxide. GHG footprint can be minimized when the local available nonedible feedstock is utilized for cleaner biodiesel production rather than to export to other regions (Fan, Klemeš, Tan, & Varbanov, 2019).

Energy efficiency and raw material cost are the major factors that lead to the total cost of biodiesel production (Chuah, Yusup, Abd Aziz, & Bokhari, 2015). Increasing the mass transfer rate between oil and alcohol with minimal cost for biodiesel production in terms of raw material's price, energy consumption, time, and scale up cost are some of the challenges faced. However, minimizing the carbon footprint of biodiesel needs to be taken into account (Lam, Varbanov, & Klemeš, 2010) and local utilization of nonedible oil is an advantage in terms of overall energy efficiency, low raw material's price, and reduced transport costs (Ng, Lam, Varbanov, & Klemeš, 2014).

Microwave, ultrasonic cavitation, reactive distillation, and hydrodynamic cavitation are the potential intensification technologies to overcome the mass transfer resistance between immiscible reactants (oil and alcohol) and resulted in high ester conversion, shorter reaction time, and high yield efficiency. Hydrodynamic cavitation is about 2-, 11-, 138-, and 253-fold more efficient compared to reactive distillation, microwave, ultrasonic cavitation, and mechanical stirring in terms of energy efficiency for biodiesel production. In a range of technologies available for intensification, the hydrodynamic cavitation-based approach could be considered as an effective one to assist and intensify the transesterification. The yield efficiency in relation to the method was in the following order: hydrodynamic cavitation > reactive distillation > microwave > ultrasonic cavitation > mechanical stirring. This paper reviews the recent achievements of the different approaches of intensification technologies. The merits and limitations of these various intensification technologies have been discussed in this paper.

3.2 Intensification technologies for biodiesel production

Process intensification of biodiesel production from sustainable feedstock nonedible oil is attaining concerns and attention due to the considerable lower energy requirement and shorter reaction time. These intensification technologies have been presented and discussed in the following sections.

3.2.1 Microwave

Microwave power started to be used in 1940, where it is generated from a very high-power generator called the magnetron. The detailed history of microwave power has been stated in Chuah, Klemeš, Yusup, Bokhari, and Akbar (2017b).

Microwave irradiation activates the smallest variance degree of polar molecules and ions (e.g., alcohol) and continuously changing the magnetic field. The oscillating microwave field causes molecules or ions to have a rapid rotation, and heat is generated due to molecular collision and friction without changing the molecular structure. The reactant is heated up by thermal energy through convection, conduction, and radiation from the reactor surface. Therefore no warm-up stages are needed as the temperature of the polar reactant with high-permittivity rises rapidly by irradiating with microwaves. Microwave irradiation accelerates the chemical reaction due to the efficiency of heat transfer and selective absorption of microwave energy by polar molecules (mixture of oil, alcohol, and KOH/NaOH contains both polar and ionic component) and results in a higher yield of product in comparison to conventional heating (transferring of energy into reactant depends on the convection and thermal conductivity of the mixture).

The wavelength and frequency of microwave irradiation are in 1–1000 mm and 0.3–300 GHz frequency. The number of publication related to biodiesel production via transesterification-assisted microwave has increased since 2007. The benefit of microwave energy in biodiesel production has drawn attention from worldwide researchers. Microwave requires about 23-fold lower energy consumption compared to mechanical stirring in biodiesel production as the energy of microwave is transmitted directly to the reactant and no require preheating step (Tan, Lim, Ong, & Pang, 2019). The major limitation of using the microwave for biodiesel production is the scaling-up from laboratory scale to industrial scale due to the small penetration depth of the microwaves in the reactive media. This small penetration depth is the reason why mechanical stirring is required to homogenize the temperature (Mazubert, Taylor, Aubin, & Poux, 2014).

Sharma, Kodgire, and Kachhwaha (2019) carried out the transesterification reaction of waste cotton-seed cooking oil in the presence of KOH and a CaO catalyst using a microwave heating system. They obtained above 90 wt% of biodiesel yield within the reaction time of 9.6 and 9.7 min, whereas 40 min reaction time was required by using mechanical stirring. Fatimah, Rubiyanto, Taushiyah, Najah, and Azmi (2019) carried out the transesterification reaction of soybean oil in the presence of ZrO_2/BLA using the microwave heating system under optimum conditions of 1:15 molar ratio of oil to methanol and 12 wt% catalyst loading. The optimum yield of 92.75 wt% was achieved in 30 min under via microwave-assisted method and 120 min with the reflux method. Lin and Chen (2017) reported that microwave heating (950 W) has effectively reduced the reaction time from 60 min using mechanical stirring (660 W) to 10 s with 90 wt% of biodiesel yield in the presence of 1.0 wt% KOH as a catalyst, 1:6 molar ratio of Jatropha oil to methanol, 200 rpm at 65°C reaction temperature. Milano et al. (2018) investigated microwave irradiation-assisted alkaline-catalyzed transesterification to produce biodiesel from the mixture of waste cooking oil and *Calophyllum inophyllum* oil (W70CI30). They found that the biodiesel yield was 97.65 wt% at 1:13 molar ratio of oil to methanol, 0.774 wt% KOH catalyst loading, 600 rpm, and 7.15 min reaction time. It is found that the oxidation stability of methyl ester W70CI30 was fourfold higher than that of methyl ester waste cooking oil alone. Nayak and Vyas (2019) carried out the transesterification reaction of papaya oil in the presence of sodium hydroxide using a microwave heating system. The optimized biodiesel yield of 99.30 wt% was achieved in the presence of 0.95 wt% NaOH, 1:9.5 molar ratio of waste cooking oil to methanol at 62.33°C reaction temperature in 5 min reaction time. Phromphithak, Meepowpan, Shimpalee, and Tippayawong (2020) investigated the microwave-heated continuous flow reactor-assisted transesterification of palm oil for biodiesel production using choline hydroxide (ChOH) as green ionic liquid catalyst under the optimum condition of 1:13.24 molar ratio of oil to methanol, the flow rate of 20 mL/min, microwave power of 800 W, catalyst loading of 6.0 wt% ChOH, 68°C, and 5 min reaction time, obtaining 89.72 wt% methyl ester content.

Silitonga et al. (2020) reported that the microwave irradiation-assisted transesterification of *Ceiba pentandra* oil in the presence of methanol and KOH. The best biodiesel yield conversion of *Ceiba pentandra* oil (95.42 wt%) was obtained using a 1:13 molar ratio of oil to methanol, 0.84 wt% catalyst loading, and 800 rpm in 6.5 min reaction time. Singh and Sharma (2017) investigated the low-cost guinea fowl bone-derived recyclable catalyst as a heterogeneous catalyst and utilized it for biodiesel production on a microwave hearing system (800 W); they found that the biodiesel conversion was 95.82 wt% at 1:18 molar ratio of custard apple seed oil to methanol, 4 wt% catalyst loading, 65°C, 700 rpm, and 20 min reaction time. El Sherbiny, Refaat, and El Sheltawy (2010) stated that the application of radio frequency microwave energy offers a fast, easy route to this valuable biofuel with the advantages of enhancing the reaction rate of 2 min instead of 150 min, which resulted in a biodiesel yield of 97.4 wt%. Martinez-Guerra and Gude (2014) investigate the simultaneous effect of microwave and ultrasound irradiations on transesterification of used vegetable oil catalyzed by the heterogeneous catalyst, that is, barium oxide (BaO). BaO has been selected due to its activity among alkaline earth oxide catalyst, and the order of activity is $BaO > SrO > CaO > MgO$. They found that the biodiesel yield was 93.5 wt% at 1:4.5 and 1:6 molar ratio of oil to methanol, 91 wt% at 1:9 molar ratio of oil to methanol, and 86 wt% at 1:12 molar ratio of oil to methanol. This could be attributed to BaO as a strong catalytic activity, and its concentration in methanol is higher at low molar ratios, which enhances the mass transfer effect and the availability of the catalyst decreases at a higher molar ratio. The biodiesel yields were higher for the simultaneous microwave/ultrasound-mediated reaction (93.5 wt%) compared to only microwave (91 wt%) and ultrasound (83.5 wt%) under the optimum condition of 1:6 molar ratio of oil to methanol and 0.75 wt% BaO as a catalyst in 2 min of reaction time.

A comparison of the microwave-assisted biodiesel production is shown in Table 3.1. Multiple microwave units could have a major impact on land and energy compared to a single large mechanical stirring reactor. It would be very useful to carry the research on several aspects of microwave design in order to increase its capacity in biodiesel production. The combination effect of microwave (eliminate the heat transfer resistance) and hydrodynamic cavitation or ultrasonic hydrodynamic (eliminate the mass transfer resistance) could overcome microwave's limitation of low penetration depth (Fig. 3.1).

Table 3.1 Microwave-assisted biodiesel production.

Oil	Catalyst	Catalyst amount (wt%)	Reactant	Oil to alcohol molar ratio	Reaction condition	Maximum yield/conversion (wt%)	Other details	Reference
Waste cotton-seed cooking oil	KOH	0.65	Methanol	1:7	50°C, 400 rpm, 9.6 min	96.55 (yield)	A pulse power (half min cycle: 8 s on and 22 s off) of 180 W (900 W) with slight stirring, 0.89 wt% FFA	Sharma et al. (2019)
	CaO	1.33	Methanol	1:9.6	50°C, 400 rpm, 9.7 min	90.41 (yield)	With stirring, 0.5 wt% FFA	Fatimah et al. (2019)
Soybean oil	ZrO_2/BLA	12.0	Methanol	1:15	30 min	92.75 (yield)	Penetration length: 27 cm, 950 W, 2.0 wt% FFA	Lin and Chen (2017)
Treated Jatropha oil	KOH	1.0	Methanol	1:6	65°C, 200 rpm, 10 s	90.0 (yield)	850 W, 9.8 wt% FFA	Milano et al. (2018)
Waste cooking oil and *Calophyllum inophyllum* oil (W70CI30)	KOH	0.774	Methanol	1:13	600 rpm, 7.15 min	97.65 (yield)	700 W, 0.8 wt% FFA, 4.7×10^{-4} g/J	Nayak and Vyas (2019)
Papaya oil	NaOH	0.95	Methanol	1:9.5	62.33°C, 5 min	99.30 (yield)	800 W, flow rate 20 mL/min, 4.2×10^{-4} g/J	Phromphithak et al. (2020)
Palm oil	ChOH	6.0	Methanol	1:13.24	68°C, 5 min	89.72 (conversion)	850 W, 2 wt% FFA	Silitonga et al. (2020)
Treated *Ceiba pentandra* oil	KOH	0.84	Methanol	1:13	800 rpm, 6.5 min	95.42 (yield)	800 W, 2.0 wt% FFA	Singh and Sharma (2017)
Custard apple seed oil	Guinea fowl bone	4.0	Methanol	1:18	65°C, 700 rpm, 20 min	95.82 (conversion)	3000 W, with stirring 200 rpm	Bokhari et al. (2015)
Treated *Ceiba pentandra* oil	KOH	2.15	Methanol	1:9.85	57.09°C, 3.29 min	98.9 (conversion)	3000 W (80 W), with stirring 200 rpm	Bokhari et al. (2019)
Canola oil	CaO	2.0	Methanol	1:5	60°C, 30 min	77.0 (yield)	10% of an exit power of 1200 W, 1% FFA, 2.8×10^{-4} g/J	Kumar, Ravi Kumar, and Chandrashekar (2011)
Pongamia pinnata seed oil	KOH NaOH	1 0.5	Methanol	1:6	60°C, 5 min	97 (yield) 96 (yield)		

Continued

Table 3.1 Microwave-assisted biodiesel production—cont'd

Oil	Catalyst	Catalyst amount (wt%)	Reactant	Oil to alcohol molar ratio	Reaction condition	Maximum yield/conversion (wt %)	Other details	Reference
Palm oil	CaO derived from waste eggshells	15	Methanol	1:18	4 min	96.7 (yield)	900 W, high content of CaO (99.2 wt%), 4.6×10^{-4} g/J	Khemthong et al. (2012)
Soybean	SrO	1.8	Methanol	1:6	60°C, 40 s	97 (conversion) 81 (conversion) 97 (conversion)	900 W, cycle mode of 21 s on and 9 s off, without stirring	Koberg, Abu-Much, and Gedanken (2011)
	KOH	1						
	Sr(OH)$_2$	2.1						
Cooked oil	SrO	1.8			60°C, 20 s	93.2 (conversion) 99.4 (conversion)	without stirring with stirring	Perin et al. (2008)
Castor oil	SiO$_2$/50% H$_2$SO$_4$	10	Methanol	1:6	60°C, 30 min	95 (conversion)	40 W	
	Al$_2$O$_3$/50% KOH				60°C, 5 min	95 (conversion)		
Karanja oil	H$_2$SO$_4$ (esterification)	3.73	Methanol	1:9.4	300 rpm, 190 s	87.5 (FFA reduction)	180 W, 8.8 wt% FFA	Venkatesh Kamath, Regupathi, and Saidutta (2011)
	KOH	1.33		1:9.3	300 rpm, 150 s	89.9 (yield)	180 W, 1.1 wt% FFA	
Waste frying palm oil	NaOH	3	Ethanol	1:12	78°C, 30 s	97 (conversion)	800 W, 4.5% FFA, continuous production, reactor coil (Teflon tubing) 0.9 cm ID × 260 cm	Lertsathapornsuk, Pairintra, Aryusuk, and Krisnangkura (2008)
Pretreated Jatropha oil	NaOCH$_3$	1	Methanol	1:6	67–78°C, 30 s	96.5 (yield)	800 W, 0.11% FFA, continuous production, reactor coil (Teflon tubing) 8 cm ID	Tippayawong and Sittisun (2012)
Used vegetable oil	KOH	1	Methanol	1:6	50°C, 30 s	97.9 (conversion)—2 L/min 98.9 (conversion)—7.2 L/min	1600 W, 0.11% FFA, continuous production, reactor coil (Teflon tubing) 8 cm ID	Barnard, Leadbeater, Boucher, Stencel and Wilhite (2007)

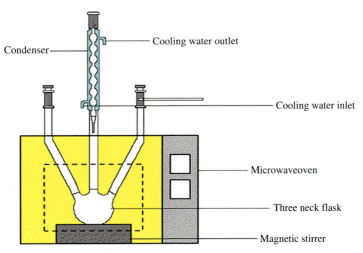

Fig. 3.1 Microwave reactor design for biodiesel production.

3.2.2 Ultrasonic cavitation

There are four types of cavitation, that is, hydrodynamic, acoustic, optical, and particle-induced (Petkovšek et al., 2013). Cavitation was discovered by Leonhard Euler in the year 1754. Cavitation is a formation, growth, and collapse of bubbles within a liquid, and releasing a large amount of energy over a small location resulting in very high densities, with the intense turbulence of liquid and chemical effects, which produce local hot spot and free radical generation (Suslick, 1990). Cavitation has increased its popularity in chemical processing due to its ability to generate high temperature and pressures under almost ambient conditions (Gogate & Pandit, 2000). The main effect of ultrasonic on chemical reactions is the formation and collapsed of microbubbles and released high energy locally in the shock waveform, through increasing the local temperature and pressure in microseconds (Suslick, 1990). This induced the molecules to generate high reactive radical species in order for product formation. High intensity of microlevel turbulence generated by oscillating cavities with a high interfacial area in the ultrasonic reactor is very effective to eliminate the mass transfer resistance during the reaction.

The sound in the audible range has no effect on the chemical reaction. The normal frequency of sound that can be responded to by human is from 0.016 to 18 kHz (Lam, Lee, & Mohamed, 2010). Ultrasound is defined as sound with a frequency above the human audibility limit, that is, more than 18 kHz. It is categorized based on its frequency into a high frequency (1000–10,000 kHz) with low power, which is normally used for diagnostic purposes, medical scanning, chemical analysis, and low frequency (20–100 kHz) with high power used for cleaning, plastic welding, and chemical reaction (Talebian-Kiakalaieh, Amin, Zarei, & Noshadi, 2013). Low frequency is commonly used for biodiesel production due to the dominant physical effects. However, methyl ester yield reduces when more than 40 kHz is applied. The power of ultrasonic can be control due to the soap formation in fast reaction (Tan et al., 2019). High frequency is not used for biodiesel production as a collapse of the cavity is weaker than the reactant's impingement. A low frequency of ultrasonic wave would force, stretch/expanse, and compress the molecular spacing of a mixture reaction in which it passes through and results in pressure variation. The molecules will be continuously vibrated to generate a huge number of cavitation bubbles. The cavities will be expanded following the violent collapse upon expanding to a certain critical size, and this asymmetric collapse of cavities generates microturbulence, which interrupts the phase boundary. These microjets from the impingement of liquid into another liquid can be reached up to a speed of 200 m/s, forming cordial mixing of the immiscible reactants at the phase boundary and then cause emulsification. This creation of emulsion continues to enhance the interfacial region and mass transfer between the immiscible reactants. Besides that, the sudden collapse of a bubble (because external pressure is higher than internal one) also resulted in high local temperature/heating ($\geq 4727°C$ and ≥ 1013 bar), increasing the intimate contact or interfacial surface of immiscible reactant and elevated pressure and thus accomplished chemical reaction in a few seconds (Bantle & Eikevik, 2014). The latter point is a generation of free radicals (OH•, HO$_2$•, and H•) from the dissociation of solvent vapor trapped in the bubble during a transient implosive collapse of a bubble (Tan et al., 2019). It can eliminate external mixing and heating due to the formation of microjets and local temperature. Ultrasonic cavitation could enhance the mass and heat transfer within the immiscible reactants and thus increase the chemical reaction rate. Consequently, it can decrease the operating cost and energy consumption as it requires lower alcohol to oil molar ratio, catalyst loading, reaction time, electricity consumption, and shorter time in phase separation compared to the mechanical stirring approached. In general, separation of the reaction mixture (biodiesel, glycerol, and alcohol) requires more than 3 h via gravity method subject to the amount of nonreacted alcohol and temperature of the mixture. Methanol ($0.79 g/cm^3$) and glycerol ($1.26 g/cm^3$) are polar compounds and the uniform phase (higher density) would be formed. In addition, the lower temperature of the reaction mixture lowers the dissolution of glycerol and methanol in the biodiesel layer. Low ultrasonic amplitude is required to avoid the formation of fine emulsion of alcohol/oil/catalyst multiphase mixture after the transesterification reaction to envisage the benefit of the ultrasonic cavitation. Low ultrasonic amplitude and lower temperature of the reaction mixture are leading to shorter separation time between biodiesel ($\approx 0.88 g/cm^3$) and byproduct layers in ultrasonic cavitation reactor.

The earliest works on the utilization of ultrasonic energy for biodiesel production has been done by Stavarache, Vinatoru, Nishimura, and Maeda (2005). They evaluated the production of biodiesel via the transesterification reaction of vegetable oil in the presence of NaOH under ultrasonic of 28 and 40 kHz. It was found that by using ultrasonic, the reaction time is much shorter (20–40 min) to obtain 98 wt% yield at both frequencies compared to mechanical stirring, which only obtains 80 wt% yield in 60 min under similar conditions of 0.5 wt% NaOH in the presence of 1:6 molar ratio of oil to methanol. Kashyap, Gogate, and Joshi (2018) investigated the ultrasonic (20 kHz, 120 W) assisted with heterogeneous acid (γ-alumina) interesterification of karanja oil for biodiesel production. They studied the effect of the catalyst loading (0.5–3.0 wt%), a duty cycle of an ultrasonic horn (40%–70%), oil to methyl acetate molar ratio (1:4–1:14), and temperature reaction (30–50°C) on the yield of biodiesel. The maximum yield of 69.3 wt% was achieved in 35 min under the following conditions of 1.0 wt% catalyst loading, oil to methyl acetate molar ratio of 1:9 at 50°C of reaction temperature. They also demonstrated the mechanical stirring, and the results show only 50.8 wt%. The author de Medeiros et al. (2019) studied the thermal effects of direct sonication on fish processing residue conversion into biodiesel without heating and stirring. This study shows that the transesterification reaction can be completed in a shorter reaction time of 2 min without any temperature control settings compared to the heating (30 min) and stirring (60 min). The high yield of biodiesel (content high concentrations of sodium, potassium, and magnesium) produced from fish waste oil was 81.02 wt% in 2 min at experimental conditions of 1:9 molar ratio of oil to methanol, 0.5 wt% NaOH catalyst, and the ultrasonic frequency of 20 kHz.

Thangarasu, Siddarth, and Ramanathan (2020) evaluated the interlaced effect of continuous flow microreactor and ultrasonic mixing on biodiesel production via the transesterification reaction of *Aegle marmelos* Correa seed oil in the presence of sodium methoxide under ultrasonic (20 kHz, 315 W). Response surface methodology experiment design was employed to study the effect of process parameters, that is, flow rate (2–10 mL/min), reaction temperature (45–65°C), catalyst amount (0.5–2.5 wt%), oil to methanol molar ratio (1:6–1:8), and ultrasonic mixing time (30–150 s). The highest biodiesel yield of 98 wt% was obtained in 0.3 mm reactor rather than 0.8 mm reactor (91.8 wt%) under the optimum condition of 6.8 mL/min flow rate, 48°C reaction temperature, 1.3 wt% catalyst loading, 1:9 molar ratio of oil to methanol, and 83 s ultrasonic mixing time. This combination of ultrasonic and microreactor was 450-fold shorter reaction time compared to without ultrasonic-assisted. Goh et al. (2020) investigated the effect of spent coffee ground oil to methanol molar ratio (1:10–1:60), catalyst loading (0.5–4.5 wt%), ultrasonic time (0.5–5 h), and ultrasonic amplitude (20%–40%) on biodiesel yield performance without employing the external heating device due to the continuous compression and rarefaction ultrasonic cavitation cycle. About 97.11 wt% of yield with 1:30 molar ratio of oil to methanol, 4 wt% KOH, and 30% ultrasonic amplitude could be achieved through ultrasonic-assisted transesterification for 3 h reaction time. Hoseini et al. (2018) investigated biodiesel production from *Ailanthus altissima* seed oil via transesterification over KOH catalyst. Response surface methodology experiment design was employed with three factors and three levels to study the effect of the amount of catalyst loading (1–2 wt%), oil to methanol molar ratio (1:2–1:12), and reaction time

(2–10 min) at a fixed reaction temperature of 50°C, 70% amplitude, and 70 pulses on the biodiesel yield. The optimum yield of 92.26 wt% was achieved under the following conditions of oil to methanol molar ratio of 1:8.5, catalyst concentration of 1.01 wt%, and a reaction time of 4.71 min.

Pascoal, Oliveira, Figueiredo, and Assuncao (2020) investigated the ultrasonic (20 kHz, 750 W) process with homogeneous alkali transesterification of *Syagrus cearensis* almond for biodiesel production. They evaluated the effect of three reaction variables in an ultrasonic-assisted reactor catalyzed by KOH, such as the molar ratio of oil to methanol (1:6–1:60), catalyst loading (1–5 wt%), and reaction time (10–30 min) at 50% amplitude. The maximum yield of 99.99 wt% was achieved under the following conditions of oil to methanol molar ratio of 1:6 and catalyst concentration of 5 wt% in 30 min reaction time. Carmona-Cabello, Sáez-Bastante, Pinzi, and Dorado (2019) studied the optimization of food waste oil via transesterification-assisted ultrasonic cavitation over the KOH catalyst. They evaluated the effect of three reaction variables in an ultrasonic-assisted reactor catalyzed by KOH, such as the molar ratio of oil to methanol (1:4.5–1:6.5), catalyst loading (0.8–2.0 wt%), and temperature reaction (40–65°C) at 50% amplitude. The results revealed that 93.23 wt% of biodiesel yield was achieved at optimum conditions of 1:6.08 molar ratio of oil to methanol in the presence of 1.28 wt% catalyst KOH at 52.5°C for 20 min reaction time. They also found that the ultrasonic cavitation was a 1.5-fold lower reaction time and 5-fold lower energy consumption compared to mechanical stirring.

Chen, Shan, Shi, and Yan (2014) investigated biodiesel production from palm oil via transesterification over the ostrich eggshell-derived CaO catalyst. They studied the effect of the amplitude of the ultrasonic power (30%–80%), reaction time (1–3 h), oil to methanol molar ratio (1:3–1:15), and amount of catalyst loading (9–10 wt%) at fixed reaction temperature of 60°C on the yield of biodiesel. The maximum yield of 92.7 wt% was achieved under the following conditions of 60% of amplitude, the reaction time of 1 h, oil to methanol molar ratio of 1:9, and catalyst concentration of 8 wt%. Maghami, Sadrameli, and Ghobadian (2015) studied the effect of catalyst concentration (0.5–2 wt%), oil to methanol molar ratio (1:3–1:12), and temperature reaction (40–60°C) on biodiesel production at 400 W ultrasonic power and conventional methods. It showed that 30 min reaction time using ultrasonic has a higher yield than 60 min when using a conventional method. The results revealed that 87 wt% of conversion and 79.86 wt% of yield was achieved at optimum conditions of 1:6 molar ratio of oil to methanol in the presence of 1 wt% catalyst KOH at 55°C reaction temperature. Sajjadi, Abdul Aziz, and Ibrahim (2015) investigated the effect of temperature (50–64°C), the molar ratio (1:6–1:12), catalyst loading (1–2 wt%), ultrasonic power (0–400 W), and reaction time (20–60 min) on biodiesel yield performance by employing central composite design. About 93.84 wt% of yield with 1:6 molar ratio of palm oil to methanol, 1 wt% KOH, 55°C, and 400 W could be achieved through ultrasonic-assisted transesterification in only 20 min, whereas only

89.09 wt% of yield was achieved by the conventional heating method under similar operating conditions. They observed that ultrasonic cavitation resulted in the generation of cavitation bubbles and the bubbles expanded to about 13–18 fold of their initial radius till the moment of collapse. This collapse resulted in the liquid experienced an energetic compression and collapsed. They reached the temperature and pressure of 486–440°C and 235.58–159.55 bar, and the reaction accelerated within the sonoluminescence bubbles due to the increasing number of energetic collisions within the cavities. This phenomenon is similar to a supercritical condition. They concluded that the oscillation velocity of these bubbles showed that cavitation plays an important role in increasing microstream, generating fine microemulsion, and hence increasing the mass transfer rate.

Pukale, Maddikeri, Gogate, Pandit, and Pratap (2015) studied the transesterification of waste cooking oil in the presence of different heterogeneous solid catalysts, such as K_3PO_4, Na_3PO_4, Na_2HPO_4, NaH_2PO_4, and KH_2PO_4 for biodiesel production. They found that K_3PO_4 exhibits high catalytic activity for the transesterification of waste cooking oil. They also investigated the influence of various operating parameters, such as catalyst concentration (1–4 wt%), oil to methanol molar ratio (1:4–1:8), and the reaction temperature (30–60°C), on the methyl ester yield progressed at a fixed 50% amplitude (supplied power of 375 W) of ultrasonic horn of frequency 22 kHz. The maximum yield of 92 wt% has been observed at optimum reaction parameters of oil to methanol molar ratio of 1:6, catalyst loading of 3 wt%, and the reaction temperature of 50°C in 90 min of reaction time.

A need for more of studies on the pulse timing and ultrasonic power in the ultrasonic system-assisted heterogeneous biodiesel production was noticed. Salamatinia, Mootabadi, Hashemizadeh, and Abdullah (2013) investigated the ultrasonic (20 kHz, 200 W) process with heterogeneous transesterification of oils, such as corn oil, canola oil, sunflower oil, and used palm oil for the production of biodiesel. They evaluated the effect of five process variables in an ultrasonic-assisted reactor catalyzed by SrO, such as pulse on (1–9 s), pulse off (1–9 s), reaction time (10–50 min), power (40–140 W), and oil amount (30–60 g) under 1:9 molar ratio of oil to methanol, 2.5 wt% catalyst loading, and 75% ultrasonic amplitude. The results indicated that the response surface methodology model could be able to predict the methyl ester yield at the lowest error. The obtained yield of 97 wt% under optimum conditions suggested by the model was at the ultrasonic pulse "on" of 9 s, pulse "off" of 2 s, the reaction time of 30.7 min, power of 130 W, and oil amount of 52 g.

The development of a continuous reactor with tubular or hexagonal geometry might be able to up scale the operation, and the use of multiple transducers with a possibility of multiple frequency operations is recommended in order to minimize the energy consumption (Fig. 3.2 and Table 3.2).

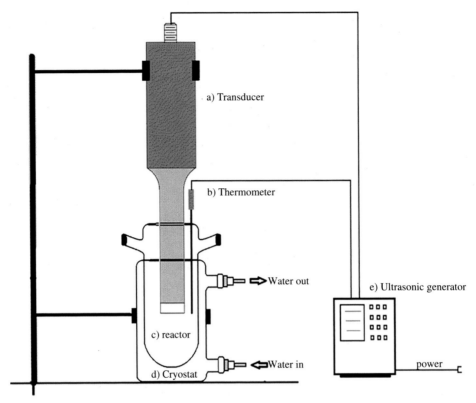

Fig. 3.2 Ultrasonic cavitation reactor design for biodiesel production Chuah, Amin, et al. (2016).

3.2.3 Hydrodynamic cavitation

Cavitation is usually divided into two classes of behavior: inertial (or transient) cavitation and noninertial cavitation. Inertial cavitation is the process where the bubbles/cavities collapsed due to recovery of pressure and released a great amount of local heating in the form of shockwaves, thus increasing local temperature (713–758°C) and pressure (159.55–235.58 bar) in microseconds and causing molecules to fragment and generate highly reactive radical species, which favors forward reaction (Sajjadi et al., 2015). The reaction speeded up within the sonoluminescence cavities due to the increasing number of energetic collisions within the oscillating cavities, which is similar to a supercritical condition. This behavior can eliminate the mass transfer limitations between two different phases in the flow system. It can be easily established by using control valves, pumps, and constriction/geometry. Noninertial cavitation is the process in which a bubble in a fluid is forced to oscillate in size or shape due to some form of energy input, such as an acoustic field. Such cavitation is often employed in ultrasonic cleaning baths and can also be observed in pumps and propellers (Shah, Pandit, & Moholkar, 1999).

Table 3.2 Ultrasonic cavitation-assisted transesterification.

Oil	Catalyst	Catalyst amount (wt%)	Reactant	Oil to alcohol molar ratio	Reaction condition	Maximum yield/conversion (wt %)	Other details	Reference
Vegetable oil	NaOH	0.5	Methanol	1:6	Room temperature, 40 min	98.00 (yield)	60% (720 W), 28 kHz	Stavarache et al. (2005)
					Room temperature, 20 min		60% (720 W), 40 kHz	
Karanja oil	γ-Alumina	1.0	Methyl acetate	1:9	50°C, 35 min	69.30 (yield)	120 W, 20 kHz, 1.9×10^{-4} g/J	Kashyap et al. (2018)
Fish processing residue	NaOH	0.5	Methanol	1:9	Without heating, 2 min	81.02 (yield)	500 W, 20 kHz, 1.6×10^{-4} g/J	de Medeiros et al. (2019)
Aegle marmelos Correa seed oil	Sodium methoxide	1.3	Methanol	1:9	48°C, 1.4 min	98.00 (yield)	315 W, 20 kHz, 1.1×10^{-4} g/J	Thangarasu et al. (2020)
Spent coffee ground oil	KOH	4.0	Methanol	1:30	Without heating, 3 h	97.11 (yield)	30% amplitude, 20 kHz	Goh et al. (2020)
Ailanthus altissima seed oil	KOH	1.01	Methanol	1:8.5	50°C, 4.71 min	92.26 (yield)	400 W, 24 kHz	Hoseini et al. (2018)
Syagrus cearensis almond	KOH	5.0	Methanol	1:6	25°C, 30 min	99.99 (yield)	750 W, 20 kHz, 50% amplitude	Pascoal et al. (2020)
Food waste oil	KOH	1.28	Methanol	1:6.08	52.5°C, 20 min	93.23 (yield)	700 W, 20 kHz, 50% amplitude	Carmona-Cabello et al. (2019)

Continued

Table 3.2 Ultrasonic cavitation-assisted transesterification—cont'd

Oil	Catalyst	Catalyst amount (wt%)	Reactant	Oil to alcohol molar ratio	Reaction condition	Maximum yield/conversion (wt %)	Other details	Reference
Treated rubber seed oil	EFB-KOH	15	Methanol	1:15	40°C, 30 min	Saponification form	500 W, 20 kHz	Chuah, Amin, et al. (2016)
Pistacia khinjuk sedd oil	SO_4^{2-}/SnO_2-SiO_2	3.5	Methanol	1:13	60°C, 50 min	88 (yield)	500 W, 20 kHz, 50% amplitude	Asif, Ahmad, et al. (2017)
Salvadora alii oil	CaO	3	Methanol	1:10	65°C, 30 min	92 (yield)	500 W, 20 kHz, 50% amplitude	Asif, Chuah, et al. (2017)
Thespesia populneoides oil		3.5		1:6	65°C, 30 min	88.6 (yield)	500 W, 20 kHz, 45% amplitude	
Jatropha oil	HPA/AC	3.5	Methanol	1:25	65°C, 40 min, 60% amplitude	91 (yield)	60% (240 W), 20 kHz	Badday, Abdullah, and Lee (2013)
Vernicia fordii	KOH	1	Methanol	1:6	20–30°C, 10 min,	91.15 (yield)	90% (270 W), 25 kHz	Manh, Chen, Chang, Chang, and Chang (2011)
Palm oil	waste ostrich eggshell-derived CaO	8	Methanol	1:9	60°C, 60 min,	92.7 (yield)	60% (120 W), 20 kHz	Chen et al. (2014)
Treated waste fish oil	KOH	1	Methanol	1:6	55°C, 30 min,	79.86 (yield)	100% (400 W), 20 kHz	Maghami et al. (2015)
Vegetable oil	KOH	1	Methanol	1:6	45°C, 10 min	85 (yield)	75% (150 W), 19.7 kHz	Ji, Wang, Li, Yu, and Xu (2006)
Palm oil	KOH	1	Methanol	1:6	55°C, 20 min	93.84 (yield)	100% (400 W), 24 kHz	Sajjadi et al. (2015)

Feedstock	Catalyst	Catalyst amount	Alcohol	Ratio	Conditions	Yield/Conversion	Power/frequency	Reference
Waste cooking oil	K_3PO_4	3	Methanol	1:6	50°C, 90 min	92 (yield)	50% (375 W), 22 kHz	Pukale et al. (2015)
Beef tallow	KOH	0.5	Methanol	1:6	60°C, 70 s	92 (yield)	100% (400 W), 24 kHz	Teixeira et al. (2009)
Vegetable oil	KOH	1	Methanol	1:6	55°C, 30 min	95 (yield)	60% (450 W), 20 kHz	Lee, Lee, and Hong (2011)
Waste cooking oil	KCH_3O	1	Methyl acetate	1:12	40°C, 30 min	90 (yield)	60% (450 W), 22 kHz	Maddikeri, Pandit, and Gogate (2013)
Palm oil	CaO SrO BaO	3	Methanol	1:9	65°C, 60 min, 50% amplitude	77.3 (yield) 95.2 (yield) 95.2 (yield)	50% (100 W), 20 kHz	Mootabadi, Salamatinia, Bhatia, and Abdullah (2010)
Jatropha oil	H_2SO_4 (esterification)	4	Methanol	1:11	60°C, 60 min	89 (FFA reduction)	100% (210 W), 20 kHz, 5.2% FFA	Deng, Fang, and Liu (2010)
	NaOH	1.4		1:6.6	60°C, 30 min	96.4 (yield)	100% (210 W), 20 kHz, 0.6% FFA	
Corn oil, canola oil, sunflower oil, used palm oil	SrO	2.5	Methanol	1:9	pulse on (9 s), pulse off (2 s), 30.7 min, 130 W, 52 g oil	97 (yield)	65% (130 W), 20 kHz	Salamatinia et al. (2013)

Hydrodynamic cavitation is the phenomenon where bubbles/cavities are created due to pressure reduction by passing the liquid through a constriction/geometry, such as an orifice, throttling valve, and venturi. According to Bernoulli's principle, incompressible fluid flows through the constriction/geometry; its velocity increases at the expense of pressure. Cavitation is initiated by the formation of bubbles/cavities when the liquid pass through the low-pressure region was below the vapor pressure of the flowing liquid, and subsequently, bubbles/cavities collapse asymmetrically when these bubbles/cavities expanse to maximum size under the isothermal condition as pressure recovers downstream of the mechanical constriction (Gogate & Pandit, 2000). The pressure recovery profile downstream of the orifice is assumed to be linear. The distance of the pressure recovery is usually in eight pipe diameters downstream of the orifice. The sudden collapse of these bubbles/cavities is due to the regaining of pressure that releases a tremendous amount of local energy, increasing the local temperature and pressure which accelerate the reaction rate (Ghayal, Pandit, & Rathod, 2013). The detailed mechanism of cavitation has been reported by Chuah, Yusup, Abd Aziz, Bokhari, and Abdullah (2016).

Cavitation number is the most mentioned term in hydrodynamic cavitation studies that represent the cavitation activities of the hydrodynamic cavitation reactor/system in terms of collapse condition as well as the ease of generation (Gole, Naveen, & Gogate, 2013). Dimensionless yet undeniably important, cavitation number summarizes the flow conditions efficiency of a hydrodynamic cavitation reactor/system by relating the velocity of flowing liquid with the properties of the liquid itself. A dimensionless parameter generally used to study hydrodynamic cavitation is the cavitation number, which is defined by the following Eq. (3.1) (Gole et al., 2013).

$$C_V = \frac{P_f - P_V}{\frac{1}{2}\rho U^2} \tag{3.1}$$

where P_f (Pa) is the fully recovered downstream pressure, P_v (Pa) is the vapor pressure of the mixture liquid, ρ (kg/m^3) is the density of the liquid mixture based on the molar ratio of oil and methanol at each of the density, and U (m/s) is the orifice velocity which can be calculated by knowing the upstream flow rate and diameter of the orifice.

Mechanical stirring is the most commonly used in the worldwide industrial application for biodiesel production. This production process requires a lot of energy, a high catalyst, a high molar ratio (alcohol to oil), and a longer time due to mass and heat transfer limitation (immiscible of alcohol and oil) to produce biodiesel. There is a strong quest to develop an efficient, time-saving, economically functional, and environmentally friendly biodiesel production intensification process at an industrial scale. Several intensification processes such as hydrodynamic cavitation, ultrasonic cavitation, supercritical methanol processes, and microwave have been developed. Hydrodynamic cavitation is one of the intensification processes that offer a substantial promise for biodiesel production in terms of methyl ester conversion, reaction rate, time, chemical consumption, energy consumption, scale-up, clean and safe (Qiu, Zhao, & Weatherley, 2010).

Table 3.3 presents the various studies that have been reported in biodiesel production derived from different feedstocks using hydrodynamic cavitation approaches. Hydrodynamic cavitation has been widely used in the shipping industry and wastewater treatment since 1990 and 2000. It is being extensively studied in biodiesel production by many researchers since 2006 due to its powerful reduction in the interfacial tension reduction between two immiscible phases such as oil and alcohol, flavor forward reaction. In addition, it is easy to operate, handle, require low energy input, and can be operated in a continuous system. The intensity of cavity collapse is subject to the physical properties of the liquid, that is, density, viscosity, vapor pressure, and so on, and also the design of the geometrical parameter of the orifice. Hydrodynamic cavitation is very sound in terms of energy and process efficiency in biodiesel production. The majority of the studies were only at laboratory scale (only a few at pilot scale), and the implementation for industrial purposes is still nowhere to be found. This may be due to the economic constraint, lack of expertise as well as technology limitation that were faced for the development and implementation of the system worldwide.

In early studies of the cavitation number, only a single hole of the orifice was used and it was found that the cavitation number was a function of orifice diameter (Yan, Thorpe, & Pandit, 1988). Then the studies expand by varying the plate geometry, including the hole size and perimeter, ratio of hole diameter and the total area to the pipe diameter, hole distribution, and inlet pressure (Yan & Thorpe, 1990). With different variables and parameters to be optimized, these studies were aimed to obtain the ideal cavitation number as possible within the range of 0.5–1. The hydrodynamic cavitation approach in biodiesel production was the pioneering work of Ji et al. (2006), and most of the studies were at laboratory scale. Hydrodynamic cavitation for biodiesel production in pilot scale (50 L) has been commenced by Chuah, Yusup, Abd Aziz, Bokhari, Klemeš, et al. (2015). Only a few studies are reporting on the design and optimization of geometrical parameters of hydrodynamic cavitation in biodiesel production. Ghayal et al. (2013) have investigated the effect of geometry and upstream pressure-assisted centrifugal pump upon methyl ester conversion derive from used frying oil in a 10 L capacity reactor. However, there is no detailed information in the literature regarding hole spacing and distribution pattern of holes in the plates. Chuah, Yusup, Abd Aziz, Bokhari, et al. (2016) reported the detail of the several newly design geometry of orifice plates upon different pressure on biodiesel production. They found that the geometrical parameters, such as α (ratio of the total hole perimeter to the total flow area) and $\beta°$ (ratio of the total flow area to the cross-section area of the pipe), are important parameters, which affects the biodiesel yield performance. Ghayal et al. (2013)—cavitation number: no reporting (α: $2\,\text{mm}^{-1}$)—and Chuah, Yusup, Abd Aziz, Bokhari, et al. (2016)—cavitation number: 0.357 (α: $4\,\text{mm}^{-1}$)—reported that the transesterification reaction-assisted hydrodynamic

Table 3.3 Comparison of hydrodynamic cavitation, microwave, ultrasonic cavitation, and mechanical stirring.

Raw material	Method	Process	Alcohol	Catalyst	Parameters Oil to alcohol molar ratio	Temperature (°C)	Catalyst (wt%)	Time (min)	Results Ester yield/conversion (wt%)	Yield efficiency 10^{-4} (g/J)	Other details	Reactor capacity (L)	Reference
Vegetable oil	HC UC	Transesterification	Methanol	KOH	1:6	45	1.0	10 30	98 (yield) 98 (yield)	15.18 0.11	7 bar 75% (150 W), 19.7 kHz	— 0.25	Ji et al. (2006)
Vegetable oil	MS HC UC	Transesterification	Methanol	NaOH	1:6	—	1 0.5	45 15 10	90 (yield) 98 (yield) 99 (yield)	0.06 33.7 0.86	900 rpm — 85 W, 20 kHz	— 10 0.1	Gogate (2008)
Fatty acid (C8–C10)	MS HC UC	Esterification	Methanol	H_2SO_4	1:10	28	0.5 1.0	180 90 120	98 (yield) 92 (conversion) 98 (conversion)	0.23 — —	— 1.5 kW 37.5% (45 W), 20 kHz	0.1 10 10	Kelkar, Gogate, and Pandit (2008)
				Caprylic acid			2.0	75	98 (conversion)	—	37.5% (45 W), 20 kHz	10	
Thumba oil	HC	Transesterification	Methanol	NaOH	1:4.5	50	1.0	30	80 (yield)	9.6	2.2 kW	10	Pal, Verma, Kachhwaha, and Maji (2010)
Frying oil	HC	Transesterification	Methanol	KOH	1:6	60	1.0	10	95 (conversion)	12.8	7.5 kW	10	Ghayal et al. (2013)
Nagchampa oil	HC UC	Transesterification	Methanol	KOH	1:6	60	1.0	20 40	92.1 (conversion) 92.5 (conversion)	8.7 0.1	1.5 kW 37.5% (45 W), 20 kHz, 350 rpm	10 0.25	Gole et al. (2013)
	MS	Transesterification						90	90.6 (conversion)	0.05	350 rpm	0.25	
Waste cooking oil	HC UC	Interesterification	Methyl acetate	CH_3OK	1:12	40	1.0	30	89.2 (yield) 90.0 (yield)	12.2 0.5	3 bar 60% (450 W), 22 kHz	15 —	Maddikeri, Gogate, and Pandit (2014)
Waste cooking oil	MS HC MS	Transesterification	Methanol	KOH	1:6	60	1	15 90	70.0 (yield) 98 (conversion) 97 (conversion)	0.3 12.5 1.5	1000 rpm 4 kW, 2 bar	— 50	Chuah, Yusup, Abd Aziz, Bokhari, et al. (2016)

Feedstock	Technology	Reaction	Alcohol	Catalyst	Molar ratio	Temp. (°C)	Catalyst (wt%)	Time (min)	Yield/Conversion (%)	Energy/Cost	Conditions	Oil (g/L)	Reference
Rubber seed oil	HC	Esterification	Methanol	H_2SO_4	1:6	55	8	30	96.4 (FFA reduction)	2.2	4 kW, 3 bar	50	Bokhari, Chuah, Yusup, Klemeš, and Kamil (2016)
	MS							90	92.8 (FFA reduction)	0.5			
Treated rubber seed oil	HC	Transesterification	Methanol	KOH	1:6	55	1	18	97.0 (conversion)	9.1	4 kW, 3 bar	50	Bokhari, Chuah, Yusup, Klemeš, Akbar, et al. (2016)
	MS							90		1.4			
Waste cooking oil	HC	Transesterification	Methanol	NaOH	1:6.8	35	1	5	99 (yield)	75	1.1 kW, 7 bar, 0.5 wt% FFA	10	Bargole et al. (2019)
Frying oil	HC	Transesterification	Methanol	KOH	1:4.5	45	0.55	20	93.6 (conversion)	27.5	1.75 kW, 1.2 wt% FFA	10	Kolhe, Gupta, and Rathod (2017)
Frying oil	MS	Transesterification	Methanol	KOH	1:6	63	1.1	60	88.5 (conversion)	N/A	1000 rpm	0.25	Chitsaz, Ghobadian, and Ardjmand (2018)
Frying oil	HC	Transesterification	Methanol	KOH	1:6	63	1.1	8	95.6 (yield)	N/A	3.27 bar	5	
Rubber seed oil	HC	Transesterification	Ethanol	KOH	1:6	40	4.5	40	92.5 (yield)	41.3	0.373 kW, 3700 g biodiesel	25	Samuel, Okwu, Amosun, Verma, and Afolalu (2019)
Schleichera Oleosa oil	HC	Transesterification	Methanol	KOH	1:6	60	0.75	30	95 (yield)	5	2.2 kW	10	Yadav, Khan, Pal, and Ghosh (2017)
Thumba oil	HC	Transesterification	Methanol	TiO_2–Cu_2O	1:6	60	1.6	60	60 (conversion)	N/A	5.5 kW, 2 bar	15	Patil, Baral, Dhanke, and Kore (2019)
Jatropha oil	MW	Transesterification	Methanol	KOH	1:7.5	65	1.5	2	97.4 (yield)	6.9	1200 W, 3.1% FFA	0.5	El Sherbiny et al. (2010)
Palm oil	MW	Transesterification	Methanol	CH_3ONa NaOH	1:6	65	0.75	3	99.5 (yield) 99.0 (yield)	6.9	1200 W, <1% FFA	0.5	Lin, Hsu, and Lin (2014)
Waste cooking oil	MW	Transesterification	Methanol	CH_3ONa NaOH	1:6	65	0.75	3	97.9 (yield) 96.2 (yield)	6.9	1200 W, <1% FFA	0.5	Chen, Lin, Hsu, and Wang (2012)
Pongamia pinnata seed oil	MW	Transesterification	Methanol	KOH NaOH	1:6	60	1 0.5	5	97 (yield) 96 (yield)	4.4	1200 W, 1% FFA	0.5	Kumar et al. (2011)

cavitation has given a shorter reaction, that is, 10 and 15 min at 60°C and 2 bar with biodiesel above 95 wt%. In addition, the reported value of α was found in the range of 0.4–4 mm^{-1}. Bokhari, Chuah, Yusup, Klemeš, and Kamil (2016) found that operating with cavitation number: 0.4 (α: 4 mm^{-1}) has successfully reduced the FFA of RSO to an acceptable level in just 30 min using an optimized orifice plate and other operating conditions reported by Chuah, Yusup, Abd Aziz, Bokhari, et al. (2016). Bokhari, Chuah, Yusup, Klemeš, Akbar, et al. (2016)—cavitation number: 0.301 (α: 4 mm^{-1})—have continued to utilize treated RSO (2.64 mg KOH/g) for cleaner rubber seed oil methyl ester production in a pilot-scale HC reactor. Bargole, George, and Saharan (2019)—cavitation number: 0.340 (α: 13.33 mm^{-1})—designed a new orifice plate with the higher value of α (1.33–13.33 mm^{-1}) in order to enhance the shear and turbulence for eliminating the mass transfer resistant and accelerate the transesterification reaction rate. It was agreed with Chuah, Klemeš, Yusup, Bokhari, Akbar, et al. (2017) that from the point of a pseudo-first-order transesterification kinetic, the hydrodynamic cavitation has about sevenfold higher rate constant compared to mechanical stirring and it is also agreed by Erdem Günay, Turker, and Tapan (2019). Bargole et al. (2019) found that 99 wt% biodiesel yield was obtained in just 5 min reaction time by using 100 circular holes of 0.3 mm diameter each at 7 bar of inlet pressure. Based on the previously mentioned studies, it was confirmed that the operation of the cavitating device at cavitation number about 0.3–0.4 could produce higher conversion/yield of biodiesel (>96.5 wt%) within 5–15 min. The reported results are in agreement with the findings of Yan and Thorpe (1990). Bargole et al. (2019) reported that the cavitational yield and reaction time have been about sixfold higher and two to threefold shorter compared to the findings of Chuah, Yusup, Abd Aziz, Bokhari, et al. (2016) and Ghayal et al. (2013) in the conditions of higher α (13.33 mm^{-1}) at lower reaction temperature (35°C), which could ease the separation process.

A few recommendations proposed later are some of the routes that could be explored as future work in biodiesel production via hydrodynamic cavitation. The effect of the orifice plate thickness on the methyl ester conversion should be further explored. Further optimization studies and process improvement by integrating microwave, hydrodynamic cavitation, and ultrasonic cavitation would be most advantageous for biodiesel production with low energy and chemical consumption. Consequently, the overall cost of biodiesel production can be substantially reduced (Fig. 3.3).

3.2.4 Reactive distillation

Reactive distillation is a promising intensification technology for esterification reaction (feedstock contents high FFA) compared to a conventional method. In addition, it could reduce the excess alcohol/alcohol usage (at least 50%) due to the continuous removal of byproduct (water). Reactive distillation has less connection between instruments, and it indirectly reduces safety issues. However, it requires a reboiler and condenser, thus

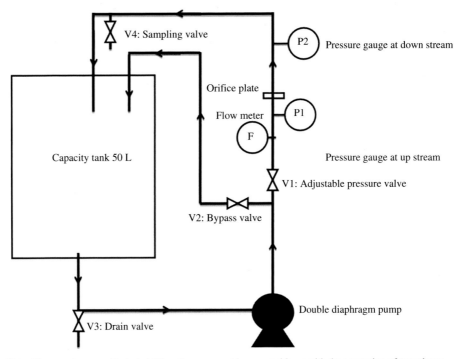

Note: The sample was collected at V4 as there was not be any problems with the separation of two phases

Fig. 3.3 Hydrodynamic cavitation reactor design for biodiesel production (Chuah, Klemeš, Yusup, Bokhari, & Akbar, 2017a, 2017b).

increase the operating cost. It is a chemical reaction and distillation separation in a single unit. Kiss (2014) have reviewed reactive distillation, and it found that it can be applied successfully to biodiesel production since the reactions leading to the end product, which controlled by chemical equilibrium. This reactive distillation has several advantages compared to the conventional process of biodiesel production, that is, (1) shorter reaction time; (2) no excess alcohol requirement; (3) lower capital cost due to the small size of a reactive distillation column and no additional separation units; (4) no neutralization and catalyst separation of the catalyst, as heterogeneous catalyst are used as a homogeneous catalyst rather; (5) simultaneous separation of reactants and products enhances the conversion and improves the selectivity by breaking the reaction equilibrium restrictions; and (6) get rid of the reheating as the heat of vaporization provides the heat of reaction in exothermic reactions. However, there are disadvantages of this reaction distillation, such as thermal degradation of the methyl esters due to the higher-temperature profile in the column. It is recommended that reactive absorption could replace reactive distillation as it can reduces the capital investment and operating cost due to the no reboiler and condenser require, higher conversion and selectivity as no products are recycled in the form of reflux or boil-up vapors, as well as no occurrence of thermal degradation of the products due to a lower-temperature profile in the column (Tan et al., 2019).

Reactive separation processes improve the biodiesel production efficiency by integrating both reaction and separation (e.g., distillation or membrane separation) into a single unit that allows the simultaneous production and removal of products and enhancing the reaction rate, improving the productivity and selectivity, reducing the energy use, eliminating the need for solvents, intensifying the mass and heat transfer, and ultimately leading to high-efficiency systems (Kiss & Bildea, 2012). It can bring important process performance and process of economic benefits.

Based on Tables 3.1–3.4, it is found that hydrodynamic cavitation was efficient, referring to time and energy consumption. The yield efficiency was shortened in the order of hydrodynamic cavitation, microwave, ultrasonic cavitation, and mechanical stirring. It can be found that hydrodynamic cavitation $(8.7–75 \times 10^{-4})$ g/J for biodiesel production in the present study is about 50–174, 39–87, 3–11, and 1–2 times more energy efficient compared to mechanical stirring $(0.05–1.5 \times 10^{-4})$ g/J, ultrasonic cavitation $(0.1–1.9 \times 10^{-4})$ g/J, microwave $(2.8–6.9 \times 10^{-4})$ g/J, and reactive distillation $(4.1–53.0 \times 10^{-4})$ g/J.

However, these comparisons are based on the different reactor capacity and feedstock's properties. In fact, the same reactor capacity of different intensification approaches should be employed for comparison in terms of energy efficiency as reported by Chuah, Yusup, Abd Aziz, Bokhari, et al. (2016). Hydrodynamic cavitation seems to be superior when compared to the other intensification processes in terms of high bubble density, scaling-up potential, reaction time, energy consumption, and biodiesel yield (Fig. 3.4).

3.2.5 Compare the costs of the different technologies

About 95% of the world biodiesel production is derived from edible oils (Tan et al., 2019). The consumption of edible oils in biodiesel production has consequently led to the price of edible oils and biodiesel to increase to 1.5–2 times the price of diesel fuel (Maddikeri et al., 2014). Therefore minimizing the carbon footprint or improving the GHG footprint of biodiesel needs to be taken into account (Lam, Varbanov, et al., 2010) so that it is an advantage to utilize it in local regions rather than export to other regions. It is more environment friendly, cleaner, and efficient at reducing energy loss and distribution cost (Ng et al., 2014). For this reason, utilizing low-cost feedstock derived from local nonedible oil such as nyamplung, Jatropha, castor, rubber, kapok, karanja, and waste cooking oils has become more attractive and a promising alternative raw material for biodiesel production, and it can also slightly reduce the raw material cost. About 75%–88% of the total cost of biodiesel is derived from raw material. The examples of the biodiesel production cost are raw materials, chemicals, utilities (water, electricity,

Table 3.4 Reactive distillation-assisted esterification and transesterification.

Raw material	Method	Process	Alcohol	Catalyst	Oil to alcohol molar ratio	Reboiler temperature (°C)	Catalyst (wt%)	Time (min)	Ester yield/conversion (wt%)	Yield efficiency 10^{-4} (g/J)	Other details	Reference
Waste cooking oil	Integrated two RDs Hybridization Conventional process RD	Transesterification	Methanol	CaO/Al_2O_3	1:4	205	3.0	60	97 (conversion) 98 (conversion) 99 (conversion)	11.11 11.36 4.05	Operated within the eight reactive stages under 3 bar pressure and reflux ratio 0.1	Petchsoongsakul, Ngaosuwan, Kiatkittipong, Aiouache, and Assabumrungrat (2017)
Oleic acid	RDC-total condenser	Esterification	Methanol	Amberlyst 15	1:4	344	1000 kg	60	99 (conversion)	53.01	Operated at 1 bar pressure and reflux ratio 1.0	Perez-Cisneros et al. (2016)
	RDC-partial condenser					343			99 (conversion)	48.51	Operated at 1 bar pressure and reflux ratio 1.75	
	RDC-NO condenser					343			99 (conversion)	52.63	Operated at 1 bar pressure and reflux ratio 2.0	
Triolein	RDC-one methanol feed	Transesterification		MgO	1:4	330			99 (conversion)	12.02	Operated at 1 bar pressure and reflux ratio 3.0	
	RDC-two methanol feeds				1:4	124			99 (conversion)	27.03	Operated at 1 bar pressure and reflux ratio 3.5	
Waste cooking oil	RD	Transesterification	Methanol	CaO/Al_2O_3	—	175	—	—	97 (yield)	—	Reflux ratio 0.1	Anantapinitwatna, Ngaosuwan, Kiatkittipong, Wongsawaeng, and Assabumrungrat (2019)
Oleic acid	RD Heat integrated RD	Esterification	Methanol	—	1:4	310	—	60	98 (conversion) 98 (conversion)	14.87 23.11	Reflux ratio 0.2	Phuenduang, Chatsirisook, Simasatikul, Paengjuntuek, and Arpornwichanop (2012)
Triglyceride	Conventional with a vac. distill	Transesterification	Methanol	NaOH	1:5	120	—	60	85 (yield)	4.31	Operated at 1 bar pressure and reflux ratio 0.1	Boon-anuwat, Kiatkittipong, Aiouache, and Assabumrungrat (2015)
	RD				1:5		—		98 (yield)	15.17		
	Conventional with vac. distill			MgO	1:15		—		93 (yield)	2.41		
	RD				1:4		—		98 (yield)	18.18		

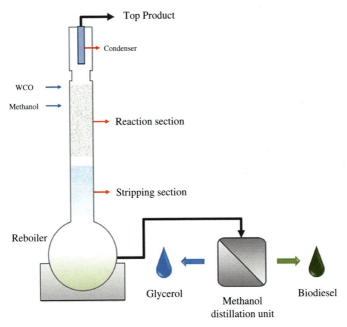

Fig. 3.4 Reactive distillation reactor design for biodiesel production.

etc.), equipment, facility construction, labor, maintenance, and separation time. According to Chuah, Amin, et al. (2016), the second major contributor cost after the raw material is the energy efficiency in biodiesel production. Consequently, yield efficiency related to energy efficiency can be used as a tool to compare the costs of the different technologies, including the conventional method. To the best of the author's knowledge, it can be seen that the yield efficiency in relation to the method was in the following order; hydrodynamic cavitation > reactive distillation > microwave > ultrasonic cavitation > mechanical stirring.

General practice, reaction mixture separation which containing methyl ester, glycerol, water, excess methanol, will require several hours via the gravity method. The amount of excess methanol used and the temperature of the reaction mixture would be the main factor to determine the separation time. Reaction mixture which containing less methanol and low temperature can result in lower separation time. It could be explained by the differences density of methanol ($0.79\,g/cm^3$) and glycerol ($1.26\,g/cm^3$) and both of it is polar compound (form uniform phase at any mixing ratio). Glycerol layer with lower excess of methanol will be resulting shorter time in during separation between methyl ester layer and glycerol layer due to significant differences of both layers. The lower temperature of the reaction mixture will lower the dissolution of glycerol and methanol in the methyl ester layer and thus resulting in a shorter separation time.

Table 3.5 Comparison of excess methanol and temperature of different intensification technologies in biodiesel production.

Method	Oil to methanol molar ratio	Temperature (°C)
Hydrodynamic cavitation	1:5–1.7	28–65
Reactive distillation	1:1–1:5	>100
Microwave	1:6–1:18	50–78
Ultrasonic cavitation	1:6–1:30	40–65
Mechanical stirring	1:5–1.7	40–65

Comparison of the oil to methanol molar ratio and temperature within the different intensification technologies is shown in Table 3.5. Reactive distillation seems to be superior (reduce the biodiesel cost especially in the separation process) when compared to the other intensification processes in terms of the methanol amount in the reaction mixture. Although the high temperature of the reaction mixture results in difficulty in the separation process, it can be solved by reducing the temperature of the mixture with external cooling assistance.

3.3 Conclusion

Today's most of the world's biodiesel production at an industrial scale is mechanical stirring with heating via transesterification by converting oil in the presence of methanol to fatty acid alkyl ester and glycerol. However, this conventional method's main drawback of the transesterification reaction is limited by mass transfer resulting in a much lower reaction rate and also much higher cost compared to diesel fuel. Different intensification technologies such as hydrodynamic cavitation, reactive distillation, ultrasonic cavitation, and microwave have been tried out to overcome these aforementioned problems, but they were only at laboratory scale and minipilot scale. It was found that microwave and ultrasonic cavitation are yet to be completely feasible for biodiesel production at an industrial scale. Hydrodynamic cavitation requires mild reaction conditions to complete the transesterification reaction. It is energy-efficient, time-saving, high product quality, eco-friendly, simple, and cheapest method of generating cavitation and easy to scale up. More studies are still needed to extend the existing information of HC in the design of the plate geometry with respect to the methyl ester conversion in order to develop a sustainable and industrially viable route for energy recovery from renewable oils. It can be concluded that the yield efficiency in relation to the method was in the following order: hydrodynamic cavitation > reactive distillation > microwave > ultrasonic cavitation > mechanical stirring. To the best of the author's knowledge, most of the produced biodiesel properties from the listed intensification technologies are in conform to both standards of EN 14214 and ASTM D 6751.

Acknowledgments

The authors gratefully acknowledge the Ministry of Higher Education Malaysia, Universiti Malaysia Terengganu, Manipal International University, University of Nottingham—Malaysia Campus, Department of Chemical and Environmental Engineering and Universiti Utara Malaysia. Heartfelt appreciation to Prof. Ir. Ts. Dr. Pau Loke Show, Chief Engineer Kevin Chin Ket Vui, Chief Mate Lim Poh Keong, Captain Shamsir Mohamed, Prof. Dr. Nor Hasni Osman (UUM), Jenna Tan Ying Min, Tan Lee Chwin, Chew Kuan Lian, Teh Bee Bee, Loh Chong Hooi, Timmy Chuah Tim Mie, and Ong Shying Weei for their support One of the authors has been supported by the EU project "Sustainable Process Integration Laboratory—SPIL," project No. CZ.02.1.01/0.0/0.0/15_003/0000456 funded by EU "CZ Operational Programme Research, Development and Education," Priority 1: Strengthening capacity for quality research, which has been gratefully acknowledged.

References

Anantapinitwatna, A., Ngaosuwan, K., Kiatkittipong, W., Wongsawaeng, D., & Assabumrungrat, S. (2019). Effect of water content in waste cooking oil on biodiesel production via ester-transesterification in a single reactive distillation. *IOP Conference Series: Materials Science and Engineering, 559*, 1–10.

Asif, S., Ahmad, M., Bokhari, A., Chuah, L. F., Klemeš, J. J., Akbar, M. M., et al. (2017). Methyl ester synthesis of *Pistacia khinjuk* seed oil by ultrasonic-assisted cavitation system. *Industrial Crops and Products, 108*, 336–347.

Asif, S., Chuah, L. F., Klemeš, J. J., Ahmad, M., Akbar, M. M., Lee, K. T., et al. (2017). Cleaner production of methyl ester from non-edible feedstock by ultrasonic-assisted cavitation system. *Journal of Cleaner Production, 161*(10), 1360–1373.

Badday, A. S., Abdullah, A. Z., & Lee, K. T. (2013). Optimisation of biodiesel production process from Jatropha oil using supported heteropolyacid catalyst and assisted by ultrasonic energy. *Renewable Energy, 50*, 427–432.

Bantle, M., & Eikevik, T. M. (2014). A study of the energy efficiency of convective drying systems assisted by ultrasound in the production of clipfish. *Journal of Cleaner Production, 65*, 217–223.

Bargole, S., George, S., & Saharan, V. K. (2019). Improved rate of transesterification reaction in biodiesel synthesis using hydrodynamic cavitating devices of high throat perimeter to flow area ratios. *Chemical Engineering and Processing Process Intensification, 139*, 1–13.

Barnard, T. M., Leadbeater, N. E., Boucher, M. B., Stencel, L. M., & Wilhite, B. A. (2007). Continuous-flow preparation of biodiesel using microwave heating. *Energy & Fuels, 21*, 1777–1781.

Bokhari, A., Chuah, L. F., Michelle, L. Z. Y., Asif, S., Shahbaz, M., Akbar, M. M., et al. (2019). Microwave enhanced catalytic conversion of canola-based methyl ester: Optimization and parametric study. In A. K. Azad, & M. Rasul (Eds.), *Woodhead publishing series in energy. Advanced biofuels: Application, technologies and environmental sustainability* (pp. 154–166). United Kingdom: Elsevier.

Bokhari, A., Chuah, L. F., Yusup, S., Ahmad, J., Shamsuddin, M. R., & Teng, M. K. (2015). Microwave-assisted methyl esters synthesis of Kapok (*Ceiba pentandra*) seed oil: Parametric and optimization study. *Biofuel Research Journal, 7*, 281–287.

Bokhari, A., Chuah, L. F., Yusup, S., Klemeš, J. J., Akbar, M. M., & Kamil, R. N. M. (2016). Cleaner production of rubber seed oil methyl ester using a hydrodynamic cavitation: Optimisation and parametric study. *Journal of Cleaner Production, 136*(B), 31–41.

Bokhari, A., Chuah, L. F., Yusup, S., Klemeš, J. J., & Kamil, R. N. M. (2016). Optimisation on pretreatment of rubber seed (*Hevea brasiliensis*) oil via esterification reaction in a hydrodynamic cavitation reactor. *Bioresource Technology, 199*, 414–424.

Boon-anuwat, N., Kiatkittipong, W., Aiouache, F., & Assabumrungrat, S. (2015). Process design of continuous biodiesel production by reactive distillation: Comparison between homogeneous and heterogeneous catalysts. *Chemical Engineering and Processing*, 1–36.

Carmona-Cabello, M., Sáez-Bastante, J., Pinzi, S., & Dorado, M. P. (2019). Optimisation of solid food waste oil biodiesel by ultrasound-assisted transesterification. *Fuel, 255*, 115–122.

Chen, K. S., Lin, Y. C., Hsu, K. H., & Wang, H. K. (2012). Improving biodiesel yields from waste cooking oil by using sodium methoxide and a microwave heating system. *Energy, 38*, 151–156.

Chen, G., Shan, R., Shi, J., & Yan, B. (2014). Ultrasonic-assisted production of biodiesel from transesterification of palm oil over ostrich eggshell-derived CaO catalysts. *Bioresource Technology, 171*, 428–432.

Chitsaz, H., Ghobadian, M. O. B., & Ardjmand, M. (2018). Optimisation of hydrodynamic cavitation process of biodiesel production by response surface methodology. *Journal of Environmental Chemical Engineering*. https://doi.org/10.1016/j.jece.2018.02.047.

Chuah, L. F., Abd Aziz, A. R., Yusup, S., Bokhari, A., Klemeš, J. J., & Abdullah, M. Z. (2015). Performance and emission of diesel engine fuelled by waste cooking oil methyl ester derived from palm olein using hydrodynamic cavitation. *Journal of Clean Technologies and Environmental Policy, 17*(8), 2229–2241.

Chuah, L. F., Abd Aziz, A. R., Yusup, S., Klemeš, J. J., & Bokhari, A. (2016). Waste cooking oil biodiesel via hydrodynamic cavitation on a diesel engine performance and greenhouse gas footprint reduction. *Chemical Engineering Transactions, 50*, 301–306.

Chuah, L. F., Amin, M. M., Yusup, S., Raman, N. A., Bokhari, A., Klemeš, J. J., et al. (2016). Influence of green catalyst on transesterification process using ultrasonic-assisted. *Journal of Cleaner Production, 136*(B), 14–22.

Chuah, L. F., Bokhari, A., Yusup, S., Klemeš, J. J., Abdullah, B., & Akbar, M. M. (2016). Optimisation and kinetic studies of acid esterification of high free fatty acid rubber seed oil. *Arabian Journal for Science and Engineering, 41*(7), 2515–2526.

Chuah, L. F., Bokhari, A., Yusup, S., & Saminathan, S. (2017). Optimisation on pretreatment of kapok seed (*Ceiba pentandra*) oil via esterification reaction in an ultrasonic cavitation reactor. *Biomass Conversion and Biorefinery, 7*(1), 91–99.

Chuah, L. F., Klemeš, J. J., Yusup, S., Bokhari, A., & Akbar, M. M. (2017a). Influence of fatty acids in waste cooking oil for cleaner biodiesel. *Journal of Clean Technologies and Environmental Policy, 3*, 1–10.

Chuah, L. F., Klemeš, J. J., Yusup, S., Bokhari, A., & Akbar, M. M. (2017b). A review of cleaner intensification technologies in biodiesel production. *Journal of Cleaner Production, 146*, 181–193.

Chuah, L. F., Klemeš, J. J., Yusup, S., Bokhari, A., Akbar, M. M., & Zhong, Z. K. (2017). Kinetic studies on waste cooking oil into cleaner biodiesel via hydrodynamic cavitation. *Journal of Cleaner Production, 146*, 47–56.

Chuah, L. F., Mohd Salleh, N. H., Osnin, N. A., Alcaide, J. I., Abdul Majid, M. H., Abdullah, A. A., et al. (2021). Profiling Malaysian ship registration and seafarers for streamlining future Malaysian shipping governance. *Australian Journal of Maritime & Ocean Affairs*. https://doi.org/10.1080/18366503.2021.1878981.

Chuah, L. F., Yusup, S., Abd Aziz, A. R., & Bokhari, A. (2015). Performance of refined and waste cooking oils derived from palm olein on synthesis methyl ester via mechanical stirring. *Australian Journal of Basic and Applied Sciences, 9*(37), 445–448.

Chuah, L. F., Yusup, S., Abd Aziz, A. R., Bokhari, A., & Abdullah, M. Z. (2016). Cleaner production of methyl ester using waste cooking oil derived from palm olein using a hydrodynamic cavitation reactor. *Journal of Cleaner Production, 112*(5), 4505–4514.

Chuah, L. F., Yusup, S., Abd Aziz, A. R., Bokhari, A., Klemeš, J. J., & Abdullah, M. Z. (2015). Intensification of biodiesel synthesis from waste cooking oil (palm olein) in a hydrodynamic cavitation reactor: Effect of operating parameters on methyl ester conversion. *Journal of Chemical Engineering and Processing: Process Intensification, 95*, 235–240.

Chuah, L. F., Yusup, S., Abd Aziz, A. R., Klemeš, J. J., Bokhari, A., & Abdullah, M. Z. (2016). Influence of fatty acids content in non-edible oil for biodiesel properties. *Journal of Clean Technologies and Environmental Policy, 18*(2), 473–482.

de Medeiros, E. F., Vieira, B. M., de Pereira, C. M. P., Nadaleti, W. C., Quadro, M. S., & Andreazza, R. (2019). Production of biodiesel using oil obtained from fish processing residue by conventional methods assisted by ultrasonic waves: Heating and stirring. *Renewable Energy, 143*, 1357–1365.

Deng, X., Fang, Z., & Liu, Y. (2010). Ultrasonic transesterification of *Jatropha curcas* L. oil to biodiesel by a two-step process. *Energy Conversion and Management, 51*, 2802–2807.

El Sherbiny, S. A., Refaat, A. A., & El Sheltawy, S. T. (2010). Production of biodiesel using the microwave technique. *Journal of Advanced Research, 1*, 309–314.

Erdem Günay, M., Turker, L., & Tapan, N. A. (2019). Significant parameters and technological advancements in biodiesel production systems. *Fuel, 250*, 27–41.

Fan, Y. V., Klemeš, J. J., Tan, R. R., & Varbanov, P. S. (2019). Graphical break-even based decision-making tool (BBDM) to minimise GHG footprint of biomass utilisation: Biochar by pyrolysis. *Chemical Engineering Transactions, 76*, 19–24. https://doi.org/10.3303/CET1976004.

Fatimah, I., Rubiyanto, D., Taushiyah, A., Najah, F. B., & Azmi, U. (2019). Use of ZrO_2 supported on bamboo leaf ash as a heterogeneous catalyst in microwave-assisted biodiesel conversion. *Sustainable Chemistry and Pharmacy, 12*, 100–108.

Fózer, D., Volantb, M., Passarini, F., Varbanov, P. S., Klemeš, J. J., & Mizsey, P. (2020). Bioenergy with carbon emissions capture and utilisation towards GHG neutrality: Power-to-gas storage via hydrothermal gasification. *Applied Energy*. https://doi.org/10.1016/j.apenergy.2020.115923.

Ghayal, D., Pandit, A. B., & Rathod, V. K. (2013). Optimisation of biodiesel production in a hydrodynamic cavitation reactor using used frying oil. *Ultrasonics Sonochemistry, 20*, 322–328.

Gogate, P. R. (2008). Cavitational reactors for process intensification of chemical processing applications: A critical review. *Chemical Engineering and Processing: Process Intensification, 47*, 515–527.

Gogate, P. R., & Pandit, A. B. (2000). Engineering design methods for cavitation reactors II: Hydrodynamic cavitation. *The American Institute of Chemical Engineers, 46*, 1641–1649.

Goh, B. H. H., Ong, H. C., Chong, C. T., Chen, W., Leong, K. Y., Tan, S. X., et al. (2020). Ultrasonic assisted oil extraction and biodiesel synthesis of spent coffee ground. *Fuel, 261*, 116–121.

Gole, V. L., Naveen, K. R., & Gogate, P. R. (2013). Hydrodynamic cavitation as an efficient approach for intensification of synthesis of methyl esters from sustainable feedstock. *Chemical Engineering and Processing: Process Intensification, 71*, 70–76.

Hoseini, S. S., Najafi, G., Ghobadian, B., Mamat, R., Ebadi, M. T., & Yusaf, T. (2018). *Ailanthus altissima* (tree of heaven) seed oil: Characterisation and optimisation of ultrasonication-assisted biodiesel production. *Fuel, 220*, 621–630.

Ji, J., Wang, J., Li, Y., Yu, Y., & Xu, Z. (2006). Preparation of biodiesel with the help of ultrasonic and hydrodynamic cavitation. *Ultrasonics, 44*, 411–414.

Kashyap, S. S., Gogate, P. R., & Joshi, S. M. (2018). Ultrasound assisted intensified production of biodiesel from sustainable source as karanja oil using interesterification based on heterogeneous catalyst (γ-alumina). *Chemical Engineering and Processing Process Intensification*. https://doi.org/10.1016/j.cep.2018.12.006.

Kelkar, M. A., Gogate, P. R., & Pandit, A. B. (2008). Intensification of esterification of acids for synthesis of biodiesel using acoustic and hydrodynamic cavitation. *Ultrasonics Sonochemistry, 15*, 188–194.

Khemthong, P., Luadthong, C., Nualpaeng, W., Changsuwan, P., Tongprem, P., & Viriya-empikul, N. (2012). Industrial eggshell wastes as the heterogeneous catalysts for microwave-assisted biodiesel production. *Catalysis Today, 190*, 112–116.

Kiss, A. A. (2014). Process intensification technologies for biodiesel production: Reactive separation processes. *Springer briefs in applied sciences and technology*. London, UK: Springer.

Kiss, A. A., & Bildea, C. S. (2012). A review of biodiesel production by integrated reactive separation technologies. *Journal of Chemical Technology & Biotechnology*, 1–19.

Koberg, M., Abu-Much, R., & Gedanken, A. (2011). Optimisation of bio-diesel production from soybean and wastes of cooked oil: Combining dielectric microwave irradiation and a SrO catalyst. *Bioresource Technology, 102*, 1073–1078.

Kolhe, N. S., Gupta, A. R., & Rathod, V. K. (2017). Production and purification of biodiesel produced from used frying oil using hydrodynamic cavitation. *Resource-Efficient Technologies, 3*, 198–203.

Kumar, R., Ravi Kumar, G., & Chandrashekar, N. (2011). Microwave assisted alkali-catalysed transesterification of *Pongamia pinnata* seed oil for biodiesel production. *Bioresource Technology, 102*, 6617–6620.

Lam, M. K., Lee, K. T., & Mohamed, A. R. (2010). Homogeneous, heterogeneous and enzymatic catalysis for transesterification of high free fatty acid oil (waste cooking oil) to biodiesel: A review. *Biotechnology Advances, 28*, 500–518.

Lam, H. L., Varbanov, P., & Klemeš, J. J. (2010). Minimising carbon footprint of regional biomass supply chains. *Resources, Conservation and Recycling, 54*, 303–309.

Lee, S. B., Lee, J. D., & Hong, I. K. (2011). Ultrasonic energy effect on vegetable oil based biodiesel synthetic process. *Journal of Industrial and Engineering Chemistry, 17*, 138–143.

Lertsathapornsuk, V., Pairintra, R., Aryusuk, K., & Krisnangkura, K. (2008). Microwave assisted in continuous biodiesel production from waste frying palm oil and its performance in a 100 kW diesel generator. *Fuel Processing Technology, 89*, 1330–1336.

Lin, J., & Chen, Y. (2017). Production of biodiesel by transesterification of Jatropha oil with microwave heating. *Journal of the Taiwan Institute of Chemical Engineers*, 1–8.

Lin, Y. C., Hsu, K. H., & Lin, J. F. (2014). Rapid palm-biodiesel production assisted by a microwave system and sodium methoxide catalyst. *Fuel, 115*, 306–311.

Maddikeri, G. L., Gogate, P. R., & Pandit, A. B. (2014). Intensified synthesis of biodiesel using hydrodynamic cavitation reactors based on the interesterification of waste cooking oil. *Fuel, 137*, 285–292.

Maddikeri, G. L., Pandit, A. B., & Gogate, P. R. (2013). Ultrasound assisted interesterification of waste cooking oil and methyl acetate for biodiesel and triacetin production. *Fuel Processing Technology, 116*, 241–249.

Maghami, M., Sadrameli, S. M., & Ghobadian, B. (2015). Production of biodiesel from fishmeal plant waste oil using ultrasonic and conventional methods. *Applied Thermal Engineering, 75*, 575–579.

Manh, D. V., Chen, Y. H., Chang, C. C., Chang, M. C., & Chang, C. Y. (2011). Biodiesel production from Tung oil and blended oil via ultrasonic transesterification process. *Journal of the Taiwan Institute of Chemical Engineers, 42*, 640–644.

Martinez-Guerra, E., & Gude, V. G. (2014). Transesterification of used vegetable oil catalyzed by barium oxide under simultaneous microwave and ultrasound irradiations. *Energy Conversion and Management, 88*, 633–640.

Mazubert, A., Taylor, C., Aubin, J., & Poux, M. (2014). Key role of temperature monitoring in interpretation of microwave effect on transesterification and esterification reactions for biodiesel production. *Bioresource Technology, 161*, 270–279.

Milano, J., Ong, H. C., Masjuki, H. H., Silitonga, A. S., Chen, W., Kusumo, F., et al. (2018). Optimisation of biodiesel production by microwave irradiation-assisted transesterification for waste cooking oil-*Calophyllum inophyllum* oil via response surface methodology. *Energy Conversion and Management, 158*, 400–415.

Mootabadi, H., Salamatinia, B., Bhatia, S., & Abdullah, A. Z. (2010). Ultrasonic-assisted biodiesel production process from palm oil using alkaline earth metal oxides as the heterogeneous catalysts. *Fuel, 89*, 1818–1825.

Nayak, M. G., & Vyas, A. P. (2019). Optimisation of microwave-assisted biodiesel production from Papaya oil using response surface methodology. *Renewable Energy, 138*, 18–28.

Ng, P. Q. W., Lam, H. L., Varbanov, P. S., & Klemeš, J. J. (2014). Waste-to-energy (WTE) network synthesis for municipal solid waste (MSW). *Energy Conversion and Management, 85*, 866–874.

Pal, A., Verma, A., Kachhwaha, S. S., & Maji, S. (2010). Biodiesel production through hydrodynamic cavitation and performance testing. *Renewable Energy, 35*, 619–624.

Pascoal, C. V. P., Oliveira, A. L. L., Figueiredo, D. D., & Assuncao, J. C. C. (2020). Optimization and kinetic study of ultrasonic-mediated in situ transesterification for biodiesel production from the almonds of *Syagrus cearensis*. *Renewable Energy, 147*, 1815–1824.

Patil, A., Baral, S. S., Dhanke, P., & Kore, V. (2019). Biodiesel production using prepared novel surface functionalised TiO_2 nano-catalyst in hydrodynamic cavitation reactor. *Materials Today: Proceedings*. https://doi.org/10.1016/j.matpr.2019.10.009.

Perez-Cisneros, E. S., Mena-Espino, X., Rodriguez-Lopez, V., Sales-Cruz, M., Viveros-Garcia, T., & Lobo-Oehmichen, R. (2016). An integrated reactive distillation process for biodiesel production. *Computers and Chemical Engineering*, 1–14.

Perin, G., Álvaro, G., Westphal, E., Viana, L. H., Jacob, R. G., & Lenardão, E. J. (2008). Transesterification of castor oil assisted by microwave irradiation. *Fuel, 87*, 2838–2841.

Petchsoongsakul, N., Ngaosuwan, K., Kiatkittipong, W., Aiouache, F., & Assabumrungrat, S. (2017). Process design of biodiesel production: Hybridization of ester-and transesterification in a single reactive distillation. *Energy Conversion and Management, 153*, 493–503.

Petkovšek, M., Zupanc, M., Dular, M., Kosjek, T., Heath, E., & Kompare, B. (2013). Rotation generator of hydrodynamic cavitation for water treatment. *Separation and Purification Technology, 118*, 415–423.

Phromphithak, S., Meepowpan, P., Shimpalee, S., & Tippayawong, N. (2020). Transesterification of palm oil into biodiesel using ChOH ionic liquid in a microwave heated continuous flow reactor. *Renewable Energy, 154*, 925–936.

Phuenduang, S., Chatsirisook, P., Simasatikul, L., Paengjuntuek, W., & Arpornwichanop, A. (2012). Heat-integrated reactive distillation for biodiesel production from Jatropha oil. In *Symposium on Process Systems Engineering, 15–19 July 2012, Singapore*.

Pukale, D. D., Maddikeri, G. L., Gogate, P. R., Pandit, A. B., & Pratap, A. P. (2015). Ultrasound assisted transesterification of waste cooking oil using heterogeneous solid catalyst. *Ultrasonics Sonochemistry, 22*, 278–286.

Qiu, Z., Zhao, L., & Weatherley, L. (2010). Process intensification technologies in continuous biodiesel production. *Chemical Engineering and Processing: Process Intensification, 49*, 323–330.

Sajjadi, B., Abdul Aziz, A. R., & Ibrahim, S. (2015). Mechanistic analysis of cavitation assisted transesterification on biodiesel characteristics. *Ultrasonics Sonochemistry, 22*, 463–473.

Salamatinia, B., Mootabadi, H., Hashemizadeh, I., & Abdullah, A. Z. (2013). Intensification of biodiesel production from vegetable oils using ultrasonic-assisted process: Optimization and kinetic. *Chemical Engineering and Processing: Process Intensification, 73*, 135–143.

Samuel, O. D., Okwu, M. O., Amosun, S. T., Verma, T. N., & Afolalu, S. A. (2019). Production of fatty acid ethyl esters from rubber seed oil in hydrodynamic cavitation reactor: Study of reaction parameters and some fuel properties. *Industrial Crops and Products, 141*, 111–125.

Shah, Y. T., Pandit, A. B., & Moholkar, V. S. (1999). *Cavitation reaction engineering*. New York: Kluwer Academic.

Sharma, A., Kodgire, P., & Kachhwaha, S. S. (2019). Biodiesel production from waste cotton-seed cooking oil using microwave-assisted transesterification: Optimisation and kinetic modeling. *Renewable and Sustainable Energy Reviews, 116*, 109–125.

Silitonga, A. S., Shamsuddin, A. H., Mahlia, T. M. I., Milano, J., Kusumo, F., Siswantoro, J., et al. (2020). Biodiesel synthesis from *Ceiba pentandra* oil by microwave irradiation-assisted transesterification: ELM modeling and optimisation. *Renewable Energy, 146*, 1278–1291.

Singh, V., & Sharma, Y. C. (2017). Low cost guinea fowl bone derived recyclable heterogeneous catalyst for microwave assisted transesterification of *Annona squamosa* L. seed oil. *Energy Conversion and Management, 138*, 627–637.

Stavarache, C., Vinatoru, M., Nishimura, R., & Maeda, Y. (2005). Fatty acids methyl esters from vegetable oil by means of ultrasonic energy. *Ultrasonics Sonochemistry, 12*, 367–372.

Suslick, K. S. (1990). Sonochemistry. *Science, 247*, 1439–1445.

Talebian-Kiakalaieh, A., Amin, N. A. S., Zarei, A., & Noshadi, I. (2013). Transesterification of waste cooking oil by heteropoly acid (HPA) catalyst: Optimization and kinetic model. *Applied Energy, 102*, 283–292.

Tan, S. X., Lim, S., Ong, H. C., & Pang, Y. L. (2019). State of the art review on development of ultrasound-assisted catalytic transesterification process for biodiesel production. *Fuel, 235*, 886–907.

Teixeira, L. S. G., Assis, J. C. R., Mendonça, D. R., Santos, I. T. V., Guimarães, P. R. B., & Pontes, L. A. M. (2009). Comparison between conventional and ultrasonic preparation of beef tallow biodiesel. *Fuel Processing Technology, 90*, 1164–1166.

Thangarasu, V., Siddarth, R., & Ramanathan, A. (2020). Modeling of process intensification of biodiesel production from *Aegle Marmelos Correa* seed oil using microreactor assisted with ultrasonic mixing. *Ultrasonics Sonochemistry, 60*, 104–119.

Tippayawong, N., & Sittisun, P. (2012). Continuous-flow transesterification of crude Jatropha oil with microwave irradiation. *Scientia Iranica, 19*, 1324–1328.

Venkatesh Kamath, H., Regupathi, I., & Saidutta, M. B. (2011). Optimisation of two step karanja biodiesel synthesis under microwave irradiation. *Fuel Processing Technology, 92*, 100–105.

Yadav, A. K., Khan, M. E., Pal, A., & Ghosh, U. (2017). Performance and emission characteristics of a stationary diesel engine fuelled by Schleichera Oleosa Oil Methyl Ester (SOME) produced through hydrodynamic cavitation process. *Egyptian Journal of Petroleum*. https://doi.org/10.1016/j.ejpe.2017.01.007.

Yan, Y., & Thorpe, R. B. (1990). Flow regime transitions due to cavitation in the flow through an orifice. *International Journal of Multiphase Flow, 16*, 1023–1045.

Yan, Y., Thorpe, R. B., & Pandit, A. B. (1988). Cavitation noise and its suppression by air in orifice flow. In *Processing of the international symposium on flow induced vibration and noise, Chicago, USA* (pp. 25–40).

CHAPTER 4

Intensified technologies for the production of renewable aviation fuel

Araceli Guadalupe Romero-Izquierdo[a] and Salvador Hernández[b]

[a]Facultad de Ingeniería, Universidad Autónoma de Querétaro, Amazcala, Querétaro, México
[b]Departamento de Ingeniería Química, División de Ciencias Naturales y Exactas, Universidad de Guanajuato, Guanajuato, Mexico

4.1 Introduction

The aviation has been recognized as an important transport way due to its speed and efficiency for promoting international transactions such as global social contacts and business (Doliente et al., 2020; Wei et al., 2019). Forecasts pointed out that the increase of the passenger's number will be twice over the next 20 years, respect to 2016 (Doliente et al., 2020); thus it is expected that the demand of aviation fuel and the carbon dioxide emissions will increase correspondingly. According to the International Air Transport Association (IATA, 2020), in 2019, the aviation sector realized 38.9 million of flights, consuming 188 billion USD of fuel, and releasing 914 million ton of carbon dioxide; these data represent increments of 2%, 4.4%, and 1% in comparison with 2018, respectively. Indeed, the International Renewable Energy Agency (IRENA, 2017) indicated that, if no strategy is implemented, the carbon dioxide emissions due to air transport could reach until 2700 million ton by 2050. This amount represents almost three times the 2019 emissions levels. Therefore the aviation sector established objectives to guarantee its sustainable development; one of them considers the reduction of 50% in carbon dioxide emissions by 2050, in relation to 2005 levels (Gutiérrez-Antonio, Romero-Izquierdo, Gómez-Castro, & Hernández, 2021). In order to achieve these goals, the four-pillar strategy has been proposed, as can be seen in Fig. 4.1, which includes technological improvements in engines and aircrafts, operational improvements by the online optimization of flight paths, market-based measures, and development of alternative fuels. It is expected that the aforementioned strategies will help to reach the proposed objectives of the aviation sector. However, each one of them will contribute in different proportions. For instance, Chiaramonti (2019) has reported that the jet fuel consumption will reach 860 million ton/year or 570 million ton/year in 2050, if air fleet renewal with technological improvements or consumption-reduction improvements measures is implemented, respectively. However, according to IATA (2020), these values are six and four times more than the fuel demand in 2019 (142 million ton/year), respectively.

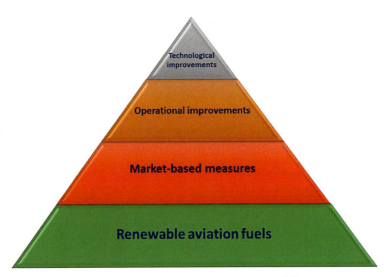

Fig. 4.1 Four-pillar strategy to reach the sustainable development of the aviation sector.

This emission reduction rate has been estimated considering an improvement of 1.5% annually through the implementation of aircraft modifications (improving the fuel efficiency) and optimizing the navigational systems (Chiaramonti, 2019; IRENA, 2017). On the other hand, it has been identified the use of alternative aviation fuel as the best strategy to decarbonize the aviation industry, due to its potential to reduce the carbon dioxide emissions in the short and medium term (Doliente et al., 2020; Gutiérrez-Antonio, Gómez-Castro, de Lira-Flores, & Hernández, 2017; IRENA, 2017).

It is important to mention that in 2020, the appearance and rapid spread of the virus SARS-CoV-2 changed the landscape of all economic sectors and swept the world (Van Fan, Jiang, Hemzal, & Klemeš, 2021). Although the overall energy demand declines, the spatial and temporal variations are complicated (Jiang, Van Fan, & Klemeš, 2021). In particular, the aviation sector experienced its worst year. The international passenger demand was 75.6% below 2019 levels; the International Air Transport Association's baseline forecast for 2021 is for a 50.4% improvement on 2020 demand, which would bring the industry to 50.6% of 2019 levels (IATA, 2021). In the new scenario outlined by the pandemic, the promotion of renewable energy has been identified as an emerging development (Klemeš, Van Fan, & Jiang, 2021). Moreover, the International Energy Agency proposed a sustainable recovery plan, where sustainable biofuels also have a potentially important part to play in reducing emissions from sectors that are challenging for low-carbon electricity to reach such as heavy-duty vehicles, aviation, and shipping (IEA, 2020). Therefore the development of renewable fuel for the aviation is crucial for its sustainable recovery.

The renewable aviation fuel is also known as synthetic paraffinic kerosene (SPK) or biojet fuel; this biofuel consists of renewable hydrocarbons from C8 to C16, mainly paraffinic and naphthenic compounds. The main difference of biojet fuel with fossil jet fuel is the amount of aromatic compounds, which depends on the production process used (Gutiérrez-Antonio et al., 2017; Gutiérrez-Antonio et al., 2021). The absence of aromatic compounds can cause wear in some engines, and they are required to swell rings and seals in engines (Liu, Yan, & Chen, 2013). In addition, the ASTM D7566 standard requires a minimum aromatic composition of 8.4% vol in the blends of renewable and fossil jet fuels (ASTM, 2019). The amount of aromatic compounds present in the biojet fuel determines the maximum biofuel blend ratios, and this percentage varies for each conversion pathway being 50% vol the maximum mixing ratio allowed (ASTM, 2019; Moreno-Gómez, Gutiérrez-Antonio, Gómez-Castro, & Hernández, 2021). It is important to mention that biojet fuel is a drop-in fuel that can be used by the current aircrafts without additional economic investment, allowing the energetic transition and reducing the GHG emissions (Doliente et al., 2020).

The renewable aviation fuel can be produced from any biomass source, such as edible crops, nonedible crops, waste biomass, micro, and macroalgae, among others (Gutiérrez-Antonio et al., 2021); however, it is suggested to avoid the consumption of edible crops, in order to ensure the food safety. According to Maity (2015), the processing routes to produce biojet fuel can be classified based on the nature of biomass, such as triglyceride conversion pathways, sugar and starchy conversion processes, and lignocellulosic conversion routes; this classification allows the processing of raw materials of different generations but with the same chemical nature. The main conversion pathways to produce biojet fuel that have received the certification by ASTM are presented next (CAAFI, 2021): Fischer-Tropsch synthesized isoparaffinic kerosene (FT-SPK), hydroprocessed ester and fatty acid synthetic paraffinic kerosene (HEFA-SPK), hydroprocessed fermented sugars to synthetic isoparaffins (HFS-SIP), Fischer-Tropsch synthetic paraffinic kerosene with aromatics (FT-SPK/A), alcohol to jet synthetic paraffinic kerosene (ATJ-SPK), catalytic hydrothermolysis synthesized kerosene (CH-SK, or CHJ), hydroprocessed hydrocarbons, esters and fatty acids synthetic paraffinic kerosene (HHC-SPK or HC-HEFA-SPK), as it can be seen in Fig. 4.2.

The pathways for the production of renewable aviation fuel have been studied in the literature (Bwapwa, Anandraj, & Trois, 2017; Goh et al., 2020; Gutiérrez-Antonio et al., 2017; Gutiérrez-Antonio et al., 2021; Shahabuddin, Alam, Krishna, Bhaskar, & Perkins, 2020; Vásquez, Silva, & Castillo, 2017; Wang et al., 2019; Wei et al., 2019; Why et al., 2019); from these works, it can be concluded that the production is feasible from the technical point of view. However, the price of biojet fuel is still not competitive respect to its fossil counterpart (Klein-Marcuschamer et al., 2013; Pearlson, Wollersheim, &

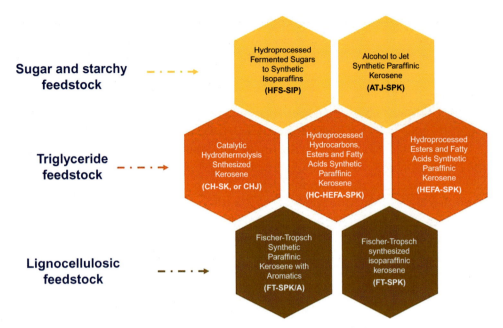

Fig. 4.2 Certified processes for the production of renewable aviation fuel.

Hileman, 2013; Wang, 2016). The prices are until almost 4 times the price of the fossil jet fuel (PEMEX, 2021). Therefore it is necessary to reduce the production costs for the production of renewable aviation fuel. In order to reach this objective, waste materials could be used, which are available and low cost (Gutiérrez-Antonio & Hernández, 2019). Another alternative is the application of energy integration or process intensification strategies, where the last ones are defined as any chemical engineering development that leads to a substantially smaller, cleaner, and more energy-efficient technology (Stankiewicz & Moulijn, 2000). In particular, process intensification has been used in several biofuel production processes, achieving interesting improvements (Athar & Zaidi, 2020; Gutiérrez Ortiz, 2020; Quiroz-Pérez, de Lira-Flores, & Gutiérrez-Antonio, 2019; Stephen & Periyasamy, 2018).

Therefore in this chapter, the use of intensified technologies for the production of renewable aviation fuel will be analyzed. The chapter is organized as follows. In Section 4.2, the modeling and design strategies for intensified technologies for biojet fuel production will be described. Later, in Section 4.3, it will be presented a revision of the literature of the different biojet fuel intensified production processes considering all types of biomasses. Moreover, in Section 4.4, the advances in the implementation of intensified technologies at an industrial scale will be discussed. After that, in Section 4.5, a technoeconomic assessment of an intensified process for the production of biojet fuel will be presented. Finally, conclusions are included in Section 4.6.

4.2 Processes intensification strategies in the production of biojet fuel

According to Stankiewicz and Moulijn (2000), process intensification concerns only engineering methods and equipment. In this section, we will focus on the methods available to design intensified equipment that can be used for the production of biojet fuel, describing those related to the reactive zone, the purification zone, and some hybrid technologies.

4.2.1 Methodologies to design intensified reactors for the production of biojet fuel

Based on the nature of the biomass used, different reactions are carried on in order to produce biojet fuel. In the case of triglyceride feedstock, the reactions involved depend on the processing pathway.

For the HEFA process, the reactions are hydrodeoxygenation, hydroisomerization, and hydrocracking. These processes involved heterogeneous reactors, since gas (hydrogen), liquid (oil), and solid (catalyst) phases are presented; moreover, the reactions occur at high pressures (20–110 bar) and temperatures (250–430°C) with catalysts based on Ni, Mo, Co, Pd, Ro, Pt, and bimetallic sulfide catalysts, such as Ni-Co-Fe, Mo-WU, $NiMoS_2$, $Ni-W/SiO_2$, $CoMoS_2$, and $NiWS_2$ supported on Al_2O_3 (Gutiérrez-Antonio et al., 2021).

For CH-SK, the reactions considered are hydrothermolysis, hydrodeoxygenation, hydroisomerization and hydrocracking. In the hydrothermolysis, the oil mixed with hot water is converted to free fatty acids and supercritical water at high pressure (210 bar) and temperature (450–475°C) (Li, Coppola, Rine, Miller, & Walker, 2010). The other reactions (hydrodeoxygenation, hydroisomerization, and hydrocracking) are performed as described previously (Gutiérrez-Antonio et al., 2021).

In the case of sugar and starchy feedstock, one of the main routes to produce biojet fuel is ATJ-SPK, which uses an alcohol as intermediate, commonly bioethanol. For ATJ-SPK process, the involved reactions are dehydration, oligomerization, and hydrogenation. The dehydration is carried out in heterogeneous reactors, since liquid (ethanol and water), gas (ethylene), and solid (acid catalyst) phases are involved. The dehydration occurs at moderated temperatures (180–300°C) on zeolites, alumina, and silica-alumina catalysts. On the other hand, in the oligomerization reaction, ethylene (gas) is converted in long-chain linear olefins (liquid) with a catalyst that consists of organic or metal-organic frameworks to support nickel or chromium; this reaction is also realized in heterogeneous reactors at low temperatures (50–150°C) and moderate pressures (15–30 bar). Finally, hydrogenation of the alkenes is performed at low temperatures (20°C) and moderate pressures (1–20 bar); in this reaction, liquid (olefins and saturated hydrocarbons), gas (hydrogen), and solid (catalyst) phases are present (Gutiérrez-Antonio et al., 2021).

For the HFS-SIP process, the reactions involved are fermentation, hydrodeoxygenation, hydroisomerization, and hydrocracking. For the fermentation reaction, farnesane is obtained through the mevalonate pathway, where *E. coli* is used; this reaction is also realized in a heterogeneous reactor that operates at low temperature (30–37°C) and environmental pressure (around 1 bar), with long residence times (24–48 h) (You et al., 2017). The other reactions (hydrodeoxygenation, hydroisomerization, and hydrocracking) are performed at as described previously (Gutiérrez-Antonio et al., 2021).

In the case of lignocellulosic feedstock, for the FT-SPK process, the necessary reactions are gasification, Fischer-Tropsch synthesis, hydrodeoxygenation, hydroisomerization, and hydrocracking. In gasification, the solid biomass is heated at 700°C in a low-oxygen atmosphere, to avoid the combustion of the biomass, to generate syngas. In the Fischer-Tropsch synthesis, syngas and hydrogen (gas) reacts to produce hydrocarbons (liquid) with a catalyst (solid). The reaction occurs at moderate temperatures (220–250°C) using iron- and cobalt-based catalysts. The other reactions (hydrodeoxygenation, hydroisomerization, and hydrocracking) are performed at as described previously (Gutiérrez-Antonio et al., 2021).

In brief, in all the conversion processes, the reactions involved in the hydroprocessing are required. Besides them, reactions such as gasification, fermentation, Fischer-Tropsch, oligomerization, hydrothermolysis, and dehydration can occur as processes for initiating the biomass transformation.

For hydrotreatment reactions, it is possible to intensify them through the one step hydroprocessing; in this alternative, all the reactions are performed with a multifunctional catalyst in the same vessel (a conventional one), along with hydrogen, at high operating conditions (temperature and pressure). In the literature, there are some experimental studies reported for the one step hydroprocessing, which are presented in Table 4.1. From Table 4.1, we can observe that the work of Scaldaferri and Pasa (2019a, 2019b) reported a good conversion without the use of hydrogen. This is an important result that will help to the establishment of the supply chain for this biofuel. On the other hand, the other studies use hydrogen, being the maximum selectivity of 78.5% to biojet fuel.

In the literature, we can find design methodologies for this type of reactors, since they are conventional ones in this aspect (Li, 2017; Ray & Das, 2020), and the intensification is realized thanks to the development of a multifunctional catalyst that allows simultaneous reactions. The synthesis of multifunctional catalysts usually considers experimentation to evaluate the effect of different catalyst concentrations, operating conditions and hydrogen flow (if required) on the production of hydrocarbons on the boiling point range of jet fuel; in some cases, response surface methodology is used to find the optimal parameters.

On the other hand, there are several intensified reactor technologies, which include multifunctional catalytic reactors, oscillatory flow reactors, microchannel reactors, spinning disk reactors, and microreactors (Quiroz-Pérez, Gutiérrez-Antonio, & Vázquez-Román, 2019; Samih, Latifi, Farag, Leclerc, & Chaouki, 2018; Tian,

Table 4.1 One-step hydroprocessing experimental works reported.

Raw material	Catalyst	Operating conditions	Biojet fuel selectivity	Reference
Soybean oil	NbOPO$_4$	10 bar H$_2$ 350°C Hydrogen	62%	(Scaldaferri & Pasa, 2019b)
Soybean oil	NbOPO$_4$	10 bar N$_2$ 350°C	58%	(Scaldaferri & Pasa, 2019a)
Jatropha oil	SDBS-Pt/ SAPO-11	5 MPa 410°C H$_2$/oil: 1000 NmL/mL	59.51%	(Li et al., 2019)
Chicken fat	NiW/ SiO$_2$-Al$_2$O$_3$	6 MPa 400°C H$_2$/oil: 450 NmL/mL	40%	(Hanafi et al., 2016)
Soybean oil	Pt/Al$_2$O$_3$/ SAPO-11	30 atm 350°C H$_2$/oil: 200 NL/L	48%	(Rabaev et al., 2015)
Jatropha oil	NiW/MSP-2	60 bar 450°C H$_2$/oil: 1500 NmL/mL	37.5%	(Verma, Rana, Kumar, Sibi, & Sinha, 2015)
Microalgae oil	NiMo-H-ZSM-5	50 bar 410°C H$_2$/oil: 1500 NmL/mL	78.5%	(Verma, Kumar, Rana, & Sinha, 2011)

Demirel, Hasan, & Pistikopoulos, 2018). These intensified reactors are of interest for the production of renewable aviation fuel since they allow to carry on reactions with lower operating conditions and major yields, respect to the conventional ones.

In the literature, there are some reports of experimental studies on microreactors for hydrogenation of 1-butyne (García Colli, Alves, Martínez, & Barreto, 2016), levulinic acid (Hommes, ter Horst, Koeslag, Heeres, & Yue, 2020), 2-methyl-3-butyn-2-ol (Okhlopkova, Prosvirin, Kerzhentsev, & Ismagilov, 2021), nitrobenzene (Liu et al., 2018), and 2-methylfuran (Liu, Ünal, & Jensen, 2012). These works represent important advances in the use of intensified technologies, but none of them allows to generate renewable aviation fuel. Regard the hydrodeoxygenation, microreactors have been reported for the conversion of acid acetic (Joshi & Lawal, 2012a) and pyrolysis oil (Joshi & Lawal, 2012b); in both cases, renewable aviation fuel is not produced. Nevertheless, Joshi and Lawal (2012b) reported that for the studied conditions the hydrogen consumption and the oxygen removal obtained in the microreactor are similar to those reported for macroreactors but with lower pressure and residence time.

The design of these equipment requires extensive knowledge of the related phenomena such as fluid flow, chemical kinetics, and heat and mass transport (Seelam, 2013). In this context, computational fluid dynamics (CFD) has been proposed as an attractive method for the design and optimization of these units (Erickson, 2005). A CFD study consists of obtaining the numerical solution of the equations that describe momentum, heat, and mass transfer in a system using computational methods. One of the most interesting characteristics of techniques based on CFD is the possibility to model and represent the aforementioned phenomena in systems with both conventional and complex geometries (Quiroz-Pérez, Gutiérrez-Antonio, & Vázquez-Román, 2019). In this type of studies, three main stages reidentified: preprocessing, processing, and postprocessing.

In the preprocessing stage, the geometry of the equipment is proposed along with the model to describe their behavior; based on the specifications defined, the model is solved in the processing stage. Finally, the results obtained in the processing stage are analyzed in the postprocessing step. An interesting characteristic is that the presentation of the results can include pictures or even small videos of the dynamic of the process. There are several computational packages where CFD studies can be realized; among them, ANSYS CFX stands out for its robustness, wide availability of models, as well as high reliability in the results. For CFD studies, it is highly desirable to have experimental results that can be used to compare the simulation results; in this way, improvements to the models can be made in order to have a better description of the phenomenon.

CFD has been used for the design of intensified equipment for the production of other biofuels (Quiroz-Pérez, de Lira-Flores, & Gutiérrez-Antonio, 2019; Quiroz-Pérez, Gutiérrez-Antonio, & Vázquez-Román, 2019). However, for the authors' knowledge there are not reported studies on the design or optimization of this type of intensified reactors for the production of renewable aviation fuel.

Respect to the gasification of biomass, several studies have been presented considering conventional technologies (He et al., 2021; Sansaniwal, Pal, Rosen, & Tyagi, 2017; Yu et al., 2021). Moreover, some intensification proposals for biomass gasification consider the integrating several functionalities into suitable fluidized bed gasifiers, such as catalytic cracking/reforming of tar, CO_2 elimination, H_2 separation, and the elimination of particles and other contaminants (Richardson, Blin, & Julbe, 2012). In addition, Chuayboon, Abanades, and Rodat (2018) proposed an experimental solar reactor to perform the gasification of woody biomass.

According to Xiong, Hong, Yu, Li, and Xu (2018), process design and intensification are most important routines to improve the final product and energy yields and overall economy for biomass gasification reactors. In this context, the use of CFD as design strategy is one of the most studied. In a recent review article, Safarian, Unnþórsson, and Richter (2019) indicated that there is significant ongoing work aimed at developing the detailed kinetic, rate laws, and CFD models to accurately describe the reaction rates and transport for biomass gasification. Thus in this area, important advances in the use of CFD as design strategy are reported.

On the other hand, conventional fermentation reactions to produce farnesane have been reported in the literature (Blanch, 2012; Liu & Khosla, 2010). This reaction could be intensified through the use of simultaneous saccharification and fermentation, which have been reported for bioethanol production (Cheng et al., 2017; Kossatz, Rose, Viljoen-Bloom, & van Zyl, 2017). However, the study of this intensified operation (from experimental and design points of view) remains as an opportunity area for the production of farnesane.

Moreover, conventional technologies for Fisher-Tropsch synthesis are described in several studies (Otun, Liu, & Hildebrandt, 2020; Panzone, Philippe, Chappaz, Fongarland, & Bengaouer, 2020). Regard the intensification of this reaction pathway, it was proposed a dual-type reactor, which is permeable to increase the conversion of carbon dioxide to carbon monoxide, and then having a higher yield to hydrocarbons (Saeidi, Amin, & Rahimpour, 2014). Also, the use of microchannel reactors was proposed for the Fischer-Tropsch synthesis, and the results indicated that the productivity increases in several orders of magnitude (from 0.004 to 1.5 gal/d) (Becker, Güttel, & Turek, 2019; Deshmukh et al., 2010; Wilhite, 2017). In addition, several studies on the use of microreactor to perform this reaction obtained conversions until 80% of carbon monoxide (Abrokwah, Rahman, Deshmane, & Kuila, 2019; Bepari et al., 2020; Bepari et al., 2021; Mohammad, Abrokwah, Stevens-Boyd, Aravamudhan, & Kuila, 2020). Regarding to the design methodologies for this type of reactor, again CFD is the most used tool. In this topic several articles can be reported where CFD is used to design conventional and intensified reactors for Fischer-Tropsch synthesis (Arzamendi et al., 2010; Guan & Yang, 2017; Shin, Park, Park, Jun, & Ha, 2013; Troshko & Zdravistch, 2009; Zhang, 2009).

The conventional technologies for the oligomerization reaction are widely studied in the literature (Alferov, Belov, & Meng, 2017; Antunes, Rodrigues, Lin, Portugal, & Silva, 2015; Budagumpi, Keri, Biffis, & Patil, 2017). Regard the use of intensified technologies, just one work was found where the dicyclopentadiene was oligomerized through a microreactor at high pressure and temperature (Yao et al., 2020); the product does not include lineal olefins, desirable to produce biojet fuel, but it is an important advance for oligomerization reaction. For the authors' knowledge, there are not CFD studies on the design or optimization of intensified reactors for the oligomerization reaction.

On the other hand, the conventional hydrothermolysis is a recent proposal for the production of free fatty acids, which later are hydroprocessed to obtain hydrocarbons in the boiling point range of jet fuel; this conversion pathway obtained the ASTM certification in 2020 (CAAFI, 2021). For the hydrothermolysis reaction just one studied was found in the literature. In that study, 91% of rapeseed cake is converted to free fatty acid and aminoacids at 246°C during 65 min (Pińkowska, Wolak, & Oliveros, 2014). For the authors' knowledge, there are not experimental or CFD studies on the design or optimization of intensified reactors for the hydrothermolysis reaction.

Finally, the dehydration reaction of ethanol has been studied in the literature (Marosz, Kowalczyk, & Chmielarz, 2020; Masih, Rohani, Kondo, & Tatsumi, 2019; Verdes, Sasca, Popa, Suba, & Borcanescu, 2021). Regard the intensification of this reaction, one study reported the use of a microreactor for the dehydration of bioethanol (Suerz et al., 2021). The results indicated a 98% of selectivity to ethylene at ambient pressure and temperatures between 225°C and 325°C. In other study, the use of a microchannel reactor reported an ethylene selectivity of 99.4% on the $TiO_2/\gamma\text{-}Al_2O_3$ catalyst (Chen, Li, Jiao, & Yuan, 2007). For the authors' knowledge, there are no CFD studies on the design or optimization of intensified reactors for the dehydration reaction.

4.2.2 Methodologies to design-intensified distillation columns for the production of biojet fuel

In spite of the conversion pathway used, renewable hydrocarbons are generated from the reactive zone, which usually are separated through distillation. Typically, these hydrocarbons can be grouped in four products: light gases (C1–C4), naphtha (C5–C7), biojet fuel (C8–C16), and green diesel (>C16). In order to have a distillation train with reduced energy consumption, light gases must be separated in the first column; in this way, the use of a refrigerant in the condenser is minimized. Then, this column must be fixed, leaving two possible combinations for the separation of the rest of the products: direct and indirect conventional distillation sequences. For the design of conventional distillation columns, the shortcut methodology Fenske-Underwood-Gilliland is one of the most used (Gadalla, Jobson, & Smith, 2003; Gutiérrez-Antonio, Gómez-Castro, Hernández, & Briones-Ramírez, 2015). Distillation is the most used unit operation for the separation of fluid mixtures worldwide; nevertheless, its thermodynamic efficiency is considerably low. Due to this, new proposals that keep the advantages of distillation but improving its efficiency have been presented.

Among the intensified process, those that can be applied for the purification of the renewable aviation fuel are the thermally coupled distillation columns, reactive distillation, and reactive thermally coupled distillation columns, as can be seen in Fig. 4.3.

The design of these intensified equipment can be realized through methodologies based on the design of the conventional trains or optimization strategies. It is desirable to have experimental data in order to validate the results of the design. However, for the author's knowledge, there is no experimental studies reported for the purification of the renewable aviation fuel through intensified equipment.

The thermally coupled distillation sequences consist of distillation columns linked between them through vapor and/or liquid interconnection flows; these flows are

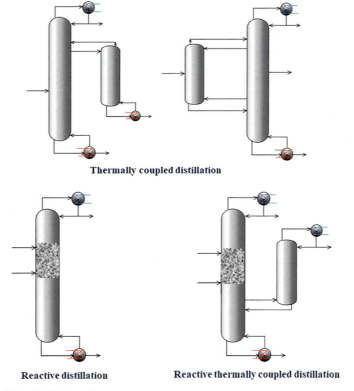

Fig. 4.3 Intensified distillation schemes.

located in specific stages in order to eliminate the remixing observed in distillation columns. As consequence, energy savings between 30% and 50% can be obtained with the use of thermally coupled distillation sequences, respect to the conventional distillation sequences (Gómez-Castro et al., 2016; Jiang & Agrawal, 2019; Kiss & Smith, 2020; Waltermann, Sibbing, & Skiborowski, 2019). In addition, the vapor and liquid interconnection flows allow the elimination of a reboiler and a condenser in the column; thus a reduction in capital costs (*CCs*) can also be obtained in the implementation of these intensified schemes in comparison with the conventional ones. Finally, several studies have shown that the control properties of thermally coupled distillation sequences are better than those observed in conventional distillation schemes (Cabrera-Ruiz, Santaella, Alcántara-Ávila, Segovia-Hernández, & Hernández, 2017; Kaymak, 2021; Nikačević, Huesman, Van den Hof, & Stankiewicz, 2012; Ramírez-Márquez et al., 2016).

The design of thermally coupled distillation columns can be realized through the analogy section methodology (Hernández & Jiménez, 1996) or the methodology of Rong

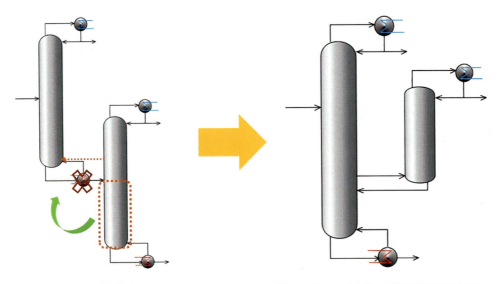

Fig. 4.4 Generation of intensified direct distillation sequence through the methodology of Hernández and Jiménez (1996).

and Errico (2012); in both cases, the design needs to be adjusted by recursive simulation, in Aspen Plus for instance, in order to reach the purities and recoveries specified.

In the methodology of Hernández and Jiménez, the conventional distillation trains are designed using a shortcut methodology. Based on that designs, the sections are moved in order to generate the intensified schemes. For instance, in order to generate the thermally coupled direct sequence, the stripping section of the second column is moved below the stripping section of the first column, as is shown in Fig. 4.4. In addition, the reboiler of the first column is eliminated and replace by vapor and liquid interconnection flows between the columns. The number of total stages remain unchanged, and then the interconnection flows are manipulated in order to minimize the heat duty, satisfying the purities and recoveries specified.

On the other hand, the methodology of Rong and Errico (2012) is also based on the design of the conventional distillation sequences (Gutiérrez-Antonio, Gómez-De la Cruz, Romero-Izquierdo, Gómez-Castro, & Hernández, 2018; Gutiérrez-Antonio, Romero-Izquierdo, Gómez-Castro, Hernández, & Briones-Ramírez, 2016; Moreno-Gómez et al., 2021). In this methodology the conventional distillation columns are taken as basis to perform some movements of sections as well as eliminations in order to generate the thermally coupled distillation schemes. This methodology is schematically represented by the next steps:

Step 1. Create the subspace of simple column configurations (SC).

Step 2. Select the simple column configuration with less energy consumption, from the previous subspace.

Step 3. Generate the original thermally coupled configurations, replacing systematically reboilers or condensers (OTC), and select the OTC with less energy consumption.

Step 4. Generate the thermodynamically equivalent structures (TES) from OTCs, using the sections rearrangement methodology.

Step 5. Identify the TESs that contain columns with a unique lateral transport section, keeping the structure of simple columns; then, the TES with less energy consumption is selected.

Step 6. Generate the intensified configurations (ISC) by eliminating the lateral transport section in the identified TES.

Step 7. If you do not select the structure with less energy consumption in steps 2–6, repeat these steps until all simple column configurations have been examined.

Step 8. Summarize all the ISC configurations.

On the other hand, reactive distillation is an intensified operation where the reaction and the separation are performed simultaneously in the same vessel. One of the advantages of reactive distillation is that the conversion increases due to the continual removal of the products of the reaction, respect to the conversion reported in conventional reactors. In addition, in the reactive distillation columns, the operation conditions are lower than the required in the conventional reactors; this also allows to have a safer process with lower energy requirements. These characteristics are relevant and they could help to improve the profitability of the production of renewable aviation fuel.

The design of reactive distillation columns has been explored in the literature. In general, these methodologies can be classified in: graphical methods (Avami, Marquardt, Saboohi, & Kraemer, 2012; Barbosa & Doherty, 1988; Carrera-Rodríguez, Segovia-Hernández, & Bonilla-Petriciolet, 2011; Carrera-Rodríguez, Segovia-Hernández, Hernández-Escoto, Hernández, & Bonilla-Petriciolet, 2014; Muthia et al., 2018; Thery, Meyer, Joulia, & Meyer, 2005) and optimization procedures (González-Rugerio, Fuhrmeister, Sudhoff, Pilarczyk, & Górak, 2014; Liñán, Bernal, Gómez, & Ricardez-Sandoval, 2021; Miranda-Galindo, Segovia-Hernández, Hernández, Gutiérrez-Antonio, & Briones-Ramírez, 2011). Nevertheless, none of these methodologies is useful to handle multireactive and multicomponent systems. In these cases, parametric analysis is the suggested methodology when there are more than five components (Segovia-Hernández & Bonilla-Petriciolet, 2016). In the parametric analysis, the objective is to obtain a feasible design of the reactive distillation column that satisfy the energy and mass balances as well as reactive and phase equilibrium. For this, the following variables are manipulated: operation pressure, number of reactive stages, number of separation stages, reflux ratio, feed stage of reactive, condenser type, and temperature of feed flow streams (Errico et al., 2020; Gutiérrez-Antonio, Soria Ornelas,

Gómez-Castro, & Hernández, 2018). The chemical reactions are assumed to take place in the liquid phase; the catalyst is assumed to be located on all the reactive trays, and the liquid phase holdup in the reactive stages is considered as the holdup of the reactor in the conventional process (Gutiérrez-Antonio, Soria Ornelas, et al., 2018).

The reactive thermally coupled distillation columns, similar to reactive distillation, are intensified equipment where the reaction and the separation is performed in the same vessel, but the separation considers thermally coupled distillation columns. In these intensified schemes, major conversions are expected and also lower operating conditions, respect to those required in the conventional reactor. In addition, lower energy requirements are involved due to the remixing effect is avoided. The design of reactive thermally coupled distillation is usually performed through parametric analysis (Barroso-Muñoz, Hernández, & Ogunnaike, 2007; Cossio-Vargas, Barroso-Muñoz, Hernandez, Segovia-Hernandez, & Cano-Rodriguez, 2012). Only one study employed an optimization procedure for the design of reactive thermally coupled distillation, but few components and reactions are involved (Miranda-Galindo et al., 2011).

Other intensified technologies are membrane distillation and microchannel distillation. Membrane distillation is a new technology where the separation efficiency depends on the volatility of the components as well as the structure of the porous membrane (Nagy, 2019). Applications for the purification of renewable aviation fuel have not been reported in the literature; however, it could be an interesting alternative for its reduced energy consumption. On the other hand, microchannel distillation consists of a distillation column with dimensions in the submillimeters or submicron dimensions, where surface tension forces exceed gravitational and hydrodynamic forces (Yang, Liu, Wang, Hou, & Fu, 2017); as consequence, the efficiency is limited, and due to this the distillation, application at microscale is still very limited (Kenig, Su, Lautenschleger, Chasanis, & Grünewald, 2013). The design of these novel technologies usually is performed with CFD, artificial neural networks, or empirical models based on design from experiments (Hitsov, Maere, De Sitter, Dotremont, & Nopens, 2015; Macedonio, Ali, & Drioli, 2017; Yang et al., 2017).

4.3 Intensified processes to produce biojet fuel

In this section, we present a revision of the literature focus on the intensified processes for the production of renewable aviation fuel, considering complete processes, which involves conditioning, reactive stages, and separation zones.

In 2015, it was proposed the intensification of the hydrotreating process for the production of renewable aviation fuel (Gutiérrez-Antonio et al., 2015). In this work, the intensification was performed in the separation zone. The analyzed schemes included direct and indirect thermally coupled distillation sequence, Petlyuk distillation column, and the dividing wall distillation column. The designs were obtained through a multiobjective algorithm genetic coupled to Aspen Plus process simulator. The results indicated that the direct thermally coupled distillation sequence presented an energy consumption 21% minor than the best conventional scheme (direct). Moreover, it was reported the production of electrical energy through a turbine incorporated for the conditioning to the separation zone. The reported yield for renewable aviation fuel was 22.05%.

Later, the hydroprocessing of *Jatropha curcas* oil was improved through the application of process intensification and energy integration strategies (Gutiérrez-Antonio et al., 2016). For the separation zone, direct and indirect distillation schemes, both conventional and intensified, were designed with a multiobjective genetic algorithm coupled to Aspen Plus. The energy integration considers the use of the released energy in the hydrodeoxygenation reaction to generate vapor that is provided for the distillation columns and heat exchanger of the conditioning zone. The results show that the total annual costs ($TACs$) are almost the same, since the savings in utilities are compensated by the increase in CCs for the energy integration; however, a decreasing of 87% in carbon dioxide emissions is observed when energy integration and an intensified scheme are used in the process. The reported yield for renewable aviation fuel was 18.61%.

The use of microalgae for the production of renewable aviation fuel was presented by Gutiérrez-Antonio, Gómez-De la Cruz, et al. (2018). The modeling was realized in Aspen Plus, and it considers a multifunctional reactor to convert the microalgae oil as well as conventional and intensified distillation sequences. Results indicated that the use of an intensified distillation scheme allows a reduction of 34% in the carbon dioxide emissions; in spite of the TAC are slightly higher in the intensified processes, respect to the conventional ones, the price of biojet fuel is competitive with its fossil counterpart due to the low price of the raw material. Moreover, the reported yield to biojet fuel was 76.61%.

Later, the use of reactive distillation for the production of renewable aviation fuel was presented (Gutiérrez-Antonio, Soria Ornelas, et al., 2018). The intensification was proposed between the hydroisomerization and hydrocracking reactor with the first distillation column of the train, taking as basis of the conventional process reported by Gutiérrez-Antonio et al. (2016) for the production of biojet fuel from *jatropha curcas*. The design of the reactive distillation column was performed with parametric analysis through simulation, since it is a multicomponent and multireaction system. Results indicated that the intensified process obtained a yield to biojet fuel 5% higher that the non-intensified processing, with a reduction in 23.5% in carbon dioxide emissions, as well as a reduction in the operating conditions from 80 to 10 bar; this will help to decrease the energy consumption and to have a safer process. Also, it is possible to generate all the electricity required in the process, when the reactive distillation is used in the hydroprocessing.

Recently, Moreno-Gómez et al. (2021) presented the intensification of the hydroprocessing of chicken fat for the production of renewable aviation fuel. The intensified schemes were generated through the application of the synthesis methodology of Rong and Errico (2012). Results show that the best scenario in terms of economic (TAC) and environmental indicators (carbon dioxide emissions) is the process that include conditioning and reactive zones along with a direct intensified sequence. The authors conclude that the price of aviation biofuel and the carbon dioxide emissions can be decreased if the energy released in the hydrodeoxygenation reactor is integrated in the process.

In the same year, Romero-Izquierdo, Gómez-Castro, Gutiérrez-Antonio, Hernández, and Errico (2021) proposed the intensification of the alcohol to jet process. The intensification is applied to the separation zone, through the application of the methodology of Rong and Errico (2012). The results show that the intensification of the separation zone allows a reduction of 5.37% in energy consumption. Based on this scheme, the energy integration was applied. As consequence, the TAC decreased by 4.83%, while the carbon dioxide emissions diminished by 4.99%, in comparison with the conventional process. In addition, they incorporated a turbine that completely satisfies the electrical energy requirement of the process.

4.4 Implementation to industrial scale

Based on the information presented in the previous sections, it can be established that the renewable aviation fuel can be produced form almost any type of biomass through several conversion pathways; however, only eight production processes are certified by ASTM to produce the biofuel that can be used in mixture until 50% vol maximum in commercial passenger flights (ASTM, 2019).

In the last two decades, the number of projects of research and development of technologies to produce biojet fuel has increased, mainly due to the goals of sustainability established by the aviation authorities (Wang, Lee, Olarte, & Zacher, 2016). According to the International Renewable Energy Agency (IRENA, 2017), in 2017, there were about 100 initiatives to produce biojet fuel; pointing out that in the period 2009–15, the new ventures increased by 1020 new projects. Worldwide, at 2020, there were reported 19 facilities for the production of biojet fuel; 9 of them were located in United States, as can be seen in Table 4.2.

From Table 4.2, it can be observed that the most used technology production is HEFA-SPK with 15 facilities, followed by FT-SPK with 3, and just one with HFS-SIP technology. These facilities constitute an important advance for the supply of biojet fuel for the aviation sector. Based on the information provided in the patents about these technologies, all the processes involve conventional technologies.

Thus in order to impulse the use of intensified technologies for the production of aviation biofuel, it is necessary to work proactively to obtain ASTM certification. For this, it is desirable to have a pilot plant of the process, in order to generate 900,000 L

Table 4.2 Facilities that produce biojet fuel (Gutiérrez-Antonio et al., 2021).

Country	Number of facilities	Production technologies	Total production capacity (mm gal/year)
United States	9	HEFA-SPK (7) FT-SPK (2)	412
Brazil	1	HFS-SIP	13
Finland	2	HEFA-SPK	161
Sweden	1	HEFA-SPK	271
United Kingdom	1	FT-SPK	17
The Netherlands	1	HEFA-SPK	271
France	1	HEFA-SPK	169
Italy	1	HEFA-SPK	155
United Arab Emirates	1	HEFA-SPK	288
Singapore	1	HEFA-SPK	271

of this biofuel, volume that is required during the certification process that takes between 3 and 5 years (Gutiérrez-Antonio et al., 2021).

On the other hand, there are several opportunity areas to apply the intensification in the process for the production of biojet fuel. Based on the revision of the literature, most of the works have been focused on the intensification of the separation of the renewable hydrocarbon; just one study considers the use of reactive distillation using conventional columns, but the use of thermally coupled reactive distillation is still missing in the literature. Few works have been found related the use of intensified reactors (such as microchannel reactors or microreactors) for the realization of the reactions involved in the different certified pathways; based on these studies, it can be expected a considerable increasing in the yields to biojet fuel and a considerable reduction in the operating conditions to produce them. Thus minor production cost and a safer process can be expected from these intensified processes. Moreover, the production processes must consider the use of mixture of raw materials; this will help to guarantee a supply for the industrial facility, in spite of variations in weather conditions and even price in the market. Finally, the use of intensified technologies in the platform conversion of the biorefineries could be an interesting alternative for the production of biojet fuel. In a biorefinery, a wide portfolio of products can be obtained, which include bioproducts (for applications in the food, pharmaceutical, cosmetic, chemical, and biotechnology industries) as well as bioenergy (biofuels, electrical energy, thermal energy). In this way, the wide variety of production volume and prices of all these products help to increase the profitability of the biomass conversion processes. Indeed, it is desirable design the biorefinery schemes as zero residues, in order to produce biojet fuel under a circular economy model.

Regard the biomass, it is desirable to consider the use of waste biomass generated in the agriculture, agroindustry, food, as well as other production processes. This type of biomass is usually available all the year, in high volumes; indeed, the accumulation of this type of residues represents a pollution problem. Thus the use of waste biomass for the production of biofuel allows solving this problem and generate a new product that can be reinserted in the market. Moreover, it is desirable to use waste biomass that is not used for the production of animal feed; in this way, the food safety is not affected for both humans and animals. In addition, in most of the cases the cost of waste biomass is only associated with its recollection from the generation sites to the industrial facilities. In this context, the proposal of local/regional supply chains must be ensured, in order to keep the costs low as well as the carbon dioxide emissions.

The consideration of all the aspects presented before will contribute to the development and establishment of the industrial production of sustainable aviation fuel.

4.5 Case of study

This section presents a case study of the design, modeling, and simulation of the intensification of the processing of bio-oil from pyrolysis to produce biojet fuel. This case study is structured by subsections. In the first one the problem statement is described. Next, the second and third sections describes the modeling of the intensified hydrotreating process of bio-oil and the intensification applied on the separation zone, respectively. Then, in the fourth subsection the simulation of the overall intensified process is presented. Finally, the last two sections showed the economic and environmental assessment of the processing.

4.5.1 Problem statement

Mexico has a high potential to produce energetic crops, such as *Jatropha curcas* and castor bean. In order to maximize the energetic recovery of these cultures and producing biofuels, energy, and chemicals economically competitive, all the crop fractions must be used (Ng, Ng, & Ng, 2017). For instance, the crop residues can be converted into bio-oil through pyrolysis, which is later hydroprocessed to obtain renewable aviation fuel as well as other hydrocarbons. In this work, we consider bio-oils from *Jatropha curcas* and castor bean as raw materials, which are hydroprocessed to generate biojet fuel and other hydrocarbons; moreover, based on this conventional process, the intensification is applied to reactive and separation zones. Table 4.3 presents the chemical composition of bio-oil from *Jatropha curcas* shell pyrolysis and castor bean residue pyrolysis (Romero-Izquierdo, 2020). According to Romero-Izquierdo (2020), the flows of bio-oil from *Jatropha curcas* shell and castor bean residue are 40,470,729.50 and 671,515.14 kg/h, respectively.

The bio-oil has high oxygen content, hence, high corrosively, high chemical reactivity, low energy content, and poor thermal instability; thus its use as nonupgraded conventional transportation fuel is not feasible (Wang et al., 2020; Wei et al., 2019). The

Table 4.3 Chemical composition of bio-oil from *Jatropha curcas* shell and castor bean residue pyrolysis (Romero-Izquierdo, 2020).

Bio-oil from *Jatropha curcas* shell		Bio-oil from castor bean residue	
Component	%wt	Component	%wt
Octadecanoic-6-acid ($C_{18}H_{34}O_2$)	54.82	Phenol (C_6H_6O)	12.69
		3-Methyl-cyclopentanone ($C_6H_{10}O$)	66.52
Hexadecenoic acid ($C_{16}H_{32}O_2$)	45.18	Syringol ($C_8H_{10}O_3$)	20.79

upgrading of bio-oil from pyrolysis to produce biojet fuel and other biofuels involves a reactive stage of two steps and a separation zone, often distillation (Wei et al., 2019). In the first reactive step, the bio-oil is hydrotreated at high temperature (350–500°C) with hydrogen at high pressure (until 7 MPa), using a solid catalyst (Ni, NiMo, CoMo, Pt, Ru supported on Al_2O_3, SiO_2, ZrO_2, etc.). In the first reactive step, chemical pathways such as hydrodeoxygenation, decarboxylation, and decarbonylation are carrying out to convert the bio-oil into oxygen-free simpler components, such as CO, CO_2, and lighter hydrocarbons (Gupta, Mondal, Borugadda, & Dalai, 2021). While in the second reactive step, the conventional catalytic hydrocracking process is realized to obtain naphtha, diesel, and jet fuel range hydrocarbons (Wang et al., 2020), this step is usually called fractionation (Wei et al., 2019). Fig. 4.5 presents the conventional processing to convert bi-oil from pyrolysis into biojet fuel. Other process to upgrade the bio-oil is the catalytic cracking, which is considered as a cheaper route to convert bio-oil into oxygen-free hydrocarbons (Wang et al., 2015; Gupta et al., 2021). This process implies the thermal breakdown of larger bi-oil compounds into simpler at high temperature (350–600°C) and atmospheric pressure with a solid catalyst, like zeolites.

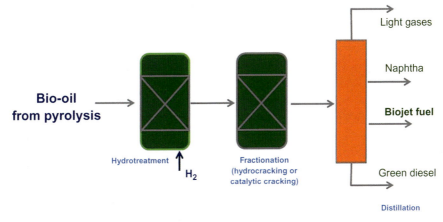

Fig. 4.5 Conventional processing of bio-oil from pyrolysis into biojet fuel.

Nevertheless, the hydrocarbons obtained by this chemical route have low carbon chains, and they do not meet the requirements of jet fuels (Gupta et al., 2021; Wang et al., 2015).

The conventional processing of bio-oil from pyrolysis of lignocellulosic materials has been reported in few works. Wang et al. (2015) studied the transformation of bio-oil from pyrolysis of straw stalk into jet and diesel hydrocarbons. The complete process involved three reaction steps: catalytic cracking of bio-oil to obtain aromatics and light olefins; then, the alkylation of low-carbon aromatics with light olefins to synthetize C8-C15 aromatic hydrocarbons, and finally, the production of C8-C15 cyclic alkanes by a hydrogenation step. The alkylation reactions were carried out using ionic liquid (bmim-Cl_2AlCl_3). According to experimental results, 88.4% selectivity was achieved for C8–C15 aromatic compounds. Then, Chen, Lin, and Wang (2020) analyzed the experimental production of jet fuel hydrocarbons from rice husk, using three reactive stages. In the first one, the pyrolysis of rice husk is performed in a fluidized bed fast reactor; in the second one, the hydroprocessing of pyrolytic oil into hydroprocessed oil is carried out, while in the last one, the hydrocracking/isomerization of the oil from the second stage is transformed into jet fuel hydrocarbons. Based on their results, the composition of jet fuel obtained is closer to the traditional jet fuel. Other works can be found in the reviews of Hansen, Mirkouei, and Diaz (2020) and Gupta et al. (2021).

A novel report about the experimental conversion of the bio-oil from pyrolysis into biojet fuel was published by Wang et al. (2020). In the research of Wang et al. (2020), the total cost of the process is improved by eliminating the hydrocracking step, recycling the last fraction separated by distillation train with boiling point >338°C, called residue stream and coprocessing it at the hydrotreating step; this last one process operates at mild conditions of temperature and hydrogen pressure, using $NiMo(S)/SiO_2$-Al_2O_3 as a catalyst. According to results, parallel hydrotreating and hydrocracking reactions occurred at a hydrotreating reactor, achieving until 19% yield of biojet fuel; likewise, up 57% of the *CC* reduction and 43% of operating cost (*OC*) reduction were reached. For this case of study, the experimental report published by Wang et al. (2020) is taken as basis in the modeling of the conversion of bio-oil from pyrolysis into biojet fuel. The mixture of bio-oil from pyrolysis of *Jatropha curcas* shell and castor bean residue are the raw materials for the production of biojet fuel, obtained as main fuel. Additionally, the Rong and Errico (2012) methodology, discussed in Section 4.2.2, is applied at the separating zone of the process. The modeling, simulation, and assessment (economic and environmental) of this case of study are presented next.

4.5.2 Modeling of the reactive stage

According to Wang et al. (2020), in order to achieve a better heteroatom removal, higher fuel-range hydrocarbons yield, and diminishing the high chemical reactivity from bio-oil, it is necessary to add a stabilizer stage, which converts the bio-oil compounds into more stable alcohols. This stage improves the bio-oil stability and enables hydrotreating without fouling the catalyst due to polymer formation, avoiding the plugging into the reactor. This stage operates at 140°C, 12.4 bar of hydrogen pressure with a ratio of H_2 (L) per bio-oil feed (L) equal to 2000, using Ru/TiO_2 as a catalyst. The reaction pathway for typical chemical groups contained in bio-oil from lignocellulosic materials pyrolysis, into the stabilizer reactor, has been reported by Wang et al. (2016). For this work, based on the experimental data provided by Wang et al. (2016) and Wang et al. (2020), the proposed chemical reactions and conversion data are presented in Table 4.4, taking into account the composition of bio-oil mixture shown in Table 4.3.

After the stabilizer reactor, the hydrotreatment is carrying out. The hydrotreating reactor operates at 420°C, 12.4 bar of hydrogen pressure, and $NiMo\ (S)/SiO_2\text{-}Al_2O_3$ as a catalyst (Wang et al., 2020). The ratio of H_2 (L) per bio-oil feeding (L) is 2100. In this reactor, the stabilized bio-oil is converted in gaseous products (CH_4, C_2H_6, C_3H_8, C_4H_{10}, CO, CO_2), naphtha (boiling point range <184°C), biojet fuel (boiling point range 150–250°C), green diesel (boiling point range 184–338°C), and residual hydrocarbons (boiling point >338°C). The yield for each hydrocarbon fraction defined by Wang et al. (2020), and the specific compounds obtained after hydrotreatment stage reported by Patel, Arcelus-Arrillaga, Izadpanah, and Hellgardt (2017), allows us to estimate the product distribution for this case of study, due to the absence of a kinetic model to describe the reaction system. Table 4.5 presents the product distribution for this case of study, along with the yield reported by Wang et al. (2020). It is important to mention that the yield and product distribution take into account the recycling stream, which belongs to residue fraction in a ratio of 6/100 (residue mass/stabilized bio-oil mass).

Table 4.4 Chemical reactions and conversion data for a stabilizer reactor (Wang et al., 2016; Wang et al., 2020).

No. Reaction	Reaction	% Conversion (Regarding to bio-oil compound)
1	$C_6H_6O + 3H_2 \rightarrow C_6H_{12}O$ (cyclohexanol)	99
2	$C_6H_{10}O + H_2 \rightarrow C_6H_{12}O$ (3-methyl-cyclopentanol)*	99
3	$C_8H_{10}O_3 + 5H_2 \rightarrow C_6H_{12}O_3 + 2CH_4$ (1,2,3-cyclohexanetriol)	99
4	$C_{18}H_{34}O_2 + 3H_2 \rightarrow C_{18}H_{38}O + H_2O$ (3-octadecanol)	99
5	$C_{16}H_{32}O_2 + 2H_2 \rightarrow C_{16}H_{34}O + H_2O$ (1-hexadecanol)	99

C_6H_6O=BIOIL-1H/$C_6H_{10}O$=BIOIL-2H/$C_8H_{10}O_3$=BIOIL-3H/$C_{18}H_{34}O_2$=BIOIL-1J/$C_{16}H_{32}O_2$=BIOIL-2J/ $C_6H_{12}O$=STAB-1H/$C_6H_{12}O$=STAB-2H*/$C_6H_{12}O_3$=STAB-3H/$C_{18}H_{38}O$=STAB-1J/$C_{16}H_{34}O$=STAB-2J.

Table 4.5 Product distribution from a hydrotreating reactor with a recycling stream (Patel et al., 2017; Wang et al., 2020).

Hydrocarbon fraction	Compounds	% Yield
Light gases (A)	CH_4	2.0
	C_2H_6	
	C_3H_8	
	C_4H_{10}	
Naphtha (B)	C_5H_{12}	36.52
	C_6H_{14}	
	C_7H_{16}	
Biojet fuel (C)	C_8H_{16}[a]	19.09
	C_8H_{18}	
	C_9H_{20}	
	$C_{10}H_{22}$	
	$C_{11}H_{24}$	
	$C_{13}H_{28}$	
	$C_{14}H_{30}$	
	$C_{16}H_{34}$[a]	
	$C_{15}H_{32}$	
	$C_{16}H_{34}$	
Green diesel (+residue) (D)	$C_{19}H_{40}$	27.39
	$C_{20}H_{42}$	
	$C_{21}H_{44}$	
	$C_{23}H_{48}$[a]	
	$C_{23}H_{48}$	
	$C_{26}H_{54}$[a]	
	$C_{26}H_{54}$[a]	

[a]C_8H_{16} Cyclopentane, 1,2,3-trimethyl/$C_{16}H_{34}$ dodecane, 5,8-diethyl/$C_{23}H_{48}$ heptadecane, 9-hexyl/$C_{26}H_{54}$ octadecane, 3-ethyl-5-(2-ethylbutyl)/$C_{26}H_{54}$ docosane, 9-butyl.

Finally, the hydrocarbon stream is separated in a distillation train, which involves three distillation columns in a sequence direct conventional, in order to separate the four grouped products: light gases (A), naphtha (B), biojet fuel (C), and green diesel (D). Fig. 4.6 presents the overall processing of bio-oil mixture for this case of study. The conventional distillation scheme will be intensified in the next section through the methodology of Rong and Errico (2012).

4.5.3 Modeling of separation zone: Process intensification

According to the methodology of Rong and Errico (2012), presented in Section 4.2.2, it is necessary to develop, analyze, and evaluate the energetic requirements of each subspace proposed, selecting the train with less energy consumption in each case: simple column configurations (SC), originally thermally coupled configurations (OTC), TES, and finally the ISC; each one of these configurations are constructed based on the previous subspace.

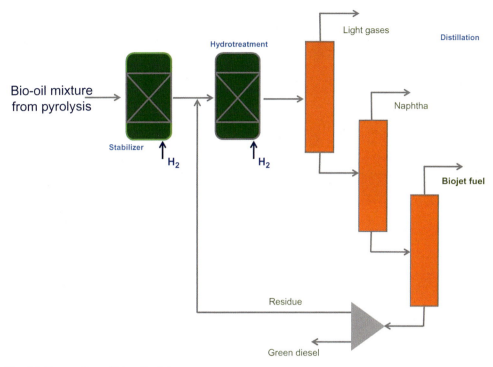

Fig. 4.6 Processing of bio-oil mixture.

For this case of study, light gases have a low boiling point; thus they must be separated in the first distillation column. In this context, each generated design will have a fixed column with partial-vapor condenser, which uses 314-A refrigerant. Assuming this consideration, Fig. 4.7 presents the possible configurations for the distillation train with three columns and four components to separate, from conventional direct sequence. It is important to note that the ISC's configurations cannot be designed from the TES's configurations, due to absence of lateral transport sections. In the next section is presented the evaluation of each configuration, and the selection of the best one, according to its energetic requirements.

4.5.4 Simulation of the process

Based on the modeling and conceptual design of bio-oil mixture from pyrolysis of *Jatropha curcas* shell and castor bean residue presented in previous sections, the simulation of the overall processing through Aspen Plus is describing in this section. The simulation starts with the feed of the components involved in the process, which are defined in Tables 4.3 and 4.4. Likewise, the suitable thermodynamic method to model the process is chosen. For this case, Peng-Robinson and BK10 were selected for the reactive and

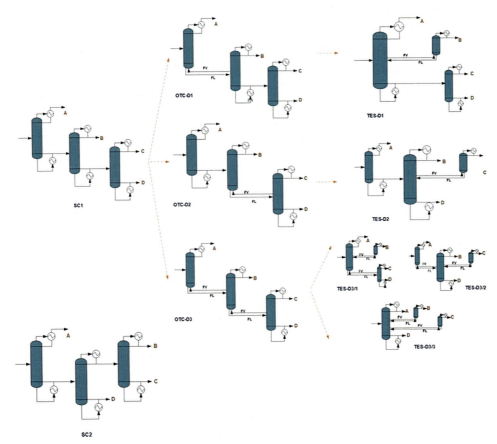

Fig. 4.7 Application of Rong and Errico methodology (Rong & Errico, 2012): SC: SC1, SC2//OTC: OTC-D1, OTC-D2, OTC-D3//TES: TES-D1, TES-D2, TES-D3, TES-D3/1, TES-D3/2, TES/D3/3.

separation zone, respectively (Carlson, 1996). The flowsheet diagram constructed in Aspen Plus is presented in Fig. 4.8.

The process starts with the bio-oil streams, whose composition and material amount were presented in Table 4.3 from Section 4.5.1, at 25°C and 1 bar. Regarding to hydrogen stream, the entrance ratio for a stabilizer and hydrotreatment reactors is defined in Section 4.5.2. It is important to mention that due to excessive hydrogen entrance for each reactive zone, the output hydrogen from a stabilizer reactor is only adjustment to fulfill the hydrotreatment requirement. The hydrogen entrance conditions are defined as 10 bar and vapor fraction equal to 1 (Gutiérrez-Antonio et al., 2016). The bio-oil and hydrogen streams are conditioning to reach the operative conditions of a stabilizer reactor. The conditioning of feed streams starts with the increase until 12.4 bar of the pressure using the *Pump* and *Compr* modules, called *PUMP-1* and *COMP-1*, for bio-oil mixture

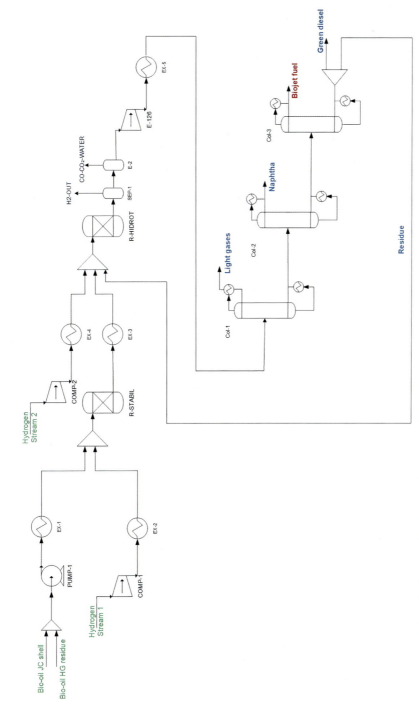

Fig. 4.8 Flowsheet diagram for bio-oil processing into biojet fuel.

and hydrogen stream, respectively. Then, the temperature is adjusted until 140°C, using a *Heater* module, called *EX-1* and *EX-2* for bio-oil and hydrogen streams, respectively. Later, the bio-oil and hydrogen streams are mixed at a *MIX* module (*MIX-2*) to be fed to the stabilizer reactor. The stabilizer reactor is modeling by a *RStoic* module, called *R-STABIL*, with the reactions and conversion data presented in Table 4.4 from Section 4.5.2. The product distribution from *R-STABIL* is presented in Table 4.6. The output stream from *R-STABIL* is conditioning until 420°C. Also, the additional hydrogen stream is conditioning using a *COMP* module and *Heater* module, called *COMP-2* and *EX-4*, respectively, in order to reach the operative conditions of hydrotreatment reactor. The hydrotreatment reactor is modeling by a *Ryield* module called *R-HIDROT*, according to product distribution and data yield presented in Table 4.5 from Section 4.5.2. It is important to mention that due to incomplete conversion from *R-STABIL*, the yield data for hydrotreatment stage must be tuned. Table 4.7 shows the product distribution from *R-HIDROT*. The remaining hydrogen and other gaseous products, such as CO, CO_2 and water (as vapor), are separated before distillation train, using *SEP* modules (*SEP-1* and *SEP-2*).

The hydrocarbon stream is conditioning by a *PUMP* module (TURBI-1) and Heater (EX-5) module to diminish its operation conditions until 1 bar and 25°C, respectively. The conditioned stream enters to distillation train with three columns in order to separate the grouped products defined in Table 4.5: (A) light gases, 72,898.56 kg/h; (B) naphtha, 1,601,572.19 kg/h; (C) biojet fuel, 2,620,555.48 kg/h; and (D) green diesel, 49,265,354.23 kg/h. The conventional design presented in Fig. 4.8 is the direct conventional train. The design procedure for the distillation configuration is performed with the short methods, based on Winn, Underwood, and Gilliland equations, programmed in *DSTW* module; in this module, we assume a reflux ratio of −1.33 times the minimum value, 10 psia pressure drop along the column, and 99% of recovery for the key components. The results from *DSTW* module are the number of stages, feed stage, actual reflux

Table 4.6 Product distribution from a stabilizer reactor (*R-STABIL*).

Components	Flow (kg/s)	Components	Flow (kg/s)
BIOIL-1H	852.35	STAB-1H	89,804.75
BIOIL-2H	4466.58	STAB-2H	451,274.22
BIOIL-3H	1396.22	STAB-3H	118,495.09
BIOIL-1J	221,862.39	STAB-1J	21,033,783.08
BIOIL-2J	182,844.91	STAB-2J	17,114,530.83
WATER	2,672,574.82	C1	28,768.10
H2	6,826,311.31		

Table 4.7 Product distribution from a hydrotreatment reactor (*R-HIDROT*).

Components	Flow (kg/s)	Components	Flow (kg/s)
BIOIL-1H	1.46	STAB-1H	334.79
BIOIL-2H	6.56	STAB-2H	1681.24
BIOIL-3H	2.19	STAB-3H	441.28
BIOIL-1J	311.45	STAB-1J	78,378.52
BIOIL-2J	256.74	STAB-2J	63,773.99
WATER	16,803.63	CO	16,803.63
H2	1,267,699.58	CO2	16,803.63
C1	16,803.63	C13	187,122.91
C2	16,803.63	C14	187,122.91
C3	16,803.63	C12-ISO	187,122.91
C4	16,803.63	C15	402,721.11
C5	536,961.23	C16	402,721.11
C6	536,961.23	C19	402,721.11
C7	536,667.29	C20	402,721.11
C8-CLC5	147,032.12	C21	147,032.12
C8	536,961.23	C17-ISO	147,032.12
C9	1871,22.91	C23	47,731,644.22
C10	187,122.91	C18-ISO	147,032.12
C11	187,122.91	C22-ISO	147,032.12

Table 4.8 Results from Col-1, Col-2, Col-3, and direct conventional train.

	Col-1	Col-2	Col-3
Number of stages	30	40	34
Q condenser (kW)	2774.98	504,892.24	711,129.31
Q reboiler (kW)	6,094,841.19	6,194,547.91	3,644,943.51
Reflux ratio	0.13	2.0	1.085

ratio, among others, whose allow the rigorous design of columns; this last task is performed with the *RadFrac* module with 99% of recovery. Table 4.8 provides the results for columns 1–3 involved in the conventional direct distillation train.

According to the report by Wang et al. (2020), the recycle stream is separated from the bottom at the last distillation column. The ratio between the recycle stream and the bio-oil mixture after the stabilizer reactor is provided in Section 4.5.2. Thus in Aspen Plus, an *FSplit* module (*SPLIT-1*) is used to separate the stream; this stream enters to *MIX-3*, before *R-HIDROT*.

As mentioned before, the methodology of Rong and Errico (2012) is used to apply process intensification on the separation zone. It is important to mention that the design in Aspen Plus of each configuration presented in Fig. 4.7, from Section 4.5.3, was constructed by the procedure described previously for conventional direct configuration. The methodology starts with the design of SC's alternatives. Each alternative is analyzed in terms of energy requirements, selected the one with minimum energy consumption. Table 4.9 presents the results for SC configurations. As can be seen, the SC2 has 66 stages more than SC1 (direct conventional), and its energy requirements are 4.30 and 6.37% higher than SCI for the condenser and reboiler, respectively. Thus the SC1 is chosen as a basis for the design of the OTC's.

The energetic assessment for each OTC's design is presented in Table 4.10. Based on the simulation results, the OTC-D1 has the minimum energy requirements for the

Table 4.9 Results of SC configurations.

	Col-1	Col-2	Col-3	TOTAL
SC1				
No. Stages	30	40	34	104
Q condenser (kW)	2774.98	504,892.24	711,129.31	1,218,796.53
Q reboiler (kW)	6,094,841.19	6,194,547.91	3,644,943.51	15,934,332.61
SC2				
No. Stages	30	99	41	170
Q condenser (kW)	2774.98	398,746.11	306,044.32	1,273,617.73
Q reboiler (kW)	6,094,841.19	8,899,741.30	477,329.45	17,019,103.13

Table 4.10 Results of OTC configurations.

	Col-1	Col-2	Col-3	TOTAL
OTC-D1				
No. Stages	30	40	34	104
Q condenser (kW)	33,970.55	504,866.92	710,829.80	1,249,667.27
Q reboiler (kW)	–	12,475,967.86	3,643,015.84	16,118,983.70
OTC-D2				
No. Stages	30	40	34	104
Q condenser (kW)	2774.97	1,052,281.80	715,794.40	1,770,851.17
Q reboiler (kW)	6,248,555.24	–	10,391,775.22	16,640,330.46
OTC-D3				
No. Stages	30	40	34	104
Q condenser (kW)	57,464.23	2,287,356.34	715,794.71	3,060,615.28
Q reboiler (kW)	–	–	18,130,092.20	18,130,092.20

Table 4.11 Results of TES configurations.

	TES-D1			
	Col-1	Col-2	Col-3	TOTAL
No. Stages	49	21	34	104
Q condenser (kW)	2767.24	198,198.07	713,486.57	914,451.88
Q reboiler (kW)	12,120,245.69	–	3,662,355.52	15,782,601.21

condenser and reboiler, with 2.47 and 1.15%, regarding to SC1. In comparison, the OTC-D2 is 31.17 and 4.24% higher than SC1 for the condenser and reboiler, respectively. Likewise, the OTC-D3 consumes 60.18 and 12.11% more energy than SC1 for the condenser and reboiler, respectively. Thereby, the OTC-D1 design is the basis for the design for the TES configuration.

Table 4.11 shows the TES results. Due to that the OTC-D1 is the best configuration, one design TES is constructed. As can be seen, the condenser and reboiler energy is 24.97 and 0.95% less than energy requirements for SC1. Thus TES-D1 is the selected configuration to separate the renewable hydrocarbons stream from bio-oil mixture processing.

Including the TES-D1 configuration, the intensified flowsheet diagram for bio-oil mixture processing is presented in Fig. 4.9.

4.5.5 Economic assessment

The economic evaluation of intensified processing of bio-oil from pyrolysis to produce biojet fuel is carried out through the estimation of TAC, defined by the CC and the OC, according to Eq. (4.1).

$$TAC\ (USD/year) = \frac{CC}{n} + OC \quad (4.1)$$

The CC is associated with the total equipment cost, obtained by Aspen Economics tool, the payback period (n) is assumed as 5 years, and 8500 h per year of process operation (Romero-Izquierdo et al., 2021). Also, according to Guthrie's method to estimate the TAC, the factors A1 (18%) and A2 (61%) are added to the cost obtained from Aspen Economics in order to include contingencies, fees related to installation cost, and cost associated with machinery, equipment, and maintenance, respectively (Turton, Bailie, Whiting, Shaeiwitz, & Bhattacharyya, 2012). On the other hand, the OC is the sum of steam, cooling water, refrigerant (314-A), hydrogen and electricity costs, and utilities required by the processing. Table 4.12 presents the prices for OC estimation, taken from Romero-Izquierdo et al. (2021). It is important to mention that the bio-oil mixture from pyrolysis of *Jatropha curcas* shell and castor bean residue are considered for free, while the catalysts costs are not considered.

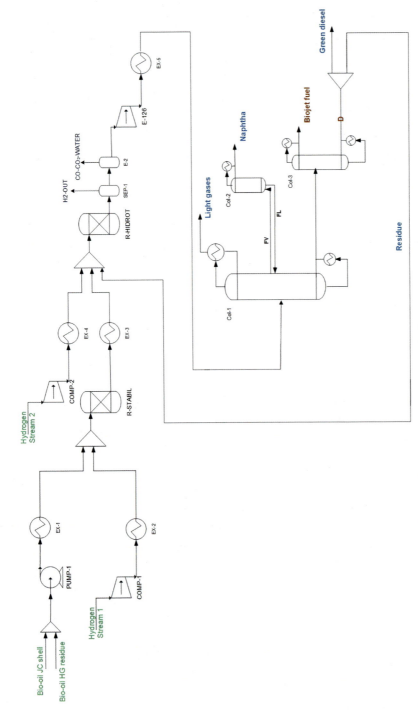

Fig. 4.9 Flowsheet diagram for bio-oil-intensified processing into biojet fuel.

Table 4.12 Prices for estimation of *OC*.

Item	Price	Reference
Steam	1 USD/GJ	(Romero-Izquierdo et al., 2021)
Cooling water	14.8 USD/1000 m^3	
Refrigerant (R314A)	7.76 USD/kg	
Hydrogen	1.80 USD/kg	
Electricity	1.4721 MX/kWh	

Table 4.13 presents the detailed costs for the bio-oil conversion into biojet fuel. As it can be seen, the hydrogen and refrigerant cost represents 94.64 and 3.48% of TAC, respectively, which are the higher values. In the case of steam and cooling water, they affect the TAC by 1 and 0.0042%, respectively. Likewise, the electricity cost represents only 0.0514% of TAC, due to the power generation (31,656.47 kWh) by the turbine before separation zone. Thus the *OC* is the major factor affecting the TAC, which represents 99.17%. It is important to emphasize that the hydrogen consumption is the major contributor to *OC*, while the process intensification allows to improve the steam and cooling water requirements, regarding to conventional direct sequence; thus its effect reduces the *OC*.

4.5.6 Environmental assessment

For this case of study, the environmental evaluation involves the estimation of CO_2 emissions due to steam and electricity generation required in the processing, provided by external sources. In the case of CO_2 emissions due to steam generation, a detailed calculation procedure has been presented by Gutiérrez-Antonio et al. (2021), which involves the burned fuel mass (gas natural) to reach the steam requirement, associated with emission factor due to the combustion process. On the other hand, the CO_2 emissions released by the electricity generation take account the emission factor estimated by the national electric system in Mexico, which is 0.582 ton of CO_2 per MWh (Romero-Izquierdo et al., 2021). In the case of refrigerant consumption, the related emissions are not considered for this case of study. Table 4.14 presents the total CO_2 emissions released by this processing.

Based on the results, 99.99% of CO_2 emissions released by the processing are due to steam requirements. The power generation by the turbine allows to diminish the external electricity consumption and the CO_2 emissions. However, in order to minimize the total CO_2 emissions, energy integration tools can be used.

Table 4.13 Overview of costs involved in the TAC estimation.

Operating Cost (USD/year)					
Heating Cost	Cooling cost	Hydrogen cost	Refrigerant cost	Electricity	Total operating cost (USD/year)
$1,745,738.03	$7269.04	$165,414,854.61	$6,082,805.28	$89,805.57	$173,340,472.53

Annual Capital Cost (USD/year)

Equipment Cost $4,052,678.95	A1 $729,482.21	A2 $2,472,134.16	Total annual cost (USD/year) $1,450,859.06
TAC (USD/year) $174,791,331.59			

Table 4.14 CO_2 emissions released by the processing due to steam and electricity requirements.

Emissions by steam generation (Mton CO_2/year)	Emissions by electricity (Mton CO_2/year)	Total emissions (Mton CO_2/year)
316,421.44	0.414	316,421.85

4.6 Conclusions

The renewable aviation fuel is one alternative that can help to guarantee the sustainable recovery of the aviation sector. In order to do that, the biojet fuel must be competitive from technical and economic points of views. In this context, process intensification is a powerful tool to decrease the operating and *CCs* associated in the production of biokerosene, contributing to having a competitive price respect to its fossil counterpart. There are several opportunity areas to intensify the production pathways of this biofuel, involving high efficiency and compact reactors as well reactive thermally coupled distillation. As can be noted in the case of study, the application of process intensification tools in the production of biojet fuel allows to diminish the *CC*, the energy requirements, and the CO_2 emissions. In this way, the intensification strategies could lead to the development of sustainable production process for renewable aviation fuel.

Acknowledgments

Financial support provided by CONACyT, grant 239765, for the development of this project is gratefully acknowledged. Also, Araceli Guadalupe Romero-Izquierdo was supported by a scholarship from CONACYT for the realization of her graduate studies.

References

Abrokwah, R. Y., Rahman, M. M., Deshmane, V. G., & Kuila, D. (2019). Effect of titania support on Fischer-Tropsch synthesis using cobalt, iron, and ruthenium catalysts in silicon-microchannel microreactor. *Molecular Catalysis, 478*, 110566.

Alferov, K. A., Belov, G. P., & Meng, Y. (2017). Chromium catalysts for selective ethylene oligomerization to 1-hexene and 1-octene: Recent results. *Applied Catalysis A: General, 542*, 71–124.

Antunes, B. M., Rodrigues, A. E., Lin, Z., Portugal, I., & Silva, C. M. (2015). Alkenes oligomerization with resin catalysts. *Fuel Processing Technology, 138*, 86–99.

Arzamendi, G., Diéguez, P. M., Montes, M., Odriozola, J. A., Falabella Sousa-Aguiar, E., & Gandía, L. M. (2010). Computational fluid dynamics study of heat transfer in a microchannel reactor for low-temperature Fischer-Tropsch synthesis. *Chemical Engineering Journal, 160*, 915–922.

ASTM. (2019). *ASTM D7566—20c standard specification for aviation turbine fuel containing synthesized hydrocarbons*. ASTM International. https://www.astm.org/Standards/D7566.htm. (Accessed 1 April 2021).

Athar, M., & Zaidi, S. (2020). A review of the feedstocks, catalysts, and intensification techniques for sustainable biodiesel production. *Journal of Environmental Chemical Engineering, 8*, 104523.

Avami, A., Marquardt, W., Saboohi, Y., & Kraemer, K. (2012). Shortcut design of reactive distillation columns. *Chemical Engineering Science, 71*, 166–177.

Barbosa, D., & Doherty, M. F. (1988). Design and minimum-reflux calculations for single-feed multicomponent reactive distillation columns. *Chemical Engineering Science, 43*, 1523–1537.

Barroso-Muñoz, F. O., Hernández, S., & Ogunnaike, B. (2007). Analysis of design and control of reactive thermally coupled distillation sequences. In V. Pleşu, & P.Ş. Agachi (Eds.), *17th European symposium on computer aided process engineering, computer aided chemical engineering* (pp. 877–882). Elsevier.

Becker, H., Güttel, R., & Turek, T. (2019). Performance of diffusion-optimised Fischer-Tropsch catalyst layers in microchannel reactors at integral operation. *Catalysis Science & Technology, 9*, 2180–2195.

Bepari, S., Li, X., Abrokwah, R., Mohammad, N., Arslan, M., & Kuila, D. (2020). Co-Ru catalysts with different composite oxide supports for Fischer-Tropsch studies in 3D-printed stainless steel microreactors. *Applied Catalysis A: General, 608*, 117838.

Bepari, S., Stevens-Boyd, R., Mohammad, N., Li, X., Abrokwah, R., & Kuila, D. (2021). Composite mesoporous SiO_2-Al_2O_3 supported Fe, FeCo and FeRu catalysts for Fischer-Tropsch studies in a 3-D printed stainless-steel microreactor. *Materials Today: Proceedings, 35*, 221–228.

Blanch, H. W. (2012). Bioprocessing for biofuels. *Current Opinion in Biotechnology, 23*, 390–395.

Budagumpi, S., Keri, R. S., Biffis, A., & Patil, S. A. (2017). Olefin poly/oligomerizations by metal precatalysts bearing non-heterocyclic N-donor ligands. *Applied Catalysis A: General, 535*, 32–60.

Bwapwa, J. K., Anandraj, A., & Trois, C. (2017). Possibilities for conversion of microalgae oil into aviation fuel: A review. *Renewable and Sustainable Energy Reviews, 80*, 1345–1354.

CAAFI. (2021). *Fuel qualification*. https://www.caafi.org/focus_areas/fuel_qualification.html. (Accessed 1 April 2021).

Cabrera-Ruiz, J., Santaella, M. A., Alcántara-Ávila, J. R., Segovia-Hernández, J. G., & Hernández, S. (2017). Open-loop based controllability criterion applied to stochastic global optimization for intensified distillation sequences. *Chemical Engineering Research and Design, 123*, 165–179.

Carlson, E. (1996). Don't gamble with physical properties for simulations. *Chemical Engineering Progress*, 35–46.

Carrera-Rodríguez, M., Segovia-Hernández, J. G., & Bonilla-Petriciolet, A. (2011). Short-cut method for the design of reactive distillation columns. *Industrial and Engineering Chemistry Research, 50*, 10730–10743.

Carrera-Rodríguez, M., Segovia-Hernández, J. G., Hernández-Escoto, H., Hernández, S., & Bonilla-Petriciolet, A. (2014). A note on an extended short-cut method for the design of multicomponent reactive distillation columns. *Chemical Engineering Research and Design, 92*, 1–12.

Chen, G., Li, S., Jiao, F., & Yuan, Q. (2007). Catalytic dehydration of bioethanol to ethylene over TiO_2/γ-Al_2O_3 catalysts in microchannel reactors. *Catalysis Today, 125*, 111–119.

Chen, Y.-K., Lin, C.-H., & Wang, W.-C. (2020). The conversion of biomass into renewable jet fuel. *Energy, 201*, 117655.

Cheng, N., Koda, K., Tamai, Y., Yamamoto, Y., Takasuka, T. E., & Uraki, Y. (2017). Optimization of simultaneous saccharification and fermentation conditions with amphipathic lignin derivatives for concentrated bioethanol production. *Bioresource Technology, 232*, 126–132.

Chiaramonti, D. (2019). Sustainable aviation fuels: The challenge of decarbonization. *Energy Procedia, 158*, 1202–1207.

Chuayboon, S., Abanades, S., & Rodat, S. (2018). Experimental analysis of continuous steam gasification of wood biomass for syngas production in a high-temperature particle-fed solar reactor. *Chemical Engineering and Processing Process Intensification, 125*, 253–265.

Cossio-Vargas, E., Barroso-Muñoz, F. O., Hernandez, S., Segovia-Hernandez, J. G., & Cano-Rodriguez, M. I. (2012). Thermally coupled distillation sequences: Steady state simulation of the esterification of fatty organic acids. *Chemical Engineering and Processing Process Intensification, 62*, 176–182.

Deshmukh, S. R., Tonkovich, A. L. Y., Jarosch, K. T., Schrader, L., Fitzgerald, S. P., Kilanowski, D. R., et al. (2010). Scale-up of microchannel reactors for Fischer–Tropsch synthesis. *Industrial and Engineering Chemistry Research, 49*, 10883–10888.

Doliente, S. S., Narayan, A., Tapia, J. F. D., Samsatli, N. J., Zhao, Y., & Samsatli, S. (2020). Bio-aviation fuel: A comprehensive review and analysis of the supply chain components. *Frontiers in Energy Research, 8*, 110.

Erickson, D. (2005). Towards numerical prototyping of labs-on-chip: Modeling for integrated microfluidic devices. *Microfluidics and Nanofluidics, 1*, 301–318.

Errico, M., Madeddu, C., Flemming Bindseil, M., Dall Madsen, S., Braekevelt, S., & Camilleri-Rumbau, M. S. (2020). Membrane assisted reactive distillation for bioethanol purification. *Chemical Engineering and Processing Process Intensification, 157*, 108110.

Gadalla, M., Jobson, M., & Smith, R. (2003). Shortcut models for retrofit design of distillation columns. *Chemical Engineering Research and Design, 81*(8), 971–986.

García Colli, G., Alves, J. A., Martínez, O. M., & Barreto, G. F. (2016). Development of a multi-layer microreactor: Application to the selective hydrogenation of 1-butyne. *Chemical Engineering and Processing Process Intensification, 105*, 38–45.

Goh, B. H. H., Chong, C. T., Ge, Y., Ong, H. C., Ng, J.-H., Tian, B., et al. (2020). Progress in utilisation of waste cooking oil for sustainable biodiesel and biojet fuel production. *Energy Conversion and Management, 223*, 113296.

Gómez-Castro, F. I., Ramírez-Vallejo, N. E., Segovia-Hernández, J. G., Gutiérrez-Antonio, C., Errico, M., Briones-Ramírez, A., et al. (2016). Energy consumption maps for quaternary distillation sequences. In Z. Kravanja, & M. Bogataj (Eds.), *26th European symposium on computer aided process engineering, computer aided chemical engineering* (pp. 121–126). Elsevier.

González-Rugerio, C. A., Fuhrmeister, R., Sudhoff, D., Pilarczyk, J., & Górak, A. (2014). Optimal design of catalytic distillation columns: A case study on synthesis of TAEE. *Chemical Engineering Research and Design, 92*, 391–404.

Guan, X., & Yang, N. (2017). CFD simulation of pilot-scale bubble columns with internals: Influence of interfacial forces. *Chemical Engineering Research and Design, 126*, 109–122.

Gupta, S., Mondal, P., Borugadda, V. B., & Dalai, A. K. (2021). Advances in upgradation of pyrolysis bio-oil and biochar towards improvement in bio-refinery economics: A comprehensive review. *Environmental Technology and Innovation, 21*, 101276.

Gutiérrez Ortiz, F. J. (2020). Techno-economic assessment of supercritical processes for biofuel production. *Journal of Supercritical Fluids, 160*, 104788.

Gutiérrez-Antonio, C., Gómez-Castro, F. I., de Lira-Flores, J. A., & Hernández, S. (2017). A review on the production processes of renewable jet fuel. *Renewable and Sustainable Energy Reviews, 79*, 709–729.

Gutiérrez-Antonio, C., Gómez-Castro, F. I., Hernández, S., & Briones-Ramírez, A. (2015). Intensification of a hydrotreating process to produce biojet fuel using thermally coupled distillation. *Chemical Engineering and Processing Process Intensification, 88*, 29–36.

Gutiérrez-Antonio, C., Gómez-De la Cruz, A., Romero-Izquierdo, A. G., Gómez-Castro, F. I., & Hernández, S. (2018). Modeling, simulation and intensification of hydroprocessing of micro-algae oil to produce renewable aviation fuel. *Clean Technologies and Environmental Policy, 20*, 1589–1598.

Gutiérrez-Antonio, C., & Hernández, S. (2019). Process intensification applied to waste-to-energy production. Chapter 4 In E. Jacob-López, L. Queiroz-Zepka, & M. I. Queiroz (Eds.), *Waste to energy* Nova Science Publishers. ISBN; 978-1-53614-431-4.

Gutiérrez-Antonio, C., Romero-Izquierdo, A. G., Gómez-Castro, F. I., & Hernández, S. (2021). *Production processes of renewable aviation fuel*. Elsevier.

Gutiérrez-Antonio, C., Romero-Izquierdo, A. G., Gómez-Castro, F. I., Hernández, S., & Briones-Ramírez, A. (2016). Simultaneous energy integration and intensification of the hydrotreating process to produce biojet fuel from jatropha curcas. *Chemical Engineering and Processing Process Intensification, 110*, 134–145.

Gutiérrez-Antonio, C., Soria Ornelas, M. L., Gómez-Castro, F. I., & Hernández, S. (2018). Intensification of the hydrotreating process to produce renewable aviation fuel through reactive distillation. *Chemical Engineering and Processing Process Intensification, 124*, 122–130.

Hanafi, S. A., Elmelawy, M. S., Shalaby, N. H., El-Syed, H. A., Eshaq, G., & Mostafa, M. S. (2016). Hydrocracking of waste chicken fat as a cost effective feedstock for renewable fuel production: A kinetic study. *Egyptian Journal of Petroleum, 25*, 531–537.

Hansen, S., Mirkouei, A., & Diaz, L. A. (2020). A comprehensive state-of-technology review for upgrading bio-oil to renewable or blended hydrocarbon fuels. *Renewable and Sustainable Energy Reviews, 118*, 109548.

He, Q., Guo, Q., Umeki, K., Ding, L., Wang, F., & Yu, G. (2021). Soot formation during biomass gasification: A critical review. *Renewable and Sustainable Energy Reviews, 139*, 110710.

Hernández, S., & Jiménez, A. (1996). Design of optimal thermally-coupled distillation systems using a dynamic model. *Chemical Engineering Research and Design, 74*, 357–362.

Hitsov, I., Maere, T., De Sitter, K., Dotremont, C., & Nopens, I. (2015). Modelling approaches in membrane distillation: A critical review. *Separation and Purification Technology, 142*, 48–64.

Hommes, A., ter Horst, A. J., Koeslag, M., Heeres, H. J., & Yue, J. (2020). Experimental and modeling studies on the Ru/C catalyzed levulinic acid hydrogenation to γ-valerolactone in packed bed microreactors. *Chemical Engineering Journal, 399*, 125750.

IATA. (2020). *Industrial statistics (fact sheet)*. https://www.iata.org/en/iata-repository/pressroom/factsheets/industry-statistics/. (Accessed 2 April 2021).

IATA. (2021). *IATA—2020 worst year in history for air travel demand*. https://www.iata.org/en/pressroom/pr/2021-02-03-02/. (Accessed 1 April 2021).

IEA. (2020). *Sustainable recovery, fuels*. https://www.iea.org/reports/sustainable-recovery/fuels#abstract. (Accessed 1 April 2021).

IRENA. (2017). *Biofuels for aviation: Technology brief*. https://www.irena.org/-/media/Files/IRENA/Agency/Publication/2017/IRENA_Biofuels_for_Aviation_2017.pdf. (Accessed 1 April 2021).

Jiang, Z., & Agrawal, R. (2019). Process intensification in multicomponent distillation: A review of recent advancements. *Chemical Engineering Research and Design, 147*, 122–145.

Jiang, P., Van Fan, Y., & Klemeš, J. J. (2021). Impacts of COVID-19 on energy demand and consumption: Challenges, lessons and emerging opportunities. *Applied Energy, 285*, 116441.

Joshi, N., & Lawal, A. (2012a). Hydrodeoxygenation of acetic acid in a microreactor. *Chemical Engineering Science, 84*, 761–771.

Joshi, N., & Lawal, A. (2012b). Hydrodeoxygenation of pyrolysis oil in a microreactor. *Chemical Engineering Science, 74*, 1–8.

Kaymak, D. B. (2021). Design and control of an alternative intensified process configuration for separation of butanol-butyl acetate-methyl isobutyl ketone system. *Chemical Engineering and Processing Process Intensification, 159*, 108233.

Kenig, E. Y., Su, Y., Lautenschleger, A., Chasanis, P., & Grünewald, M. (2013). Micro-separation of fluid systems: A state-of-the-art review. *Separation and Purification Technology, 120*, 245–264.

Kiss, A. A., & Smith, R. (2020). Rethinking energy use in distillation processes for a more sustainable chemical industry. *Energy, 203*, 117788.

Klein-Marcuschamer, D., Turner, C., Allen, M., Gray, P., Dietzgen, R. G., Gresshoff, P. M., et al. (2013). Technoeconomic analysis of renewable aviation fuel from microalgae, *Pongamia pinnata*, and sugarcane. *Biofuels, Bioproducts and Biorefining, 7*, 416–428.

Klemeš, J. J., Van Fan, Y., & Jiang, P. (2021). COVID-19 pandemic facilitating energy transition opportunities. *International Journal of Energy Research, 45*, 3457–3463.

Kossatz, H. L., Rose, S. H., Viljoen-Bloom, M., & van Zyl, W. H. (2017). Production of ethanol from steam exploded triticale straw in a simultaneous saccharification and fermentation process. *Process Biochemistry, 53*, 10–16.

Li, S. (2017). Chapter 7—Analysis and design of heterogeneous catalytic reactors. In S. Li (Ed.), *Reaction engineering* (pp. 311–368). Boston: Butterworth-Heinemann.

Li, X., Chen, Y., Hao, Y., Zhang, X., Du, J., & Zhang, A. (2019). Optimization of aviation kerosene from one-step hydrotreatment of catalytic Jatropha oil over SDBS-Pt/SAPO-11 by response surface methodology. *Renewable Energy, 139*, 551–559.

Li, L., Coppola, E., Rine, J., Miller, J. L., & Walker, D. (2010). Catalytic hydrothermal conversion of triglycerides to non-ester biofuels. *Energy and Fuels, 24*, 1305–1315.

Liñán, D. A., Bernal, D. E., Gómez, J. M., & Ricardez-Sandoval, L. A. (2021). Optimal synthesis and design of catalytic distillation columns: A rate-based modeling approach. *Chemical Engineering Science, 231*, 116294.

Liu, T., & Khosla, C. (2010). Genetic engineering of Escherichia coli for biofuel production. *Annual Review of Genetics, 44*, 53–69.

Liu, X., Ünal, B., & Jensen, K. F. (2012). Heterogeneous catalysis with continuous flow microreactors. *Catalysis Science & Technology, 2*, 2134–2138.

Liu, G., Yan, B., & Chen, G. (2013). Technical review on jet fuel production. *Renewable and Sustainable Energy Reviews, 25*, 59–70.

Liu, M., Zhu, X., Chen, R., Liao, Q., Ye, D., Zhang, B., et al. (2018). Tube-in-tube hollow fiber catalytic membrane microreactor for the hydrogenation of nitrobenzene. *Chemical Engineering Journal, 354*, 35–41.

Macedonio, F., Ali, A., & Drioli, E. (2017). 3.10 membrane distillation and osmotic distillation. In E. Drioli, L. Giorno, & E. Fontananova (Eds.), *Comprehensive membrane science and engineering (second edition)* (pp. 282–296). Oxford: Elsevier.

Maity, S. K. (2015). Opportunities, recent trends and challenges of integrated biorefinery: Part I. *Renewable and Sustainable Energy Reviews, 43*, 1427–1445.

Marosz, M., Kowalczyk, A., & Chmielarz, L. (2020). Modified vermiculites as effective catalysts for dehydration of methanol and ethanol. *Catalysis Today, 355*, 466–475.

Masih, D., Rohani, S., Kondo, J. N., & Tatsumi, T. (2019). Catalytic dehydration of ethanol-to-ethylene over rho zeolite under mild reaction conditions. *Microporous and Mesoporous Materials, 282*, 91–99.

Miranda-Galindo, E. Y., Segovia-Hernández, J. G., Hernández, S., Gutiérrez-Antonio, C., & Briones-Ramírez, A. (2011). Reactive thermally coupled distillation sequences: Pareto front. *Industrial and Engineering Chemistry Research, 50*, 926–938.

Mohammad, N., Abrokwah, R. Y., Stevens-Boyd, R. G., Aravamudhan, S., & Kuila, D. (2020). Fischer-Tropsch studies in a 3D-printed stainless steel microchannel microreactor coated with cobalt-based bimetallic-MCM-41 catalysts. *Catalysis Today, 358*, 303–315.

Moreno-Gómez, A. L., Gutiérrez-Antonio, C., Gómez-Castro, F. I., & Hernández, S. (2021). Modelling, simulation and intensification of the hydroprocessing of chicken fat to produce renewable aviation fuel. *Chemical Engineering and Processing Process Intensification, 159*, 108250.

Muthia, R., Reijneveld, A. G. T., van der Ham, A. G. J., ten Kate, A. J. B., Bargeman, G., Kersten, S. R. A., et al. (2018). Novel method for mapping the applicability of reactive distillation. *Chemical Engineering and Processing Process Intensification, 128*, 263–275.

Nagy, E. (2019). Chapter 19—Membrane distillation. In E. Nagy (Ed.), *Basic equations of mass transport through a membrane layer (second edition)* (pp. 483–496). Elsevier.

Ng, D. K. S., Ng, K. S., & Ng, R. T. L. (2017). Integrated biorefineries. In *Encyclopedia of sustainable technologies* (pp. 299–314).

Nikačević, N. M., Huesman, A. E. M., Van den Hof, P. M. J., & Stankiewicz, A. I. (2012). Opportunities and challenges for process control in process intensification. *Chemical Engineering and Processing Process Intensification, 52*, 1–15.

Okhlopkova, L. B., Prosvirin, I. P., Kerzhentsev, M. A., & Ismagilov, Z. R. (2021). Capillary microreactor with PdZn/(Ti, Ce)O2 coating for selective hydrogenation of 2-methyl-3-butyn-2-ol. *Chemical Engineering and Processing Process Intensification, 159*, 108240.

Otun, K. O., Liu, X., & Hildebrandt, D. (2020). Metal-organic framework (MOF)-derived catalysts for Fischer-Tropsch synthesis: Recent progress and future perspectives. *Journal of Energy Chemistry, 51*, 230–245.

Panzone, C., Philippe, R., Chappaz, A., Fongarland, P., & Bengaouer, A. (2020). Power-to-liquid catalytic CO2 valorization into fuels and chemicals: Focus on the Fischer-Tropsch route. *Journal of CO_2 Utilization, 38*, 314–347.

Patel, B., Arcelus-Arrillaga, P., Izadpanah, A., & Hellgardt, K. (2017). Catalytic hydrotreatment of algal biocrude from fast hydrothermal liquefaction. *Renewable Energy, 101*, 1094–1101.

Pearlson, M., Wollersheim, C., & Hileman, J. (2013). A techno-economic review of hydroprocessed renewable esters and fatty acids for jet fuel production. *Biofuels, Bioproducts and Biorefining, 7*, 89–96.

PEMEX. (2021). *Precio al público de productos petrolíferos. Estadísticas Petroleras*. Petróleos Mex. https://www.pemex.com/ri/Publicaciones/IndicadoresPetroleros/epublico_esp.pdf. (Accessed 1 April 2021).

Pińkowska, H., Wolak, P., & Oliveros, E. (2014). Hydrothermolysis of rapeseed cake in subcritical water. Effect of reaction temperature and holding time on product composition. *Biomass and Bioenergy, 64*, 50–61.

Quiroz-Pérez, E., de Lira-Flores, J. A., & Gutiérrez-Antonio, C. (2019). Microreactors: Design methodologies, technology evolution, and applications to biofuels production. In D. Gruyter (Ed.), *Process intensification—Design methodologies*.

Quiroz-Pérez, E., Gutiérrez-Antonio, C., & Vázquez-Román, R. (2019). Modelling of production processes for liquid biofuels through CFD: A review of conventional and intensified technologies. *Chemical Engineering and Processing Process Intensification, 143*, 107629.

Rabaev, M., Landau, M. V., Vidruk-nehemya, R., Koukouliev, V., Zarchin, R., & Herskowitz, M. (2015). Conversion of vegetable oils on Pt/Al_2O_3/SAPO-11 to diesel and jet fuels containing aromatics. *Fuel, 161*, 287–294.

Ramírez-Márquez, C., Cabrera-Ruiz, J., Segovia-Hernández, J. G., Hernández, S., Errico, M., & Rong, B.-G. (2016). Dynamic behavior of the intensified alternative configurations for quaternary distillation. *Chemical Engineering and Processing Process Intensification, 108*, 151–163.

Ray, S., & Das, G. (2020). Chapter 15—Reactors and reactor design. In S. Ray, & G. Das (Eds.), *Process equipment and plant design* (pp. 527–559). Elsevier.

Richardson, Y., Blin, J., & Julbe, A. (2012). A short overview on purification and conditioning of syngas produced by biomass gasification: Catalytic strategies, process intensification and new concepts. *Progress in Energy and Combustion Science, 38*(6), 765–781.

Romero-Izquierdo, A. G. (2020). *Diseño, modelado y simulación de un esquema de biorefinería para el aprovechamiento integral de mezclas de materias primas renovables*. Univ. Guanajuato PhD Thesis.

Romero-Izquierdo, A. G., Gómez-Castro, F. I., Gutiérrez-Antonio, C., Hernández, S., & Errico, M. (2021). Intensification of the alcohol-to-jet process to produce renewable aviation fuel. *Chemical Engineering and Processing Process Intensification, 160*, 108270.

Rong, B.-G., & Errico, M. (2012). Synthesis of intensified simple column configurations for multicomponent distillations. *Chemical Engineering and Processing Process Intensification, 62*, 1–17.

Saeidi, S., Amin, N. A. S., & Rahimpour, M. R. (2014). Hydrogenation of CO2 to value-added products—A review and potential future developments. *Journal of CO_2 Utilization, 5*, 66–81.

Safarian, S., Unnþórsson, R., & Richter, C. (2019). A review of biomass gasification modelling. *Renewable and Sustainable Energy Reviews, 110*, 378–391.

Samih, S., Latifi, M., Farag, S., Leclerc, P., & Chaouki, J. (2018). From complex feedstocks to new processes: The role of the newly developed micro-reactors. *Chemical Engineering and Processing Process Intensification, 131*, 92–105.

Sansaniwal, S. K., Pal, K., Rosen, M. A., & Tyagi, S. K. (2017). Recent advances in the development of biomass gasification technology: A comprehensive review. *Renewable and Sustainable Energy Reviews, 72*, 363–384.

Scaldaferri, C. A., & Pasa, V. M. D. (2019a). Hydrogen-free process to convert lipids into bio-jet fuel and green diesel over niobium phosphate catalyst in one-step. *Chemical Engineering Journal, 370*, 98–109.

Scaldaferri, C. A., & Pasa, V. M. D. (2019b). Production of jet fuel and green diesel range biohydrocarbons by hydroprocessing of soybean oil over niobium phosphate catalyst. *Fuel, 245*, 458–466.

Seelam, P. K. (2013). Microreactors and membrane microreactors: Fabrication and applications. In *Handbook of membrane reactors reactor types and industrial applications*. https://www.researchgate.net/publication/257942264_Microreactors_and_membrane_microreactors_fabrication_and_applications. (Accessed 1 April 2021).

Segovia-Hernández, J. G., & Bonilla-Petriciolet, A. (2016). *Process intensification in chemical engineering: Design optimization and control*. Springer International Publishing.

Shahabuddin, M., Alam, M. T., Krishna, B. B., Bhaskar, T., & Perkins, G. (2020). A review on the production of renewable aviation fuels from the gasification of biomass and residual wastes. *Bioresource Technology, 312*, 123596.

Shin, M.-S., Park, N., Park, M.-J., Jun, K.-W., & Ha, K.-S. (2013). Computational fluid dynamics model of a modular multichannel reactor for Fischer-Tropsch synthesis: Maximum utilization of catalytic bed by microchannel heat exchangers. *Chemical Engineering Journal, 234*, 23–32.

Stankiewicz, A. I., & Moulijn, J. A. (2000). Process intensification: Transforming chemical engineering. *Chemical Engineering Progress*. https://www.aiche.org/sites/default/files/docs/news/010022_cep_stankiewicz.pdf (accessed 4.1.21).

Stephen, J. L., & Periyasamy, B. (2018). Innovative developments in biofuels production from organic waste materials: A review. *Fuel, 214*, 623–633.

Suerz, R., Eränen, K., Kumar, N., Wärnå, J., Russo, V., Peurla, M., et al. (2021). Application of microreactor technology to dehydration of bio-ethanol. *Chemical Engineering Science, 229*, 116030.

Thery, R., Meyer, X. M., Joulia, X., & Meyer, M. (2005). Preliminary design of reactive distillation columns. *Chemical Engineering Research and Design, 83*, 379–400.

Tian, Y., Demirel, S. E., Hasan, M. M. F., & Pistikopoulos, E. N. (2018). An overview of process systems engineering approaches for process intensification: State of the art. *Chemical Engineering and Processing Process Intensification, 133*, 160–210.

Troshko, A. A., & Zdravistch, F. (2009). CFD modeling of slurry bubble column reactors for Fisher-Tropsch synthesis. *Chemical Engineering Science, 64*, 892–903.

Turton, R., Bailie, R. C., Whiting, W. B., Shaeiwitz, J. A., & Bhattacharyya, D. (2012). *Analysis, synthesis, and design of chemical processes* (4th ed.).

Van Fan, Y., Jiang, P., Hemzal, M., & Klemeš, J. J. (2021). An update of COVID-19 influence on waste management. *Science of the Total Environment, 754*, 142014.

Vásquez, M. C., Silva, E. E., & Castillo, E. F. (2017). Hydrotreatment of vegetable oils: A review of the technologies and its developments for jet biofuel production. *Biomass and Bioenergy, 105*, 197–206.

Verdes, O., Sasca, V., Popa, A., Suba, M., & Borcanescu, S. (2021). Catalytic activity of heteropoly tungstate catalysts for ethanol dehydration reaction: Deactivation and regeneration. *Catalysis Today, 366*, 123–132.

Verma, D., Kumar, R., Rana, B. S., & Sinha, A. K. (2011). Aviation fuel production from lipids by a single-step route using hierarchical mesoporous zeolites. *Energy & Environmental Science, 4*, 1667–1671.

Verma, D., Rana, B. S., Kumar, R., Sibi, M. G., & Sinha, A. K. (2015). Diesel and aviation kerosene with desired aromatics from hydroprocessing of jatropha oil over hydrogenation catalysts supported on hierarchical mesoporous SAPO-11. *Applied Catalysis A: General, 490*, 108–116.

Waltermann, T., Sibbing, S., & Skiborowski, M. (2019). Optimization-based design of dividing wall columns with extended and multiple dividing walls for three- and four-product separations. *Chemical Engineering and Processing Process Intensification, 146*, 107688.

Wang, W.-C. (2016). Techno-economic analysis of a bio-refinery process for producing hydro-processed renewable jet fuel from Jatropha. *Renewable Energy, 95*, 63–73.

Wang, J., Bi, P., Zhang, Y., Xue, H., Jiang, P., Wu, X., et al. (2015). Preparation of jet fuel range hydrocarbons by catalytic transformation of bio-oil derived from fast pyrolysis of straw stalk. *Energy, 86*, 488–499.

Wang, M., Dewil, R., Maniatis, K., Wheeldon, J., Tan, T., Baeyens, J., et al. (2019). Biomass-derived aviation fuels: Challenges and perspective. *Progress in Energy and Combustion Science, 74*, 31–49.

Wang, H., Lee, S.-J., Olarte, M. V., & Zacher, A. H. (2016). Bio-oil stabilization by hydrogenation over reduced metal catalysts at low temperatures. *ACS Sustainable Chemistry & Engineering, 4*, 5533–5545.

Wang, H., Meyer, P. A., Santosa, D. M., Zhu, C., Olarte, M. V., Jones, S. B., et al. (2020). Performance and techno-economic evaluations of co-processing residual heavy fraction in bio-oil hydrotreating. *Catalysis Today, 365*, 357–364.

Wei, H., Liu, W., Chen, X., Yang, Q., Li, J., & Chen, H. (2019). Renewable bio-jet fuel production for aviation: A review. *Fuel, 254*, 115599.

Why, E. S. K., Ong, H. C., Lee, H. V., Gan, Y. Y., Chen, W.-H., & Chong, C. T. (2019). Renewable aviation fuel by advanced hydroprocessing of biomass: Challenges and perspective. *Energy Conversion and Management, 199*, 112015.

Wilhite, B. A. (2017). Unconventional microreactor designs for process intensification in the distributed reforming of hydrocarbons: A review of recent developments at Texas A&M University. *Current Opinion in Chemical Engineering, 17*, 100–107.

Xiong, Q., Hong, K., Yu, X., Li, T., & Xu, F. (2018). Editorial overview—Process design and intensification of biomass pyrolysis and gasification reactors: Experimental and modeling studies. *Chemical Engineering and Processing Process Intensification, 131*, 161–163.

Yang, R.-J., Liu, C.-C., Wang, Y.-N., Hou, H.-H., & Fu, L.-M. (2017). A comprehensive review of micro-distillation methods. *Chemical Engineering Journal, 313*, 1509–1520.

Yao, Z., Xu, X., Dong, Y., Liu, X., Yuan, B., Wang, K., et al. (2020). Kinetics on thermal dissociation and oligomerization of dicyclopentadiene in a high temperature & pressure microreactor. *Chemical Engineering Science, 228*, 115892.

You, S., Yin, Q., Zhang, J., Zhang, C., Qi, W., Gao, L., et al. (2017). Utilization of biodiesel by-product as substrate for high-production of β-farnesene via relatively balanced mevalonate pathway in Escherichia coli. *Bioresource Technology, 243*, 228–236.

Yu, J., Guo, Q., Gong, Y., Ding, L., Wang, J., & Yu, G. (2021). A review of the effects of alkali and alkaline earth metal species on biomass gasification. *Fuel Processing Technology, 214*, 106723.

Zhang, W. (2009). A review of techniques for the process intensification of fluidized bed reactors. *Chinese Journal of Chemical Engineering, 17*, 688–702.

CHAPTER 5

Opportunities in the intensification of the production of biofuels for the generation of electrical and thermal energy

Noemí Hernández-Neri[a], Julio Armando de Lira-Flores[b], Araceli Guadalupe Romero-Izquierdo[a], Juan Fernando García-Trejo[a], and Claudia Gutiérrez-Antonio[a]

[a]Facultad de Ingeniería, Universidad Autónoma de Querétaro, Amazcala, Querétaro, Mexico
[b]Facultad de Química, Universidad Autónoma de Querétaro, Centro Universitario, Querétaro, Mexico

5.1 Introduction

The energy demand forecasted for 2040 was 25% higher regarding to consumption levels for 2019, accompanied with a population growth of 1.7 billion for reaching 9.2 billion in 2040 (BP Energy Outlook, 2019). At the same time, this historic growth of energy demand has been sustained for the use of nonrenewable fuels, which have generated the climate change problem. Thus, in order to achieve the sustainability in the energy production, several alternative and renewable energy sources were proposed and studied (Fagbohungbe, Komolafe, & Okere, 2019; Guo, Liu, Sun, & Jin, 2018; Krishnan & McCalley, 2016; Oumer, Hasan, Baheta, Mamat, & Abdullah, 2018; Pfeifer, Dobravec, Pavlinek, Krajačić, & Duić, 2018; Powell, Rashid, Ellingwood, Tuttle, & Iverson, 2017; Shuba & Kifle, 2018; Wei et al., 2019; Won, Kwon, Han, & Kim, 2017). However, due to the pandemic caused by COVID-19, in 2020 the global economies collapsed, alongside the health systems and the societies ordinary life style; likewise, the energetic sector was harshly affected. According to the International Energy Agency (IEA, 2020), for the countries in full lockdown, their energy demand declined by 25% per week, while the countries in partial lockdown declined by 18%; globally, the energy demand declined 3.8% in the first quarter of 2020. Regarding to fossil oil demand, it fell roughly 5% in the first quarter of 2020, due to containment measures; the road mobility and aviation were 50% and 60% below in relation to the 2019 behavior. This same effect has been experienced by the power sector, which has diminished its demand until 20% or more, due to reductions in commercial and industrial operations. Despite this, the renewable energies have been the only source with demand growth, boosted by their resilience and low impact. In this context, the International Energy Agency (IEA,

2021) has identified three areas related to the energy systems and clean energy transitions, as consequence of the pandemic:
(1) the energy security is the cornerstone of the economies.
(2) the electricity security and resilient energy systems are indispensable for modern societies.
(3) the clean energy transitions must be at center of the economic recovery plans.

Thereby, the renewables energies are a solid strategy to achieve the sustainable recovery. There are several sources of renewables energies, such as the solar, wind, tidal, geothermal, and bioenergy. In particular, the bioenergy proceeds from biomass, and it has gained interest due to its wide availability alongside the earth. The biomass is defined as the organic matter derived from plants and available on a renewable basis, including edible, nonedible energy crops, and waste organic material from harvesting or processing agricultural products, including animal waste and rendered animal fat, forestry products, wood waste, and sewage (Basu, 2013). Due to its broad variety, there are several conversion processes to obtain bioenergy; nevertheless, it is important to emphasize that the edible organic matter is not considered into the bioenergy production, in order to ensure the food safety. The biomass can be classified considering its chemical nature as triglyceride, lignocellulosic, sugar, and starchy materials (Maity, 2015). The triglyceride material involves oil and fats and its wastes, while the lignocellulosic feedstock includes wood, husk, straws, kernels, among others. Finally, into the sugar biomass is considered sugarcane and sugar beet, while into the starchy materials are included corn and wheat (Gutiérrez-Antonio, Romero-Izquierdo, Gómez-Castro, & Hernández, 2021). These biomasses can be converted in biofuels, which can be found as liquid, gaseous, and solid. As part of liquid biofuels are included the bioethanol, biobutanol, biodiesel, biojet fuel and green diesel, among others; into the gaseous biofuels, the syngas, biogas, biomethane, and biohydrogen are considered the most common. Finally, into the solid biofuels are founded the biochar from pyrolysis and fuel pellets. Biofuels as bioethanol, biobutanol, biodiesel, green diesel, and inclusive biogas are used as substitutes, additives, or mixed with road vehicles fuel; while biomethane and biogas are the renewable counterpart of natural gas. Syngas is commonly used to produce electricity, and sometimes the biogas is used for the same goal. On the other hand, biohydrogen is the substitute of hydrogen obtained from petrochemical plants; the biochar and fuel pellets are involved into the heating activities, as a replace of mineral carbon and nonsustainable firewood. Overall, the biofuels have been broadly studied. In accordance to the scientific browser "Sciencedirect," in the period 2016–21, with the words "(biofuel name) production process" and "(biofuel name) production," the number of research articles for common biofuels is presented in Fig. 5.1. As it can be seen, biogas, biodiesel, and fuel pellets have the majority of research papers, while biojet fuel, biobutanol, and biohydrogen have the minority.

Although the number of research papers is extensive, there are few papers focused to the process intensification applied on the biofuel production. The process intensification

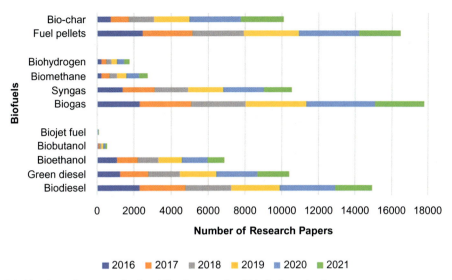

Fig. 5.1 Number of papers obtained for the search of biofuel production processes.

is a chemical engineering tool related to the development of production technologies and equipment in order to substantially enhance the manufacturing and processing. This tool involves decreasing the unit operation number, the equipment size, the energy consumption and waste production, and improving the process economy and the production-capacity ratio (Gutiérrez-Antonio, Romero-Izquierdo, et al., 2021). Consequently, its application on biofuel production allows improving its market competitivity, its energy consumption, and thus, its environmental impact. Fig. 5.2 presents the research paper search related to process intensification on the biofuel production, using "Sciencedirect" as browser with the keywords "(name of biofuel) process intensification" and "(name of biofuel) intensification".

As it can be seen in Fig. 5.2, the biogas, biodiesel, and fuel pellets have the highest number of research papers related to intensification, and biojet fuel, biobutanol, and biohydrogen have the lowest number. However, the majority of these papers about the biogas and fuel pellets are focused on the improvement through different ways to use or convert their bioenergy; thus the number of research papers really involved with the application of process intensification on the production processes is reduced. In the literature, we can found several reviews related to the application of process intensification for the production of liquid fuels (Quiroz-Pérez, Gutiérrez-Antonio, & Vázquez-Román, 2019); for instance, biodiesel (Natarajan et al., 2019; Wong, Ng, Chong, Lam, & Chong, 2019), bioethanol (Khalid et al., 2019; Ko, Lee, Jung, & Lee, 2020), biobutanol (Morone & Pandey, 2014; Rochón et al., 2020), biogasoline (Hassan, Sani, Abdul Aziz, Sulaiman, & Daud, 2015), green diesel (Ameen, Azizan, Yusup, Ramli, & Yasir, 2017), and biojet fuel (Gutiérrez-Antonio et al., 2021;

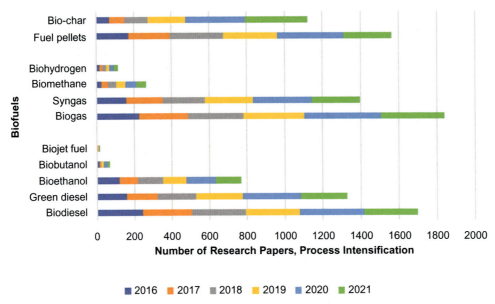

Fig. 5.2 Number of papers obtained for the search of biofuel-intensified production processes.

Gutiérrez-Antonio, Gómez-Castro, de Lira-Flores, & Hernández, 2017). Nevertheless, for the author's knowledge, a compilation of the application of the intensification strategies for the production processes of biofuels that can be used for the generation of thermal and electrical energy is still missing; these biofuels are biogas, renewable hydrogen, and fuel pellets. Thus it is important to identify the advances regard the intensification of biofuels production processes, as well as the opportunity areas.

Therefore in this work, an extensively research about the process intensification applied on the production of biogas, renewable hydrogen, and fuel pellets is presented. The chapter is organized as follows. In Section 5.2, the conventional production processes for biogas, renewable hydrogen, and fuel pellets are presented, while in Section 5.3, the proposal of process intensification on these biofuels are discussed, according to the literature revision. Then, in Section 5.4 a case of study about the process intensification on the production of fuel pellets is presented. Finally, in Sections 5.5 and 5.6 the future trends as well as the concluding remarks derived of this investigation are included.

5.2 Conventional production processes

In this section, the definition, properties, and conventional processes involved in the production of biogas, biohydrogen, and fuel pellets reported in the literature are presented.

5.2.1 Biogas

Biogas is a gaseous mixture of methane, carbon dioxide, hydrogen, carbon monoxide, hydrogen sulfide, ammonia, and oxygen traces (Matheri, Ndiweni, Belaid, Muzenda, & Hubert, 2017); the standard composition is around 50%–75% methane and 25%–50% carbon dioxide (Kainthola, Kalamdhad, & Goud, 2019). This gaseous mixture is used as fuel for electricity or heating, and contributes to reduce greenhouse gases emissions, the eutrophication, as well as the depletion of dissolved oxygen, among others (Kainthola et al., 2019). Biogas can be produced from the organic fraction of any material, mainly organic wastes. These feedstocks are easily digestible for the microbiological community (distinct bacterial and archaeal consortia), called inoculum, responsible of its transformation into biogas (Patinvoh, Osadolor, Chandolias, Sárvári Horváth, & Taherzadeh, 2017). The quality and quantity of the biogas is commonly represented by the methane content, and this last one strongly depends on the feedstock composition and concentration, for instance, the carbon-nitrogen ratio or the presence of micro and macro elements as nutrients (Tabatabaei et al., 2020). The biogas production is carried out in series of four biochemical fundamentals steps in the absence of oxygen: hydrolysis, acidogenesis, acetogenesis, and methanogenesis, called the overall process anaerobic digestion. In the hydrolysis stage, large polymers such as proteins, carbohydrates, and lipids are enzymatically degraded into their respective monomers by anaerobic bacterial consortium, such as *Cellulomonas*, *Clostridium*, *Bacillus*, *Thermomonospora*, *Ruminococcus* *Bacteroides*, *Erwinia*, *Acetovibrio*, *Microbispora*, and *Streptomyces*. On the other hand, the acidogenesis step uses acidogenic bacteria such as *Streptococcus*, *Lactobacillus*, *Bacillus*, among others to convert the monomers into organic acids, for instance acetic acid, alcohols, propionic, and butyric acids. Then, in the acetogenic step, these acids are transformed into hydrogen, carbon dioxide, and acetic acid. In this step, the responsible bacteria are *Clostridium*, *Acetobacterium*, *Syntrophomonas*, *Thermosyntropha*, among others. Finally, in the methanogenic step, methane is obtained from the products of the last stage, by reduction of carbon dioxide with hydrogen or cleaving acetic acid (Kainthola et al., 2019; Matheri et al., 2017; Tabatabaei et al., 2020). The anaerobic digestion process has as key operating parameters:

- The temperature: psychrophilic condition <10°C, mesophilic condition, 20°C–45°C, and thermophilic condition 50°C–55°C.
- The pH: hydrolysis: 5.5–6; acidogenesis: 6–7; acetogenesis: 6–7; methanogenesis: 6.5–7.5
- The hydraulic retention time (HRT, time required to degrade completely the organic material): commonly 15–30 days under mesophilic conditions
- The organic load rate (OLR, this value represents the amount of volatile solids (VSs) fed into the reactor per day)
- The carbon to nitrogen ratio (C/N reflects the nutrients level of raw materials): optimal production 25 ± 5.

In addition, the control of nutrients and minerals are important to avoid inhibitors, when inorganic elements are insufficient and the macronutrients are excessive (Gonçalves Neto, Vidal Ozorio, Campos de Abreu, Ferreira dos Santos, & Pradelle, 2021; Matheri et al., 2017). There are several studies about the anaerobic digestion of a broad variety of feedstocks, proving different operating conditions (since inoculum type until the digestion reactor) in order to improve the yield and quality of biogas obtained. In 2017, Huang et al. (2017) studied the biogas production using various lignocellulosic hydrolysates, such as bagasse, rice straw, and corncob; it was used a semipilot scale reactor, with the objective of increasing the methane yield and improving the instability of a conventional lignocellulosic system. According to the results, up to $0.549\,m^3$ of biogas with $0.381\,m^3$ of methane, per kg of raw material, were obtained when the system was stable, converting more than 85% of feedstock. These results are important to demonstrate that different kinds of lignocellulosic substrates allow to obtain biogas with high quality and quantity, and also with a great potential of the industrialization. In the same year, Onthong and Juntarachat (2017) carried out the experimental evaluation of biogas production from soybean residues, papaya peels, sugarcane bagasses, rice straws, and galangals, using separately a batch and continuous reactor. Based on their results, the soybean residue generates the highest biogas yield of 560.47 mL at 25 days of HRT; likewise, this same residue produces 63.01 L of biogas per day into the 200 L continuous reactor between 60 days. However, in relation the methane yield, the galangal waste produces 73.50%, the highest yield. Then, Ning et al. (2019) proved the anaerobic codigestion to improve the associate problems with the low biogas yield from agricultural wastes, using a mixture of pig manure and corn straw for simultaneous biogas and biogas slurry production. Also, the effect of C/N ratio, OLR, and total solid (TS) content was experimentally studied. They found that the highest biogas production was obtained at a C/N ratio of 25 or 35 for biogas slurry performance, with five VSs per liter per day of OLR and 20% of TS. On the other hand, Adekunle et al. (2019) investigated the biogas production using cow dung, jatropha fruit exocarp, cattle dung, and the mixture of those previously mentioned; during 5 weeks, 50 mL of inoculum were added to compensate the dead or weak bacteria. According to the results, the raw material mixture reached the high volume of biogas collected of 586 mL, while 77 mL were obtained for cow dung, as the lowest value. Also, for the mixture on the fifth week was obtained six times emission of total produced biogas, it means 350 mL of biogas per kg of mixture on the fifth week. An interesting review about the biogas production from vinasse was presented by Parsaee, Kiani Deh Kiani, and Karimi (2019). Vinasse is a waste in the bioethanol production industry from sugarcane, and it is considered as a strong source of contamination; however, it is one of the sources with high potential for biogas production when it is complemented with substances rich in macronutrients and micronutrients, for instance animal manure, organic industrial waste, and lime fertilizers. Their research revealed that 22.4 GL of vinasse are produced in the world, having a potential to produce until 407.68 GL of biogas. In 2020, Abraham et al. (2020) presented a work compilation about the inclusion of pretreatment stage in the anaerobic digestion of lignocellulosic biomass, which allow to enhance the biogas yield promoting the lignin removal and the destruction of their complex structures. The conclusion of that study is that the pretreatment methods can be grouped in physical, chemical, physicochemical, and biological, and they can be used based on the type of the lignocellulosic biomass; in addition, it is possible to reach until 1200% in the biogas production when ionic liquids are used as pretreatment. Then, Vijin Prabhu, Sivaram, Prabhu, and Sundaramahalingam (2021) reported a comprehensive overview on the enhancement of biogas production from various organic wastes using techniques as codigestion, two-phase digestion, recirculation of slurry, thermal, alkaline, acid, ultrasound, hydrothermal, and milling pretreatments; all focused on improving the biogas production. According to this study, the codigestion greatly improves the biogas production, regarding the other strategies for cow manure and food wastes, while any pretreatment method allows to increase the biogas yield, for instance, the hydrothermal method improves the biogas yield almost twice, regarding the conventional digestion. At the end of 2021, Singh, Hariteja, Sharma, Raju, and Prasad (2021) published a report about the codigestion of human feces with poultry litter and cow dung into a laboratory scale digester, during 52 days at room temperature. Their results indicated that the codigestion of human feces with cow dung allows to obtain the better gas production. In that study the substrate proportion and pH were varied. Other works about the innovative strategies to enhance the biogas production from agricultural residues, food waste, and microalgal biomass can be found in Abdul Aziz, Hanafiah, and Mohamed Ali (2019), Yadav and Vivekanand (2021), Shamurad et al. (2020), and Oleszek and Krzemińska (2021), respectively.

From the revision of the literature, it can be noted that the use of lignocellulosic wastes has been widely studied for the production of biogas, specially its mixtures. The results are promising, since the yields can be duplicated with the new coprocessing technologies.

5.2.2 Renewable hydrogen

The biohydrogen, or renewable hydrogen, is the low carbon hydrogen obtained from renewable sources such as the biomass (Saratale, Saratale, & Chang, 2013). This is a clean biofuel that can be generated with low greenhouse gases emissions, and it is considered a great source of heating and transport fuel due to its high energy content (Materazzi, Taylor, & Cairns-Terry, 2019; Saratale et al., 2013). The energy content of hydrogen is 120 MJ/kg, which is higher than other energy sources, for instance, liquefied natural gas with 54.4 MJ/kg, ethanol with 29.6 MJ/kg, or methane with 50 MJ/kg (Tian et al., 2019). There are three conventional pathways to produce biohydrogen (Materazzi et al., 2019; Tian et al., 2019):

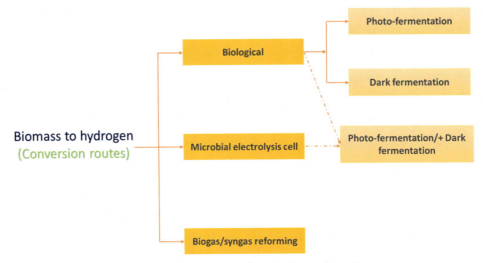

Fig. 5.3 Conventional processing pathways to produce hydrogen from biomass.

(1) biological, in which are included the photo and dark fermentation.
(2) microbial electrolysis cell (MEC), which uses power obtained from renewable sources, such as wind and solar.
(3) biogas/syngas reforming, wherein the gaseous raw material is commonly produced from anaerobic digestion or biomass gasification.

Fig. 5.3 presents the conventional routes to produce biohydrogen. The biological processes have low yield of biohydrogen, due to their poor efficiency on direct microbial assimilation of biomass (Kumar & Chowdhary, 2016; Saratale et al., 2013). In the photo-fermentation, biohydrogen is produced through photosynthetic bacteria, in the presence of light (solar or artificial) and carbon source (simple sugar or volatile fatty acids), and the absence of oxygen and nitrogen (Tian et al., 2019); however is slower than dark fermentation.

A recent report about photo fermentation was published by Lu et al. (2020), wherein the biohydrogen production was realized from alfalfa, using active/passive saccharification plus photo fermentation (by photosynthetic bacteria called HAU-M1). Some variables were studied to improve the biohydrogen yield, such as initial pH, substrate concentration, and cellulase loading. According to the results, 55.8 mL per g alfalfa were obtained as the highest yield, using passive saccharification and photo fermentation. Usually, the photo-fermentation can be used with dark fermentation to increase the biohydrogen yield (Chen, Yang, Yeh, Liu, & Chang, 2008). On the other hand, in the dark fermentation, the microorganisms are responsible to produce hydrogen under anaerobic and dark condition. The hydrogen formation is a consequence of the protons reduction by the electrons generated during the degradation of the carbon source (Materazzi et al.,

2019; Tian et al., 2019). The dark fermentation of pure carbohydrate, for instance the glucose and starch, is an efficient hydrogen source; however, these pure compounds are so expensive, thus the use of organic waste materials rich in carbohydrates are a cheap, accessible, and excellent source for hydrogen production (Tian et al., 2019). In 2019, Lu et al. (2019) studied the biohydrogen production in a $3\,m^3$ pilot scale fermented using glucose through dark fermentation, to determine the effects of hydraulic retention time ad substrate concentration. The optimal hydraulic retention time was 24 h, being concluded that the hydrogen production rate increased with the substrate concentration increase. Other work about dark fermentation was published by Varanasi and Das (2020). In their work, it was developed a single-chamber MEC to develop the electrohydrogenesis, after dark fermentation. This two-stage arrangement improves the overall hydrogen yield, reaching until 67.69 L hydrogen per kg consumed organic matter and 46% energy recovery. Then, in 2021, Rambabu et al. (2021) presented the augmented biohydrogen production by dark fermentation assisted for nanometal oxides (NiO and CoO), using rice mill wastewater as raw material. Based on the results, the hydrogen yield improved until 109% for NiO and 90% for CoO. The feedstock reduction efficiency was 77.6% and 69.5% for NiO and CoO addition, respectively, regarding 57.5% for conventional processing. In the same year, Jamali, Md Jahim, Mumtaz, and Abdul (2021) presented an experimental work about the biohydrogen production from palm oil mill effluent as a carbon source through a dark fermentation, investigating the performance of a specific immobilized bacterial consortium (*Caldicellulosiruptor saccharolyticus*) on granular activated carbon. In their study, the performance of consortium inside the robust environment was tested, reaching until 38.8 mL hydrogen per liter, with 3 mol hydrogen per mol sugar as maximum hydrogen yield. Other novel reports about dark fermentation are presented by Kovalev, Kovalev, Litti, and Katraeva (2020), Akhbari, Chuen, and Ibrahim (2021), and Panin, Setthapun, Elizabeth Sinsuw, Sintuya, and Chu (2021). More information about dark fermentation and its main involved variables can be found in Preethi, Usman, Rajesh Banu, Gunasekaran, and Kumar (2019).

The MEC is a system that combines the production of electricity through electrochemically active bacteria, which fed from waste organic and mineral nutrients (Florio et al., 2019). The MEC can include two electrodes: anode and cathode collocated in the same chamber (single-chamber MEC) or separated in two chambers (two-chamber MEC). With the two-chamber MEC, high purity of hydrogen is obtained and the interference between two electrodes is diminished. In this scheme, membranes are used to separate the two chambers, which can be proton or cation/anion exchange membrane, charge-mosaic or bipolar membranes. It is important to mention that during the process, there is a potential loss of the membrane, and due to its high cost, the single-chamber MEC is preferred. However, both schemes have the same operating principle; the organic material into the anaerobic system (sensitive to oxygen) is oxidized in the anode to produce electrons, which are transported to the anode, and consequently to the cathode, being accepted by the protons in the cathode in order to produce hydrogen (Habashy, Ong, Abdeldayem, Al-Sakkari, & Rene, 2021; Tian et al., 2019). Florio et al. (2019) studied a two-step process to produce hydrogen, using an MEC system as pretreatment with anaerobic digestion from municipal organic waste. For the MEC system, a single chamber, air cathode, and membranes with graphite plates as electrodes were used. The substrate from MEC was tested in a batch experiment for the production of biohydrogen and biomethane trough anaerobic digestion. According to the results, the biohydrogen production was 276%, regarding to that produce from the anaerobic digestion using the same fresh untreated municipal organic waste. Then, in 2021, Keruthiga, Mohamed, Gandhi, and Muthukumar (2021) investigated the biohydrogen production from rice mill wastewater using an artificial photoassisted MEC with an inexpensive anode prepared from carbonaceous material, recovered of sugar industry. Based on their experiments, the hydrogen production on the fifth fermentation day was 220 mL, having a production rate of 3.6 mL per liter of waste material per hour. These results were obtained at pH of 6 and acid concentration of 1.5%, which are the main variables affecting the processing. In the same year, Ndayisenga et al. (2021) reviewed the all possible pretreatment methods of the lignocellulosic agricultural residues, operational parameters, and the main drawbacks for the integrated system MEC-dark fermentation. They concluded that the MEC system significantly upgrades hydrogen production in a dark fermentation reactor by promoting a deep decomposition of the lignocellulosic material. An interesting review about the biohydrogen production using photofermentation and dark fermentation, their main microbial mediators, operational variables, and conditions from food waste can be founded in Habashy et al. (2021).

Finally, biohydrogen can be produced also through the biogas or syngas reforming. The reforming technology uses steam, at high temperature and mild pressure, to react with methane, from biogas or CO from syngas, to obtain hydrogen and carbon dioxide (Tian et al., 2019). The steam reforming includes the water-gas-shift reaction, which involves a high-temperature step, in order to accelerate the reaction, and a low temperature step, with the objective of achieve high conversion rate. In the water-gas-shift reaction, the catalyst is usually required to promote the hydrogen production (Materazzi et al., 2019). It is important to mention that the reformation of methane from biogas is a usually carry on when the biogas has high methane content, due to its similarity to gas natural reforming; likewise, in the case of the reforming of syngas from gasification, the production of hydrogen can be oriented into the processing, varying the reaction temperature, gasification agent, catalyst, feedstock composition, and properties, among others (Materazzi et al., 2019; Tian et al., 2019). Some works about the biogas and syngas reforming to produce biohydrogen are presented next.

In 2018 Gao et al. presented a deep investigation about the catalyst, the optimization of operation conditions, the influence of biogas impurities, and reactor type into the biohydrogen production by gas reforming. They found that the process is highly affected by temperature, pressure, calcination and reduction conditions, gas hourly space velocity, and reactor type, such as conventional fixed bed and a fluidized bed reactor, or novel as membrane reactors, microreactors, or solar thermal flow reactors. Then, Rosha, Mohapatra, Mahla, and Dhir (2019) studied the catalytic reforming of biogas for biohydrogen production, using Ni supported in $ZnO-CeO_2$ as a catalytic system. In their study, the hydrogen production with respect to various performance parameters was tested, such as the temperature. Results showed that when the temperature is increased between 650°C and 900°C, the reactant conversion and the product yield also increased, reaching until 78.5% and 35.7%, respectively. Also, the carbon deposition rate in a mixed support catalyst is less in comparison with the single support (Ni supported on ZnO). In 2021, García, Gil, Rubiera, Chen, and Pevida (2021) investigated thermodynamically and experimentally the hydrogen production from biogas by sorption enhanced steam reforming into a fluidized bed reactor. This system combines the catalytic reforming reaction of biogas using Pd/Ni-Co hydrotalcite-like material, with simultaneous CO_2 removal by dolomite as a CO_2 sorbent, in a single step. Some variables were tested on the process performance, such as temperature, steam-methane molar ratio, and gas hourly space velocity. Based to their results, 98.4% vol of hydrogen purity and 92.7% of hydrogen yield were reached at 550°C–600°C range. Recent advances on catalyst development for methane (biogas) reforming into syngas have been presented in Abdulrasheed et al. (2019), including reaction pathways, catalytic activities, metal/support deactivation, and the interactions of them are reviewed. On the other hand, regarding to the biohydrogen production from syngas, in 2019 Materazzi et al. collect the available work to design a commercial waste to hydrogen plant, involving an assessment of future hydrogen markets, the identification of an appropriate scale for the plants, and the development of specifications for process design and output streams. A thermodynamic study to investigate the feasibility of syngas biohydrogen production from palm oil mill effluent using steam reforming was presented by Cheng et al. (2019). In their study, it was tested the effect of reaction temperature on hydrogen yield, through the minimization of the total Gibbs free energy and experimental information. They found that the steam reforming of the organic effluent is viable at 773 K as minimum temperature; also, it is obtained a hydrogen-carbon monoxide ratio between 25 and 3457, higher hydrogen production, compared with other studies. More information about the biohydrogen production from syngas reforming is presented by Materazzi et al. (2019).

In the production of renewable hydrogen, the electro-based technologies are widely used; however, the yields are still low for its industrial scale implementation. Similar to biogas production, the focus is given to the use of lignocellulosic wastes mainly. In spite of hydrogen can be produced from syngas, the production costs could be not competitive.

5.2.3 Fuel pellets

The use of biomass as an energetic source has some drawbacks, such as low bulk density, the seasonality of supply, and scattered distribution into a geographic region, as well as the high cost associated with handling, transportation, and storage (Méndez-Vázquez et al., 2016). Thus the densification of biomass as fuel pellets or briquettes, with regular geometry and standard size and weight, allows compact storage, improving handling, and best feeding in large-scale unit operations (Pradhan, Mahajani, & Arora, 2018). It is important to mention that the main difference between fuel pellets and briquettes is their dimensions as well as the final application. Fuel pellets and briquettes are cylindrical solids fuels, but the briquettes have 50–90 mm of diameter and 75–300 mm length, while the pellets have 6–8 mm of diameter and less than 40 mm of length. Due to its size, briquettes are applied on large and medium scale thermal industry, while pellets are used into smaller equipment, usually for residential stove, gasifier, among others (Pradhan, Mahajani, & Arora, 2018). These renewable fuels allows high energy conversion efficiency, generating less particles emissions, respect to the original biomass state, or other conventional biomass fuels (Méndez-Vázquez et al., 2016). Regarding the feedstock to produce pellets, woody waste biomass is the most common, but other agricultural and forest wastes, such as energy crops, husk, straws, leaves, and so on, are used too. The conventional processing to convert the low-density biomass into densified pellets involves three steps: (1) pretreatment (s), (2) pelletization, and (3) conditioning of fuel pellets (Pradhan, Arora, & Mahajani, 2018; Kato, 2021). It is important to point out that the collection and storage of biomass are the main bottlenecks in the pellets commercialization, due to its seasonality supply, the scattered distribution of biomass, and the scarce supply chains for this biofuel, mainly in countries highly dependent on crude oil (Méndez-Vázquez et al., 2016). On the other hand, regarding to pretreatment stage, it varies on the initial biomass characteristics. According to Pradhan, Arora, and Mahajani (2018), the biomass pretreatment improves bulk density, energy density, storage, and handling of biomass. Into the pretreatment methods are included size reduction, drying, torrefaction, steam explosion, hydrothermal carbonization, and biological treatments. Commonly, the fuel pellets have less than 10% of moisture content; thus an efficient drying stage is important to reach high energy conversion efficiency (Kato, 2021). A detailed information about the pretreatment methods into the fuel pellets production can be found in Pradhan, Arora, and Mahajani (2018) and Kato (2021).

Respect to the pelletization stage, it consists of a perforated hard steel die with a couple of rollers; when the die and rollers are rotating, the raw material is forced alongside the perforations to form densified pellets. Into the pelletization process, the main variables are feedstock composition, moisture biomass content, particle size, presence of binders, pressure machine gap, die diameter, channel length, die speed, among others (Méndez-Vázquez et al., 2016; Kato, 2021 ; Pradhan, Arora, & Mahajani, 2018). Finally, after the pelletization stage, the typical conditioning of pellets involves the pellet cooler in order to achieve an appropriate temperature for the bagging and storage. Some works about the pellets production have been published. Ståhl and Berghel (2011) investigated the pilot-scale production of pellets from mixtures of sawdust and rapeseed cake. During the experimentation, the load current, the die pressure, and the die temperature were measured, in order to examine how the mixture affected the energy consumption of the pelletizing machine and also the mechanical properties of pellets obtained. They found that the energy consumption decreased and the fines amount increased with high proportion of rapeseed cake into the raw material mixture; however, the pellet durability increases with this material; thus, the production of pellets from this mixture must be at an equilibrium point between the energy consumption and the pellet durability. Then, Méndez-Vázquez et al. (2016) presented a mathematical programming model to determinate the optimal locations of the pellet production plants, the collection centers, together with the optimal distribution logistics in order to have a truly environmentally friendly fuel, minimizing the global environmental impact associated with the supply chain and the production process into the pellet generation. This multiobjective optimization problem includes the minimization of total annual cost and carbon dioxide emissions for the whole supply chain. It is important to mention that the optimal supply chain for the fuel pellets allows to overcome the challenges involved into its production and distribution. In 2018, Pradhan et al. investigated the pilot scale assessment of the fuel pellet production from garden waste biomass without an additional binder. The main parameters affecting the pelletization and the pellet quality were feedstock moisture content, milling size, and die size; in addition, the authors developed appropriate regression models for each quality attribute. According to the results, the average durability value decreased of 95% until 92.5, using garden wastes with moisture content of 5% and 15%, respectively. Also, in relation with a combustion test, the garden waste pellets may be used in a residential cookstove. Then, Ríos-Badrán, Luzardo-Ocampo, García-Trejo, Santos-Cruz, and Gutiérrez-Antonio (2020) presented the production and characterization of pellets from rice husk, wheat straws, and their mixtures. The characterization procedure was aligned with the ISO-17225-6 standard, which stablishes the main parameters for high-quality pellets from biomass. Results indicated that the mixture of both biomasses exhibits the highest calorific value (until 4573.5 kcal/kg), regarding the individual biomasses, having a reduced ash amount up 13.06%. Nevertheless, the pellets from both biomass mixture exceed the moisture, ashes, and nitrogen content established in the normativity, but its diameter, length, and durability are fulfilled. In 2021, Pradhan, Mahajani, and Arora (2021) investigated the effect of biomass milling size on the pelletization process and pellet quality using garden wastes (leaves). In their study, three milling sizes were used (<25.4, 25.4, 6.25 mm) with a moisture content of 10% and a die size of 15 mm for each case. Based on the results, an increased throughput capacity of 29.5 kg/h until 60 kg/h was observed for the smaller and greater particle sizes, respectively. Also, the specific energy consumption value was reduced from 141.2 to 100.2 kWh/ton by reducing biomass milling size, for the intermediate and bigger particle sizes. Thereby, the production cost was reduced by 23% for the raw material with higher particle size. Other technical aspects about the pellet production from any material are presented in Kato (2021).

As can be seen, the production of pellets has captured the attention of the scientific community in the last years. The emphasis has been given in the use of residues and mixtures of them in order to fulfill the ISO standard. In this biofuel, the use of binder is another aspect to be considered, since usually is water.

5.3 Proposals for the intensification of the production processes

In this section, the proposal on process intensification strategies applied on the production of biogas, biohydrogen, and fuel pellets is presented.

5.3.1 Biogas

In the biogas production, there are a lot of reports to enhance the anaerobic digestion performance, focused on optimization of operating parameters, pretreatment methods, additives addition, codigestion, reactor design, genetic technology to improve the bacterial consortium, among others (Zhang, Loh, & Zhang, 2019). To the author's knowledge, the application of process intensification to improve the processing is scarce; next, it is provided a revision of works about the process intensification or intensification into the biogas production. Patil, Gogate, Csoka, Dregelyi-Kiss, and Horvath (2016) investigated the effect of the hydrodynamic cavitation as a pretreatment method for the biogas production form wheat straw. They proved three speeds rotor, raw material-water ratios, and treatment times. The results indicated that the methane yield obtained by using the pretreatment increased twice, regarding to the process without pretreatment; thus the biogas production was intensified. In 2020, Pessoa, Sobrinho, and Kraume (2020) presented a report about the use of a sonication pretreatment alongside an induced magnetic field to an enzyme-substrate complex to improve the biogas production from sugar beet pulp. As result, the methane production increased 62% with the magnetic field, reaching until 79% with the sonication plus magnetic field. Other reports about the application of pretreatment stages to intensify the biogas production are from Lizama, Figueiras, Gaviria, Pedreguera, and Ruiz Espinoza (2019), You, Pan, Sun, Kim, and Chiang (2019), and Joshi and Gogate (2019). An interesting review about the application of nanoparticles of zero-valent iron, metallic and metal oxides, and carbon bases to enhance the biodigestion of organic wastes was published by Dehhaghi, Tabatabaei, Aghbashlo, Kazemi Shariat Panahi, and Nizami (2019). They concluded that the zero-valent iron nanoparticles are the most promising for increase the biogas yield, due to the stabilizer effect and improving the microorganism growth into the anaerobic digestion. Hassaneen et al. (2020) studied two types of nanocomposite formulations: zinc ferrite with 10% carbon nanotubes and zinc ferrite with 10% C76 fullerene, in order to boost the biogas production from cattle manure. Their experimental work was developed into a lab-scale biodigesters containing organic slurry, operating by 50 days. Their results showed that the zinc ferrite with 10% carbon nanotubes improved the methane yield by 183%, regarding the conventional process. On the other hand, regarding to the reactor design for anaerobic digestion, Gaballah, Abdelkader, Luo, Yuan, and El-Fatah Abomohra (2020) reported a solar heating system for household biogas digesters to improve the digestion of cattle manure in cold winter. The experimental reactor used is a low-cost tubular digester, which was heated by two modes: using the solar greenhouse integrated with a solar water heating system and a capillary heat exchanger, or heating using only solar greenhouse. According to their results, the temperature increased 9.5°C and 4.9°C, and achieving until 247 and 181 L biogas per kg VSs for proposed designs, respectively. Other novel solar design was reported by Zaied et al. (2020) for the production of biogas from a mixture of palm oil mil effluent and cattle manure; the biogas yield was improved by the used a semicontinuously solar-assisted bioreactor and ammonium bicarbonate to provide the nitrogenous substrate and buffering potential. In their system, the solar radiation was collected by a solar panel and converted into electricity, which is used to heat the substrate. The results indicated that 29.80% and 42.30% more of cumulative biogas and methane production are obtained, regarding the conventional process. Then, in 2021, Duarte et al. (2021) presented the biogas recirculation in a biogas-lift reactor in order to rise the waste frying oil degradation. For their study, two bioreactors were used: a biogas-lift bioreactor with gas and liquid recirculation and a control reactor with liquid recirculation, adding to waste oil glycerol and volatile fatty acids. They obtained that 79% of methane content was obtained for first design, while 67% for the second one. Thus the biogas recirculation facilitates the waste oil degradation. In this same year, a novel two-layered reactor to produce biogas from oil-palm empty fruit bunches and palm oil mill effluents was reported by Wadchasit, Suksong, O-Thong, and Nuithitikul (2021). In their system, excess heat from the outer liquid-state anaerobic digestion of first biomass is transferred to the inner solid-state anaerobic digestion of second biomass for promoting the digestion. The temperature range used was 55°C–70°C, identifying bacteria and archaea consortium. They observed that the system increases the methane yield by 6.4%. Also, if is added a NaOH pretreatment on first biomass, the methane yield increases by 14.1%. An extensive review about the recent innovations in biogas production, since reactor designs alongside operative ways and conditions, is provided by Tabatabaei et al. (2020).

It can be observed that the use of nanoparticles, hydrodynamic cavitation, and solar bioreactors are the process intensification approaches that has been more explored in the literature. In particular, the use of solar bioreactor could help to reduce operating costs and make more viable from the economic point of view.

5.3.2 Renewable hydrogen

In the biohydrogen production, there are several opportunities areas in order to enhance the biohydrogen yield, the energy efficiency, and the process economy. In this context, as it can be seen in previous Section 2.2, several reports have been published. Also, recent approaches about the reactor design such as substrate pretreatments, inhibitors removal, bioaugmentation, catalyst development, among others, have been investigated (Banu et al., 2021; Gao, Jiang, Meng, Yan, & Aihemaiti, 2018; Habashy et al., 2021). However, to the authors' knowledge, the process intensification applied into the biohydrogen production is still scarce. In this subsection, a revision about the process intensification approach applied into the biohydrogen production is presented. In 2013, Bakonyi, Nemestóthy, and Bélafi-Bakó (2013) summarized the simultaneous biohydrogen production and separation using nonporous, polymeric, and ionic liquid membranes. The proposal is a promising alternative to be coupled with hydrogen producing bioreactors, giving the chance for biohydrogen concentration in situ. In their study, the recent applications of these membranes for biohydrogen recovery, focusing on the operational conditions affecting their behavior and performance, are provided. Then, Ugarte et al. (2017) studied the simultaneous dry reforming of methane, catalyst regeneration, and biohydrogen separation into a single fluidized bed. The reactor involves two zones with hydrogen selective membranes called TZFBR and MB, and the presence of a catalyst. The experiment was developed proving different catalysts such as Ni/Al_2O_3, $Ni-Ce/Al_2O_3$, $Ni-Co/Al_2O_3$, and a membrane combination (TZFBR, TZFBR + MB); also, the system was tested through the use of CO_2 for in situ catalyst regeneration. For all catalysts, the use of the membrane combination and CO_2 for catalyst regeneration are the best alternatives to produce high yield of biohydrogen. Other interesting contribution about membranes system was published by Saleem, Lavagnolo, and Spagni (2018); in their investigation, the use of dynamic membranes as a solid-liquid separation medium for biohydrogen production into a dark fermentation reactor was tested, under mild conditions. Particularly, the fouling of dynamic membrane was assessed in response to the biohydrogen production. It was observed that high feed concentration affects the filtration behavior of dynamic membrane, accumulating volatile fatty acids and inhibiting the biohydrogen production. Then, in 2019, de Jesús Montoya-Rosales et al. (2019) proposed the use of binary enzymatic hydrolysates of agave bagasse into a continuous stirred-tank reactor and a trickling bed reactor, in order to improve the biohydrogen production compared to individual hydrolysates. The results of the experiments indicate that the substrate availability is the limiting factor to biohydrogen production into a trickling bed reactor; however, the type of reactor and hydrolysate affected strongly the reaction performance. In order to obtain high hydrogen yield, binary hydrolysates use as a best alternative for scaling-up potential into the biohydrogen production. In 2020, Ghasemi, Yaghmaei, Abdi, Mardanpour, and Haddadi (2020) reported the fabrication and the use of a new cathode electrode assembly, constructed from polyaniline and graphene on a stainless-steel mesh. This cathode replaces the conventional and expensive cathode of MEC. Also, the assessed of different catalysts to elucidate the potential of the new cathode for larger-scale MEC was also presented. According to the experiments, the use of new cathode allows 82% of wastewater degradation and $0.805\,m^3$ of hydrogen per m^3 anolyte per day; these values are 7% and 20% lower than the use of conventional cathode, respectively. However, the fabrication cost of the new cathode was 50% lower than the conventional cathodes. In the same year, Chen, Lu, Tran, Lin, and Naqvi (2020) reported a novel catalytic tube reactor for efficient hydrogen production with low cost, from ethanol steam reforming. The design was developed using computational fluid dynamics, but the behavior of ethanol steam reforming over a nickel-based catalyst was conducted from experimental values. Based on their results, the high reaction pressure, the increase of diameter to channel width ratio, or catalyst thickness to tube diameter ratio can improve the reaction conversion and hydrogen yield. In 2021, Renaudie, Clion, Dumas, Vuilleumier, and Ernst (2021) presented a new reactor design to improve the biohydrogen production using dark fermentation. A hollow fiber membrane module was coupling to combine biohydrogen production, in situ liquid-gas separation and hydrogen producing bacteria retention in a single reactor. At the operation, no washout of hydrogen-producing bacteria was noted at 2 h; thus further hydrogen production rate enhances using an optimized organic loading rate. The membrane configuration used in their study enables the testing and selection of fermentation strategies in order to simplify the implementation of the dark fermentation process by addressing its main bottlenecks. A recent review about the use of sonochemistry technology to improve the technical challenges involved in the biohydrogen production, storage, and usage has been presented by Zore, Yedire, Pandi, Manickam, and Sonawane (2021). Sonochemistry is a novel technology and intensification technique for promoting acoustic cavitation phenomena caused by ultrasound, where higher reaction rates occur locally. Thus in their work the sonochemical routes for developing fuel cell catalysts, fuel refining, biofuel production, chemical processes for hydrogen production, as well as the physical, chemical, and electrochemical hydrogen storage techniques are presented. More information about a successful catalyst for enhancing the biohydrogen production are summarized in Nabgan et al. (2021).

In the production of renewable hydrogen, some of the strategies of process intensification that has been applied include simultaneous dry reforming, use of MECs, membranes as well as sonochemistry. One alternative that must be improved is the transformation of carbon dioxide through photocatalysis, which can lead to neutral carbon hydrogen production.

5.3.3 Fuel pellets

In the pellet production, there are some opportunity areas, mainly regarding to diminish the energy consumption involved in the pelletization stage. However, to the author's knowledge, the application of process intensification strategies to any fuel pellets production stage is negligible. In this sense, in this subsection are presented the process intensification reports about specific operations involved into the pellet production, such as drying and milling/grinding, which are also oriented to solid handling.

The drying is typically the water removal or other solvent from solids. Novel process intensification technologies and methods to enhance the solids drying are the fluidized bed, TORBED, screw conveyor, microwave, ultrasound, impinging stream (IS), and a vertical thin film dryer. A screw conveyor and IS have higher drying efficiency, regarding to the conventional fluidized bed reactor; however, TORBED technology is more convenient due to its design. On the other hand, microwaves and ultrasound are unconventional drying methods, but they are more time efficient with uniform drying. Extensive details about each one can be founded in Wang et al. (2017). Regarding to milling/grinding, these are physical methods to reduce the size particles. Impinging steam reactor, wet stirred media milling (WSMM) for nanoparticle production, gas-in-solid IS reactor, and high speed stirred ball mills are the process intensification technologies to develop the milling/grinding. However, the impinging steam reactor is the intensified technology accepted by some industries due to its practicalities. The other ones are in the experimental or assessment stage. A detailed information about the milling/grinding intensified technologies consult Wang et al. (2017).

The production of fuel pellets implies many mechanical operations, where the process intensification has not been widely studied. To the author's knowledge, there are no proposals regard the intensification of fuel pellets production, neither simulation nor experimental studies. Indeed, the first study simulation where process intensification is applied to fuel pellet production is presented in the next section of this chapter.

5.4 Intensification of the production of fuel pellets

5.4.1 Case of study

As case of study, it is selected the intensification of the production process to obtain fuel pellets including the production of heat and power through a power plant. The conventional process is described next.

The conventional process considers a small-scale biomass power plant for combined heat and power production (CHP), with the objective of valorizing the residual agricultural biomass available in Mexico. There are some reports about the power generation technologies using biomass as fuel (Megwai, 2014). Examples of these technologies are internal combustion engine power plants, gas turbine power plants, and steam power plants. In the process, pellets from agricultural residues (mixture: 90% wt of bean straw and 10% wt of rice husk) are produced and burned. The generated heat is transferred to preheated water and producing steam, which is the working fluid for electricity generation through a turbine. This process can be divided into three steps: (1) pellet production, (2) combustion, and (3) power generation. The next section describes the modeling and simulation of conventional and intensified processes for combined heat and power generation as well as the analysis of power generation's economy to compare both processes.

On the other hand, the intensified process considers the use of solar energy for the production of electrical energy that is used in the production process. Thus the effect of the incorporation of photovoltaic panels as a renewable energy source is also presented. It is important to mention that the cost and sizing analysis of the involved equipment is not considered in this work, excluding the equipment cost of some critical units.

5.4.2 Modeling and simulation of conventional and intensified pellet production processes

5.4.2.1 Conventional pellet production process

The modeling starts with the chemical definition of bean straw and rice husk. The chemical components of these biomasses vary due to different growing locations, climate, soil, fertilizer used, among others factors (Zou & Yang, 2019). The proximate and ultimate analysis for bean straw and rice husk are provided by Jenkins and Ebeling (1985) and Mansaray and Ghaly (1997), respectively, which are presented in Table 5.1. The moisture content was taken from experimental data reported by Hernández-Neri (2020). In regard to the particle size distribution (PSD), Table 5.2 presents rice husk (Mansaray & Ghaly, 1997). In this case of study, we assume that both biomasses have cylindrical geometry; thus, an equivalent volume is calculated to obtain an approximate diameter. The particle size estimated for bean straw is between 14 and 16 mm.

The modeling of these biomasses is based on the information previously presented. Also, in order to simulate both biomasses in Aspen Plus, they are assumed as solid components. The simulation of nonconventional solids takes into account the solid property models, such as HCOALGEN and DCOALIGT for calculating enthalpy and density of these lignocellulosic biomasses, using the stream class MCINCPSD.

The flowsheet into the simulation space starts when a feed stream of 1350 kg/h of bean straw is mixed with a stream of 150 kg/h of rice husk. The resulting mixture is fed to a crusher to reduce the particle size at 25°C and 1 bar. In the crusher model, the outlet PSD was based on the selected equipment method to spell out the crusher type, breakage function parameters, rotor velocity, and some of its sizing parameters. In the operating parameters section, Gyratory crusher type was selected. The specified maximum particle diameter was 10 mm as a breakage function parameter to calculate the power consumption.

Table 5.1 Proximate, ultimate, and sulfate analysis of bean straw and rice husk.

	Bean straw (Jenkins & Ebeling, 1985)	Rice husk (Mansaray & Ghaly, 1997)
Proximate analysis (wt%, dry basis)		
Volatile	75.30	66.42
Fixed carbon	18.77	10.23
Ash	5.93	23.35
Moisture (wet basis)	6.45	5.58
Ultimate analysis (wt%, dry basis)		
Carbon	42.77	37.602
Hydrogen	5.59	5.421
Nitrogen	0.83	0.381
Chlorine	0.13	0.011
Sulfur	0.01	0.034
Oxygen	44.74	33.201
Sulfate analysis (wt%, dry basis)		
Pyritic	0	0
Sulfate	0	0
Organic	0.01	0.10
Calorific value (MJ/kg)	16.89[b]	15.06[a]
Bulk density (kg/m^3)	120[b]	92[a]

[a] Hernández-Neri (2020).
[b] Trejo-Zamudio (2018).

Table 5.2 Weight, length, width, and diameter distribution of rice husk (Mansaray & Ghaly, 1997).

No.	Mesh	Weight	Length average (mm)	Width average (mm)	Equivalent diameter (mm)
1	7	2.06	11.6–12.4	2.8–4.0	5.1–6.7
2	8	14.22	10.2–11.4	2.4–2.8	4.5–5.1
3	10	23.6	8.8–10.0	2.0–2.4	3.7–4.5
4	12	31.34	6.2–8.0	1.7–2.0	3–3.7
5	14	18.72	4.0–6.0	1.4–1.7	2.3–3
6	18	7.37	3.2–4.0	1.0–1.4	1.6–2.3
7	20	1.17	1.5–2.4	0.8–1.0	1.1–1.6
8	25	0.67	0.8–1.2	0.7–0.9	0.8–1.1
Pan	–	0.85	dust	dust	dust

A granulator unit operation was used to model a pelletizer, due to the absence a predefined module for pelletization available in Aspen Plus (Manouchehrinejad & Mani, 2019). Regarding to the granulator model, the user specifies the variations between PSD, temperature and moisture content. These values are based on the results presented

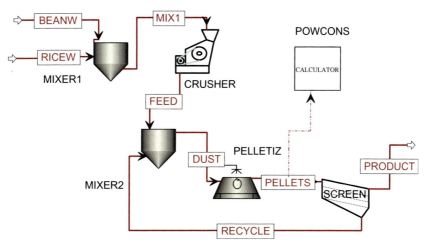

Fig. 5.4 Flowsheet of pellets processing in Aspen Plus.

in Hernández-Neri (2020). This process unit generates 65% of the feed: pellets with 8 mm of diameter and 30 mm length (the equivalent diameter size for Aspen Plus is 14.228 mm ± 0.3 mm), while the 35% of the dust (not pelletized) is recovery by the screen and recirculated. The biomass moisture content is around of 9.78% for water spraying. For this processing stage, 50°C was considered as result of steam injection and frictions involved into the pelletization processing. The pelletizer unit has an efficiency of 84% and the power engine of 5 hp. The energy consumption was considered as 62.142 kWh/Mg, applying a calculation block. Fig. 5.4 presents the flowsheet for pellets processing by Aspen Plus.

The process continues with the combustion stage. The obtained pellets at stream PRODUCT are combusted into a furnace-boiler system to produce steam and hot water, as can be seen in Fig. 5.5. In the combustion process, a ratio of 7.5 kg of air/kg of biomass is feed at 25°C and 1 bar. The air is compressed to 1.1 bar in the AIR-COMP module, before entering the combustion system; the last one consists of an operation unit set: RYIELD and RGIBBS reactors and a heat exchanger as an evaporator. Also, an ash filter, as a solid separator, is included in this system. The RYIELD model decomposes the biomass to its basic constituents, H_2O, C (solid), H_2, N_2, S, O_2, and ASH. The procedure starts defining the temperature of the biomass at approximately 200°C and 1 bar, using any initial values for these yields; these values will be calculated in a calculator block based on the 90% of straw beans and 10% of rice husk attributes. This evaluation requires to define variables in the Calculator block CAL1, including import (mass flow stream, compattr-vec streams) and export (mass yields) variables related to the flows and mass yields.

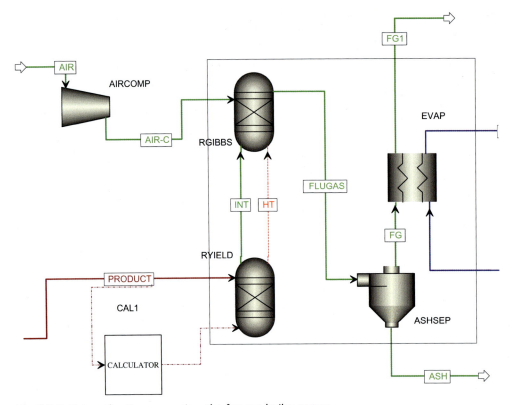

Fig. 5.5 Pellet combustion processing, the furnace-boiler system.

On the other hand, Table 5.3 shows the defined equations in order to estimate the molar fraction composition in the reactor. The definition of each variable is presented next.
- YCOMP is the mass fraction of the bean straw in the pellet biomass (90%).
- FACT1 and FACT2 are the factors to recalculate the ultimate analysis composition to wet basis for bean straw and rice husks attributes, respectively.
- The remaining equations are the mass balance for each element.

On the Sequence page, it is necessary to specify the calculator block execution sequence as Before-Unit operation RYIELD.

The RGIBBS unit requires only specifying as calculation option: Calculate phase equilibrium, chemical equilibrium, and the pressure of 1 bar, due to energy stream (HT) and mass stream (INT) from the RYIELD. In the Tab Product selecting: Identify possible products option and select all the components that can be present in the flue gas.

Table 5.3 Equations proposed to estimate the molar fraction composition in the reactor.

Equation
YCOMP = BIOMASS1/(BIOMASS1 + BIOMASS2)
FACT1 = (100 − WATER1)/100
FACT2 = (100-WATER2)/100
H20 = (WATER1 * YCOMP + WATER2 − (1 − YCOMP))/100
ASH = (ULTI(1) * FACT1 * YCOMP + ULT2(1) * FACT2 * (1 − YCOMP))/100
C = (ULT1(2) * FACT1 * YCOMP + ULT2(2) * FACT2 * (1 − YCOMP))/100
H2 = (ULT1(3) * FACT1 * YCOMP + ULT2(3) * FACT2 * (1 − YCOMP))/100
N2 = (ULT1(4) * FACT1 * YCOMP + ULT2(4) * FACT2 * (1 − YCOMP))/100
CL2 = (ULT1(5) * FACT1 * YCOMP + ULT2(5) * FACT2 * (1 − YCOMP))/100
S = (ULT1(6) * FACT1 * YCOMP + ULT2(6) * FACT2 * (1 − YCOMP))/100
Q2 = (ULT1(7) * FACT1 * YCOMP + ULT2(7) * FACT2 * (1 − YCOMP))/100

The solid separator ASHSEP (SSplit module) is specified with the value of 1.0, for a split fraction in the specification column and for the FG stream in the substream MIXED.

The reboiler unit EVAP is modeled with a MHeatX module. This unit is used to preheat and evaporate the water and superheat the steam into a boiler. A HeatX block or two heater blocks can be used for the same purpose. In the flue gas, FG1, the outlet temperature is fixed to 530°C. A design specification is implemented to satisfy this condition in the Rankine cycle described later.

Next, the power generation is simulated using a steam Rankine cycle. The Steam Rankine includes four critical devices: a boiler (heat exchanger), a steam turbine, a condenser (heat exchanger), and a pump (Dincer & Demir, 2018). The process is shown in Fig. 5.6. In this case, a recuperation unit (RECU) is included. Drescher and Brüggemann (2007) explain that the implementation of a recuperation unit increases the efficiency considerably. The pump and the turbine (isentropic efficiency of 94%) have a 14 and 1 bar discharge pressure, respectively. The operation conditions of the heat exchangers are presented in Table 5.4. The Rankine working fluid is water, using the IAPWS-95 steam tables. According to Aspen Help documentation, these tables can calculate any thermodynamic property of water.

The mass flow of water, at 80°C and 1 bar, for this system was specified in the low-pressure water (LPW) stream. However, any value was specified because a design specification determined the final flow value to have the HPS (high-pressure steam) at 500°C. Another design specification is used to get the cool water mass flow for the H2O-IN stream, to set water temperature at the outlet to 90°C (H2O-OUT).

Finally, in Fig. 5.7 is presented the complete conventional pellets production process, described previously.

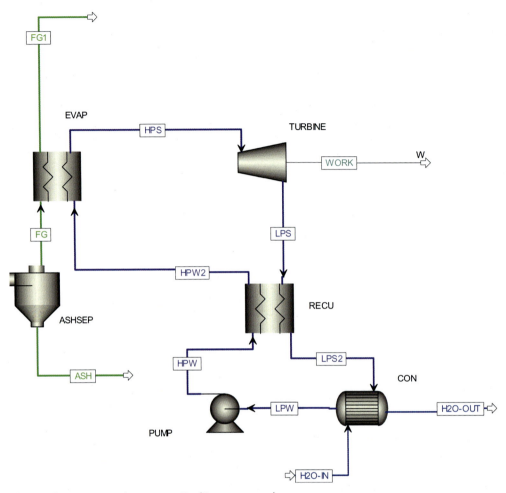

Fig. 5.6 Power generation system, Rankine steam cycle.

Table 5.4 Operating conditions for the power generation system.

Variable	Condenser	Evaporator	Recuperation
Pressure (bar)	1	14	1
Hot stream outlet temperature (°C)	80	530	100

5.4.2.2 Intensified pellet production process

The intensified process consists of the same process units. However, photovoltaic (PV) panels are used to supply the required electricity to the network, instead of the electricity produced with fossil resources.

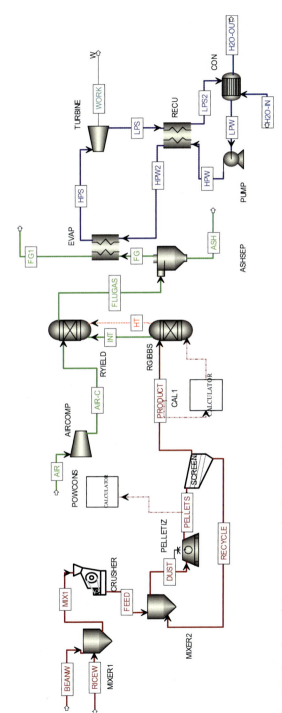

Fig. 5.7 Flowsheet of the conventional pellet production process.

Table 5.5 Retail price for components (USD/Wp) (GIZ, 2020).

System component	Reference capacity for prices (kWp)							
	0–25	2.5–5	5–15	15–30	30–50	50–100	100–250	250–500
Panels	0.45	0.35	0.4	0.43	0.43	0.55	0.49	0.31
Investors	0.28	0.24	0.22	0.19	0.19	0.23	0.21	0.1
Structures	0.13	0.21	0.19	0.17	0.13	0.41	0.23	0.24
Electric material	0.11	0.19	0.17	0.1	0.1	0.11	0.11	0.1
Workforce	0.13	0.2	0.18	0.13	0.1	0.1	0.11	0.08
Accessories	0.07	0.13	0.12	0.08	0.06	0.03	0.06	0.02
Additional (margin, indirect costs, contingency)	0.09	0.27	0.3	0.62	0.24	0.15	0.34	0.16
Electrical installations verification unit (UVIE)	2.00	0.57	0.24	0.45	0.03	0.04	0.23	0.03
Electrical Inspection Unit (UIE)	1.08	0.48	0.26	0.57	0.05	0.04	0.23	0.03
Total (USD/Wp)	4.34	2.64	2.08	2.74	1.33	1.66	2.01	1.07

The installation of a photovoltaic system has several components: panels, investors, structures, electric material, workforce, accessories, and, also, electrical installations verification unit (UVIE), electrical inspection unit (UIE); these last ones are considered as indirect costs. The costs for the installation, operation and maintenance of each distributed generation system varies according to their power range; Table 5.5 shows the retail price for these components (USD/Wp) for Mexico industry (GIZ, 2020). The study provides by GIZ-Mexico (GIZ, 2020) concludes that the costs for panels and inverters decrease as the size of the system increases. However, the total cost is increased because it is needed more specialized structures. Then, it is possible to determine a tentative cost of the installation of the photovoltaic system according to retail prices in Mexico, through the determination of the amount of electrical energy required by the processing.

The prices provided by Table 5.5 are in USD/Wp. Wp is the peak power of a PV system or panel, and it refers to the maximum power that a solar panel can deliver under standard conditions. The effective production in kWh depends on many other factors; for instance, the orientation of the panels, the weather conditions, the shading, the outside temperature, or the cleanliness of the panels are some factors that influence the production of a solar panel. Thus, kWp is different to kWh.

According with Ru, Kleissl, and Martinez (2013), the equation to calculate the power generated from solar panels is described in Eq. (5.1).

$$P_{pv} = GHT \cdot S \cdot \eta \tag{5.1}$$

where GHI (W/m^2) is the global horizontal irradiation at the location of solar panels, S (m^2) is the total area of solar panels, and η is the solar conversion efficiency of the PV. This model does not account for PV panel temperature effects. The output power, P_s, for a PV cell/module with the temperature effects is obtained by using the Eqs. (5.2)–(5.6), provided by Atwa, El-Saadany, Salama, and Seethapathy (2010).

$$P_s = FF \cdot V \cdot I \tag{5.2}$$

$$FF = \frac{V_{MPP} \cdot I_{MPP}}{V_{oc} \cdot I_{sc}} \tag{5.3}$$

$$Tc = T_A + s_a \left(\frac{N_{OT} - 20}{0.8} \right) \tag{5.4}$$

$$I = s_a [I_{sc} + K_i (Tc - 25)] \tag{5.5}$$

$$V = V_{oc} - K_v \cdot Tc \tag{5.6}$$

where V and I are the voltage and amperage in the module, respectively. FF is the fill factor, I_{sc} is the short circuit current in amperes, V_{oc} is the open-circuit voltage in volts, I_{MPP} is the current at maximum power point in amperes, V_{MPP} is the voltage at maximum power point in volts, Tc and T_A are the cell and ambient temperature in °C, respectively. K_v and K_i are the voltage (V/°C) and current (A/°C) temperature coefficient, respectively. N_{OT} is the nominal operating temperature of cell in °C, and, finally, s_a is the average solar irradiance. These parameters are reported in the product datasheet of photovoltaic panels, which includes specifications, mechanical properties, and temperature properties. The Table 5.6 shows the parameters for a specific module (Sunceco, 2021).

Next, the procedure to estimate the cost of the photovoltaic system is pointed out:
1. Determine the required electrical power.
2. Calculate the amount of energy per day.
3. Establish the percentage of photovoltaic solar electricity production
4. Identify the number of hours of peak sun for the inclined plane.
5. Obtain capacity by dividing daily consumption by peak hours.
6. Obtain the capacity by compensating losses (dividing by an efficiency factor).
7. Specify the peak power of the chosen panel.
8. Determine the number of modules to satisfy the required capacity.
9. Obtain the occupied area (horizontal and vertical assembly of the rows in which the modules are organized).
10. Calculate the final capacity.
11. Determine the savings in electricity consumption.

Table 5.6 Features of 300–320 W polycrystallin solar module (Sunceco, 2021).

Module characteristics	Units	Polycrystalline solar module				
Maximum power (P_{max})	Wp	300	305	310	315	320
Open circuit voltage (V_{oc})	V	44.71	44.72	44.76	44.82	44.84
Short circuit current (I_{sc})	A	8.947	9.094	9.234	9.371	9.515
Voltage at maximum power	V	37.23	37.24	37.32	37.46	37.62
Current at maximum power	A	8.06	8.19	8.31	8.41	8.51
Maximum system voltage	1000 V (iec), 600 V (UL)					
Cell efficiency	%	17.46	17.75	18.05	18.34	18.63
Module efficiency	%	15.46	15.72	15.98	16.23	16.49
Number of by-pass diodes	6					
Maximum series fuse	15 A					
Temperature coefficient of P_{max}	−0.46%/°C					
Temperature coefficient of V_{oc}	−0.34%/°C					
Temperature coefficient of I_{sc}	−0.05%/°C					
Nominal operating cell temperature	47°C ± 2°C					
Operating temperature	−40°C to 85°C					
Dimensions	1956 × 992 × 50 mm					

Table 5.7 Particle size distribution of biomass.

PSD (mm)	Inlet	Outlet
0.42–6.803	0.1	0.7687
−13.19	0	0.2155
−19.57	0	0.0157
−25.95	0	0
⋮	⋮	⋮
−134.5	0	0
−140.8	0.0382	0
−147.2	0.2873	0
−153.6	0.2872	0
−160	0.2872	0

5.4.3 Results and discussion

In this case of study has been modeled and simulated the pellets production from the mixture of bean straw and rice husk, in order to produce heat and power. According to the simulation results, the produced pellets have a calorific value of 3414.59 kcal/kg, 9.50% of ashes in dry weight, absolute humidity of 9.69%, bulk density of 613.89 kg/m^3, and durability of 99.81%. Also, the size reduction ratio between the feed and the outlet stream from the crusher is 32.5674. The Sauter mean diameter (diameter of spheres with the same volume/area ratio as the particle mixture) is also reduced from 4.2485 to 29.6718 mm. Table 5.7 presents the PSD, obtaining 98.43% of the total particles have a diameter size between 0.42 and 13.19 mm; being the power consumption negligible. Also, as can be seen, the pelletizer-screen system produces 1500 kg/h of pellets with 14.6594 mean diameter particle size.

The stream results for the biomass combustion process are shown in Table 5.8. The combustion reaction of 1500 kg/h of biomass with 12,000 kg/h of air produces 13,392.04 kg/h of flue gas at 1302°C, mainly, N_2, CO_2, H_2O, and O_2. The final temperature is 530°C; assuming negligible heat losses for this system, the reached temperature could cover these requirements, and the final value could be smaller. The transferred heat is 3.657 MW from FG stream to water in the evaporator, increasing 386.2°C the temperature of the water to get superheated steam at 500°C and 14 bar, according to the results presented in Table 5.9. The turbine generates 805.92 kW of energy, reducing the temperature and steam pressure at 169°C and 1 bar. The steam is again cooled in the recuperation unit at 100°C. The removed heat (169.32 kW) is transferred to the high-pressure water stream (HPW) as countercurrent flow direction to preheat it before entering the boiler. Then, the steam at 100°C is condensed and cooled to 80°C, using a cooling water flow in a shell and tube exchanger. Next, the water in the LPW is pumped at 14 bar to the recuperation unit. The amount of water within the work cycle was 4394.03 kg/h. With this flow, the pump consumed 5.52 kW of electrical energy. The cooling water temperature was increased from 25°C to 90°C, to use it as heating water. The mass flow was 37,792 kg/h and the transferred heat was 2857 MW.

Table 5.10 shows the power generation and consumption for some equipment. The produced network for the system is 761.8994 kW, that is, 6.679 GWh/year, which represents incomes higher than half of one US million dollars per year (12.908 kWh/USD). Also, more incomes are generated by the sale of water for heating. The ideal efficiency of the Rankine cycle, η_H, is calculated according the Eq. (5.7):

$$\eta_H = \frac{|W_T| - |W_p|}{|Q_H|} \tag{5.7}$$

where W_T is power generated in the turbine, W_p is the power consumption in the pump, and Q_H is the transferred heat in the boiler. For this case, the estimated value is

$$\eta_H = \frac{(805.9174 - 5.5226) kW}{3657.3819\ kW} = 0.2188$$

The ideal Rankine cycle efficiency is 21.88%, while the turbine generates 0.537 kWh/kg of pellets.

In the case of the intensified process, the PV system cost is determined. The Aspen Process Economic Analyzer provides the total utilities required. The required electricity is 107.524 kW (8.33 USD/h), and the fuel is 17.96 MMBTU/H (141 USD/h). Therefore the electrical energy cost is 0.0775 USD/kWh. The estimated amount of energy per day, E_{day}, is

Table 5.8 Stream results for the pellet combustion process.

	Product	Int	Air-C	Flugas	FG	FG1	Ash
T (°C)	25.22	200.00	36.40	1302.17	1302.17	530.00	1302.17
P (bar)	1.00	1.00	1.10	1.00	1.00	1.00	1.00
Mass vapor fraction	0.00	0.53	1.00	0.99	1.00	1.00	0.00
Mass solid fraction	1.00	0.47	0.00	0.01	0.00	0.00	1.00
Mass enthalpy (kJ/kg)	−6299.52	−613.68	11.53	−689.70	−700.54	−1683.71	655.64
Mass density (kg/m^3)	9.30	0.60	1.23	0.22	0.22	0.44	3000.00
Enthalpy flow (kW)	−2624.80	−255.70	38.43	−2586.37	−2606.03	−6263.41	19.66
Mass flows (kg/h)	1500.00	1500.00	12,000.00	13,500.00	13,392.04	13,392.04	107.96
Rice husks (kg/h)	150.00	0	0	0	0	0	0
Bean straws (kg/h)	1350.00	0	0	0	0	0	0
H$_2$O (kg/h)	0	95.45	0	794.53	794.53	794.53	0
CO$_2$ (kg/h)	0	0	0	2174.21	2174.21	2174.21	0
CO (kg/h)	0	0	0	0.07	0.07	0.07	0
H$_2$ (kg/h)	0	78.28	0	0	0	0	0
O$_2$ (kg/h)	0	612.06	2795.00	1196.96	1196.96	1196.96	0
C (kg/h)	0	593.41	0	0	0	0	0
N$_2$ (kg/h)	0	11.02	9205.00	9208.87	9208.87	9208.87	0
S (kg/h)	0	0.17	0	0	0	0	0
Ash (kg/h)	0	107.96	0	107.96	0	0	107.96
NH$_3$ (kg/h)	0	0	0	7.25E−10	7.25E−10	7.25E−10	0
H$_2$S (kg/h)	0	0	0	6.57E−14	6.57E−14	6.57E−14	0
Cl$_2$ (kg/h)	0	1.66	0	2.46E−05	2.46E−05	2.46E−05	0
NO$_2$ (kg/h)	0	0	0	0.06	0.06	0.06	0
NO (kg/h)	0	0	0	15.28	15.28	15.28	0
SO (kg/h)	0	0	0	1.03E−06	1.03E−06	1.03E−06	0
SO$_2$ (kg/h)	0	0	0	0.35	0.35	0.35	0
HCl (kg/h)	0	0	0	1.70	1.70	1.70	0

Table 5.9 Stream results for the Rankine cycle.

	LPW	HPW	HPW2	HPS	LPS	LPS2	H$_2$O-IN	H$_2$O-OUT
T (°C)	80.00	80.83	113.8	500	169	100	25	90
P (bar)	1	14	14	14	1	1	1	1
Mass vapor fraction	0	0	0	1	1	1	0	0
Mass enthalpy (MJ/kg)	−15.6	−15.6	−15.5	−12.5	−13.2	−13.3	−15.9	−15.6
Mass density (kg/m^3)	971.8	971.9	948.6	3.97	0.49	0.59	997	965.31
Enthalpy flow (MW)	−19.1	−19.1	−18.9	−15.3	−16.1	−16.2	−166.6	−163.7

Table 5.10 Power consumption per equipment.

Equipment	Power (kW)
Crusher	negligible
Pelletizer	0.0621
Pump	5.5226
Compressor	38.4333
Turbine	−805.9174
Network	**−761.8994**

$$E_{day} = 107.524 \text{ kW } (24 \text{ h}) = 2580.576 \text{ kWh}$$

Using PV panels, it is required to specify the percentage of the electricity that is supplied by the photovoltaic system, for instance, a maximum penetration limit. In this case, it is assumed a value of 30% of the peak load. Then, the power provided by the PV, P_{PV}, is 774.1728 kWh per day. Thus the considered hours of peak sun are 6.5, and the installation capacity is

$$Pc' = \frac{774.1728 \text{ kWh/day}}{6.5 \text{ } h_{peak}/\text{day}} = 119.1 \text{ kWp}$$

The installation capacity of the PV system is dividing by 0.70 for compensating losses (70% efficiency in converting to alternating current), as

$$Pc = \frac{Pc'}{0.7} = 170.15 \text{ kWp}$$

The polycrystalline solar module is operated with a maximum power of 310 Wp (according to Table 5.5); thus the number of modules is

$$N = \frac{170,150 \text{ Wp}}{310 \text{ Wp}} = 548.864 = 549 \text{ modules}$$

Then, these PV modules produce 170.19 kWp. One module produces approximately 1 kWp per 8 m²; thus the area required for mounting the PV array is estimated. Each PV module has 1956 × 992 × 50 mm for length, width, and thickness. The total area is estimated by the Eq. (5.8):

$$\text{Area} = 549(1.956 \text{ m})(0.992 \text{ m}) = 1065.25 \text{ m}^2 \tag{5.8}$$

The cost for 170.19 kWp is approximately 2.01 USD/Wp, according to Table 5.5. The total cost for the PV power system is

$$C_{PV} = 2010 \frac{\text{USD}}{\text{kWp}} \cdot 170.19 \text{ kWp} = 342,081.9 \text{ USD}$$

Thereby, the total savings per day, S_{day} are

$$S_{day} = 774.1728 \frac{\text{kWh}}{\text{day}} \left(0.0775 \frac{\text{USD}}{\text{kWh}}\right) = 60 \frac{\text{USD}}{\text{day}}$$

With this value, the years to recover the investment are 15.6 years, according to Eq. (5.9):

$$\#\text{years} = \frac{342,081.9 \text{ USD}}{60 \frac{\text{USD}}{\text{day}}} = 5701.365 \text{ day} = 15.6 \text{ years} \qquad (5.9)$$

Georgitsioti, Pearsall, Forbes, and Pillai (2019) mentioned that the operation period of a PV system is assumed to be at least 25 years since the PV module warranties, provided by the PV manufacturers, are usually around 20–25 years. Therefore in 10 years, the process plan is savings for the 30% of the total cost of the electricity required. However, these savings can be more significant by increasing the penetration percentage to around 45%. Hence, the required PV power is 255.44 kWh (824 PV modules). According to this information, the rate cost is 1070 USD/Wp, which is 53.23% of the initial price per Wp, so it only takes 8.3 years to recover the investment. It should be clear that these referral prices are based on the study carried out in Mexico between suppliers and clients between 2007 and 2018. Also, the size of the power plant can affect the power demand. Finally, biomass gasification is often preferred to combustion because of better efficiency and reduced emissions. There are in the literature examples with high efficiency of gasification processes with an Organic Rankine Cycle (Drescher & Brüggemann, 2007; Ishaq, Islam, Dincer, & Yilbas, 2020; Kalina, 2011; Megwai, 2014; Rentizelas, Karellas, Kakaras, & Tatsiopoulos, 2009).

It is important to mention that the profitability of the process was not reviewed. However, the purchase and operation costs are the same for both processes, except for the percentage of electricity that is supplied by the PV system. Also, the raw material only requires a transport cost from the agricultural regions. The main results show that 6679 GWh per year of electrical energy are produced, which represents 0.537 kWh per kg of biomass feed. In addition, it was determined that including a PV system requires an investment of approximately 1070 USD/Wp to satisfy 45% of the electrical energy required for the process. According to the literature, the investment expense could be recovered in the first 8.5 years of operation.

5.5 Future trends

The development and research of processes for the production of biogas, renewable hydrogen, and fuel pellets has been increased in the last years. Several studies have investigated how to improve the yields, decreasing the residence times, as well as using low-cost raw materials, mainly wastes. Regard the application of process intensification strategies, few efforts have been reported. For the production of renewable hydrogen, the main intensification alternatives include the use of membranes for the purification, as well as the use of simultaneous dry reforming, MECs, and sonochemistry for the reaction process. On the other hand, the use of hydrodynamic cavitation, nanoparticles, as well as solar bioreactors, have been explored for the production of biogas; in the case of fuel pellets production for the author's knowledge, the first application is presented in this chapter. Therefore there are many opportunities areas for the application of process intensification strategies for the production of these biofuels.

In the case of hydrogen and biogas production, the efforts must be focus on the development of solar-assisted process, which allows a significant reduction in the energy consumption as well in the carbon footprint. Another strategy is the genetic modification of microorganisms for the synthesis of precursory compounds that will facilitate the processing to generate these biofuels. Finally, both alternatives must be capable to process waste biomasses as well as mixtures of them. Other strategy recently explored is the conversion of carbon dioxide for the production of hydrogen and methane; this process is feasible but the yields are still lows. In this context, the development of more efficient catalyst is an interesting research area. In addition, this proposal would enable the production of these biofuels in a neutral carbon cycle.

Respect the fuel pellets, the use of solar energy through photovoltaic panels to satisfy the energy requirements of the process is one alternative for the intensification of the process. On the other hand, another strategy will imply the redesign of the process in order to have one vessel, where the conditioning (regard moisture content and particle size) and the pelletization can be performed in one step.

5.6 Concluding remarks

The development of efficient production processes for gaseous and solid biofuels is required in order to satisfy the increasing energy demand in a sustainable way. In this context, the application of process intensification strategies is a powerful tool to increase the yield to biofuels and reduce the energy consumption required in its production, as is analyzed by the case of study provided in this chapter, wherein up to 45% of the electrical energy consumptions are covered by the inclusion of the PV system to the small-power plant fed by pellets fuel from biomass. On the other hand, according to the revision of the literature, it can be observed that there is an interesting opportunity area for the application of process intensification strategies for the production of gaseous and solid biofuels, such as biogas, biohydrogen, and fuel pellets. Thus, it is necessary to focus the efforts in the development of sustainable process for the production of biofuels, in order to guarantee the satisfaction of the energetic demand as well as a future as society.

Acknowledgments

Financial support provided by CONACyT, through the grants of Noemí Hernández-Neri and Araceli Guadalupe Romero-Izquierdo for the realization of her graduate studies, is gratefully acknowledged.

References

Abdul Aziz, N. I. H., Hanafiah, M. M., & Mohamed Ali, M. Y. (2019). Sustainable biogas production from agrowaste and effluents—A promising step for small-scale industry income. *Renewable Energy, 132*, 363–369.

Abdulrasheed, A., Jalil, A. A., Gambo, Y., Ibrahim, M., Hambali, H. U., & Shahul Hamid, M. Y. (2019). A review on catalyst development for dry reforming of methane to syngas: Recent advances. *Renewable and Sustainable Energy Reviews, 108*, 175–193.

Abraham, A., Mathew, A. K., Park, H., Choi, O., Sindhu, R., Parameswaran, B., et al. (2020). Pretreatment strategies for enhanced biogas production from lignocellulosic biomass. *Bioresource Technology, 301*, 122725.

Adekunle, A. S., Ibitoye, S. E., Omoniyi, P. O., Jilantikiri, L. J., Sam-Obu, C. V., Yahaya, T., et al. (2019). Production and testing of biogas using cow dung, jatropha and iron filins. *Journal of Bioresources and Bioproducts, 4*, 143–148.

Akhbari, A., Chuen, O. C., & Ibrahim, S. (2021). Start-up study of biohydrogen production from palm oil mill effluent in a lab-scale up-flow anaerobic sludge blanket fixed-film reactor. *International Journal of Hydrogen Energy, 46*, 10191–10204.

Ameen, M., Azizan, M. T., Yusup, S., Ramli, A., & Yasir, M. (2017). Catalytic hydrodeoxygenation of triglycerides: An approach to clean diesel fuel production. *Renewable and Sustainable Energy Reviews, 80*, 1072–1088.

Atwa, Y. M., El-Saadany, E. F., Salama, M. M. A., & Seethapathy, R. (2010). Optimal renewable resources mix for distribution system energy loss minimization. *IEEE Transactions on Power Apparatus and Systems, 25*, 360–370.

Bakonyi, P., Nemestóthy, N., & Bélafi-Bakó, K. (2013). Biohydrogen purification by membranes: An overview on the operational conditions affecting the performance of non-porous, polymeric and ionic liquid based gas separation membranes. *International Journal of Hydrogen Energy, 38*, 9673–9687.

Banu, J. R., Usman, T. M. M., Kavitha, S., Kannah, R. Y., Yogalakshmi, K. N., Sivashanmugam, P., et al. (2021). A critical review on limitations and enhancement strategies associated with biohydrogen production. *International Journal of Hydrogen Energy, 46*, 16565–16590.

Basu, P. (2013). Appendix A—Definition of biomass. In P. Basu (Ed.), *Biomass gasification, pyrolysis and torrefaction* (2nd ed., pp. 457–459). Boston: Academic Press.

BP Energy Outlook. (2019). *BP Energy Outlook 2019 Edition*. BP p.l.c. Retrieved April 12, 2022, from https://www.bp.com/content/dam/bp/business-sites/en/global/corporate/pdfs/energy-economics/energy-outlook/bp-energy-outlook-2019.pdf.

Chen, W.-H., Lu, C.-Y., Tran, K.-Q., Lin, Y.-L., & Naqvi, S. R. (2020). A new design of catalytic tube reactor for hydrogen production from ethanol steam reforming. *Fuel, 281*, 118746.

Chen, C.-Y., Yang, M.-H., Yeh, K.-L., Liu, C.-H., & Chang, J.-S. (2008). Biohydrogen production using sequential two-stage dark and photo fermentation processes. *International Journal of Hydrogen Energy, 33*, 4755–4762.

Cheng, Y. W., Lee, Z. S., Chong, C. C., Khan, M. R., Cheng, C. K., Ng, K. H., et al. (2019). Hydrogen-rich syngas production via steam reforming of palm oil mill effluent (POME)—A thermodynamics analysis. *International Journal of Hydrogen Energy*, *44*, 20711–20724.

de Jesús Montoya-Rosales, J., Olmos-Hernández, D. K., Palomo-Briones, R., Montiel-Corona, V., Mari, A. G., & Razo-Flores, E. (2019). Improvement of continuous hydrogen production using individual and binary enzymatic hydrolysates of agave bagasse in suspended-culture and biofilm reactors. *Bioresource Technology*, *283*, 251–260.

Dehhaghi, M., Tabatabaei, M., Aghbashlo, M., Kazemi Shariat Panahi, H., & Nizami, A.-S. (2019). A state-of-the-art review on the application of nanomaterials for enhancing biogas production. *Journal of Environmental Management*, *251*, 109597.

Dincer, I., & Demir, M. E. (2018). 4.9 Combined energy conversion systems. In I. Dincer (Ed.), *Comprehensive energy systems* (pp. 312–363). Oxford: Elsevier.

Drescher, U., & Brüggemann, D. (2007). Fluid selection for the Organic Rankine Cycle (ORC) in biomass power and heat plants. *Applied Thermal Engineering*, *27*, 223–228.

Duarte, M. S., Sinisgalli, E., Cavaleiro, A. J., Bertin, L., Alves, M. M., & Pereira, M. A. (2021). Intensification of methane production from waste frying oil in a biogas-lift bioreactor. *Renewable Energy*, *168*, 1141–1148.

Fagbohungbe, M. O., Komolafe, A. O., & Okere, U. V. (2019). Renewable hydrogen anaerobic fermentation technology: Problems and potentials. *Renewable and Sustainable Energy Reviews*, *114*, 109340.

Florio, C., Nastro, R. A., Flagiello, F., Minutillo, M., Pirozzi, D., Pasquale, V., et al. (2019). Biohydrogen production from solid phase-microbial fuel cell spent substrate: A preliminary study. *Journal of Cleaner Production*, *227*, 506–511.

Gaballah, E. S., Abdelkader, T. K., Luo, S., Yuan, Q., & El-Fatah Abomohra, A. (2020). Enhancement of biogas production by integrated solar heating system: A pilot study using tubular digester. *Energy*, *193*, 116758.

Gao, Y., Jiang, J., Meng, Y., Yan, F., & Aihemaiti, A. (2018). A review of recent developments in hydrogen production via biogas dry reforming. *Energy Conversion and Management*, *171*, 133–155.

García, R., Gil, M. V., Rubiera, F., Chen, D., & Pevida, C. (2021). Renewable hydrogen production from biogas by sorption enhanced steam reforming (SESR): A parametric study. *Energy*, *218*, 119491.

Georgitsioti, T., Pearsall, N., Forbes, I., & Pillai, G. (2019). A combined model for PV system lifetime energy prediction and annual energy assessment. *Solar Energy*, *183*, 738–744.

Ghasemi, B., Yaghmaei, S., Abdi, K., Mardanpour, M. M., & Haddadi, S. A. (2020). Introducing an affordable catalyst for biohydrogen production in microbial electrolysis cells. *Journal of Bioscience and Bioengineering*, *129*, 67–76.

GIZ. (2020). *Monitor de información comercial e índice de precios de sistemas de generación distribuida de electricidad con fuentes renovables comercializados en México*. Report.

Gonçalves Neto, J., Vidal Ozorio, L., Campos de Abreu, T. C., Ferreira dos Santos, B., & Pradelle, F. (2021). Modeling of biogas production from food, fruits and vegetables wastes using artificial neural network (ANN). *Fuel*, *285*, 119081.

Guo, S., Liu, Q., Sun, J., & Jin, H. (2018). A review on the utilization of hybrid renewable energy. *Renewable and Sustainable Energy Reviews*, *91*, 1121–1147.

Gutiérrez-Antonio, C., Gómez-Castro, F. I., de Lira-Flores, J. A., & Hernández, S. (2017). A review on the production processes of renewable jet fuel. *Renewable and Sustainable Energy Reviews*, *79*, 709–729.

Gutiérrez-Antonio, C., Romero-Izquierdo, A. G., Gómez-Castro, F. I., & Hernández, S. (2021). 6—Process intensification and integration in the production of biojet fuel. In C. Gutiérrez-Antonio, A. G. Romero-Izquierdo, F. I. Gómez-Castro, & S. Hernández (Eds.), *Production processes of renewable aviation fuel* (pp. 171–199). Elsevier.

Gutiérrez-Antonio, C., et al. (2021). *Production processes of renewable aviation fuel*. Elsevier.

Habashy, M. M., Ong, E. S., Abdeldayem, O. M., Al-Sakkari, E. G., & Rene, E. R. (2021). Food waste: A promising source of sustainable biohydrogen fuel. *Trends in Biotechnology*, *39*(12), 1274–1288. https://doi.org/10.1016/j.tibtech.2021.04.001.

Hassan, S. N., Sani, Y. M., Abdul Aziz, A. R., Sulaiman, N. M. N., & Daud, W. M. A. W. (2015). Biogasoline: An out-of-the-box solution to the food-for-fuel and land-use competitions. *Energy Conversion and Management, 89*, 349–367.

Hassaneen, F. Y., Abdallah, M. S., Ahmed, N., Taha, M. M., Abd ElAziz, S. M. M., El-Mokhtar, M. A., et al. (2020). Innovative nanocomposite formulations for enhancing biogas and biofertilizers production from anaerobic digestion of organic waste. *Bioresource Technology, 309*, 123350.

Hernández-Neri, N. (2020). *Uso de cáscara de arroz y paja de frijol para la producción de biocombustibles sólidos* (Master thesis). Univ. Autónoma Querétaro.

Huang, C., Guo, H.-J., Wang, C., Xiong, L., Luo, M.-T., Chen, X.-F., et al. (2017). Efficient continuous biogas production using lignocellulosic hydrolysates as substrate: A semi-pilot scale long-term study. *Energy Conversion and Management, 151*, 53–62.

IEA. (2020). *Global Energy Review* (p. 2020).

IEA. (2021). *COVID-19—Topics—IEA*. Int. Energy Agency.

Ishaq, H., Islam, S., Dincer, I., & Yilbas, B. S. (2020). Development and performance investigation of a biomass gasification based integrated system with thermoelectric generators. *Journal of Cleaner Production, 256*, 120625.

Jamali, N. S., Md Jahim, J., Mumtaz, T., & Abdul, P. M. (2021). Dark fermentation of palm oil mill effluent by *Caldicellulosiruptor saccharolyticus* immobilized on activated carbon for thermophilic biohydrogen production. *Environmental Technology and Innovation, 22*, 101477.

Jenkins, B., & Ebeling, J. (1985). Thermochemical properties of biomass fuels. *Hilgardia, 39*(5), 14–16. https://doi.org/10.3733/ca.v039n05p14.

Joshi, S. M., & Gogate, P. R. (2019). Intensifying the biogas production from food waste using ultrasound: Understanding into effect of operating parameters. *Ultrasonics Sonochemistry, 59*, 104755.

Kainthola, J., Kalamdhad, A. S., & Goud, V. V. (2019). A review on enhanced biogas production from anaerobic digestion of lignocellulosic biomass by different enhancement techniques. *Process Biochemistry, 84*, 81–90.

Kalina, J. (2011). Integrated biomass gasification combined cycle distributed generation plant with reciprocating gas engine and ORC. *Applied Thermal Engineering, 31*, 2829–2840.

Kato, M. (2021). Fuel design and fabrication: Pellet-type fuel. In *Reference module in earth systems and environmental sciences* Elsevier.

Keruthiga, K., Mohamed, S. N., Gandhi, N. N., & Muthukumar, K. (2021). Sugar industry waste-derived anode for enhanced biohydrogen production from rice mill wastewater using artificial photo-assisted microbial electrolysis cell. *International Journal of Hydrogen Energy, 46*, 20425–20434.

Khalid, A., Aslam, M., Qyyum, M. A., Faisal, A., Khan, A. L., Ahmed, F., et al. (2019). Membrane separation processes for dehydration of bioethanol from fermentation broths: Recent developments, challenges, and prospects. *Renewable and Sustainable Energy Reviews, 105*, 427–443.

Ko, J. K., Lee, J. H., Jung, J. H., & Lee, S.-M. (2020). Recent advances and future directions in plant and yeast engineering to improve lignocellulosic biofuel production. *Renewable and Sustainable Energy Reviews, 134*, 110390.

Kovalev, A. A., Kovalev, D. A., Litti, Y. V., & Katraeva, I. V. (2020). Biohydrogen production in the two-stage process of anaerobic bioconversion of organic matter of liquid organic waste with recirculation of digister effluent. *International Journal of Hydrogen Energy, 45*, 26831–26839.

Krishnan, V., & McCalley, J. D. (2016). The role of bio-renewables in national energy and transportation systems portfolio planning for low carbon economy. *Renewable Energy, 91*, 207–223.

Kumar, G. R., & Chowdhary, N. (2016). Biotechnological and bioinformatics approaches for augmentation of biohydrogen production: A review. *Renewable and Sustainable Energy Reviews, 56*, 1194–1206.

Lizama, A. C., Figueiras, C. C., Gaviria, L. A., Pedreguera, A. Z., & Ruiz Espinoza, J. E. (2019). Nano-ferrosonication: A novel strategy for intensifying the methanogenic process in sewage sludge. *Bioresource Technology, 276*, 318–324.

Lu, C., Jing, Y., Zhang, H., Lee, D.-J., Tahir, N., Zhang, Q., et al. (2020). Biohydrogen production through active saccharification and photo-fermentation from alfalfa. *Bioresource Technology, 304*, 123007.

Lu, C., Wang, Y., Lee, D.-J., Zhang, Q., Zhang, H., Tahir, N., et al. (2019). Biohydrogen production in pilot-scale fermenter: Effects of hydraulic retention time and substrate concentration. *Journal of Cleaner Production, 229*, 751–760.

Maity, S. K. (2015). Opportunities, recent trends and challenges of integrated biorefinery: Part I. *Renewable and Sustainable Energy Reviews, 43*, 1427–1445.

Manouchehrinejad, M., & Mani, S. (2019). Process simulation of an integrated biomass torrefaction and pelletization (iBTP) plant to produce solid biofuels. *Energy Conversion and Management: X, 1*, 100008.

Mansaray, K. G., & Ghaly, A. E. (1997). Physical and thermochemical properties of rice husk. *Energy Sources, 19*, 989–1004.

Materazzi, M., Taylor, R., & Cairns-Terry, M. (2019). Production of biohydrogen from gasification of waste fuels: Pilot plant results and deployment prospects. *Waste Management, 94*, 95–106.

Matheri, A. N., Ndiweni, S. N., Belaid, M., Muzenda, E., & Hubert, R. (2017). Optimising biogas production from anaerobic co-digestion of chicken manure and organic fraction of municipal solid waste. *Renewable and Sustainable Energy Reviews, 80*, 756–764.

Megwai, G. (2014). *Process simulations of small scale biomass power plant.* (MSc thesis in Resource Recovery–Sustainable Engineering).

Méndez-Vázquez, M. A., Gómez-Castro, F. I., Ponce-Ortega, J. M., Serafín-Muñoz, A. H., Santibañez-Aguilar, J. E., & El-Halwagi, M. M. (2016). Mathematical optimization of the production of fuel pellets from residual biomass. In Z. Kravanja, & M. Bogataj (Eds.), *26th European symposium on computer aided process engineering, computer aided chemical engineering* (pp. 133–138). Elsevier.

Morone, A., & Pandey, R. A. (2014). Lignocellulosic biobutanol production: Gridlocks and potential remedies. *Renewable and Sustainable Energy Reviews, 37*, 21–35.

Nabgan, W., Tuan Abdullah, T. A., Nabgan, B., Jalil, A. A., Nordin, A. H., Ul-Hamid, A., et al. (2021). Catalytic biohydrogen production from organic waste materials: A literature review and bibliometric analysis. *International Journal of Hydrogen Energy, 46*(60), 30903–30925. https://doi.org/10.1016/j.ijhydene.2021.04.100.

Natarajan, Y., Nabera, A., Salike, S., Dhanalakshmi Tamilkkuricil, V., Pandian, S., Karuppan, M., et al. (2019). An overview on the process intensification of microchannel reactors for biodiesel production. *Chemical Engineering and Processing: Process Intensification, 136*, 163–176.

Ndayisenga, F., Yu, Z., Zheng, J., Wang, B., Liang, H., Phulpoto, I. A., et al. (2021). Microbial electrohydrogenesis cell and dark fermentation integrated system enhances biohydrogen production from lignocellulosic agricultural wastes: Substrate pretreatment towards optimization. *Renewable and Sustainable Energy Reviews, 145*, 111078.

Ning, J., Zhou, M., Pan, X., Li, C., Lv, N., Wang, T., et al. (2019). Simultaneous biogas and biogas slurry production from co-digestion of pig manure and corn straw: Performance optimization and microbial community shift. *Bioresource Technology, 282*, 37–47.

Oleszek, M., & Krzemińska, I. (2021). Biogas production from high-protein and rigid cell wall microalgal biomasses: Ultrasonication and FT-IR evaluation of pretreatment effects. *Fuel, 296*, 120676.

Onthong, U., & Juntarachat, N. (2017). Evaluation of biogas production potential from raw and processed agricultural wastes. *Energy Procedia, 138*, 205–210.

Oumer, A. N., Hasan, M. M., Baheta, A. T., Mamat, R., & Abdullah, A. A. (2018). Bio-based liquid fuels as a source of renewable energy: A review. *Renewable and Sustainable Energy Reviews, 88*, 82–98.

Panin, S., Setthapun, W., Elizabeth Sinsuw, A. A., Sintuya, H., & Chu, C.-Y. (2021). Biohydrogen and biogas production from mashed and powdered vegetable residues by an enriched microflora in dark fermentation. *International Journal of Hydrogen Energy, 46*, 14073–14082.

Parsaee, M., Kiani Deh Kiani, M., & Karimi, K. (2019). A review of biogas production from sugarcane vinasse. *Biomass and Bioenergy, 122*, 117–125.

Patil, P. N., Gogate, P. R., Csoka, L., Dregelyi-Kiss, A., & Horvath, M. (2016). Intensification of biogas production using pretreatment based on hydrodynamic cavitation. *Ultrasonics Sonochemistry, 30*, 79–86.

Patinvoh, R. J., Osadolor, O. A., Chandolias, K., Sárvári Horváth, I., & Taherzadeh, M. J. (2017). Innovative pretreatment strategies for biogas production. *Bioresource Technology, 224*, 13–24.

Pessoa, M., Sobrinho, M. A. M., & Kraume, M. (2020). The use of biomagnetism for biogas production from sugar beet pulp. *Biochemical Engineering Journal, 164*, 107770.

Pfeifer, A., Dobravec, V., Pavlinek, L., Krajačić, G., & Duić, N. (2018). Integration of renewable energy and demand response technologies in interconnected energy systems. *Energy, 161*, 447–455.

Powell, K. M., Rashid, K., Ellingwood, K., Tuttle, J., & Iverson, B. D. (2017). Hybrid concentrated solar thermal power systems: A review. *Renewable and Sustainable Energy Reviews, 80*, 215–237.

Pradhan, P., Arora, A., & Mahajani, S. M. (2018). Pilot scale evaluation of fuel pellets production from garden waste biomass. *Energy for Sustainable Development, 43*, 1–14.

Pradhan, P., Mahajani, S. M., & Arora, A. (2018). Production and utilization of fuel pellets from biomass: A review. *Fuel Processing Technology, 181*, 215–232.

Pradhan, P., Mahajani, S. M., & Arora, A. (2021). Pilot scale production of fuel pellets from waste biomass leaves: Effect of milling size on pelletization process and pellet quality. *Fuel, 285*, 119145.

Preethi, Usman, T. M. M., Rajesh Banu, J., Gunasekaran, M., & Kumar, G. (2019). Biohydrogen production from industrial wastewater: An overview. *Bioresource Technology Reports, 7*, 100287.

Quiroz-Pérez, E., Gutiérrez-Antonio, C., & Vázquez-Román, R. (2019). Modelling of production processes for liquid biofuels through CFD: A review of conventional and intensified technologies. *Chemical Engineering and Processing: Process Intensification, 143*, 107629.

Rambabu, K., Bharath, G., Thanigaivelan, A., Das, D. B., Show, P. L., & Banat, F. (2021). Augmented biohydrogen production from rice mill wastewater through nano-metal oxides assisted dark fermentation. *Bioresource Technology, 319*, 124243.

Renaudie, M., Clion, V., Dumas, C., Vuilleumier, S., & Ernst, B. (2021). Intensification and optimization of continuous hydrogen production by dark fermentation in a new design liquid/gas hollow fiber membrane bioreactor. *Chemical Engineering Journal, 416*, 129068.

Rentizelas, A., Karellas, S., Kakaras, E., & Tatsiopoulos, I. (2009). Comparative techno-economic analysis of ORC and gasification for bioenergy applications. *Energy Conversion and Management, 50*, 674–681.

Ríos-Badrán, I. M., Luzardo-Ocampo, I., García-Trejo, J. F., Santos-Cruz, J., & Gutiérrez-Antonio, C. (2020). Production and characterization of fuel pellets from rice husk and wheat straw. *Renewable Energy, 145*, 500–507.

Rochón, E., Cortizo, G., Cabot, M. I., García Cubero, M. T., Coca, M., Ferrari, M. D., et al. (2020). Bioprocess intensification for isopropanol, butanol and ethanol (IBE) production by fermentation from sugarcane and sweet sorghum juices through a gas stripping-pervaporation recovery process. *Fuel, 281*, 118593.

Rosha, P., Mohapatra, S. K., Mahla, S. K., & Dhir, A. (2019). Catalytic reforming of synthetic biogas for hydrogen enrichment over Ni supported on $ZnOCeO_2$ mixed catalyst. *Biomass and Bioenergy, 125*, 70–78.

Ru, Y., Kleissl, J., & Martinez, S. (2013). Storage size determination for grid-connected photovoltaic systems. *IEEE Transactions on Sustainable Energy, 4*(1), 68–81.

Saleem, M., Lavagnolo, M. C., & Spagni, A. (2018). Biological hydrogen production via dark fermentation by using a side-stream dynamic membrane bioreactor: Effect of substrate concentration. *Chemical Engineering Journal, 349*, 719–727.

Saratale, G. D., Saratale, R. G., & Chang, J.-S. (2013). Chapter 9—Biohydrogen from renewable resources. In A. Pandey, J.-S. Chang, P. C. Hallenbecka, & C. Larroche (Eds.), *Biohydrogen* (pp. 185–221). Amsterdam: Elsevier.

Shamurad, B., Sallis, P., Petropoulos, E., Tabraiz, S., Ospina, C., Leary, P., et al. (2020). Stable biogas production from single-stage anaerobic digestion of food waste. *Applied Energy, 263*, 114609.

Shuba, E. S., & Kifle, D. (2018). Microalgae to biofuels: 'Promising' alternative and renewable energy, review. *Renewable and Sustainable Energy Reviews, 81*, 743–755.

Singh, S., Hariteja, N., Sharma, S., Raju, N. J., & Prasad, T. J. R. (2021). Production of biogas from human faeces mixed with the co-substrate poultry litter & cow dung. *Environmental Technology and Innovation, 23*, 101551.

Ståhl, M., & Berghel, J. (2011). Energy efficient pilot-scale production of wood fuel pellets made from a raw material mix including sawdust and rapeseed cake. *Biomass and Bioenergy, 35*, 4849–4854.

Sunceco. (2021). *Technical drawings solar cells poly-crystalline 156 × 156 MM 72 PCS. (6 × 12)-4 bus bars.*

Tabatabaei, M., Aghbashlo, M., Valijanian, E., Kazemi Shariat Panahi, H., Nizami, A.-S., Ghanavati, H., et al. (2020). A comprehensive review on recent biological innovations to improve biogas production, Part 1: Upstream strategies. *Renewable Energy, 146*, 1204–1220.

Tian, H., Li, J., Yan, M., Tong, Y. W., Wang, C.-H., & Wang, X. (2019). Organic waste to biohydrogen: A critical review from technological development and environmental impact analysis perspective. *Applied Energy, 256*, 113961.

Trejo-Zamudio, D. (2018). *Producción de pellets de residuos de cultivo de frijol con máximo contenido energético* (Master thesis). Univ. Autónoma Querétaro.

Ugarte, P., Durán, P., Lasobras, J., Soler, J., Menéndez, M., & Herguido, J. (2017). Dry reforming of biogas in fluidized bed: Process intensification. *International Journal of Hydrogen Energy, 42*, 13589–13597.

Varanasi, J. L., & Das, D. (2020). Maximizing biohydrogen production from water hyacinth by coupling dark fermentation and electrohydrogenesis. *International Journal of Hydrogen Energy, 45*, 5227–5238.

Vijin Prabhu, A., Sivaram, A. R., Prabhu, N., & Sundaramahalingam, A. (2021). A study of enhancing the biogas production in anaerobic digestion. *Materials Today: Proceedings, 45*(9), 7994–7999. https://doi.org/10.1016/j.matpr.2020.12.1009.

Wadchasit, P., Suksong, W., O-Thong, S., & Nuithitikul, K. (2021). Development of a novel reactor for simultaneous production of biogas from oil-palm empty fruit bunches (EFB) and palm oil mill effluents (POME). *Journal of Environmental Chemical Engineering, 9*, 105209.

Wang, H., Mustaffar, A., Phan, A. N., Zivkovic, V., Reay, D., Law, R., et al. (2017). A review of process intensification applied to solids handling. *Chemical Engineering and Processing Process Intensification, 118*, 78–107.

Wei, H., Liu, W., Chen, X., Yang, Q., Li, J., & Chen, H. (2019). Renewable bio-jet fuel production for aviation: A review. *Fuel, 254*, 115599.

Won, W., Kwon, H., Han, J.-H., & Kim, J. (2017). Design and operation of renewable energy sources based hydrogen supply system: Technology integration and optimization. *Renewable Energy, 103*, 226–238.

Wong, K. Y., Ng, J.-H., Chong, C. T., Lam, S. S., & Chong, W. T. (2019). Biodiesel process intensification through catalytic enhancement and emerging reactor designs: A critical review. *Renewable and Sustainable Energy Reviews, 116*, 109399.

Yadav, M., & Vivekanand, V. (2021). Combined fungal and bacterial pretreatment of wheat and pearl millet straw for biogas production—A study from batch to continuous stirred tank reactors. *Bioresource Technology, 321*, 124523.

You, Z., Pan, S.-Y., Sun, N., Kim, H., & Chiang, P.-C. (2019). Enhanced corn-stover fermentation for biogas production by NaOH pretreatment with CaO additive and ultrasound. *Journal of Cleaner Production, 238*, 117813.

Zaied, B. K., Nasrullah, M., Siddique, M. N. I., Zularisam, A. W., Singh, L., & Krishnan, S. (2020). Co-digestion of palm oil mill effluent for enhanced biogas production in a solar assisted bioreactor: Supplementation with ammonium bicarbonate. *Science of the Total Environment, 706*, 136095.

Zhang, L., Loh, K.-C., & Zhang, J. (2019). Enhanced biogas production from anaerobic digestion of solid organic wastes: Current status and prospects. *Bioresource Technology Reports, 5*, 280–296.

Zore, U. K., Yedire, S. G., Pandi, N., Manickam, S., & Sonawane, S. H. (2021). A review on recent advances in hydrogen energy, fuel cell, biofuel and fuel refining via ultrasound process intensification. *Ultrasonics Sonochemistry, 73*, 105536.

Zou, Y., & Yang, T. (2019). Chapter 9—Rice husk, rice husk ash and their applications. In L.-Z. Cheong, & X. Xu (Eds.), *Rice bran and rice bran oil* (pp. 207–246). AOCS Press.

CHAPTER 6

Intensified and hybrid distillation technologies for production of high value-added products from lignocellulosic biomass

Le Cao Nhien[a,*], Junaid Haider[b,*], Nguyen Van Duc Long[c,d], and Moonyong Lee[a]
[a]School of Chemical Engineering, Yeungnam University, Gyeongsan, South Korea
[b]Sustainable Process Analysis, Design, and Engineering Laboratory, Energy and Chemical Engineering Department, Ulsan National Institute of Science and Technology (UNIST), Ulsan, South Korea
[c]School of Engineering, University of Warwick, Coventry, United Kingdom
[d]School of Chemical Engineering and Advanced Materials, University of Adelaide, Adelaide, SA, Australia

6.1 Introduction
6.1.1 Lignocellulosic biomass and high value-added products

The world economy is currently overdependent on mineral resources, which have limited reserves and are responsible for global warming. Therefore, the attention on sustainable conversion processes for producing fuels and chemicals from renewable resources has increased in recent years. Lignocellulosic biomass, the most abundant feedstock on Earth, acts as a potential candidate to replace petroleum and is currently prepared for industrial production. Manufacturing facilities that produce biofuels, power, and bio-chemicals from various biomass feedstock are called biorefineries. Conceptually, this is analogous to a petroleum refinery that produces fuels and chemicals from petroleum. However, the complex structure of lignocellulosic biomass, which consists of 30%–55% cellulose, 20%–35% hemicellulose, and 10%–25% lignin is the greatest challenge for the subsequent processing of assorted choice products with optimal recovery. A wide range of value-added products can be produced from lignocellulose. In this book chapter, we focus on the following high-value bioproducts:

➢ Cellulosic ethanol (CE), manufactured industrially by yeast fermentation of C6 sugars obtained from the hydrolysis of cellulose, is the most important biofuel that can be mixed with gasoline without modifying current engines.
➢ Biodiesel, produced from renewable energy resources, is considered a clean, safe, nontoxic, biodegradable, and environmental benign biofuel [43]. The heating value of biodiesel is less than that of petro-diesel but is feasible for diesel engines, emits less

---
* These two authors contributed equally to this work.

carbon, and is environmentally safer due to the presence of O_2 in the biodiesel product line.

- Furfural is a major bio-based platform chemical that competes with oil-based chemicals. In addition to being used as a precursor for the production of many industrial chemicals, up to 65% of all furfural is used as raw material to produce furfuryl alcohol (FOL). Furfural can only be produced from lignocellulosic biomass through the acidic hydrolysis of hemicellulose.
- Levulinic acid (LA), which is listed as one of the top 12 value-added chemicals derived from biomass, is produced through the hydrolysis of cellulose and hemicellulose in the presence of sulfuric acid. LA is used as a starting material for many industrial and pharmaceutical compounds, for example, methyltetrahydrofuran (MTHF), ethyl levulinate, and diphenolic acid (Werpy & Petersen, 2004). LA manufacturers, in particular, could face a potential demand of more than 20,000 kt by 2020 for only MTHF applications (Grand View Research, 2014).
- Biobutanol produced from biomass has gained special attention because of its prior physical and chemical properties that extend their applications as fuel additives for gasoline, diesel oil, and kerosene (Antoni, Zverlov, & Schwarz, 2007). Compared to other liquid biofuels, biobutanol has superior properties, such as a high calorific value (29.2 MJ dm^3), high energy density, and high concentration; additionally, biobutanol does not require retrofitting of vehicle engines and can converted into jet fuel (Bharathiraja et al., 2017)
- 2,3-Butanediol (2,3-BDO) is a synthetic chemical compound that also can be produced by biomass fermentation, which is gaining share in the global market as an intermediate product for numerous applications, i.e., as liquid fuel or fuel additive.

6.1.2 Characteristics of biorefineries

The chemical and biochemical industries are moving toward more sustainable production practices in response to changing consumer demands (Lutze & Gorak, 2013). In particular, during the transition from a petroleum- to a bio-based economy, distillation retains its significance as the primary method for separating mixtures; nonetheless, this workhorse of the chemical industry faces new and significant challenges (Kiss, 2013a; Long & Lee, 2017).

In petroleum refineries, oil is typically separated and purified using distillation technologies (along with liquid extraction, crystallization, absorption, adsorption, and membranes) that can account for 40%–50% of the total cost (Fig. 6.1) (Kiss et al., 2016). A biorefinery is the integral upstream, midstream, and downstream processing of biomass to convert it into a range of products; (de Jong & Jungmeier, 2015) yet, it is not a completely new concept. Many traditional bio-mass conversion industries, such as the sugar, starch, and the pulp and paper industries, can be (partly) considered as biorefineries.

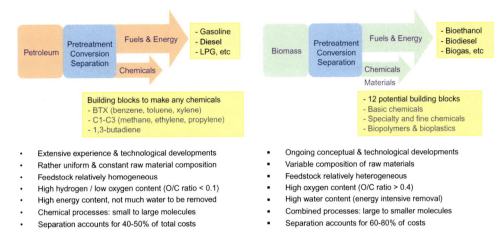

Fig. 6.1 Analogy between petroleum refinery (left) and biorefinery (right) (Kiss et al., 2016).

However, several economic and environmental drivers, such as global warming, energy conservation, supply security, and agricultural policies, have directed these industries to further improve their operations in a manner similar to that in a biorefinery.

A biorefinery has functions similar to those of a classic refinery but uses biomass as feedstock instead of oil. The four primary conversion processes are biochemical (e.g., fermentation and enzymatic conversion), thermochemical (e.g., gasification and pyrolysis), chemical (e.g., acid hydrolysis, synthesis, and esterification), and mechanical (e.g., fractionation, pressing, and size reduction) (de Jong & Jungmeier, 2015). In contrast to separation in petroleum refineries, which do not remove much water, separation in biorefineries faces problems related to high water content. This leads to high energy requirements for separation, which account for 60%–80% of total costs (Kiss et al., 2016). All separation technologies applied in biorefineries are derivatives from the petrochemical industry. As the balance in the properties of streams is different from that in the petrochemical industry, a different balance/emphasis in separation technologies is expected (Kiss et al., 2016).

6.1.3 Distillation process challenges in biorefineries

Separation and purification processes play a critical role in biorefineries and their optimal selection, design, and operation to maximize product yields and to improve overall process efficiency (Ramaswamy, Huang, & Ramarao, 2013). Separation and purification are necessary for upstream processes and for the optimization and improvement of downstream product recovery processes. Distillation is responsible for approximately 3% of the total U.S. energy consumption, more than 90% of all product recovery and purification separations in the United States, and more than 95% of chemical industry consumption worldwide (Vazquez-Castillo et al., 2009). In the change from a petroleum- to bio-based economy, distillation retains its significance as the primary method for separating mixtures, despite facing new and significant challenges (Kiss, 2013a).

Because of the aforementioned characteristics, distillation processes in biorefineries face several specific issues and challenges. One of the primary issues is excess water or a low product concentration in the feed after the conversion step, which necessitate energy-intensive distillation processes, especially when water forms azeotrope mixtures with the main product. The presence of many azeotropes in the bio-based processes make the separation and purification tasks much challenging. The vapor–liquid equilibrium (VLE) and liquid–liquid equilibrium (LLE) data play a crucial role in designing biorefinery processes. However, there is a lack of thermodynamic data of a large number of biochemical. In most commercial process simulators, UNIFAC (UNIQUAC Functional-group Activity Coefficients) has become popular method for estimating the missing binary parameters because of its large range of applicability and reliable predictions (Wittig, Lohmann, & Gmehling, 2003).

Interest in hybrid distillation processes is growing due to the increasing demand for energy-efficient processes in biorefineries. In particular, when separating nonideal multicomponent mixtures into pure components, appropriate combination of existing distillation processes with other separations can overcome existing limitations and produce substantial synergies (Górak & Stankiewicz, 2011; Long & Lee, 2017; Moulijn, Stankiewicz, Grievink, & Górak, 2008). Any single type of separation process is subject to limitations that can be overcome by using intensified hybrid designs (Beneke et al., 2012).

The separation by distillation of mixtures containing large quantities of water is particularly energy intensive because of the presence of a heterogeneous azeotrope between target components and water, which can occur in biofuel and biochemical production from biomass (Cao Nhien, Van Duc Long, Kim, & Lee, 2019; Nhien, Long, Kim, & Lee, 2016a, 2016b; Nhien, Long, & Lee, 2016). In such cases, a hybrid purification process that combines extraction and distillation to purify target components can improve the process efficiency of conventional distillation sequences. Furthermore, an evaporator, which partially removes water and/or solvent, can assist in the distillation process. This configuration can be improved by hybrid-blower-and-evaporator-assisted distillation (Van Duc Long, Hong, Nhien, & Lee, 2018) or a multi-effect-evaporation-assisted distillation configuration (Hong, Van Duc Long, Harvianto, Haider, & Lee, 2019). For removing small quantities of targeted components from a mixture, adsorption or membranes are popular because of their selectivity and attractive options for forming hybrid systems with distillation processes (Long & Lee, 2017).

Microorganisms or enzymes, which are sensitive to temperature, are typically used for conversion processes operated under mild conditions to reduce the probability of the integration of the reaction and distillation in the equipment. Additionally, the presence of numerous compounds or impurities can result in a complex separation sequence that causes many difficulties in synthesis, design, and optimization. Furthermore, insoluble solids, such as cellulose, xylan, lignin, protein, ash, microorganisms, and tar, can be present in distillation (Humbird et al., 2011; Nhien, Long, & Lee, 2017).

6.1.4 Intensified and hybrid distillation technologies for biorefineries

Biorefinery technologies largely lag far behind petrorefinery technologies. The use of biomass for producing biofuels and useful chemicals has faced extreme challenges in recent years in terms of energy consumption and production cost. Principal biorefinery technology platforms that convert biomass into high value-added products are five carbon sugars (hydrolysis of hemicellulose), six carbon sugars (hydrolysis of starch, cellulose, and hemicellulose), hydrogen (by steam reforming, water electrolysis, or dark fermentation), syngas (a mixture of CO and H_2 from gasification of Fischer-Tropsch synthesis), and biogas (methane from anaerobic digestion) (Górak Andrzej, 2018). To make biomass more competitive with fossil fuels, many hybrid and intensified techniques have been used in the production processes, especially in separation and purification. This book chapter focuses on a number of promising intensified hybrid distillation technologies that can considerably improve the process efficiency of biorefineries, such as reactive distillation (RD), dividing wall columns (DWC), hybrid extraction distillation (ED), hybrid membrane distillation (MD), and heat pump-assisted distillation (HPD).

6.1.4.1 DWC

The attention on technologies based on process intensification with complex configurations has recently increased dramatically due to economic and environmental issues. Using the appropriate configuration can achieve considerable savings in terms of energy requirements and capital cost. Remarkably, DWC is one of the best examples for illustrating the benefits of intensified techniques in distillation (Kiss, 2013b; Nhien, Long, & Lee, 2016). Fig. 6.2 shows a schematic diagram of three types of DWC, including top DWC, normal DWC, and bottom DWC. DWC allows reversible splits with no part of the separation being performed more than once and, therefore reduces operating costs by up to 30% and achieve significant capital cost savings compared to conventional sequences (Long, Minh, Nhien, & Lee, 2015; Nhien, Long, Kim, & Lee, 2016a; Nhien, Long, & Lee, 2016). Interest in DWC has increased considerably over the past many years in both academia and industrial applications. The number of industrial DWCs developed by Montz increased sharply from 1984 to 2009, as shown in Fig. 6.3 (Dejanović, Matijašević, & Olujić, 2010).

6.1.4.2 RD

RD, a well-known technique among separation processes, can be advantageously applied to overcome the difficulties and limitations encountered in several chemical processes. In

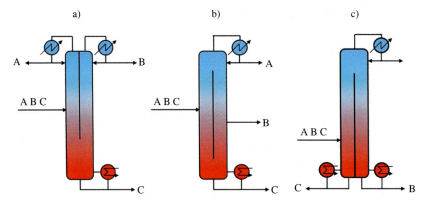

Fig. 6.2 Schematic diagram of the (A) TDWC, (B) DWC, and (C) BDWC.

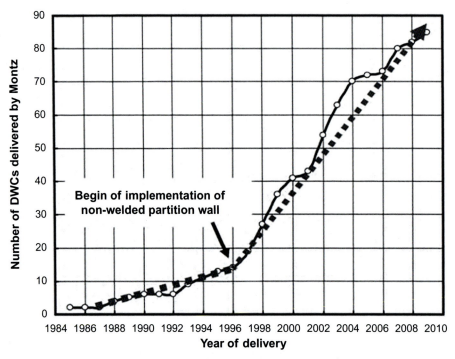

Fig. 6.3 The number of industrial DWCs developed by Montz (Dejanović et al., 2010).

particular, combining a reaction and a separation step in a distillation column can serve as a simpler process and afford higher yields by overcoming equilibrium limitations and poor selectivity of the desired products (Metkar et al., 2015). Therefore, the interest in RD as a promising technology not only to reduce investment costs but also to improve product selectivity and conversion has increased substantially in recent years (Malone &

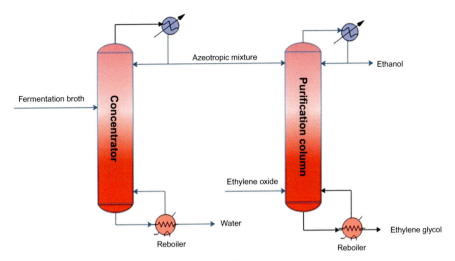

Fig. 6.4 Ethanol dehydration through reactive distillation.

Doherty, 2000). Dehydration of ethanol by RD column is an intensively discussed example of overcome the energy issues associated with azeotropic mixture with the benefit of side product and process economy. Over the years, many researchers improve RD column design by considering the energy efficiency, economics, and control of process. In this context, Kaymak (Kaymak, 2019) has provided the process design and control structure of bioethanol separation from fermentation broth, as shown in Fig. 6.4. It was observed that the ethylene oxide was provided as a component that selectively reacts with water to produce the ethylene glycol as an additional product. As a result, pure ethanol was recovered from the column with the improved economics as 19.3% of overall TAC savings were observed compared to conventional methods.

6.1.4.3 Heat pump-assisted distillation

Distillation-based separation processes are considered more reliable due to their gradual maturity and well-developed design. Thus, most processes are equipped with distillation operations despite their high energy consumption (Haider, Qyyum, Minh, & Lee, 2020). High energy consumption is directly related to greenhouse gas emissions, thus resulting in environmental hazards. In this context, many researchers have made efforts to introduce modified and intensified schemes based on specific examples. HPD has emerged rapidly in different applications. Compared to conventional schemes, HPD is generally considered a more promising and suitable approach that not only reduces energy consumption but also lowers overall emissions with a more economical process (Jana, 2014). The operational scheme of HPD is demonstrated in Fig. 6.5. Heat supplied to a reboiler through external utility contributes additional heat produced from the vapor recompression effect by installing a compressor at the top of the distillation column. The

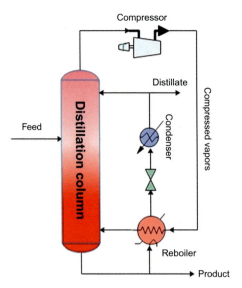

Fig. 6.5 Heat pump-assisted distillation scheme with application of vapor recompression effect.

coupling of heat within a system is a promising and innovative technique to mitigate emissions, increase thermal efficiency, and save energy. However, this innovative scheme is only possible under certain conditions. The temperature difference of vapors from the top stage and bottom stream must be minimal, the allowable temperature limit in the compressor cannot be exceeded to avoid operational limitations, and the cost of additional equipment must be considered. The effect of HPD is studied with intensified hybrid techniques for process enhancement and increased thermodynamic efficiency. Kazemi, Mehrabani-Zeinabad, and Beheshti (2018) highlighted recently developed HPD configurations. The major findings were projected for vapor recompression, bottom flashing, and external heat pump schemes. The separation of propylene from a propylene-propane mixture was performed for 18 different heat pump-assisted distillation schemes. Each technique has specific tradeoffs when increasing the heat pump effect; however, the vapor recompression system exhibits the lowest operating expenses on an annual basis with minimal total annualized cost (Kazemi et al., 2018). Nevertheless, this analysis is only used for specific applications. Other feed mixtures with more typical characteristics may have different operational results depending on the type, feed composition, objective functions, and process constraints.

6.1.4.4 Hybrid extraction-distillation

Hybrid extraction-distillation involves the integration of two commercial techniques (extraction and distillation) to enhance energy efficiency. Such integrated techniques are usually adopted when product recovery is economically infeasible due to the

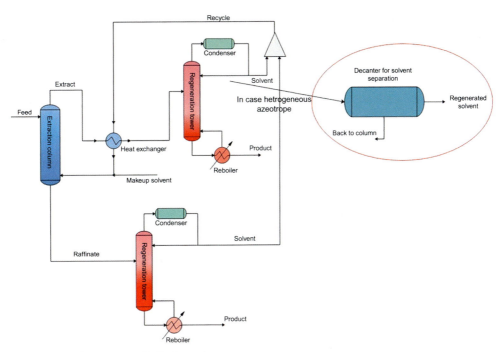

Fig. 6.6 Hybrid extraction-distillation scheme.

formation of azeotropes and low-concentration production in the feed mixture. However, the selective hybrid approach primarily depends on certain factors, such as the choice of solvent based on price, ease of regeneration, solvent-to-feed ratio, separation factor, or selectivity in the extraction section for product recovery; the purification of the product and recovery of the solvent for recycling are performed in subsequent distillation columns based on relative volatility differences. A schematic of the hybrid extraction-distillation process is shown in Fig. 6.6. Solvent properties are the most crucial factors to consider when selecting the solvent as the boiling point, toxic nature of the solvent, density, viscosity, and cost may act as operational constraints and retard the practical, large-scale implementation of solvents (Skiborowski, Harwardt, & Marquardt, 2013). Therefore, prior to process design, these major factors are first considered and experimentally validated in laboratories. Similarly, determination of the optimum solvent flow in the extraction column is of significant importance for estimating energy consumption in the solvent regeneration column, because a higher solvent rate requires high energy consumption. The solvent must not produce any azeotrope with any product or component in the mixture. Nevertheless, many industrial mixtures are complex and present a miscibility gap, thus creating heterogeneous azeotropes. The exploitation of this gap is performed by the combination of a decanter at the top of the distillation column

(Haider, Harvianto, Qyyum, & Lee, 2018) (Fig. 6.6). Practical examples of hybrid extraction–distilling-based process designs are illustrated in the next section.

6.1.4.5 Hybrid membrane-distillation

Considering energy efficiency and global warming hazard, membrane technology has attracted significant attention for commercial implementation, specifically in biorefineries. Membrane separation has several advantages over other available techniques, such as operating at normal temperature without addition of a third component, and a phase change that results in low-energy consumption (Lipnizki, Thuvander, & Rudolph, 2019). Moreover, the recovery of low-concentration products is easy and commercially viable with membrane technology, especially in biorefineries. This is the fact that the current membrane market has a massive annual share in developing the bioeconomy of $300–400 million USD (Lipnizki et al., 2019). Among available membrane technologies, vapor permeation, pervaporation, electrodialysis, and gas separation are emerging and widely used in biorefineries. Considering liquid biofuels, the membrane process is used to concentrate the final product obtained from fermentation, to break azeotropes, to recover low-concentration products, and to purify the concentrated product. However, no single technique viably meets the economical target purity of any biofuel. Hence, a hybrid of a membrane with distillation may be a possible solution. A schematic of the hybrid membrane-distillation scheme for light key separation is presented in Fig. 6.7. In this regard, the separation and purification of bioethanol, biodiesel, and biobutanol are energy and cost-effective with the selective adoption of hybrid membrane-distillation schemes (Hajilary, Rezakazemi, & Shirazian, 2019; Khalid et al., 2019). Possible combinations of membrane and distillation may vary by product, specified target purity, and product properties.

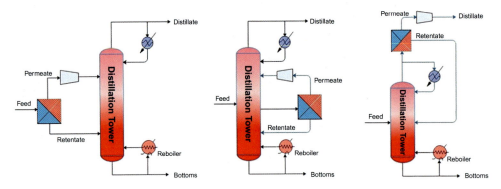

Fig. 6.7 Hybrid membrane-distillation configurations specific for light key component recovery.

6.2 Application of intensified and hybrid distillation technologies for high value-added processes from biomass

6.2.1 Cellulosic ethanol (CE)

Bioethanol is one of the most important biofuels that can be mixed with gasoline without modifying current engines to reduce carbon monoxide and other smog-causing emissions. The first generation of bioethanol is mostly produced from the sugar fermentation of food crops, such as corn and wheat. However, due to the food-versus-fuel scenario, the production of second-generation bioethanol, CE, from nonedible lignocellulose has been extensively explored. Renewable lignocellulose, such as corn stove-ethanol can reduce carbon dioxide (CO_2) emissions 37% and 86% more than corn-ethanol and oil resources, respectively, resulting in more sustainable bioethanol production (Christian, 2015; Halder, Azad, Shah, & Sarker, 2019).

Typical CE production consists of three steps: pretreatment, fermentation, and separation. First, biomass undergoes a pretreatment step before being introduced to the fermentation step to produce the fermentation broth containing 5–12 wt% ethanol. Subsequently, the diluted CE is further concentrated over 99.8 wt% to comply with all standards. Energy-intensive separation techniques proposed to overcome the azeotrope between ethanol and water (95.6 wt% ethanol) include adsorption, extractive distillation (ED), azeotropic distillation (AD), pervaporation, vapor permeation, and pressure swing distillation (Huang, Ramaswamy, Tschirner, & Ramarao, 2008; Humbird et al., 2011; Kiss & Suszwalak, 2012; Loy, Lee, & Rangaiah, 2015). Of these, the capacities for adsorption, pervaporation, and vapor permeation have reached their limits and are cost-intensive on an industrial scale (Frolkova & Raeva, 2010). ED is more energy-effective than AD and is commonly used in industry for the production of anhydrous bioethanol (Huang et al., 2008). Kiss and Ignat proposed an extractive DWC configuration to improve the energy efficiency of the bioethanol dehydration process (Kiss & Ignat, 2012a). The extractive DWC achieves high purity ethanol of 99.8 wt% by using only one distillation column. The proposed process saves 17% more energy than the conventional process—a promising case for a grassroots project but also as a retrofit for existing plants. However, the use of ethylene glycol (EG) as the extracting solvent may cause several serious environmental issues and has been restricted in several countries (Schladt, Ivens, Karbe, Ruhl-Fehlert, & Bomhard, 1998). Other entrainers such as inorganic salts and ionic liquids (ILs) were explored to enhance the separation of ethanol–water azeotrope (Hussain & Pfromm, 2013; Lei, Dai, Zhu, & Chen, 2014; Meindersma, Quijada-Maldonado, Aelmans, Hernandez, & De Haan, 2012). Hussain et al. proposed saline ED using calcium chloride ($CaCl_2$) as an economical approach to the ethanol purification process (Hussain & Pfromm, 2013). ILs, which utilized the advantages of solid salts (high separation ability) and liquid solvents (easy operation) have received considerable attention (Lei et al., 2014). Meindersma et al. carried out the pilot plant

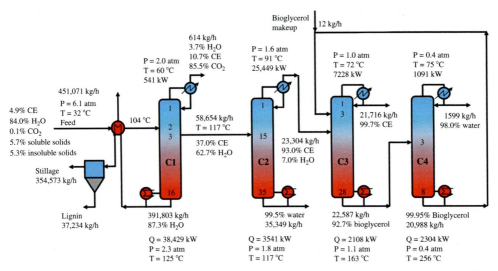

Fig. 6.8 Schematic diagram of the conventional CE process from actual fermentation broth (Nhien, Long, & Lee, 2017).

experiments with both ionic liquid 1-ethyl-3-methylimidazolium dicyanamide, [C_2mim]N(CN)$_2$, and EG (Meindersma et al., 2012). The results showed that the [C_2mim]N(CN)$_2$ process can reduce 16% energy requirement compared to the EG process after heat integration.

AD techniques are also an efficient way to overcome the ethanol-water azeotrope. An azeotropic DWC with an n-pentane entrainer was reportedly feasible for bioethanol dehydration, reducing the specific energy demand from 1.78 to 1.42 kWh/kg compared with the azeotropic conventional process (Kiss & Suszwalak, 2012).

However, most studies on the bioethanol separation process assume that the feed from the fermentation broth is comprised of only bioethanol and water. This assumption leads to a simpler process but does not represent the real lignocellulose-based feed composition. In fact, breaking down the cellulose–hemicellulose–lignin structure of lignocellulosic biomass leads to the presence of more compounds in fermentation output, resulting in a more complex separation step. The National Renewable Energy Laboratory (NREL) reported an excellent study on CE production from actual lignocellulose of corn stover (Humbird et al., 2011), which was converted to CE through acidic hydrolysis, enzymatic saccharification, and fermentation processes. The fermentation broth containing 5–12 wt% CE was introduced to the recovery and dehydration processes to achieve commercial-grade purity. Based on the NREL process, Nhien, Long, and Lee (2017) designed the CE dehydration process using ED with bioglycerol as the extracting solvent. Fig. 6.8 depicts the CE production process from actual fermentation broth with key stream information and process design parameters.

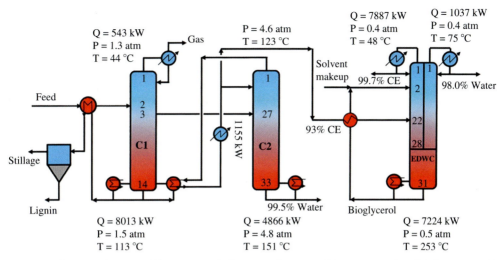

Fig. 6.9 Schematic diagram of the proposed CE process from actual fermentation broth (Nhien, Long, & Lee, 2017).

The actual fermentation broth, which comprises gas and soluble and insoluble solids, makes the recovery process more challenging. The feed was first sent to C1 to remove CO_2 in the overhead, insoluble solids and approximately 90 wt% of water in the bottom. Subsequently, the function of C2 is to concentrate the raw CE stream close to the azeotrope composition and deliver it to the ED process. The ED uses bioglycerol as the extracting solvent and consists of two columns: extractive column C3 to separate the CE and solvent recovery column C4 to recover bioglycerol, which is recycled back into C3. Notably, the use of a green bioglycerol solvent—a major by-product of biodiesel production—can achieve a high degree of integration in a biorefinery context. To increase the energy efficiency of CE purification, several improvements were proposed: optimization of the CE concentration after the recovery step, double-effect heat integration of the recovery process, and extractive DWC for the dehydration process. Fig. 6.9 shows the improved CE process with the key design parameters. Remarkably, the extractive DWC configuration reduces energy requirements by 11.1% compared to the ED process, whereas the proposed process saves up to 47.6% of the total annual cost (TAC) compared to the conventional CE process.

6.2.2 Biodiesel

Produced from renewable energy resources, biodiesel is considered a clean, safe, nontoxic, biodegradable, and environmental benign biofuel (Knothe, Krahl, & Van Gerpen, 2010). Consequently, market demand for biodiesel is gradually rising. However, drawbacks encountered during processing include pumping, atomization, and combustion issues due to the high viscosity of biodiesel. Similarly, fuel injectors coking and engine wear are critical concerns for biologically produced diesel fuel. Moreover, product purity at the industrial level is another aspect that depends on the raw material, catalyst and reactor type, and production method used to produce the biodiesel. Biodiesel produced from biomass (e.g., agricultural) sources generates particularly serious criticism for potentially increasing food and animal prices and food insecurity. In this context, biomass sources and associated land that are not feasible for growing food have less impact on sustainability. Nevertheless, feedstock type, production process, and volume have significant importance in comprising the biodiesel price, competitive market rates with petro-diesel, and sustainability. The heating value of biodiesel is less than that of petro-diesel but is feasible for diesel engines, emits less carbon, and is environmentally safer due to the presence of O_2 in the biodiesel product line.

Biodiesel production requires crucial details on feedstock, reaction kinetics, process parameters, process equipment, and detailed purification, energy, economic, and environmental analyses. The synthesis of biodiesel involves a reaction called transesterification of glycerides, whereby the product is generally obtained as glycerol. Methanol, the preferred alcohol for the reaction, has the dual advantages of being cheaper and producing lightweight esters as by-products, which are easily separable from biodiesel (Dhawane, Kumar, & Halder, 2018). Because glycerol is a byproduct, the presence of unreacted alcohol in the product mixture may largely reduce the combustion quality of biodiesel; whereas other impurities, such as water, can undergo production of free fatty acids and cause corrosion (Berrios & Skelton, 2008), setting problems, low oxidation stability, and reduced engine efficiency (Berrios & Skelton, 2008). Hence, the purification of biodiesel is a mandatory step before delivery and end use. The concentration of the final product varies by source, and other effective parameters are associated with the methodology and catalyst. The key factors for upgrading biodiesel are directly dependent on the amount of water, the type of catalyst, and methanol.

In the conventional approach, biodiesel is produced and upgraded in a series of columns sequenced as a reaction chamber, separator, biodiesel wash column, alcohol recovery column, and glycerol recovery column (Fjerbaek, Christensen, & Norddahl, 2009) (Fig. 6.10). However, biodiesel purification is conducted through either wet or dry washing techniques. Despite being most adopted approach, wet washing presents challenges, such as increased overall processing time and investment (Berrios & Skelton, 2008). Additionally, the separation of biodiesel from water is critical, and excess wastewater is difficult to process (Saleh, Tremblay, & Dubé, 2010). Dry washing of biodiesel is also commercialized and employed through ion-exchange resins as a substitute for water. Both technologies are commercialized and working well with the specified targets; however, the challenges corresponding to energy and processing costs persist, as the sustainable development of renewable resources still face numerous challenges compared to

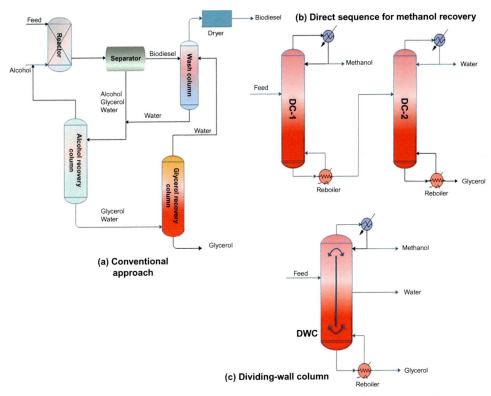

Fig. 6.10 Biodiesel separation and purification steps: (A) conventional approach, (B) methanol recovery in intensified direct sequence, (C) intensified dividing-wall column for enhanced recovery of methanol in biodiesel separation.

conventional fossil fuel resources. In this respect, intensified energy and cost-efficient approaches are continuously sought by practitioners and researchers.

6.2.2.1 DWC

Purification of the biodiesel mixture is usually performed in a decanter for the separation of the biodiesel- and glycerol-rich phases; the excess alcohol remaining at the end of each phase is further processed to recover and recycle back as a reactant. Therefore, a ternary mixture of alcohol-water-glycerol must be separated. The conventional approach for biodiesel purification requires multiple pieces of equipment, and each associated column has a high energy requirement, which adds additional equipment cost; additionally, an enormous amount of energy is required to recover each constituent. In this context, DWC has been identified as an energy and cost-effective solution for multicomponent feed mixtures compared to conventional distillation schemes. Additionally, DWC is the only recognized commercial-scale process intensification model that saves investment cost (30%) and reduces energy consumption (by up to 40%). A similar approach was introduced by Kiss and Ignat (2012b) to upgrade biodiesel through the conventional direct sequence of distillation columns and further intensification in a single divided wall column (DWC) (Fig. 6.10). The authors present a detailed design of glycerol–water–methanol separation from a biodiesel product line leading from a steady state to a dynamic simulation using Aspentech software and concluded that the energy consumption and TAC compared to those of the direct sequence were reduced to 27% and 25%, respectively; additionally, carbon emissions were reduced by up to 26.7%. Moreover, the SEC was reduced to 26.7% compared to the base case (Kiss & Ignat, 2012b). Despite the optimal design claimed by Kiss and Ignat (2012b), distillation is an energy-intensive operation and a single process is usually insufficient or energetically inefficient for biofuel separation and purification (Harvianto et al., 2018). Therefore, a hybrid approach combining two or more techniques for the efficient recovery and purification of biofuels is evaluated as energy and cost effective (Haider, Harvianto, et al., 2018; Nhien, Long, Kim, & Lee, 2016b; Novita, Lee, & Lee, 2018).

6.2.2.2 Reactive DWC

An intensified hybrid scheme to simultaneously produce and separate biodiesel through RD yielded more biodiesel and separation in a single unit, unlike conventional schemes (Kolah, Lira, & Miller, 2013). Other advantages include less equipment, decreased alcohol recovery unit, increased the energy efficiency, and overall process investment reduction. Moreover, glycerol is proven to be an advantageous entrainer for breaking the water/ethanol azeotrope; thus, in a result, the cost of production of high purity ethanol within a system can be reduced (Navarrete-Contreras, Sánchez-Ibarra, Barroso-Muñoz, Hernández, & Castro-Montoya, 2014). Consequently, coupling the integrated system largely reduced the overall energy requirement and total investment.

Hernández et al. (Cossio-Vargas, Barroso-Muñoz, Hernandez, Segovia-Hernandez, & Cano-Rodriguez, 2012; Cossio-Vargas, Hernandez, Segovia-Hernandez, & Cano-Rodriguez, 2011) designed complex reactive and thermally coupled distillation schemes to estimate the feasibility of biodiesel production and purification in a single step using sulfuric acid as a homogenous catalyst. The results of the study indicated that the process is more favorable in terms of cost and energy. Similarly, Kiss et al. (Kiss, 2011) studied the effect of reaction, production, and the separation mechanism of biodiesel in an RD column and concluded that the heat-integrated RD design has the potential to replace the conventional scheme with a specific load of 108.8 kWh/ton ester and the additional benefits of zero catalyst loss, the absence of soap formation, and reduced TAC and energy consumption. However, RD configurations are capital-intensive compared to simple distillation operations. Further, the purification of water and biodiesel requires additional units. Hence, based on thermodynamic equilibrium, the RD column would be further intensified to a complex reactive divided wall column (R-DWC), resulting in high purities

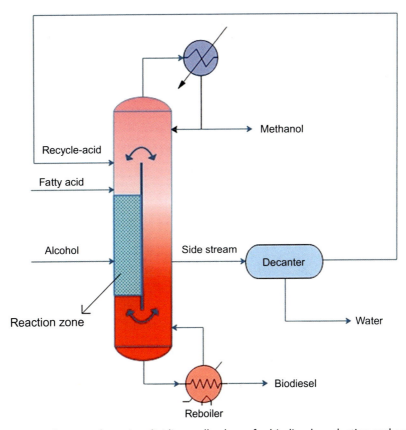

Fig. 6.11 Process diagram of reactive-dividing wall column for biodiesel production and separation.

of all three streams (water-alcohol-biodiesel). The authors (Kiss, Segovia-Hernández, Bildea, Miranda-Galindo, & Hernández, 2012) proposed an R-DWC design with the products with methanol from the top of the column, water as the side product, and biodiesel as the bottom product, depending on relative volatility differences (Fig. 6.11). Consequently, excess methanol of 15% was determined to be required for the complete conversion of feedstock to biodiesel, and energy consumption was reduced to 25% compared to conventional biodiesel processes. Gómez-Castro et al. (Gómez-Castro et al., 2012) designed and optimized thermally coupled RD (TCRD) column configurations. R-DWC and TCRD showed a significant reduction in energy consumption (23%–25%) after optimization of proposed schemes. However, concerns related to economic aspects in the proposed designs may be serious, as reactions occur at high temperatures and pressures. Radu et al. (Ignat & Kiss, 2013) studied the dynamics and control structure of R-DWC for biodiesel production using Aspen Plus commercial software and dynamics to perform simulations and analysis and concluded that the vapor feed of alcohol is best suited for the desired product specification. Scientists perform simulation analysis and present energy-saving opportunities for intensified reactive separation approaches. However, the commercialization of R-DWC (Fig. 6.11) remains an open issue due to its complexity, control issues, economic viability, handling, and large-scale upgradation.

6.2.3 Furfural

Furfural, which was listed as one of the top 30 potential chemicals from biomass by NREL, is an important renewable resource for producing biofuels and biochemicals (Werpy & Petersen, 2004). Furfural is used as an industrial solvent, and approximately 65% of the furfural produced is used for FOL production (Zeitsch, 2000). The first furfural plant was built by Quaker Oats in 1921, and the same basic process remains in use today. Currently, a large proportion of furfural in the world market is produced in China, based on small-scale fixed-bed reactor technology with a low yield of approximately 50% (De Jong & Marcotullio, 2010). Furfural technologies have shown little enhancement over a century since most existing furfural industrial processes are based on the original Quaker Oats process. Accordingly, a significant technological improvement is necessary to upgrade furfural production to compete with petroleum-based products.

The typical furfural production process from agricultural residues rich in hemicellulose (xylan) involves two steps: reaction and purification. In the reaction step, xylan is hydrolyzed to xylose by aqueous acid catalysis, and the xylose is dehydrated to furfural via a single process step. The raw furfural stream then undergoes purification to achieve commercial-grade purity. Recently, with a strong push for sustainable chemistry worldwide, many studies were carried out to improve furfural yield and reduce production costs. The Escher Wyss process using fluid bed reactor technology reported the residence time of 45 min at a reaction temperature of 170°C (Zeitsch, 2000). In the Stake process, a feeder gun system was applied, reducing the residence time to 6.3 min at 230°C (Zeitsch, 2000). Notably, this process could achieve 66% furfural yield without using foreign acid as a catalyst.

6.2.3.1 RD

For those processes, including both reaction and purification steps, RD, which can combine the reaction and distillation into a single unit, is the most promising intensified technology to improve process efficiency. A batch RD was proposed to produce furfural from various types of biomass feeds (Mandalika & Runge, 2012). In the RD column, the furfural product was instantly removed from the reacting liquid phase by the vapor upflow. As a result, the furfural loss reactions could be prevented, and the yield of 85% was achieved. Recently, a continuous RD process was developed by Metka et al. using solvent sulfolane and solid catalysts. In this process, furfural produced was instantly removed from the liquid phase by steam injected from the RD bottom. The furfural yield of 75% was obtained at an operating temperature of 175°C. To make those processes become

Intensified and hybrid distillation technologies   215

more feasible on industrial scale, further optimization is necessary to solve technical issues, e.g., solvent regeneration, catalyst deactivation, and regeneration.

6.2.3.2 DWC

It is worth noting that most previous research focused on reaction steps to improve yield, although the furfural purification step is also energy-intensive. Typically, the raw furfural stream from the top of furfural reactor is introduced into the purification step to achieve a furfural purity target. As shown in Fig. 6.12, the conventional furfural purification consists of three distillation columns and there are potential to combine those columns with the intensified technique DWC (Nhien, Long, Kim, & Lee, 2016a).

Fig. 6.13 shows the intensified furfural purification process with detailed key design parameters (Nhien, Long, Kim, & Lee, 2016a). In this process, light component, methanol, was delivered at top product, wastewater as bottom stream, and the azeotrope as a side stream. The bottom DWC combines the C1 and C2 in one distillation column. The heterogeneous azeotrope between furfural and water was separated naturally through a decanter. Furfural was collected as the C3 bottom product while the C3 top distillate was recycled to the decanter. Heat pump and preheating were also applied to improve the

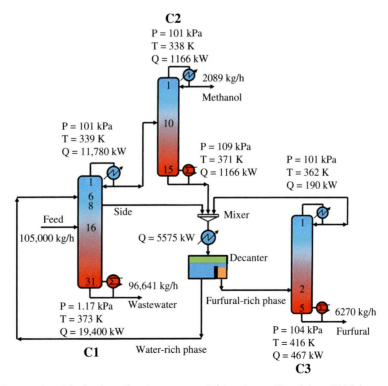

Fig. 6.12 Conventional furfural purification process (Nhien, Long, Kim, & Lee, 2016a).

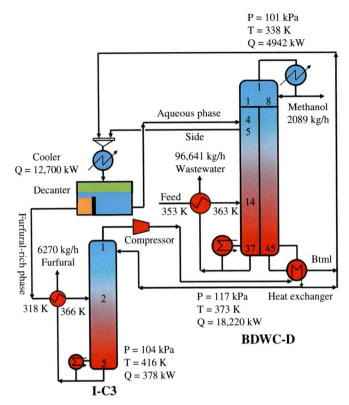

Fig. 6.13 Intensified furfural purification process with DWC (Nhien, Long, Kim, & Lee, 2016a).

heat recovery. The intensified process can reduce 11.6% of operating cost, 10.1% of TAC, and 11.6% CO_2 emission compared to the conventional process. The proposed DWC structure is promising in both grassroots and retrofit projects.

6.2.3.3 Hybrid ED

For the separation of furfural from the bulk water and several minor components, liquid–liquid extraction may be promising. However, finding an effective solvent and designing a solvent regeneration part make an extraction method more challenging. A comprehensive procedure for solvent selection was proposed to find the most favorable solvent as shown in Fig. 6.14 (Nhien, Long, Kim, & Lee, 2017): screening numerous solvents from literature, examining separation feasibility, designing processes, estimating costs.

Butyl chloride was the most suitable solvents as its process can save up to 44.7% of TAC compared to the toluene process. Besides, compared with the purification process

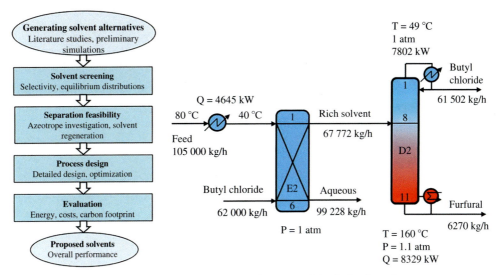

Fig. 6.14 A comprehensive procedure for solvent selection and the furfural purification process using butyl chloride solvent (Nhien, Long, Kim, & Lee, 2017).

using the distillation technique, the hybrid ED can reduce 19.2% and 58.3% in terms of TAC and CO_2 emissions, respectively.

6.2.4 Biobutanol and Biobutanediol

Among available commercialized biofuels, biobutanol produced from biomass has gained special attention because of its prior physical and chemical properties that extend their applications as fuel additives for gasoline, diesel oil, and kerosene (Antoni et al., 2007). Compared to other liquid biofuels, biobutanol has superior properties, such as a high calorific value (29.2 MJ dm^3), high energy density, and high concentration; additionally, biobutanol does not require retrofitting of vehicle engines and can converted into jet fuel (Bharathiraja et al., 2017). Commercial-scale production of biobutanol is generally obtained through Clostridium species acting as a biocatalyst. Other species used for production are *Escherichia coli*, Pseudomonas sp., Saccharomyces sp., and *Bacillus subtilis* for the biosynthesis of biobutanol. However, acetone-butanol-ethanol (ABE) fermentation for the production of butanol depends on multiple factors, including the type of substrate, fermentation conditions, biocatalysts, process systems engineering aspects, and life-cycle assessment.

Similarly, owing to the high octane number and heating value comparable to that of bioethanol, 2,3-butanediol (BDO) has better drop-in-fuel properties. Further, BDO has many other industrial applications; a blend of acetone-BDO-ketal is used as octane booster, and BDO dehydration produces industrial chemical methyl ethyl ketone used as an intermediate in the rubber industry (Ji, Huang, & Ouyang, 2011; Wang et al., 2013). Consequently, BDO and its derivatives are valued at approximately $43 billion on the global market (Köpke et al., 2011). However, the production of BDO from biomass presents numerous challenges, including complex product mixtures, low product quality, and expensive downstream processing. The fermentation product has a limited range of biobutanol and BDO productivity and purity, and the recovery and purification steps must be considered to ensure availability for commercial use.

6.2.4.1 Intensified ABE separation schemes

Conventional methods to achieve high-purity products from ABE fermentation include distillation, membrane separation, adsorption, solvent extraction, and membrane solvent extraction. Distillation is the most common and oldest technique for product recovery; however, in the case of ABE separation, energy consumption is usually very high, owing to the excess water evaporation and water-butanol azeotrope formation at 92.7°C (Oudshoorn, 2012). In this context, Vane et al. (Vane, Alvarez, Rosenblum, & Govindaswamy, 2013) estimated the energy requirements to obtain 99.5 wt% pure butanol from the ABE fermentation product and observed that conventional distillation decanter schemes require 14.5 MJ/kg energy. Similarly, Green et al. (Green, 2011) reported that distillation is an energy intensive yet robust and well-proven technique that requires 12 tons of steam to produce 1 ton of pure solvents from the ABE product. Similar aspects can be seen in other available techniques, such as adsorption; adsorbent regeneration requires several separation methods prior to reuse, and bacterial action can harm the adsorbent and decrease its efficiency (Kujawska, Kujawski, Bryjak, & Kujawski, 2015). Compared to other separation methods, high selectivity can be achieved in solvent extraction, but the primary disadvantages associated with this method are emulsion formation and extractant fouling; additionally, recovery of the solvent requires an additional separation column, and the product obtained after extraction requires further purification at the industrial/commercial level (Kujawska et al., 2015). Severe issues with membrane separation can be observed due to the presence of multiple components in the fermentation product, the low mass transfer rate, membrane swelling, membrane solvent extraction due to the high extractant viscosity, and pressure loss (Kujawska et al., 2015). The production of pure butanol obtained from low-concentration fermentation broth requires either high energy consumption or high investment cost, and the single technique is not sufficiently economical to efficiently recover biobutanol. Process enhancement, integration, and intensification are approaches to potentially overcome these economic challenges. Many researchers (Blahušiak et al., 2018) have recently introduced intensified hybrid designs that are energy efficient and competitive in commercial markets.

In this regard, Patrașcu et al. (Patrașcu, Bîldea, & Kiss, 2017) presented an intensified eco-efficient biobutanol separation from ABE fermentation. The authors claimed that conventional schemes require a multiequipment energy-intensive distillation configuration compared to heat-integrated DWC. The results showed that the specific energy consumption decreased from 2.28–1.71 kWh/kg of butanol in the case of three product recoveries (ABE), whereas a 45% reduction was achieved for butanol compared to the base case (14.5–79.5 to 4.46 MJ/kg). Furthermore, the total annualized cost savings obtained were 20% greater compared to those of the conventional approach (Patraşcu et al., 2017). The authors (Patraşcu, Bîldea, & Kiss, 2018) further proposed an intensified vapor recompression-assisted DWC to minimize energy expenditure and claimed that the proposed scheme saves up to 58% of total energy, increases the intensification factor (energy basis) by 2.33, and obtains TAC savings >24% compared to those of the base case. However, the inclusion of a solvent-assisted extraction column with conventional and advanced distillation schemes is advantageous for minimizing energy consumption and breaking homogeneous and heterogeneous azeotropes in ABE-water feed (Errico, Sanchez-Ramirez, Quiroz-Ramìrez, Rong, & Segovia-Hernandez, 2017). Errico et al. (Errico et al., 2017) studied alternative separation configurations and optimized the process by maintaining the triple objective functions of TAC, the environmental assessment (eco indicator 99), and the controllability behavior of the process evaluation of ABE separation. The authors claimed that the scheme based on an extractor integrated with a DWC containing two reboilers saved 22% of TAC, improved the environmental evaluation by 18%, and improved controllability. Similarly, Segovia-Hernández et al. (2020) analyzed intensified schemes developed for biobutanol purification. Based on hybrid and intensified schemes, the authors evaluated the risk quantification, the green degree factor, and the dynamic process. The results indicated that the green properties were enhanced as the degree of intensification increased for biobutanol separation processes. Similar effects were observed in the evaluation of dynamic properties.

Although the degree of effectiveness improved in each selective process, the implementation of the proposed scheme remains a challenging issue, as many other factors exist, such as the degree of uncertainty, rigorous economic efficiency, detailed control analysis, practical process efficiency, production rate, and robust design.

6.2.4.2 Process intensification schemes for 2,3-butanediol

The recovery of BDO from fermentation broth requires intensive separation techniques because of the complex product mixture that includes light and heavy components. A mixture of fermentation broth generally consists of excess water and other trace acidic mixtures with solid impurities. Solid impurities are removed through filtration, and the remaining liquid stream is rich in >87 wt% water, 4–9 wt% BDO, and liquid impurities (Jeon et al., 2014). BDO does not have any azeotrope with water, and the large boiling temperature difference between water (100°C) and BDO (180°C) makes separation easy and compatible with a simple distillation column. However, the distillation operation for this specific feed is highly energy intensive, as a huge amount of water is evaporated, increasing the load on the reboiler, which requires excess thermal energy to complete

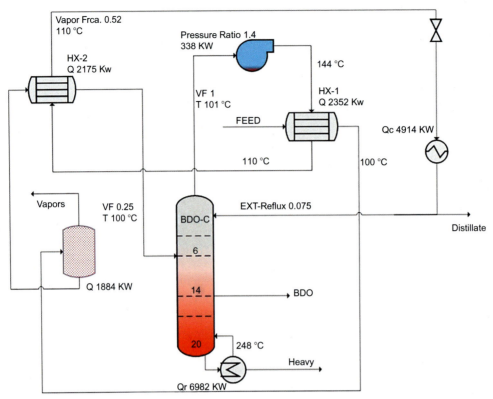

Fig. 6.15 Hybrid-evaporator assisted vapor recompression distillation scheme for 2,3-butanediol separation and purification.

the process. Haider et al. (Haider, Qyyum, Hussain, Yasin, & Lee, 2018) proposed different schemes based on distillation and estimated the energy requirements for each design. The authors claimed that pure BDO could be obtained as a product from the side stream of a single distillation column. They further intensified the distillation schemes based on hybrid evaporator-assisted distillation and blower-assisted vapor recompression distillation schemes to enhance the energy efficiency as shown in Fig. 6.15 (Haider et al., 2020; Haider, Qyyum, et al., 2018; Van Duc Long et al., 2018). On the one hand, the results showed that the vapor recompression scheme using novel blower-assisted distillation saved energy costs by up to 51% and reduced TAC expenses by up to 55%. On the other hand, the hybrid evaporator-assisted configuration reduced the amount of energy and the cost by 66.6% and 61.2%, respectively. A similar group (Haider et al., 2020) further enhanced the process configuration and estimated the operating and annual expenses based on hybrid DWC designs. The results revealed operating cost and TAC savings up to 31% and 25%, respectively.

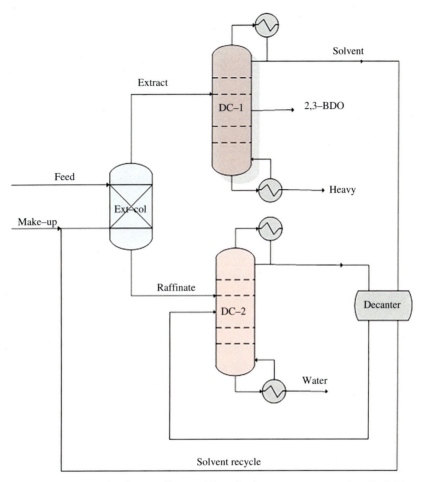

Fig. 6.16 Hybrid extraction-distillation of butanol-based solvent to recover and purify 2,3-butanediol.

Harvianto et al. (2018) introduced a hybrid extraction-distillation approach for BDO recovery and presented a detailed design based on rigorous solvent selection, process design, and economic evaluation. Oleyl alcohol was chosen as the leading solvent because of its superior properties, such as distribution coefficient, selectivity, boiling temperature differences, and density. In conclusion, the thermal load and TAC were reduced by 54.8% and 25.8%, respectively, compared to the conventional distillation approach. However, oleyl alcohol is viscous and costly. Therefore, further intensification (Haider, Harvianto, et al., 2018) was performed to select butanol-based solvents to overcome the cost and viscosity issues as shown in Fig. 6.16. Because butanol forms heterogeneous azeotropes with water, solvent recovery was considered in the decanter-assisted distillation column. The process evaluation showed that the thermal load decreased to

27%–32.9%; operating expenses and TAC savings reached 27%–34% and 24%–31.3%, respectively.

6.2.5 LA

LA was listed as one of the top 12 potential candidates by the NREL and PNNL (Pacific Northwest National Laboratory) based on the potential markets, derivatives, and the complexity of the synthetic pathways (Werpy & Petersen, 2004). However, LA production on an industrial scale is limited due to the expensive feedstock, low synthetic yield, and lack of detailed process design (Seibert, 2010). LA is typically converted from C6 sugar by acidic hydrolysis and introduced into a recovery process to achieve purity target.

6.2.5.1 Hybrid ED

Liquid–liquid extraction is promising to separate LA from an aqueous stream obtained from the acid hydrolysis (Nhien, Long, Kim, & Lee, 2016b). Nhien et al. proposed an comprehensive procedure of solvent selection for LA recovery process (Nhien, Long, Kim, & Lee, 2016b). This procedure consists of five steps: generating solvent alternatives from literature, screening solvent based on selectivity, properties, examining separation feasibility, designing processes, and calculating costs. The results showed that furfural is the most favorable solvent for the LA recovery process as the furfural process can reduce 30.7% of TIC, 30.6% TAC, and 27% TAE compared to the octanol process.

Fig. 6.17 depicts the hybrid ED configuration using furfural solvent for LA recovery process. The aqueous feed was fed into the extractor to produce a furfural-rich stream at the bottom before introduced to a series of distillation columns to achieve commercial-grade purity of LA. Furthermore, the advantage of this hybrid ED process is that one of the products, furfural, can be used as an extraction solvent.

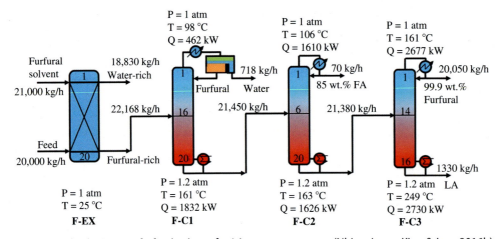

Fig. 6.17 Hybrid ED using furfural solvent for LA recovery process (Nhien, Long, Kim, & Lee, 2016b).

6.2.5.2 DWC

To make the LA recovery process more energy-efficient, the intensified technique top DWC was applied to replace the F—C1 and F—C2 distillation columns. Fig. 6.18 shows the intensified LA recovery process with top DWC—decanter configuration with key design parameters. The use of innovative top DWC—decanter has many benefits, such as increasing the furfural recovery, increasing separation efficiency, and significantly decreasing the TAC. Incorporating a decanter into the top DWC is a very attractive option for improving the separation efficiency and energy savings, especially for the separation of heterogeneous azeotrope mixtures. The proposed process can save 21.6% and 20.6% in terms of TOC and TAC, respectively, as compared to the conventional process. The use of top DWC—decanter for LA production, which has the limitations at commercial scale, is technically feasible and particularly interesting in constructing a new plant.

6.2.6 Industrial biorefineries

Biorefinery has been encouraged by many countries for a sustainable economy. Currently, most commercial biorefineries used sugarcane, corn, or soybeans as feedstocks to produce biofuels and value-added products (Hossain, Liu, & Du, 2017a). Biodiesel production from soybean oils in the United States has increased dramatically from 0.5 million gallons in 1999 to 1.07 billion gallons in 2011 (Hossain, Liu, & Du, 2017b). The attention has now shifted to the second-generation biofuels from lignocellulosic biomass such as agricultural residues, woody biomass, and municipal solid wastes. The first commercial CE plant in the world was operated by Beta Renewables in 2012 (Rosales-Calderon, Arantes, Biotechnol Biofuels, Rosales-Calderon, & Arantes, 2019).

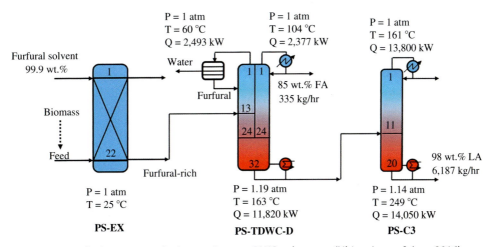

Fig. 6.18 Intensified LA process by innovative top DWC—decanter (Nhien, Long, & Lee, 2016).

A CE plant with a capacity of 20 MMgy was built by GranBio in 2014 in Brazil while Abengoa celebrated the opening of a 25 Mmgy CE plant in Kansas, USA in 2015 (Rosales-Calderon et al., 2019). However, many projects were experimented with financial difficulties, resulting in a decrease in global new investment in biofuels in recent years (Rosales-Calderon et al., 2019). To make biorefinery more competitive with petrorefinery, much more advanced techniques such as hybrid, integrated, and intensified techniques must be applied and industrially demonstrated into biorefinery processes to reduce the production costs and environmental impacts.

6.3 Conclusions

Biofuels or biochemicals are typically obtained by acidic hydrolysis or fermentation as diluted aqueous streams (such as ethanol, butanol, furfural, or LA), requiring the removal of large quantities of water. Combining other separation techniques in hybrid processes, such as extraction/distillation and distillation/decanting can efficiently overcome the water removal issue. The other primary drawback of biorefinery is the mass yield of product to feed, which is typically much less than that of petroleum-based refineries. Therefore, reducing investment and operating costs is of utmost important. Separation and purification processes, which account for a large proportion of production costs, are crucial factors in determining biorefinery accomplishments. Applying intensified techniques, such as RD and DWC, is the key to reducing total costs. Moreover, utilizing more bioproducts (e.g., glycerol, a major by-product of biodiesel production used as an extracting solvent) is needed to obtain a high degree of integration in a biorefinery context.

References

Antoni, D., Zverlov, V. V., & Schwarz, W. H. (2007). Biofuels from microbes. *Applied Microbiology and Biotechnology*. https://doi.org/10.1007/s00253-007-1163-x.

Beneke, D., Peters, M., Glasser, D., Hildebrandt, D., Beneke, D., Peters, M., et al. (2012). *Understanding distillation using column profile maps*. Hoboken, NJ, USA: John Wiley & Sons, Inc. https://doi.org/10.1002/9781118477304.

Berrios, M., & Skelton, R. L. (2008). Comparison of purification methods for biodiesel. *Chemical Engineering Journal, 144*, 459–465. https://doi.org/10.1016/j.cej.2008.07.019.

Bharathiraja, B., Jayamuthunagai, J., Sudharsanaa, T., Bharghavi, A., Praveenkumar, R., Chakravarthy, M., et al. (2017). Biobutanol—An impending biofuel for future: A review on upstream and downstream processing tecniques. *Renewable and Sustainable Energy Reviews*. https://doi.org/10.1016/j.rser.2016.10.017.

Blahušiak, M., Kiss, A. A., Babic, K., Kersten, S. R. A., Bargeman, G., & Schuur, B. (2018). Insights into the selection and design of fluid separation processes. *Separation and Purification Technology*. https://doi.org/10.1016/j.seppur.2017.10.026.

Cao Nhien, L., Van Duc Long, N., Kim, S., & Lee, M. (2019). Novel reaction-hybrid-extraction-distillation process for furfuryl alcohol production from raw bio-furfural. *Biochemical Engineering Journal, 148*, 143–151. https://doi.org/10.1016/j.bej.2019.05.005.

Christian, S. (2015). Is cellulosic ethanol the next big thing in renewable fuels? *Earth Island Journal, 5*. https://www.earthisland.org/journal/index.php/articles/entry/is_cellulosic_ethanol_the_next_big_thing_in_renewable_fuels/#:~:text=Despite%20the%20uncertainty%2C%20cellulosic%20ethanol,America's%20homegrown%20renewable%20fuel%20sources.

Cossio-Vargas, E., Barroso-Muñoz, F. O., Hernandez, S., Segovia-Hernandez, J. G., & Cano-Rodriguez, M. I. (2012). Thermally coupled distillation sequences: Steady state simulation of the esterification of fatty organic acids. *Chemical Engineering and Processing Process Intensification*. https://doi.org/10.1016/j.cep.2012.08.004.

Cossio-Vargas, E., Hernandez, S., Segovia-Hernandez, J. G., & Cano-Rodriguez, M. I. (2011). Simulation study of the production of biodiesel using feedstock mixtures of fatty acids in complex reactive distillation columns. *Energy*. https://doi.org/10.1016/j.energy.2011.10.005.

de Jong, E., & Jungmeier, G. (2015). Biorefinery concepts in comparison to petrochemical refineries. In *Industrial biorefineries and white biotechnology* (pp. 3–33). Elsevier. https://doi.org/10.1016/B978-0-444-63453-5.00001-X.

De Jong, W., & Marcotullio, G. (2010). Overview of biorefineries based on co-production of furfural, existing concepts and novel developments. *International Journal of Chemical Reactor Engineering, 8*.

Dejanović, I., Matijašević, L., & Olujić, Ž. (2010). Dividing wall column—A breakthrough towards sustainable distilling. *Chemical Engineering and Processing Process Intensification, 49*, 559–580. https://doi.org/10.1016/j.cep.2010.04.001.

Dhawane, S. H., Kumar, T., & Halder, G. (2018). Recent advancement and prospective of heterogeneous carbonaceous catalysts in chemical and enzymatic transformation of biodiesel. *Energy Conversion and Management*. https://doi.org/10.1016/j.enconman.2018.04.073.

Errico, M., Sanchez-Ramirez, E., Quiroz-Ramìrez, J. J., Rong, B. G., & Segovia-Hernandez, J. G. (2017). Multiobjective optimal acetone-butanol-ethanol separation systems using liquid-liquid extraction-assisted divided wall columns. *Industrial and Engineering Chemistry Research*. https://doi.org/10.1021/acs.iecr.7b03078.

Fjerbaek, L., Christensen, K. V., & Norddahl, B. (2009). A review of the current state of biodiesel production using enzymatic transesterification. *Biotechnology and Bioengineering*. https://doi.org/10.1002/bit.22256.

Frolkova, A. K., & Raeva, V. M. (2010). Bioethanol dehydration: State of the art. *Theoretical Foundations of Chemical Engineering, 44*, 545–556. https://doi.org/10.1134/S0040579510040342.

Gómez-Castro, F. I., Rico-Ramírez, V., Segovia-Hernández, J. G., Hernández-Castro, S., González-Alatorre, G., & El-Halwagi, M. M. (2012). Simplified methodology for the design and optimization of thermally coupled reactive distillation systems. *Industrial and Engineering Chemistry Research*. https://doi.org/10.1021/ie201397a.

Górak Andrzej, A. S. (2018). Intensification of biobased processes. *Green Chemistry Series No. 55*. Royal Society of Chemistry.

Górak, A., & Stankiewicz, A. (2011). Intensified reaction and separation systems. *Annual Review of Chemical and Biomolecular Engineering, 2*, 431–451. https://doi.org/10.1146/annurev-chembioeng-061010-114159.

Grand View Research. (2014). *Levulinic acid market analysis and segment forecasts to 2020*. https://doi.org/ISBN 978-1-68038-068-2.

Green, E. M. (2011). Fermentative production of butanol-the industrial perspective. *Current Opinion in Biotechnology*. https://doi.org/10.1016/j.copbio.2011.02.004.

Haider, J., Harvianto, G. R., Qyyum, M. A., & Lee, M. (2018). Cost- and energy-efficient butanol-based extraction-assisted distillation designs for purification of 2,3-butanediol for use as a drop-in fuel. *ACS Sustainable Chemistry & Engineering, 6*, 14901–14910. https://doi.org/10.1021/acssuschemeng.8b03414.

Haider, J., Qyyum, M. A., Hussain, A., Yasin, M., & Lee, M. (2018). Techno-economic analysis of various process schemes for the production of fuel grade 2,3-butanediol from fermentation broth. *Biochemical Engineering Journal, 140*, 93–107. https://doi.org/10.1016/j.bej.2018.09.002.

Haider, J., Qyyum, M. A., Minh, L. Q., & Lee, M. (2020). Purification step enhancement of the 2,3-butanediol production process through minimization of high pressure steam consumption. *Chemical Engineering Research and Design*. https://doi.org/10.1016/j.cherd.2019.11.005.

Hajilary, N., Rezakazemi, M., & Shirazian, S. (2019). Biofuel types and membrane separation. *Environmental Chemistry Letters*. https://doi.org/10.1007/s10311-018-0777-9.

Halder, P., Azad, K., Shah, S., & Sarker, E. (2019). Prospects and technological advancement of cellulosic bioethanol ecofuel production. In *Advances in eco-fuels for a sustainable environment* (pp. 211–236). Elsevier. https://doi.org/10.1016/b978-0-08-102728-8.00008-5.

Harvianto, G. R., Haider, J., Hong, J., Van Duc Long, N., Shim, J.-J., Cho, M. H., et al. (2018). Purification of 2,3-butanediol from fermentation broth: Process development and techno-economic analysis. *Biotechnology for Biofuels*, *11*, 18. https://doi.org/10.1186/s13068-018-1013-3.

Hong, J., Van Duc Long, N., Harvianto, G. R., Haider, J., & Lee, M. (2019). Design and optimization of multi-effect-evaporation-assisted distillation configuration for recovery of 2,3-butanediol from fermentation broth. *Chemical Engineering and Processing Process Intensification*, *136*, 107–115. https://doi.org/10.1016/j.cep.2019.01.002.

Hossain, G. S., Liu, L., & Du, G. C. (2017a). Industrial bioprocesses and the biorefinery concept. In *Current developments in biotechnology and bioengineering* (pp. 3–27). Elsevier. https://doi.org/10.1016/B978-0-444-63663-8.00001-X.

Hossain, G. S., Liu, L., & Du, G. C. (2017b). Industrial bioprocesses and the biorefinery concept. In *Current developments in biotechnology and bioengineering* (pp. 3–27). Elsevier. https://doi.org/10.1016/B978-0-444-63663-8.00001-X.

Huang, H.-J., Ramaswamy, S., Tschirner, U. W., & Ramarao, B. V. (2008). A review of separation technologies in current and future biorefineries. *Separation and Purification Technology*, *62*, 1–21. https://doi.org/10.1016/j.seppur.2007.12.011.

Humbird, D., Davis, R., Tao, L., Kinchin, C., Hsu, D., Aden, A., et al. (2011). *Process design and economics for biochemical conversion of lignocellulosic biomass to ethanol: Dilute-acid pretreatment and enzymatic hydrolysis of corn Stover*. Colorado: National Renewable Energy Laboratory.

Hussain, M. A. M., & Pfromm, P. H. (2013). Reducing the energy demand of cellulosic ethanol through salt extractive distillation enabled by electrodialysis. *Separation Science and Technology*, *48*, 1518–1528. https://doi.org/10.1080/01496395.2013.766211.

Ignat, R. M., & Kiss, A. A. (2013). Optimal design, dynamics and control of a reactive DWC for biodiesel production. *Chemical Engineering Research and Design*, *91*, 1760–1767.

Jana, A. K. (2014). Advances in heat pump assisted distillation column: A review. *Energy Conversion and Management*, *77*, 287–297. https://doi.org/10.1016/j.enconman.2013.09.055.

Jeon, S., Kim, D.-K., Song, H., Lee, H. J., Park, S., Seung, D., et al. (2014). 2,3-Butanediol recovery from fermentation broth by alcohol precipitation and vacuum distillation. *Journal of Bioscience and Bioengineering*, *117*, 464–470. https://doi.org/10.1016/j.jbiosc.2013.09.007.

Ji, X.-J., Huang, H., & Ouyang, P.-K. (2011). Microbial 2,3-butanediol production: A state-of-the-art review. *Biotechnology Advances*, *29*, 351–364. https://doi.org/10.1016/j.biotechadv.2011.01.007.

Kaymak, D. B. (2019). Design and control of an alternative bioethanol purification process via reactive distillation from fermentation broth. *Industrial and Engineering Chemistry Research*, *58*, 1675–1685. https://doi.org/10.1021/acs.iecr.8b04832.

Kazemi, A., Mehrabani-Zeinabad, A., & Beheshti, M. (2018). Recently developed heat pump assisted distillation configurations: A comparative study. *Applied Energy*, *211*, 1261–1281. https://doi.org/10.1016/j.apenergy.2017.12.023.

Khalid, A., Aslam, M., Qyyum, M. A., Faisal, A., Khan, A. L., Ahmed, F., et al. (2019). Membrane separation processes for dehydration of bioethanol from fermentation broths: Recent developments, challenges, and prospects. *Renewable and Sustainable Energy Reviews*, *105*, 427–443. https://doi.org/10.1016/j.rser.2019.02.002.

Kiss, A. A. (2011). Heat-integrated reactive distillation process for synthesis of fatty esters. *Fuel Processing Technology*. https://doi.org/10.1016/j.fuproc.2011.02.003.

Kiss, A. A. (2013a). *Advanced distillation technologies: Design, control and applications*. John Wiley & Sons.

Kiss, A. A. (2013b). Novel applications of dividing-wall column technology to biofuel production processes. *Journal of Chemical Technology and Biotechnology*, *88*, 1387–1404. https://doi.org/10.1002/jctb.4108.

Kiss, A. A., & Ignat, R. M. (2012a). Innovative single step bioethanol dehydration in an extractive dividing-wall column. *Separation and Purification Technology*, *98*, 290–297. https://doi.org/10.1016/j.seppur.2012.06.029.

Kiss, A. A., & Ignat, R. M. (2012b). Enhanced methanol recovery and glycerol separation in biodiesel production—DWC makes it happen. *Applied Energy, 99*, 146–153. https://doi.org/10.1016/j.apenergy.2012.04.019.

Kiss, A. A., Lange, J.-P., Schuur, B., Brilman, D. W. F., van der Ham, A. G. J., & Kersten, S. R. A. (2016). Separation technology–making a difference in biorefineries. *Biomass and Bioenergy, 95*, 296–309. https://doi.org/10.1016/j.biombioe.2016.05.021.

Kiss, A. A., Segovia-Hernández, J. G., Bildea, C. S., Miranda-Galindo, E. Y., & Hernández, S. (2012). Reactive DWC leading the way to FAME and fortune. *Fuel, 95*, 352–359. https://doi.org/10.1016/j.fuel.2011.12.064.

Kiss, A. A., & Suszwalak, D. J. P. C. (2012). Enhanced bioethanol dehydration by extractive and azeotropic distillation in dividing-wall columns. *Separation and Purification Technology, 86*, 70–78. https://doi.org/10.1016/j.seppur.2011.10.022.

Knothe, G., Krahl, J., & Van Gerpen, J. (2010). *The biodiesel handbook: Second edition*. https://doi.org/10.1016/C2015-0-02453-4.

Kolah, A. K., Lira, C. T., & Miller, D. J. (2013). Reactive distillation for the biorefinery. In *Separation and purification technologies in biorefineries*. https://doi.org/10.1002/9781118493441.ch16.

Köpke, M., Mihalcea, C., Liew, F., Tizard, J. H., Ali, M. S., Conolly, J. J., et al. (2011). 2,3-butanediol production by acetogenic bacteria, an alternative route to chemical synthesis, using industrial waste gas. *Applied and Environmental Microbiology, 77*, 5467–5475. https://doi.org/10.1128/aem.00355-11.

Kujawska, A., Kujawski, J., Bryjak, M., & Kujawski, W. (2015). ABE fermentation products recovery methods—A review. *Renewable and Sustainable Energy Reviews*. https://doi.org/10.1016/j.rser.2015.04.028.

Lei, Z., Dai, C., Zhu, J., & Chen, B. (2014). Extractive distillation with ionic liquids: A review. *AICHE Journal, 60*, 3312–3329. https://doi.org/10.1002/aic.14537.

Lipnizki, F., Thuvander, J., & Rudolph, G. (2019). Membrane processes and applications for biorefineries. In *Current trends and future developments on (bio-) membranes: Membranes in environmental applications* (pp. 283–301). Elsevier Inc. https://doi.org/10.1016/B978-0-12-816778-6.00013-8.

Long, N. V. D., & Lee, M. (2017). *Advances in distillation retrofit* (1st ed.). Singapore: Springer Singapore. https://doi.org/10.1007/978-981-10-5901-8.

Long, N. V. D., Minh, L. Q., Nhien, L. C., & Lee, M. (2015). A novel self-heat recuperative dividing wall column to maximize energy efficiency and column throughput in retrofitting and debottlenecking of a side stream column. *Applied Energy, 159*, 28–38. https://doi.org/10.1016/j.apenergy.2015.08.061.

Loy, Y. Y., Lee, X. L., & Rangaiah, G. P. (2015). Bioethanol recovery and purification using extractive dividing-wall column and pressure swing adsorption: An economic comparison after heat integration and optimization. *Separation and Purification Technology, 149*, 413–427. https://doi.org/10.1016/j.seppur.2015.06.007.

Lutze, P., & Gorak, A. (2013). Reactive and membrane-assisted distillation: Recent developments and perspective. *Chemical Engineering Research and Design, 91*, 1978–1997.

Malone, M. F., & Doherty, M. F. (2000). Reactive distillation. *Industrial and Engineering Chemistry Research, 39*, 3953–3957. https://doi.org/10.1021/ie000633m.

Mandalika, A., & Runge, T. (2012). Enabling integrated biorefineries through high-yield conversion of fractionated pentosans into furfural. *Green Chemistry, 14*, 3175–3184. https://doi.org/10.1039/C2GC35759C.

Meindersma, G. W., Quijada-Maldonado, E., Aelmans, T. A. M., Hernandez, J. P. G., & De Haan, A. B. (2012). Ionic liquids in extractive distillation of ethanol/water: From laboratory to pilot plant. *ACS Symposium Series, 1117*, 239–257. https://doi.org/10.1021/bk-2012-1117.ch011.

Metkar, P. S., Till, E. J., Corbin, D. R., Pereira, C. J., Hutchenson, K. W., & Sengupta, S. K. (2015). Reactive distillation process for the production of furfural using solid acid catalysts. *Green Chemistry, 17*, 1453–1466. https://doi.org/10.1039/C4GC01912A.

Moulijn, J. A., Stankiewicz, A., Grievink, J., & Górak, A. (2008). Process intensification and process systems engineering: A friendly symbiosis. *Computers and Chemical Engineering, 32*, 3–11. https://doi.org/10.1016/j.compchemeng.2007.05.014.

Navarrete-Contreras, S., Sánchez-Ibarra, M., Barroso-Muñoz, F. O., Hernández, S., & Castro-Montoya, A. J. (2014). Use of glycerol as entrainer in the dehydration of bioethanol using extractive batch distillation: Simulation and experimental studies. *Chemical Engineering and Processing Process Intensification, 77*, 38–41. https://doi.org/10.1016/j.cep.2014.01.003.

Nhien, L. C., Long, N. V. D., Kim, S., & Lee, M. (2016a). Design and optimization of intensified biorefinery process for furfural production through a systematic procedure. *Biochemical Engineering Journal, 116*, 166–175. https://doi.org/10.1016/j.bej.2016.04.002.

Nhien, L. C., Long, N. V. D., Kim, S., & Lee, M. (2016b). Design and assessment of hybrid purification processes through a systematic solvent screening for the production of levulinic acid from lignocellulosic biomass. *Industrial and Engineering Chemistry Research, 55*, 5180–5189. https://doi.org/10.1021/acs.iecr.5b04519.

Nhien, L. C., Long, N. V. D., Kim, S., & Lee, M. (2017). Techno-economic assessment of hybrid extraction and distillation processes for furfural production from lignocellulosic biomass. *Biotechnology for Biofuels, 10*. https://doi.org/10.1186/s13068-017-0767-3.

Nhien, L. C., Long, N. V. D., & Lee, M. (2016). Design and optimization of the levulinic acid recovery process from lignocellulosic biomass. *Chemical Engineering Research and Design, 107*, 126–136. https://doi.org/10.1016/j.cherd.2015.09.013.

Nhien, L. C., Long, N. V. D., & Lee, M. (2017). Novel heat–integrated and intensified biorefinery process for cellulosic ethanol production from lignocellulosic biomass. *Energy Conversion and Management, 141*, 367–377. https://doi.org/10.1016/j.enconman.2016.09.077.

Novita, F. J., Lee, H.-Y., & Lee, M. (2018). Reactive distillation with pervaporation hybrid configuration for enhanced ethyl levulinate production. *Chemical Engineering Science, 190*, 297–311. https://doi.org/10.1016/j.ces.2018.06.024.

Oudshoorn, A. (2012). *Recovery of bio-based butanol (Ph.D. thesis)*. Biotechnology Department, Delft University of Technology.

Patraşcu, I., Bîldea, C. S., & Kiss, A. A. (2017). Eco-efficient butanol separation in the ABE fermentation process. *Separation and Purification Technology*. https://doi.org/10.1016/j.seppur.2016.12.008.

Patraşcu, I., Bîldea, C. S., & Kiss, A. A. (2018). Eco-efficient downstream processing of biobutanol by enhanced process intensification and integration. *ACS Sustainable Chemistry & Engineering*. https://doi.org/10.1021/acssuschemeng.8b00320.

Ramaswamy, S., Huang, H.-J., & Ramarao, B. V. (2013). *Separation and purification technologies in biorefineries*. United Kingdom: John Wiley & Sons.

Rosales-Calderon, O., Arantes, V., Biotechnol Biofuels, A., Rosales-Calderon, O., & Arantes, V. (2019). A review on commercial-scale high-value products that can be produced alongside cellulosic ethanol. *Biotechnology for Biofuels, 12*, 240. https://doi.org/10.1186/s13068-019-1529-1.

Saleh, J., Tremblay, A. Y., & Dubé, M. A. (2010). Glycerol removal from biodiesel using membrane separation technology. *Fuel*. https://doi.org/10.1016/j.fuel.2010.04.025.

Schladt, L., Ivens, I., Karbe, E., Ruhl-Fehlert, C., & Bomhard, E. (1998). Subacute oral toxicity of tetraethylene-glycol and ethylene-glycol administered to Wistar rats. *Experimental and Toxicologic Pathology, 50*, 257–265.

Segovia-Hernández, J. G., Sánchez-Ramírez, E., Alcocer-García, H., Quíroz-Ramírez, J. J., Udugama, I. A., & Mansouri, S. S. (2020). Analysis of intensified sustainable schemes for biobutanol purification. *Chemical Engineering and Processing Process Intensification*. https://doi.org/10.1016/j.cep.2019.107737.

Seibert, F. (2010). *A method of recovering levulinic acid*. WIPO.

Skiborowski, M., Harwardt, A., & Marquardt, W. (2013). Conceptual design of distillation-based hybrid separation processes. *Annual Review of Chemical and Biomolecular Engineering, 4*, 45–68. https://doi.org/10.1146/annurev-chembioeng-061010-114129.

Van Duc Long, N., Hong, J., Nhien, L. C., & Lee, M. (2018). Novel hybrid-blower-and-evaporator-assisted distillation for separation and purification in biorefineries. *Chemical Engineering and Processing Process Intensification, 123*, 195–203. https://doi.org/10.1016/j.cep.2017.11.009.

Vane, L. M., Alvarez, F. R., Rosenblum, L., & Govindaswamy, S. (2013). Hybrid vapor stripping-vapor permeation process for recovery and dehydration of 1-butanol and acetone/butanol/ethanol from dilute aqueous solutions. Part 2. Experimental validation with simple mixtures and actual fermentation broth. *Journal of Chemical Technology and Biotechnology*. https://doi.org/10.1002/jctb.4086.

Vazquez-Castillo, J. A., Venegas-Sánchez, J. A., Segovia-Hernández, J. G., Hernández-Escoto, H., Hernández, S., Gutiérrez-Antonio, C., et al. (2009). Design and optimization, using genetic algorithms, of intensified distillation systems for a class of quaternary mixtures. *Computers and Chemical Engineering, 33*, 1841–1850. https://doi.org/10.1016/j.compchemeng.2009.04.011.

Wang, X., Lv, M., Zhang, L., Li, K., Gao, C., Ma, C., et al. (2013). Efficient bioconversion of 2, 3-butanediol into acetoin using Gluconobacter oxydans DSM 2003. *Biotechnology for Biofuels, 6*, 155.

Werpy, T., & Petersen, G. (2004). *Top value added chemicals from biomass volume I—Results of screening for potential candidates from sugars and synthesis gas*. National Renewable Energy Laboratory (NREL).

Wittig, R., Lohmann, J., & Gmehling, J. (2003). Vapor−liquid equilibria by UNIFAC group contribution. 6. Revision and extension. *Industrial and Engineering Chemistry Research, 42*, 183–188. https://doi.org/10.1021/ie020506l.

Zeitsch, K. J. (2000). *The chemistry and technology of furfural and its many by-products*. Netherlands: Elsevier Science.

CHAPTER 7

Intensification of biodiesel production through computational fluid dynamics

Harrson S. Santana[a], Marcos R.P. de Sousa[a], and João L. Silva Júnior[b]
[a]University of Campinas, School of Chemical Engineering, Campinas, SP, Brazil
[b]Federal University of ABC, CECS—Center for Engineering, Modeling and Applied Social Sciences, Alameda da Universidade, São Bernardo do Campo, SP, Brazil

7.1 Introduction

The growing concern about the oil dependence leads to a constant search for alternative energy sources to substitute the fossil material in a renewable and economically feasible fashion. This concern can be quantified. According to a report from the U.S. Energy Information Administration (EIA), the worldwide energy usage will increase from 524×10^{12} BTU (5.5×10^{17} J or 9.5×10^{10} equivalent oil barrels) in 2010 to 630×10^{12} BTU in 2020 and to 820×10^{12} BTU in 2040 (EIA, 2013). From this total energy, the consumption of 80% was derived from fossil fuels, e.g., oil, gas and coal, which released 31 billion of metric tons of carbon dioxide to the atmosphere, according to the EIA report. According to recent data, global fossil fuel consumption in 2017 was 132,514 TWh (4.77×10^{20} J) (Ritchie & Roser, 2017).

The fossil fuel dependence, especially from oil, presents consequences such as strong climate changes inherent from the greenhouse effects emissions (GEE) and the oil reserves depletion is followed by a scarce of energy (Pang, 2016). Therefore, efforts have been made by the scientific community to explore the feasibility of alternative fuels with lower pollutant emission levels, also contributing to minimize the greenhouse effects. From the substitute energy sources, the biomass appears as a suitable renewable alternative (Gent, Twedt, Gerometta, & Almberg, 2017; Pang, 2016).

According to Bonechi et al. (2017), the term biomass refers to organic matter from vegetable or animal source, from natural origin or cultivated by man, in land or sea, produced direct or indirectly by photosynthesis process involving chlorophyll. Hence, biomass can be understood as any element with an organic matrix. Some of the main advantages of biomass as energy source includes: global availability of biomass, reduction of demand and consumption of fossil fuels and the reduction of greenhouse effect gases emissions (Bonechi et al., 2017). Fig. 7.1 presents some processes of energy extraction from biomass.

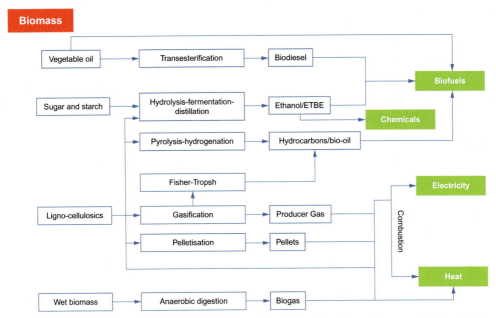

Fig. 7.1 Energy extraction processes from biomass. *(Credit: Reprinted from Bioenergy Systems for the Future, 1st ed., Bonechi, C., Consumi, M., Donati, A., Leone, G., Magnani, A., Tamasi, G., Rossi, C. (2017) Biomass: An overview, pp. 3–42, Copyright (2017), with permission from Elsevier.)*

In Fig. 7.1 is observed that several types of biomass, e.g., vegetable oil, sugar and starch, wet biomass, etc., can be converted in different forms of energy, using distinct chemical, physical or biological processes. From the energy forms, we have the biofuels that can be used in transportation, electricity generation, heat supply, electronics charging, cleaning of oil and grease spilling, cooking, lubricants, paint, and adhesive removal (SGB, 2016). Among these, biodiesel is usually considered a viable alternative to the use of diesel, due to the inherent biomass advantages already pointed out. Biodiesel have been received great attention of scientific community. From the Scopus platform, we found up to 2020, more than 40,212 documents related to the topic "Biodiesel", as shown in Fig. 7.2. Biodiesel can be obtained from vegetable oils or animal fats, and is renewable and biodegradable, nontoxic, providing several environmental benefits as the reduction of greenhouse effect gases emission, lower levels of air, water, and soil pollution regarding the use of diesel (Santana, Silva, & Taranto, 2019).

In 2015, the ASTM D6751-15 standard defined biodiesel as a fuel composed by mono-alkyl esters of long-chain fatty acid, derived from vegetable oils and/or animal fats, commonly produced by transesterification method (ASTM, 2015). This process consists in a chemical reaction between alcohol and triglycerides from the organic matrix, producing esters and glycerol, being catalyzed by acid, alkali, or enzymatic media. The

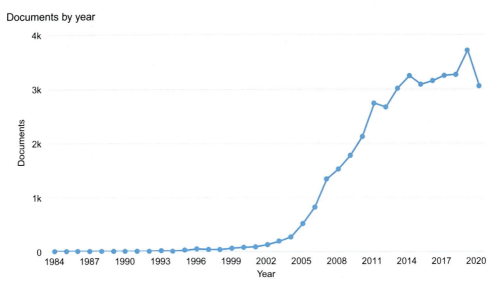

Fig. 7.2 Number of published documents per year considering the topic: biodiesel. *(Credit: Source: Scopus. Accessed in October 15, 2020.)*

chemical reaction corresponds to the transformation of an ester in another one, by the exchange of the acyl group between ester and acid, ester and other ester, or ester and alcohol. Also, one of the goals is to reduce the vegetable oil viscosity and then to use the biodiesel directly in the combustion engines designed for diesel, without the requirements of major modifications (Freedman, Pryde, & Mounts, 1984; Santana, Silva, & Taranto, 2019).

Commonly, the transesterification reaction is performed in batch reactors (BRs). The batch process can take from minutes to hours to achieve high yield (Santana, Silva, & Taranto, 2019). Therefore, to improve the biodiesel synthesis efficiency and to reduce the process costs, several technologies were proposed to intensify the biodiesel production. In this context, some advantages have been highlighted from the use of microreactors regarding the traditional BRs, especially, the shorter residence times to reach higher conversion levels (Mohadesi, Aghel, Maleki, & Ansari, 2020; Rahimi, Aghel, Alitabar, Sepahvand, & Ghasempour, 2014; Santana, Silva, & Taranto, 2019; Sun, Ju, Zhang, & Xu, 2008; Wen, Yu, Tu, Yan, & Dahlquist, 2009). The topics explored in these research studies include the influences of operating conditions (oil type, temperature, catalyst concentration, alcohol/oil molar ratio), reactor design, mass transfer-chemical kinetics correlation, and reactor scale-up in the biodiesel synthesis performance.

Microreactors or microchannel reactors are a class of microfluidic devices composed by interconnected microchannels, where low amounts of reactants are handled and reacts for a determined time (Zhang, Wiles, Painter, Watts, & Haswell, 2006). The main

characteristics of microreactors reported in literature are (Fogler, 2006; Santana, Silva, & Taranto, 2019; Whitesides, 2006):
- Laminar flow regime;
- Low amount of reactants and samples;
- Higher surface area-to-volume ration (over $10,000\,m^2\,m^{-3}$);
- Lower resistances to mass and heat transfer;
- Narrower residence time distribution;

One of the outstanding characteristic of microreactors is the laminar flow regime. In most of applications microdevices operate under very low Reynolds numbers, i.e., the viscous effects dominate the fluid flow behavior. The laminar flow regime is marked by the absence of macromixing. Such characteristics generate the behavior defined as the segregated flow, i.e., a symmetric flow pattern depicted by the continuous parallel flow along the channel. However, depending on the flow velocity or the presence of mixers when two immiscible phases are feed, passive mixers, i.e., static elements such as internal baffles; or active mixers, i.e., the use of external electrical or magnetic field, can be employed to generate chaotic advection, a flow pattern marked by vortex generation. This flow regime is marked by recirculation/vortex zones, and it is preferable over the segregated in microchannels used for chemical reactions, mostly for that the mixing extension greatly affects the reaction rate, i.e., the process is dominated by reactants mixing (Santana, Silva, & Taranto, 2015; Santana, Silva, & Taranto, 2019).

Biodiesel synthesis in microreactors is strongly influenced by the reactants mixing degree. Hence, in the microreactor design for biodiesel synthesis one of the main elements is the micromixer, the element responsible to promote the fluid mixing. In this context, the reactor design can be very complex, once one needs to evaluate the effects of micromixer design, microreactor dimensions, flow rate, type of vegetable oil, alcohol and catalyst, temperature, reactants molar ration, and catalyst concentration on reactor performance, leading to an enormous number of experimental runs. So, how to consider all these influent variables on the microreactor design?

Firstly, Design of Experiments tools can be employed to reduce the number of experimental runs. Several microreactors composed by different micromixer designs can be fabricated. Also, these devices can be tested in distinct operating conditions. After all these runs, the measurement of reactants mixing or biodiesel yield will provide data to decide which design exhibited superior performance. These procedures involve numerous human, physical and financial resources that are not always available. An alternative to experimental runs are the numerical runs using the computational fluid dynamics (CFD). CFD is a powerful tools used in several knowledge areas, including, chemical processing, physics, mechanical, aerospace and biomedical engineering, and recently in Microfluidics.

Silva (2019) and Santana et al. (2020) summarized some advantages of CFD usage in the development of microdevices: reduction of time and costs in projects—design and analysis of several geometric layouts can be accomplished before the manufacturing of physical prototypes; accurate prediction of flow field in severe conditions in which experimental measurements are complex or even unfeasible.

Accordingly, the present chapter aims to discuss the application of CFD in the intensification of biodiesel production. Firstly, a brief introduction to CFD was provided, approaching the governing equations of fluid motion, solving procedure, and boundary conditions. Also, the most important discretization methods were presented, followed by mesh generation aspects and quality metrics. After this, a literature review focusing the application of CFD in traditional macroscale reactor and microreactors for biomass processing was detailed. Finally, a study case was provided approaching CFD simulations of traditional batch scale reactors and micro and millichannel reactors for biodiesel synthesis. The CFD application in micromixer development and optimization was reported and a performance comparison between BR and micro and millireactors was performed considering production rate, energy consumption and excess reactant recovery, scale-up, and numbering-up aspects. Finally, conclusions and future perspectives were provided.

7.2 Brief introduction to CFD concepts
7.2.1 What is CFD?

CFD is the area of fluid mechanics that applies computational tools to solve fluid flow problems. This approach of fluid flow analysis is very useful to simulate cases in which there is two- or three-dimensional flow, which is eminently described through partial differential equations. Thus, in such conditions, it may be too difficult to obtain an analytical solution for the governing equations, and computers can provide numerical solutions to these equations with good accuracy. CFD techniques are therefore also applied to simulate slightly complicated fluid flow cases as turbulent or multiphase flows (Pulliam, Lomax, & Zingg, 1999).

Whatever the CFD tool used to solve a proposed problem, some validation is required. This initial step is the verification if the implemented model in such CFD tool provides results that reasonably match the physical observations in a fluid flow. This comparison is typically made in wind tunnels, although it can also be done by verifying an analytical solution or using an empirical analysis.

With the increasing of computers capacity and performance in the past decades, simulation tools and techniques have come to play a key role to development of science and technology, being essential in some particular areas. Some industries that adopt intensive design solutions are heavy CFD users nowadays, such as aeronautics and aerospace—those being the pioneers in CFD usage—automotive, marine, chemical, and nuclear (Moukalled, Mangani, & Darwish, 2016).

In this context, our main goal in this chapter is to present how modeling and simulation can be a powerful tool to develop process intensification (PI) solutions in biodiesel production processes.

7.2.2 Using CFD to solve a problem

The general steps to solve a fluid flow problem applying CFD are described in the flowchart presented in Fig. 7.3. The starter step when developing a CFD model is the observation and analysis of a given physical process. It consists not only of phenomena investigation, but also of assumptions survey based on observer's experience in the process.

The following step is the physical modeling, i.e., establishing the control volume by imposing control surfaces in a given flow domain, and applying appropriate assumptions to these surfaces. Basically, the physical model will be represented by a flow domain geometry, which can be generated in CAD software. Subsequently a mathematical model will be used to represent how fluid flow parameters vary over the flow domain. It consists of applying conservation laws to the control volume, so there are dependent field variables that describe the fluid flow. This step results in a set of partial differential

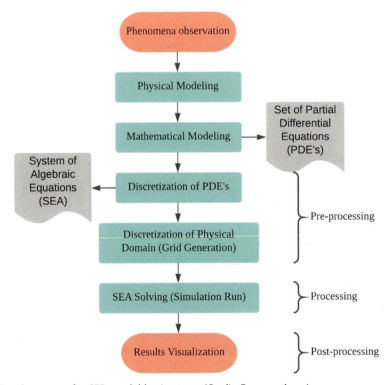

Fig. 7.3 Development of a CFD model basic steps. *(Credit: From authors.)*

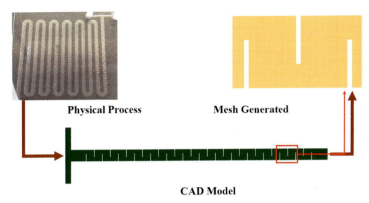

Fig. 7.4 Transforming a physical process geometry into a CAD model and then into a mesh. *(Credit: From authors.)*

equations (PDEs), where the field variables can usually be, e.g., pressure and velocity, and independent variables can be domain dimensions and time, if there is a transient problem.

Since the fluid domain is usually assumed as a continuum, independent variables vary continuously. Consequently, field variables are also continuous variables. In order to solve the set of PDEs that describes the fluid flow, it is needed to discretize such variables, so it can be possible to implement the model in a digital computer, and then solve it with applicable numerical methods. Therefore, the set of PDEs is discretized using an adequate technique—it can be a finite-difference, finite-element, or finite-volume method. The physical domain is also discretized, i.e., a geometry that contains an infinite number of points is transformed into a domain with a finite number of points, elements or cells, which is the so-called mesh generation process. The discretization of PDEs and Mesh Generation then results in a System of Algebraic Equations (SAE), where the field variables are now discrete variables, and the system of PDEs for all domain are now a SAE for each domain element. Thus, discretization of PDEs and Mesh Generation are preprocessing steps, which means preparing the problem for the computer to solve it. All the above described steps means transforming a continuous flow domain in a discrete flow domain as described in Fig. 7.4.

A simulation run then consists of a SAE solving for each element of the domain. There are several well-known numerical methods to solve these SAE, even though the variables are coupled, and they eminently are in a fluid flow. Usually, this SAE can be composed of linear equations, but numerical methods can be applied also to nonlinear problems, which are frequently present in chemical process problems, e.g., if there is a second or higher order kinetics reaction. If the solving process converges then acquired simulation data can be postprocessed in an adequate software to obtain graphical details of fluid flow (Moukalled et al., 2016).

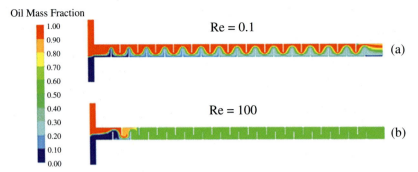

Fig. 7.5 Postprocessing results of oil mass fraction in a biodiesel synthesis via alkaline ethanolysis. (Credit: Reprinted from Santana, H. S., Sanchez, G. B., Taranto, O. P., 2017. Evaporation of excess alcohol in biodiesel in a microchannel heat exchanger with Peltier module. Chemical Engineering Research & Design, 124, 20–28 with permission from Elsevier.)

Results obtained can be graphically represented for a simulation run of a CFD model for any given field variable if there is a suitable postprocessing software for it. Fig. 7.5 shows sunflower oil mass fraction in a biodiesel synthesis via alkaline ethanolysis using CAD geometry presented in Fig. 7.4 (Santana, Sanchez, & Taranto, 2017).

7.2.3 Fluid flow mathematical modeling

7.2.3.1 Governing equations

As above mentioned, fluid dynamics is described in terms of partial differential equations. According to flow velocity relative to the speed of sound, these equations can be: elliptic, if there is a subsonic flow; parabolic, if flow velocity is equal to the speed of sound; and hyperbolic, if there is a supersonic flow (Chung, 2002). Since some chemical processes as biofuels production, do not involve gas flow, the governing equations of a fluid flow are usually elliptic.

There are two different approaches when developing a mathematical model of a fluid flow. In the Lagrangian approach, the observer is interested in describing the path of a specific fluid particle in the domain. On the other hand, in the Eulerian approach, all points are specified and the observer is interested in describing how field variables vary over the domain. From this point on we specify other fluid dynamics concepts based in a Eulerian approach.

Building a mathematical model usually consists of applying the governing equations to all points of a flow domain. A differential analysis so represents applying these conservation equations to an infinitesimal dimensions' control volume, then such equations turn into a set of PDEs. Solving these PDEs then gives field variables details, such as velocity, pressure, density, etc.

In a simple case of transient three-dimensional fluid flow, independent variables to be set can be, e.g., dimensions in each direction and time. Note that these are scalar variables.

In this example the field variables to be determined can usually be velocity and pressure. In this as well in all cases, velocity fields are vector fields, and pressure is described by a scalar field, by definition (Moukalled et al., 2016). Still considering the previous example, it is sufficient to solve a set of two differential equations, those for mass and momentum conservation. The mass conservation equation is a scalar equation given as:

$$\frac{\partial \rho}{\partial t} + \nabla \cdot (\rho U) \tag{7.1}$$

where ρ is fluid density and U is the velocity. Eq. (7.1) is also called the continuity equation, since its formulation is based on the hypothesis of a continuum for a given domain. This assumption considers that all flow variables vary continuously with space and time. The first term on the left-hand side of Eq. (7.1) represents the mass variation in the control volume while the second term expresses the balance of mass flux across control surfaces.

This equation can be rewritten in terms of material derivative, or Lagrangian derivative. This derivative can be defined as the variation of a transport variable with time but also considering the mean velocity. The material derivative of ρ is given in Eq. (7.2) (Bird, Stewart, & Lightfoot, 2006):

$$\frac{D\rho}{Dt} = \frac{\partial \rho}{\partial t} + u_x \frac{\partial \rho}{\partial x} + u_y \frac{\partial \rho}{\partial y} + u_z \frac{\partial \rho}{\partial z} \tag{7.2}$$

Rearranging Eq. (7.1) using the material derivative given in Eq. (7.2), the continuity equation can also be written as:

$$\frac{D\rho}{Dt} + \rho \nabla \cdot U = 0 \tag{7.3}$$

The momentum conservation equation is derived from Newton's second law of motion. When it is applied to a case of an incompressible, isothermal, and laminar flow of a Newtonian fluid, then the equation obtained is what is known as the Navier-Stokes equation, commonly acknowledge as the foundation stone of fluid mechanics. This is a vector equation given as:

$$\rho \frac{DU}{Dt} = \rho g - \nabla p + \nabla \left[\mu \left(\nabla U + {}^T \nabla U \right) \right] \tag{7.4}$$

where the left-hand side term is the material derivative, which can be understood as the advective term, the terms of the right-hand side of Eq. (7.4) represent the gravitational force, the most usually field force present in flow cases, the pressure gradient and the momentum diffusion of a Newtonian fluid, respectively.

Here, we presented the concept of advection and diffusion of transport variables. Basically, advection means the transport due to a global fluid motion and diffusion is the transport mechanism due to molecular interaction. Note that this equation is written for a laminar flow. If there is a turbulent flow case, a turbulent shear stress term needs to be added to the momentum flux term.

According to assumptions that can be made in order to simplify a mathematical model, Navier-Stokes equations can also be simplified to a particular case so that: if there is an inviscid flow, i.e., viscosity can be neglected, then the diffusion term of Eq. (7.4) can be neglected; if this flow is also incompressible, then both diffusion and density variations can be neglected (Bird et al., 2006).

In most chemical processes other governing equations have to be written to describe a case, such as the energy conservation and species conservation. Here, we emphasize this last one, since biofuels production are related to processes that eminently involve transfer, generation, and consumption of chemical species. Species transport as momentum transport is represented by equations that contain advective and diffusive terms. In most cases, species conservation can be described as:

$$\frac{\partial C_i}{\partial t} + \nabla \cdot (U C_i) = \nabla \cdot [D(\nabla C_i)] + r_i \tag{7.5}$$

where the first term on the left-hand side is the concentration variation of a species i over time while the second term is the advective term. On the right-hand side of Eq. (7.5), the first and second term represent, respectively, the diffusive transport and the chemical species mass balance considering a chemical reaction, commonly expressed by the reaction kinetics, given as:

$$r_i = r_{i,generation} - r_{i,consumption} \tag{7.6}$$

All these mentioned governing equations are naturally coupled, so all equations need to be solved simultaneously. If there is a transient flow, then the solution is obtained for field variables varying not only in space, but also in time.

7.2.3.2 Boundary conditions

The physical model is related to the mathematical model also by the boundary conditions. They are constraints presented in the problem adopted at the control surfaces of the flow domain, always represented as equations. Such equations are naturally requirements when solving differential equations. To understand what is a boundary condition and its types, we establish here a general ordinary differential equation given as:

$$\frac{\partial \varphi}{\partial x} + \varphi = 0 \tag{7.7}$$

where φ is the dependent variable and x is the independent variable given in the closed domain $[a, b]$. As above mentioned, an additional equation, i.e., boundary condition, needs to be stated in order to well establish this problem. There are five types of boundary conditions, each one suitable to a given real physics situation: Dirichlet, Neumann, Robin, Mixed, and Cauchy. The Dirichlet, Neumann, and Robbin are also called first-type, second-type, and third-type conditions, respectively.

Dirichlet boundary condition is that when there is a prescribed value for the dependent variable at a control surface. Thus, possible Dirichlet boundary conditions for Eq. (7.7) are:

$$\varphi(a) = A \quad (7.8)$$

$$\varphi(b) = B \quad (7.9)$$

where A and B are prescribed values, usually constants. In a fluid flow case, for example, a Dirichlet boundary condition can be adopted if pressure or velocity are known in a given inlet or outlet of a duct (Chung, 2002).

Neumann boundary condition is that when the derivative of a dependent variable is prescribed at a control surface. Thus, possible Neumann boundary conditions for Eq. (7.7) are:

$$\varphi'(a) = \alpha \quad (7.10)$$

$$\varphi'(b) = \beta \quad (7.11)$$

where α and β are prescribed values that can be both, a constant or a function.

On the other hand, the Robbin boundary condition consists in an equation weighted by Dirichlet and Neumann conditions, so possible Robbin conditions for Eq. (7.7) are:

$$w_1 \cdot \varphi(a) + w_2 \cdot \varphi'(a) = \gamma_1 \quad (7.12)$$

$$w_1 \cdot \varphi(b) + w_2 \cdot \varphi'(b) = \gamma_2 \quad (7.13)$$

where w_1 and w_2 are constants that express the respective weights.

A problem established with mixed boundary conditions refers to cases when Dirichlet condition are prescribed in some control surfaces while Neumann are prescribed in others. Thus, a set of mixed boundary conditions for Eq. (7.7) are:

$$\varphi(a) = A \quad (7.14)$$

$$\varphi'(b) = \beta \quad (7.15)$$

Finally, even this is a less usual than others, Cauchy boundary conditions are also used, especially when a problem is stated in second-order differential equations. In such boundary condition, both, a function value and its derivative are prescribed for the same point at the domain (Chung, 2002).

$$\varphi(a) = A \quad (7.16)$$

$$\varphi'(a) = \alpha \qquad (7.17)$$

Therefore, Cauchy boundary conditions accord to imposing Dirichlet and Neumann conditions, simultaneously, for the same point.

7.2.4 Discretization of a mathematical model

Once the mathematical model that describes physics of a flow case is established, it is needed to discretize the set of PDEs so that a digital computer can solve the problem. There are several techniques to discretize equations in a CFD model according to some case specificity, even so they can be summarized into three types: finite-difference, finite-element, and finite-volume methods.

7.2.4.1 Finite-difference methods

Finite-difference methods are also understood as purely mathematical techniques of discretization, i.e., in these situations there is no concern to maintain physics characteristics during equations discretization. These methods are a straightforward approach to discretize a set of PDEs. It consists in assuming a point in space where the continuum can be considered for the original set of PDEs and then these equations are turned into discrete equations, the so-called finite-difference equations.

Such methods are usually adopted for regular geometries, in which they can be very efficient for solving a fluid flow case. However, these methods are not commonly applied to irregular domain shapes. So, a finite-difference method is a suitable discretization option if the flow domain is rectangular, box-shaped, or can be represented in blocks.

In finite-difference methods, discretization is made for both, the mathematical and physical model, dimension by dimension. Therefore, it is easier in these methods to increase the order of discrete elements in order to obtain a response with higher order accuracy. This means that if a high accuracy computational simulation is needed, then by this method, it only takes to increase the number of points that represents the physical domain.

On the other hand, finite-difference methods do not handle well with variables discontinuities due to geometry. Furthermore, if it is necessary to increase the number of discrete elements in a certain region of a domain, such as in too complex shapes, this method is too difficult to implement (Pulliam et al., 1999).

7.2.4.2 Finite-element methods

Finite-element methods are those based in subdividing the physical model in very small defined parts of simple shape elements. All these small parts combined form a mesh of finite elements. After this step, the set of PDEs that describes fluid flow in terms of field variables are formulated not for the entire domain, but for each element. These techniques so consist in approximate field variables in simple low-degree polynomial functions, as linear or quadratic. It results in a local approximation into a SAE for each element. When this formulation is extrapolated for all domain elements, then a SAE is obtained, usually represented by a sparse matrix that can be solved through any well-known sparse matrix solver.

In these methods, as in finite-difference methods, it is also possible to increase the order of elements in order to obtain more accurate data of field variables. However, finite-element methods are more advantageous if it is needed to increase response precision in specific regions of a domain, such as corners. Usually, increasing the number of elements, i.e., refining the mesh, gives the simulation results more accuracy. Furthermore, these methods also allow local mesh refining, i.e., increasing the number of elements only in a corner, e.g., not in all domain. This technique is called as adaptive mesh refinement. Thus, the finite-element methods are well suitable to curvy or irregular shape geometries in a natural way. On the other hand, finite-element methods require a relatively advanced mathematical expertise for its implementation. Transient cases are even more complex (Pulliam et al., 1999).

7.2.4.3 Finite-volume methods

CFD implementations are naturally developed applying finite-volume methods, since they consist in applying conservation laws to the elements of a domain, here called cells. Thus, finite-volume methods are similar to finite-element methods when both are based in dividing the geometry in very small simple shape elements, although the approaches adopted over governing equations are quite different.

In finite-volume methods, a flux variable that enters in a cell's face have to leave the other side's face of the same cell. Therefore, implementing these methods results in a set of flux conservation equations defined for mean variables at the cells. Since most of governing equations of a fluid flow are based in conservation laws, as described before, these methods are successful to solve such problems.

These methods handle well with nonlinearities, which usually appear in transport phenomena problems, e.g., in cases that involve mass transfer with high order kinetics chemical reaction. As in the finite-elements methods, in the finite-volume methods more accurate results can be obtained by refining a mesh locally (Hirt & Nichols, 1981; Pulliam et al., 1999).

7.2.5 Mesh generation

Up to this point, we mentioned how governing equations can represent fluid flow through a flow domain, so that computer can read and solve them. However, as well as the continuum equations have to be discretized, also the continuous flow domain needs to. Thus, meshing consists in discretizing the physical model, i.e., to generate a discrete representation of a problem geometry. This mesh unfolds which are the points

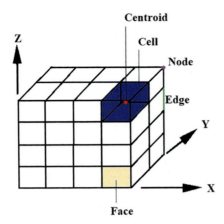

Fig. 7.6 Representation of mesh terminology concepts. *(Credit: From authors.)*

to which the equations will be solved, and which points belong to the interior and to the borders of the domain.

In a simulation, meshing strongly influence convergence rate or even the lack of convergence of a solution, results accuracy, and time required for the computer to solve the problem. Hence, CFD designers attempt mesh quality parameters toward good simulation results. Some of these parameters, which are further up discussed, are mesh density, nearing cells volume/length ratio, cell skewness, if there are tetrahedral or hexahedral elements, mesh aspect near flow boundary layer, and if there is mesh adaptive refinement (Moukalled et al., 2016).

Mesh geometry can be developed following top-down or bottom-up approach. A top-down approach means that the computational domain is generated from logic operations on primitive shapes as cylinders, prisms, or spheres. On the other hand, bottom-up means that firstly vertices are designed over the domain and then such points are linked to build lines, these lines are linked to generate faces, and faces are combined to create volume elements.

Volume elements of a mesh are usually called cells, which constitute each control volume of a 3D grid. Vertices are usually called nodes, and a centroid is a point referring to the center of a cell, the point to which flow properties are calculated during a simulation. Lines are called edges, which consist in the boundaries of a face. So that, a face is a cell boundary. Nodes, faces and cells may be part of a zone. Finally, a flow domain is made of nodes, faces, and cell zones (Bakker, 2006). The described mesh terminology is represented in Fig. 7.6.

Regarding to mesh types, there are structured and unstructured meshes. A structured mesh is one where there is indexing of cells in order to locate the nearing cells of one for which the problem is being solved. Thus, structured meshes can only be used in

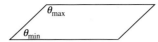

Fig. 7.7 Quadrilateral with the angles used when calculating skewness. *(Credit: From authors.)*

noncomplex geometries, so that all of a 3D domain may be addressed by three index variables i, j, and k. On the other hand, unstructured mesh is that in which cells are arranged without elements indexing, so there is no limitation on the geometry. However, using an unstructured mesh implies in memory and CPU overhead due cells referencing. Summarily, there is no logical representation of cells addresses in such mesh type.

Regarding to mesh quality, comparing meshes with similar cells quantity, meshes made of hexahedral elements tend to produce more accurate solutions, especially if there is an alignment between edges and flow direction. Mesh density, i.e., the amount of elements per area, have to be high enough to capture all relevant features of a flow. Also, mesh adjacent to the walls of a domain has to be sufficiently finer to solve the boundary layer issue. Other three main mesh quality parameters are skewness, smoothness, and aspect ratio (Moukalled et al., 2016).

Skewness of a cell can be determined as in Eq. (7.18) based on the deviation from a normalized equilateral angle according to presented in Fig. (7.7), for a quadrilateral. Such definition can be applied either to all cells or a face shape, even if the elements are prisms and pyramids.

$$\text{skewness} = \max\left(\frac{\theta_{max} - 90 \text{ degrees}}{90 \text{ degrees}}, \frac{90 \text{ degrees} - \theta_{min}}{90 \text{ degrees}}\right) \tag{7.18}$$

Skewness can also be determined as in Eq. (7.19) when referring to equiangle skew, where θ_e is the angle for an equiangular face or cell. For both skewness definitions presented, the range of such parameter varies from 0 to 1, where 0 is the best value, when there is no mesh skewness, and 1 is the worst (Bakker, 2006).

$$\text{equiangle skewnss} = \max\left(\frac{\theta_{max} - \theta_e}{180° - \theta_e}, \frac{\theta_e - \theta_{min}}{\theta_e}\right) \tag{7.19}$$

Smoothness in a mesh refers to the change of cell sizes between nearing cells. The aspect ratio is ratio between the longest edge length and the shortest edge length. Such definitions are represented in Fig. 7.8.

Assurance of a good mesh resolution may be an issue when flow features are not clearly known. Therefore, CFD designers usually apply solution-based adaption, from which mesh can be refined or coarsened based on solution values, field variable gradients, boundaries, or specific characteristics of a certain flow domain, as vertex formation. Mesh adaption basically consists on adding cells where is needed to resolve a flow field.

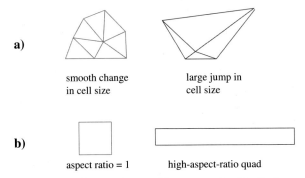

Fig. 7.8 Representation of: (A) mesh smoothness; (B) aspect ratio. *(Credit: From authors.)*

Even so, errors in a simulation may appear related to meshing. It usually occurs when: mesh is coarser; there is high size jumps between nearing cells; high screening ratios; interpolation errors in nonconformal surfaces; and unsuitable mesh for boundary layer.

Summarily, a good mesh design and development is critical for the success of a CFD analysis. This way, choosing a convenient mesh depends on the geometry complexity, field flow features, and cell shapes supported by the solver used in the simulation (Bakker, 2006).

7.2.6 The advantages of applying CFD in processes analysis

In the past decades some great disadvantages of applying CFD were the difficult to obtain solution for problems with complex geometry, strong nonlinearities, turbulent and time-dependent cases, due to computational limitations. However, with rapidly increase of computational resources at low cost, such as processing capacity and memory, now available even in personal computers, the abovementioned disadvantages are no longer observed. A CFD approach can be adopted when analyzing a production process considering: low cost, speed, flexibility to analyze extreme process conditions, and detailed domain information.

The major advance provided by CFD is the low cost when compared to an experimental investigation. If a proposed model to apply in a CFD simulation has already been validated, then analyzing a process through computational runs is an approach which cost is extremely lower than experimental tests. Thus, CFD analysis saves not only financial resources, but also human resources and material that would be spent in experiments. Besides that, unlike in lab tests, where human and undetermined errors may influence in results obtained, simulation runs when properly conducted tends to give always accurate results (Anderson, 1995).

Concerning the design of new equipment and processes, CFD is also a powerful tool to speed up this process. A designer can study, for example, several different configurations of a reactor in order to choose the optimum design that could give the higher reaction yield in a biodiesel synthesis. Furthermore, it is possible to simulate cases when extreme variables are applied, for example, fluid flows with reaction at very high temperatures, or that involves toxic or flammable reactants, and even in processes that are too fast or take too much time.

CFD also allows obtaining variables information in process where experimental data acquisition is too difficult. For example, in process that involves mass transfer, or even more complex, mass transfer with reaction cases, experimental data could not be assessed at real-time, since compounds qualitative and quantitative analysis techniques, such as spectrometry or chromatography, would be needed. CFD simulations, differently, allows visualizing variables such as species concentration or a phase fraction, in multiphase flows, in real time when considering a time-dependent process.

Finally, one of the most prominent advantages of a computer solution is that it gives complete and detailed information for a given domain. All variables, such as velocity, pressure, density, viscosity, temperature, species concentration, can be obtained throughout the domain of interest, i.e., in all points computed. On the other hand, in an experimental approach, just a few locations are accessible to visualize such variables. Also, the more locations assessed, the more flow disturbance is caused by the probes. Thus, even when experiments are made to analyze a process, a computer solution is always a relevant supplement to the experimental information (Chung, 2002; Pulliam et al., 1999).

7.3 Literature review
7.3.1 CFD studies in macroscale reactors

Koyunoğlu and Karaca (2019) used CFD to evaluate the biomass liquefaction, i.e., the production of liquid fuels from solid biomass. The main goals were to determine the flow patterns and its effects on the gas-liquid mixture and to estimate the gas retention. The authors used the Ansys Fluent software with a Multiphase Eulerian model. The Multiple Reference Frame (MRF) approach was employed to model the gas distribution. The MRF is an approximation of steady-state conditions were individual cell zones can be attributed to different rotation speed and/or translation (ENEA, 2009). The authors used a traditional BR with four baffles, two impellers, and a ring sparger. The maximum gas-liquid contact was defined by the fluids properties. In the gas retention analysis, the authors observed a superior product yield for longer contact time, however, shorter contact time resulted in higher conversions.

A multiphase Eulerian CFD model was used by Liu et al. (2017) to simulate a lab-scale fluidized bed reactor of fast pyrolysis of biomass, i.e., chemical reaction occurring by the action of high temperatures. The authors integrated the 3-parameter particle shrinking model by Di Blasi to the Eulerian framework, to consider the particle size evolution along the process. The Ansys Fluent was used and the Quadrature Method of Moments (QMOM) was employed to solve the population balance equation of particles. Such approach allowed accounting the particle diameter variation due to the biomass devolatilization. The approach allowed ranging the particle contraction parameters to represent different process patterns, aiming to evaluate the shrinking effects on particle motion, heat transfer and product yield. The computational domain consisted in a 40 mm × 340 mm 2D representation of a lab-scale cylindrical fluidized bed reactor operating at 300 g/h. The pressure-velocity coupling was solved by SIMPLE algorithm and the Volume-of-Fraction (VOF) approach was used for solid phase. The authors reported a decrease on apparent density due to the continuous transfer of particle mass and formation of pores. In the nonshrinking model, the mass transfer from biomass particle was totally accounted by the pore formation keeping the particle size constant. In the other cases, this behavior was partially explained by the formation of pores and was dependent on particle shrinking process. The numerical results also showed that the shrinking patterns also affected the biomass devolatilization process, directly influencing the conversion, and coal distribution. The bio-oil yield, about 63%, was not affected by the shrinking patterns. In contrast, the charcoal yield slightly increased with the particle shrinking, due to the increment on the residence time of biomass particles.

Park and Choi (2019) evaluated the biomass pyrolysis in a gas-solid flow in a spouted bed reactor. The Computational Particle Fluid Dynamics (CPFD) using the MP-PIC method, an Eulerian-Lagrangian approach, was used to numerically study the reactive multiphase flow field. The CPFD allowed the comprehension of the complex fluid dynamics, heat transfer and chemical kinetics characteristics. The CPFD solves the momentum balance equations for fluid and particles. The fluid motion is described by the volume-averaged Navier-Stokes equations, while the particles are approached by the Lagrangian MP-PIC method, by ordinary differential equations and a bidirectional coupling with the continuous fluid phase. The numerical approach of the particle is similar to the finite control volume, where the spatial region has a unique property of the fluid. In the CPFD, the particle collisions are modeled by the normal tension of particle, derived from its volume fraction determined from the particle volume mapped from the grid. The authors used a 3D bed with dimensions: internal diameter cylindrical section of 160 mm, bottom diameter of 50 mm, and gas inlet section diameter of 25 mm. The total bed height was 400 mm with a conical section angle of 28 degrees. The computational analysis was assessed by the CPFD code BARRACUDA 17.0.3 (CPFD Software LLC). The authors' conclusions include: the mean deviations between CPFD prediction and experimental data of yield were 15.2%, 7.3%, and 29.3% for the oil, char, and gas, respectively; the increment of inlet gas velocity caused a significative change on the product yield that was related to the enhancement on mass and heat transfer rates between bed materials and biomass particles. The obtained results showed increment on tar yield from 55.6% to 64.8% (w/w) with the inlet gas velocity increased from 4 to 6 m/s, and then a decrease to 60.7% for a further increment to 7 m/s. The raise on temperature from 400°C to 450°C lead to an increase on tar yield from 55.6% to 58.7% (w/w), achieving an upper limit at 450°C.

Jin, Wang, Wu, Ren, and Ou (2019) used CFD to investigate a supercritical water fluidized bed (SCWFB), a reactor type used to achieve efficient and clean biomass gasification. Due to the inherent flow complexity of such reactor and experimental limitations of the SCWFB, is a hard task to determine accurately the mass and heat transfer rates between the solid particles and the supercritical water (SCW). Accordingly, the authors used a simplified two particle model to study the fluid flow pattern of the biomass particles with SCW. The study focused on the drag coefficient and biomass particles flow characteristics aiming to obtain the influent factors and interaction mechanisms of the particles in the range of $10 < Re < 200$. The OpenFOAM package was used and the model was considered at steady-state, incompressible, and laminar flow regime conditions. The computational domain was a cuboid of $40D \times 20D \times 20D$ where D is the equivalent spherical diameter of the particle based on its volume. The particles were perpendicularly located at a distance of $10D$ from the inlet at the centers of Y and Z directions. In the two particle model, the sphere diameter was considered 200 μm, with a particle fixed at the origin position, P_1 and the other particle defined at P_2. The distance between P_1 and P_2 was defined as L and the angle between the centerline of particles and the inlet flow direction was defined as α, thus the relative position changes with L and α. The bed was operated at 22.1 MPa and 670 K. The authors concluded that the particle interactions affected the drag coefficient, and at lower Re, particles presented a larger area of influence, perpendicular to the flow field, and the particle interactions can be transmitted for longer distances. The numerical predictions also showed similar wake zones of the SCW around the particles regarding water at room conditions.

A two-dimensional CFD modeling in a palladium membrane reactor (MR) was performed by Ghasemzadeh, Ghahremani, Amiri, and Basile (2019). The hydrogen production by glycerol vapor reforming was assessed. The authors compared the performance of MR to a traditional reactor, under distinct operating conditions. The mathematical model, after the validation, was used to predict the concentration changes of the chemical species along the reactor and also to quantify the glycerol conversion, selectivities, and the hydrogen membrane permeation. The study evaluated the effects of the most influential operating parameters: pressure, temperature, and feed flow rate. The geometry of the MR consisted in a fine tube with a zone of catalyst particles, with a length about 120 mm and internal diameter of 10 mm. The COMSOL Multiphysics 5.3 software was used with the SIMPLE pressure-velocity coupling. The authors noticed that temperature and pressure increments mitigated the CO formation, improved the hydrogen recovery, and the glycerol conversion. The countercurrent flow was preferable than the co-current arrangement.

7.3.2 CFD studies in microscale reactors

One of the first CFD codes used to study the biodiesel synthesis in microreactors was developed by Han, Charoenwat, and Dennis (2011) using COMSOL and Matlab software to solve chemical kinetics and transport equations in a two-dimensional model. The authors evaluated the soy oil transesterification with methanol, considering the Lagrangian approach with two phases. The computational domain was a rectangular segment, representing a slice of the capillary reactor. The reaction yield was analyzed in function of the residence time that exhibited a significative influence in the chemical species variation. The superior yield was observed for longer residence times. The authors did not approach the effects of temperature, alcohol/oil molar ratio, and catalyst concentration on the chemical reaction behavior. These three operating variables (temperature, reactants molar ratio, and catalyst concentration) are directly related to the biodiesel synthesis performance. In addition, other aspects, such as the reactor design and mixing degree of reactants should be also considered in the design of a microdevice for biodiesel synthesis. Recently, research studies have been encompassed these two aspects (device design and mixing degree) and are discussed following.

One of the major difficulties to the usage of microreactor in biodiesel synthesis is the reactants mixing, i.e., between vegetable oil and short chain alcohol, due to physicochemical characteristics of the chemical species and the laminar regime flow, marked by the absence of macromixing between the fluid layers and consequently, the molecular diffusion dominance in mass transfer. The reactants mixing degree is intimately related to the synthesis performance, thus, a main element of microreactor is the micromixer that should promote a fast and efficient mixture of the reactants.

Santana, Silva, and Taranto (2015) performed a numerical analysis of the effects of different configurations of reactor inlet on biodiesel synthesis behavior. The evaluated configurations were: T-shape (two inlets, (1) alcohol, (2) oil), Cross-shape (three inlets: inlets 1 and 3—alcohol, inlet 2—oil), double T-shape (four inlets: inlets 1 and 3—alcohol, inlets 2 and 4—oil). The Ansys CFX software was used considering a multicomponent single-phase flow under steady-state laminar regime conditions. All configuration exhibited high mixing degrees at low Reynolds numbers. Ranging Re from 10 to 100, the cross-shape design presented superior mixing and reaction yield. The behavior was attributed to the fluid dynamic patterns induced by the cross-shape inlet. In such design, the triglyceride molecules are induced to diffuse in two directions instead of only one, resulting in a shorter time required to complete the mass transfer. However, for longer residence times, the reactants molecules experience enough time to complete the diffusion process along the microchannels and the oil conversions was virtually similar in all tested micromixer configurations.

After this study, Santana, Silva, and Taranto (2015) evaluated the effects of circular obstruction as micromixer elements. The numerical model was the same from the previous study. The internal obstacles allowed a good mixing performance in larger channels. From literature was found superior performance in smaller microreactors, however at the expense of higher pressure drops and production costs and also, operating difficulties (Xie, Zhang, & Zu, 2012). The use of microchannels with internal obstacles allowed to enlarge the microreactor, once the micromixer elements promoted a reduction of diffusion path and increased the specific interfacial area. Accordingly, two main advantages were highlighted regarding traditional channels without obstacles: (1) increment on operating flow rate; (2) reduction of residence time.

From these results, an experimental study was performed for the transesterification of sunflower oil with ethanolic solution of sodium hydroxide in two microreactor configurations: conventional microreactor (without obstacles) (Santana, Tortola, Reis, Silva, & Taranto, 2016) and microreactor with static elements (MSE) (Santana, Sanchez, & Taranto, 2017). Both geometries have dimensions of $1500\,\mu m \times 200\,\mu m \times 411\,mm$ (channel width x channel height x longitudinal length) and were made by PDMS. For the traditional layout, without obstacles or CFD optimization, the superior oil conversion was 95.8% at 50°C, alcohol/oil molar ratio of 5, 0.85% of catalyst concentration and residence time of 60 s. For the MSE layout, a design optimized by CFD, the superior performance was given by an oil conversion of 99.53% at 50°C, alcohol/oil molar ratio of 9, 1.0% of catalyst concentration, and residence time of 12 s. Accordingly, high biodiesel synthesis efficiency can be accomplished from larger devices using effective and optimized micromixer elements. A second conclusion highlights the minimization of the residence time with the increment on oil conversion, from 95.8% to 99.53%, from 60 to 12 s of residence time, when using static elements to enhance the reactants mixing. The significative optimization observed for the MSE configuration can be mostly attributed to the mixing quality promoted by the design. For example, for a traditional 35 mm long microchannel, the superior mixing index was 0.31, in contrast with a mixing index of 0.99 noticed from the microchannel with static elements. These results highlight the potentiality for micromixer performance optimization using CFD.

López-Guajardo, Ortiz-Nadal, Montesinos-Castellanos, and Nigam (2017) performed a comparative study between a tubular microreactor (TMR—made by stainless steel, inner diameter of 710 μm and length of 5 m) and an ideal mixing BR. Oil conversions up to 99% were noticed in the TMR for a residence time of 4 min. A CFD analysis was carried out to visualize and comprehend the interaction between the two phases. The methanol/oil molar ratio used was 6. One of the main goals of the synthesis in the TMR was to achieve a slug flow pattern. The COMSOL Multiphysics code was used considering a laminar regime flow and a Lagrangian field modeling. Dirichlet boundary conditions were used and the slug flow pattern, defining the oil and the methanol regions were fixed to emulate the internal motion of the slug as well its motion along the reactor. Moving wall and slip boundary conditions were specified. All simulations were solved in steady-state conditions by the PARDISO solver. The results showed superior oil conversion in shorter residence times (99% of conversion at 4 min of residence time) concerning the BR. Distinct flow pattern can arise in the TMR, directly impacting the reaction performance. The slug flow pattern increased the mass transfer due to the internal motion of the slugs, providing reliable results without the excessive spent of energy and reactants molar ratio.

Coming back to micromixer research and development, Sabry, El-Emam, Mansour, and Shouman (2018) proposed a uniflow passive micromixer based on multilamination principle using CFD. The main goal was to achieve high mixing quality under low pressure drops. The methanol/oil molar ratio was 6 at 60°C. The channel cross section has dimensions of 240 μm × 240 μm. The mathematical model consisted in a multicomponent (two chemical species), incompressible, Newtonian, isothermal, and laminar flow regime. The Ansys Fluent code was used and the SIMPLE algorithm was used along the QUICK scheme. The numerical predictions showed a high surface area to volume ratio, allowing a higher mixing quality at relative low pressure drops.

Santana, Silva, and Taranto (2019) employed the mathematical model from previous studies (Santana, Sanchez, & Taranto, 2017) to propose a new design named micromixer with triangular baffles and circular obstructions (MTB). The main difference of such design was the combination of three mass transfer enhancement mechanisms: reduction of diffusion path, split and recombination of streams and vortex generation. After the geometric parameters optimization, the device was use to promote the mixing in two binary systems (vegetable oil/ethanol and water/ethanol), for the Reynolds number range from 0.01 to 200. Also the transesterification of sunflower oil with ethanolic solution of sodium hydroxide was assessed. High mixing indexes were noticed for several channel heights (200–2000 μm) and widths (1500–3000 μm). The geometry W3000H400, i.e., MTB with a channel width of 3000 μm and channel height of 400 μm, was employed as a millireactor. The maximum oil conversion observed was 92.67% for a residence time of 30 s. For the water/ethanol system the geometry W1500H200 was used. High mixing index ($M=0.99$) was observed for very low Reynolds number ($Re=0.1$) and also at higher Reynolds numbers (50 and 100). The MTB device, in contrast with reported in literature, could be used as microdevice operating under very low throughputs, e.g., in sensors, and also as a larger millidevice, to operate under higher flow rates, e.g., as millireactor for modular micro-chemical plants, allowing an easy scale-up and numbering-up to achieve the production demand. The development of larger microdevices for chemical synthesis appears as an interesting alternative to apply Microfluidics devices at commercial scale. Clearly, the CFD appears as a fundamental tool for design and optimization, as highlighted by Silva, Haddad, Taranto, and Santana (2020).

Shah et al. (2020) proposed three new micromixer designs, evaluating the biodiesel synthesis performance by numerical simulation and experimental runs. The COMSOL software was used along with the laminar flow and the dilute chemical species transport modules. No-slip conditions were employed at walls. The results showed high mixing indexes and the flow patterns observed from numerical simulations were similar to the experimental runs.

Using the Ansys Fluent with the VOF to track the liquid-liquid interface, Laziz et al. (2020) evaluated the effect of KOH catalyst and residence time on the vegetable oil conversion with methanol into biodiesel in a T-junction microchannel, taking advantage of the high mass transfer rates inherent from the slug flow regime provides. The increase on reaction efficiency, performed at room temperature, was investigated based on hydrodynamic factors, including the interfacial area and chemical species mixing. Numerical and experimental runs were carried out using a transparent capillary microchannel made by fluorinated ethylene propylene (FEP) (inner diameter of 690 μm, 1 m of longitudinal length). The authors used a pressure-based solver to solve transient simulations in the three-dimensional domain. The chemical reactions were not considered in the numerical simulations. The continuity equation was solved with the interface tracking considering the volume fraction of each phase and the surface tension effects were considered by the continuous surface force (CSF) model by source terms in the momentum balance equations. The results showed that the KOH concentration and the residence time presented a negligible effect on the total interfacial area. The effects of residence time in the biodiesel yield did not show sigmoidal tendency that indicated no mass transfer limitations at the transesterification beginning. However, a relative high catalyst concentration, 5% (w/w) of KOH, was necessary to achieve 98.6% of oil conversion. The numerical prediction allowed observing the torus recirculation patterns inside the slugs, increasing the mixing and the mass transfer between the reactants, improving the global process performance.

Laziz et al. (2020) used the VOF with CFD model to evaluate the oil-alcohol flow without chemical reactions. Souza, Santana, and Taranto (2020) used VOF along with CSF and Continuum Species Transfer (CST) to solve the oil-alcohol two-phase flow considering the mass transfer between the phases and the chemical reaction. The model was implemented in the OpenFOAM to simulate the mixing between sunflower oil and ethanol with KOH catalyst. The CST model was used to solve the concentration field evolution in space and time, using a condition analogue to the Henry's Law at the liquid-liquid interface for miscible components and continuous mass fluxes across the interface. The microreactor geometry was the same of Santana, Sanchez, and Taranto (2017). The numerical results highlighted that VOF with CSF and CST provided an accurate method to track the two-phase flow interface. The numerical model also captured precisely the vortex generate in the flow field due to the geometry design, resulting in high mixing rate and consequently, high oil conversion. The authors observed for the first time that glycerol tends to aggregated in reactor dead zones. Finally, the authors concluded that the VOF-based model implemented on the OpenFOAM is more effective to predict the biodiesel synthesis regarding the pseudo-multiphase approach.

7.4 Case study

PI is a fundamental strategy employed by engineers to enhance the process performance. In this context, as already highlighted, Microfluidics appears as an interesting alternative in PI. Accordingly, in this section a comparison between BR and continuous micro and millireactors is presented. In order to assess the performance of these reactors and the scale-up features, the vegetable oil conversion, the mass flow rate of biodiesel and unreacted ethanol by the equivalent number of microreactor, n_{MR} (Eq. (7.20), Pinheiro et al., 2018) were evaluated. Furthermore, the required power input (mechanical agitation in batch process and pumping of continuous micro and millidevices) were estimated from CFD results. The sBR results were converted to a rate-basis considering the Eq. (7.21).

$$n_{MR} = \frac{\dot{m}_{BR}}{\dot{m}_{MR}} \quad (7.20)$$

$$\dot{m}_{BR} = \frac{C_i t_f}{V_{BR}} \quad (7.21)$$

$$\dot{m}_{MR} = C_{i,out} \dot{V} \quad (7.22)$$

where the $\dot{m}_{BR}$ and $\dot{m}_{MR}$ are the molar flow rate in the BR and micro/millireactor, respectively, C_i is the ith chemical species molar concentration at the final time of batch, t_f, V_{BR} is the BR volume, $C_{i,out}$ is the ith chemical species molar concentration at the micro/millireactor outlet and $\dot{V}$ is the volumetric flow rate of the micro/millireactor.

Two BR volumes were analyzed: 1 L and 10 m³, corresponding to lab-scale and industrial scale reactors, as detailed in Fig. 7.9. A standard stirred tank reactor configuration equipped with a 4-pitched (45 degrees) blade turbine was used (Fig. 7.9B). This configuration was chosen due to the relative low power requirements, since the pitched-blade tends to achieve a constant nondimensional power number of about 1.3 in turbulent agitation regime.

The microreactor designs were based on CFD analysis. Fig. 7.10 shows the device design evolution achieved by our research group in order to overcome oil-ethanol mixing issues (Fig. 7.10A and B). Santana, Silva, and Taranto (2015) evaluated two microchannel designs using circular obstacles: uniform disposed obstacles and alternate small/larger circular obstacles (Fig. 7.10C). The effects of channel width scale-up from 200–1500 μm were evaluated. The CFD results indicated that the obstacles caused a slight improvement on oil conversion. Larger microdevices exhibited oil conversions comparable to the narrower microchannel (200 μm). Furthermore, no significative differences on oil conversion were observed between the two micromixer designs. Santana, Sanchez, and Taranto (2017) proposed the Micromixer with Static baffle Elements (MSE) (Fig. 7.10D). The proposed design generate high chaotic advection zones,

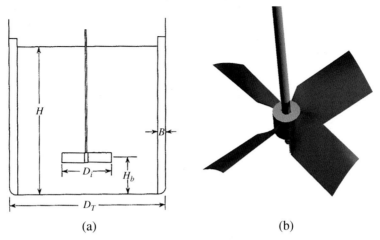

Fig. 7.9 (A) Geometric parameters of the stirred tank reactor: $D_T = 0.038$ m (1 L), $D_T = 0.816$ (10 m^3); $D_i/D_T = 1/3$, $H/D_T = 1$; $H_b/D_T = 1/3$, $B/D_T = 1/10$ (McCabe, Smith, & Harriott, 1993); (B) pitched-blade turbine composed by 4 blades (45 degrees). *(Credit: From authors.)*

enhancing the biodiesel yield to 99.53% at 50°C, ethanol/oil molar ratio of 9, catalyst concentration of 1%, for a residence time about only 12 s.

Based on previous observation of fluid flow pattern from CFD results, Santana, Silva, and Taranto (2019) proposed the micromixer with triangular baffles (MTB). This design combined three mass transfer mechanisms: split and recombination of streams, vortex generation, and reduction of molecular diffusion path (Fig. 7.10E). The optimal design was assessed by CFD analysis considering a CCDR-2^3 of geometric parameters and their effects on oil-ethanol mixing. The MTB exhibited a good performance, providing a higher mixing index (about 0.9) already at the first mixing unit (Fig. 7.10F). The predicted oil conversion was 92.7% at 50°C, 1% (w/w) of NaOH catalyst and ethanol/oil molar ratio of 9 for a scaled channel with dimensions of 400 μm × 3000 μm. The scale-up effects on oil-ethanol mixing and biodiesel synthesis were evaluated and the results obtained from MTB were used in the performance comparison with the BR.

Two operating conditions were evaluated. The first one was based on experimental work of Reyero, Arzamendi, Zabala, and Gandía (2015) that obtained high levels of oil conversion and was used for the BR. Reyero et al. (2015) provided a complete detailed chemical kinetics mathematical model including parallel saponification steps. The second operating condition was based on previous studies (Santana et al., 2016; Santana, Silva, & Taranto, 2015) on continuous micro and millidevices. The mathematical model was based on Marjanovic, Stamenkovic, Todorovic, Lazic, and Veljkovic (2010) due to the good agreement with experimental data found on previous studies (Santana et al., 2016; Santana, Sanchez, & Taranto, 2017).

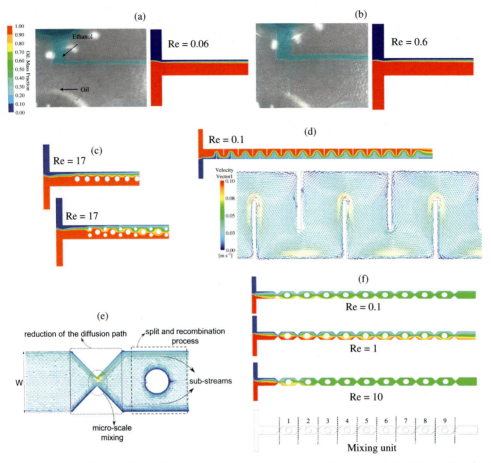

Fig. 7.10 Vegetable oil distribution in microdevices: experimental pictures and CFD results of oil distribution in the traditional T-shape microchannel at (A) $Re=0.06$ and (B) $Re=0.6$ (Santana, Silva, & Taranto, 2015), (C) CFD predictions at $Re=17$ micromixer designs proposed by Santana, Silva, and Taranto (2015) based on circular obstacles arrangements, (D) CFD predictions of MSE design at $Re=0.1$ (Santana, Sanchez, & Taranto, 2017), (E) MTB design with illustrative details of the mixing enhancement mechanisms proposed, (F) CFD results for the MTB device at different Reynolds numbers (0.1, 1 and 10) (Santana, Silva, & Taranto, 2019). *(Credit: Reprinted from Santana, H. S., Silva, J. L., Taranto, O. P., 2015. Numerical simulations of biodiesel synthesis in microchannels with circular obstructions.* Chemical Engineering and Processing, 98, *137–146; Santana, H. S., Sanchez, G. B., Taranto, O. P., 2017. Evaporation of excess alcohol in biodiesel in a microchannel heat exchanger with Peltier module.* Chemical Engineering Research & Design, 124, *20–28; Santana, H. S., Silva, J. L., Taranto, O. P., 2019. Optimization of micromixer with triangular baffles for chemical process in millidevices.* Sensors and Actuators B: Chemical, 281, *191–203) with permission of Elsevier.)*

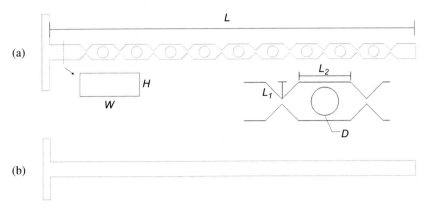

Fig. 7.11 (A) MTB design, (B) traditional T-shape microchannel. $H \times W \times L$: channel height, channel width, longitudinal length. MTB static elements dimensions: $L_1 = 650\,\mu m$; $L_2 = 2100\,\mu m$; $D = 1100\,\mu m$, from the optimal design (Santana, Silva, & Taranto, 2019). *(Credit: From authors.)*

- Operating condition 1 (Batch process): $T = 50°C$, E:O molar ratio of 12 and 0.3% (w/w) of NaOH catalyst, $N = 400$ rpm, providing $Re_i > 10^6$ to ensure turbulent mixing.
- Operating condition 2 (Continuous process): $T = 50°C$, E:O molar ratio of 9 and 1% (w/w) of NaOH catalyst, residence time, τ, ranging from 2 to 60 s. Three devices were considered: T-shape (traditional microchannel, $H \times W \times L = 200\,\mu m \times 1500\,\mu m \times 35.85 \times 10^3\,\mu m$), MTB ($H \times W \times L = 400\,\mu m \times 3000\,\mu m \times 75.50 \times 10^3\,\mu m$) microdevice and MTB millidevice ($H \times W \times L = 800\,\mu m \times 6000\,\mu m \times 151 \times 10^3\,\mu m$). $H \times W \times L$: channel height, channel width, longitudinal length. Further details are provided in Fig. 7.11.

7.4.1 Mathematical model

The reactive flow system consisted in the transesterification reaction of vegetable oil with ethanol solution containing sodium hydroxide catalyst. Three-dimensional flow was solved with the main considerations of: multicomponent Newtonian fluid under isothermal (50°C) and incompressible conditions. The flow regime was considered laminar in the micro and millireactors. For the BR, a turbulent flow regime was considered. The RANS approach was employed by using the Shear Stress Transport (SST) turbulence model. Accordingly, the total mass (continuity), the Navier-Stokes and the chemical species mass conservations were solved, as given by Eqs. (7.23)–(7.25), respectively:

$$\nabla \cdot U = 0 \quad (7.23)$$

$$\rho\left(\frac{\partial U}{\partial t} + U \cdot \nabla U\right) = -\nabla p + \mu \nabla^2 U + \rho g \quad (7.24)$$

$$\rho\left(\frac{\partial Y_i}{\partial t} + U \cdot \nabla Y_i\right) = \rho D_i \nabla^2 Y_i + S_i \quad (7.25)$$

where ρ is the specific mass (kg m^{-3}), U is the velocity vector (m/s), P is the pressure (kg m^{-1} s^{-1}), μ is the dynamic viscosity (kg m^{-1} s^{-1}), g is the gravity acceleration (m s^{-2}), the subscript i denotes the chemical species, Y is the mass fraction of chemical species, D is the mass diffusion coefficient (m^2 s^{-1}) estimated from Wilke-Chang correlation, S is the mass source term due to the chemical reaction, detailed following. The fluid properties ρ and μ where weighted by the mass fraction of chemical species. The BR was modeled as transient, in contrast with the steady-state assumption for the continuous micro and millidevices.

7.4.2 Batch reactor modeling: Chemical kinetics of vegetable oil transesterification with parallel saponification reactions

The reactive flow system consisted in the transesterification reaction of vegetable oil with ethanol solution containing sodium hydroxide catalyst. The mathematical model for the first operating condition was based on based on Reyero et al. (2015) accounting for saponification parallel reactions. This chemical kinetics is valid for NaOH catalyst concentration up to 0.3% (w/w) and consists in three transesterification steps (Eqs. 7.26–7.28), one hydroxide/ethoxide step (Eq. 7.29) and four saponification reactions (Eqs. 7.35–7.38):

$$TG + EtOH \underset{k_{-1}}{\overset{k_1}{\rightleftarrows}} DG + E \tag{7.26}$$

$$DG + EtOH \underset{k_{-2}}{\overset{k_2}{\rightleftarrows}} MG + E \tag{7.27}$$

$$MG + EtOH \underset{k_{-3}}{\overset{k_3}{\rightleftarrows}} GL + E \tag{7.28}$$

$$EtOH + OH^- \underset{k_y}{\overset{k_x}{\rightleftarrows}} EtO^- + H_2O \tag{7.29}$$

where TG, DG, MG, GL, E, $EtOH$, OH^-, EtO^- and H_2O denotes for triglycerides, diglycerides, monoglycerides, glycerol, ethyl esters (biodiesel), ethanol, hydroxide, ethoxide, and water, respectively. The reaction rates were modeled as:

$$r_1 = k_1 C_{TG} C_{EtOH} - k_{-1} C_{DG} C_E \tag{7.30}$$

$$r_2 = k_2 C_{DG} C_{EtOH} - k_{-2} C_{MG} C_E \tag{7.31}$$

Table 7.1 Chemical kinetics and equilibrium constant data for 50°C.

Reaction step	Kinetic rate constant, k	Equilibrium constant, K_{eq}
Triglyceride transesterification (k_1)	0.110 ($L^2\,mol^{-2}\,s^{-1}$)	3.21
Diglyceride transesterification (k_2)	0.171 ($L^2\,mol^{-2}\,s^{-1}$)	3.18
Monoglyceride transesterification (k_3)	0.076 ($L^2\,mol^{-2}\,s^{-1}$)	72.77
Hydroxide/ethoxide reaction (k_x)	3.1 ($L\,mol^{-2}\,s^{-1}$)	0.73
Glycerides saponification ($k_{sap,1} - k_{sap,3}$)	0.482 ($L\,mol^{-2}\,s^{-1}$)	–
Biodiesel saponification ($k_{sap,4}$)	0.062 ($L\,mol^{-2}\,s^{-1}$)	–

$$r_3 = k_3 C_{MG} C_{EtOH} - k_{-3} C_{MG} C_E \qquad (7.32)$$

$$r_4 = k_x C_{EtOH} C_{OH^-} - k_y C_{EtO^-} C_{H_2O} \qquad (7.33)$$

The direct and reverse reactions constant factors are related to equilibrium relationship as given by Eq. (7.34):

$$K_{eq} = \frac{k_{dir}}{k_{rev}} \qquad (7.34)$$

where K_{eq} is the equilibrium constant, the subscripts *dir* and *rev* denote for direct (1,2,3 and x) and reverse (−1, −2, −3 and y) reactions.

$$TG + NaOH \xrightarrow{k_{sap,1}} DG + Soap \qquad (7.35)$$

$$DG + NaOH \xrightarrow{k_{sap,2}} MG + Soap \qquad (7.36)$$

$$MG + NaOH \xrightarrow{k_{sap,3}} GL + Soap \qquad (7.37)$$

$$E + NaOH \xrightarrow{k_{sap,4}} EtOH + Soap \qquad (7.38)$$

The saponification reaction rates were given by Eqs. (7.39)–(7.42):

$$r_{sap,1} = k_{sap,1} C_{TG} C_{NaOH} \qquad (7.39)$$

$$r_{sap,2} = k_{sap,2} C_{DG} C_{NaOH} \qquad (7.40)$$

$$r_{sap,3} = k_{sap,3} C_{MG} C_{NaOH} \qquad (7.41)$$

$$r_{sap,4} = k_{sap,4} C_E C_{NaOH} \qquad (7.42)$$

The employed kinetic constant and equilibrium constant data at 50°C are summarized in Table 7.1.

Considering the reaction rate expressions given by Eqs. (7.25), (7.30)–(7.33), (7.39)–(7.42) and the stoichiometric coefficients of chemical Eqs. (7.26)–(7.29) and (7.35)–(7.38), the chemical species mass transfer solved were:

$$\rho\left(\frac{\partial Y_{TG}}{\partial t} + U \cdot \nabla Y_{TG}\right) = \rho D_{TG} \nabla^2 Y_{TG} - (r_1 + r_{sap,1}) M_{TG} \qquad (7.43)$$

$$\rho\left(\frac{\partial Y_{DG}}{\partial t} + U \cdot \nabla Y_{DG}\right) = \rho D_{DG} \nabla^2 Y_{DG} + (r_1 - r_2 - r_{sap,1} - r_{sap,2}) M_{DG} \qquad (7.44)$$

$$\rho\left(\frac{\partial Y_{MG}}{\partial t} + U \cdot \nabla Y_{MG}\right) = \rho D_{MG} \nabla^2 Y_{MG} + (r_2 - r_3 - r_{sap,2} - r_{sap,3}) M_{MG} \qquad (7.45)$$

$$\rho\left(\frac{\partial Y_{EtOH}}{\partial t} + U \cdot \nabla Y_{EtOH}\right) = \rho D_{EtOH} \nabla^2 Y_{EtOH} - (r_1 + r_2 + r_3 - r_{sap,4}) M_A \qquad (7.46)$$

$$\rho\left(\frac{\partial Y_{GL}}{\partial t} + U \cdot \nabla Y_{GL}\right) = \rho D_{GL} \nabla^2 Y_{GL} + (r_3 + r_{sap,3}) M_{GL} \qquad (7.47)$$

$$\rho\left(\frac{\partial Y_E}{\partial t} + U \cdot \nabla Y_E\right) = \rho D_E \nabla^2 Y_E + (r_1 + r_2 + r_3 - r_{sap,4}) M_E \qquad (7.48)$$

$$\rho\left(\frac{\partial Y_{NaOH}}{\partial t} + U \cdot \nabla Y_{NaOH}\right) = \rho D_{NaOH} \nabla^2 Y_{NaOH} - \sum_{j=1}^{4} r_{sap,j} M_{NaOH} \qquad (7.49)$$

$$\rho\left(\frac{\partial Y_{Soap}}{\partial t} + U \cdot \nabla Y_{Soap}\right) = \rho D_{Soap} \nabla^2 Y_{Soap} + \sum_{j=1}^{4} r_{sap,j} M_{Soap} \qquad (7.50)$$

Also, inert oil with similar transport properties of sunflower oil was added in system as a constraint to ensure the mass fraction restriction:

$$\sum_{i}^{n_i} Y_i = 1 \qquad (7.51)$$

7.4.2.1 Numerical details for the batch reactor

The domain was initially considered filled with ethanol-oil-catalyst mixture, considering the ethanol-oil molar ratio of 12 and 0.3% of NaOH. The no-slip condition was applied at wall zones. The Multiple Reference Frame approach was adopted to simulate the rotor motion. Also, a symmetry condition was used and the domain consisted in 1/4 of the total reactor. High resolution schemes were used in the numerical solution procedure. A RMS target of 1×10^{-4} was employed with 10 iterations at internal loop. The time step was ranged from 0.05 to 0.25 s that provided good convergence on the solution. A total of 45 min were simulated.

7.4.3 Continuous micro and millireactors modeling: Chemical kinetic considering a global oil transesterification mechanism

For the second operating condition evaluated, the biodiesel synthesis was considered by the overall reaction given by Eq. (7.52) (Freedman et al., 1984):

$$TG + 3A \underset{k_{inv}}{\overset{k_{dir}}{\rightleftarrows}} GL + 3E \tag{7.52}$$

where TG, A, GL, and E denotes for triglycerides, alcohol, glycerol, and ethyl esters (biodiesel), respectively. The reaction rate, written as the triglyceride consumption rate at equilibrium state, as proposed by Marjanovic et al. (2010):

$$r_{TG} = -k_{dir} C_{TG} C_A + k_{inv} C_{GL} C_E \tag{7.53}$$

where $-r$ is the reaction rate (mol m^{-3} s^{-1}), k_{dir} and k_{inv} are the reaction rate constants for direct and reverse reactions (m^3 mol^{-1} s^{-1}), respectively, C is the molar concentration (mol m^{-3}). The reactive flow simulations were carried out at optimal conditions, according to Santana, Sanchez, and Taranto (2017) with $k_{dir} = 15.68 \times 10^{-6}$ m^3 mol^{-1} s^{-1} and $k_{inv} = 9.42 \times 10^{-8}$ m^3 mol^{-1} s^{-1} at 50°C, 1% (w/w) of NaOH and ethanol/oil molar ratio of 9. Accordingly, the chemical species mass balances solved were:

$$\rho\left(\frac{\partial Y_{TG}}{\partial t} + U \cdot \nabla Y_{TG}\right) = \rho D_{TG} \nabla^2 Y_{TG} - r_{TG} M_{TG} \tag{7.54}$$

$$\rho\left(\frac{\partial Y_A}{\partial t} + U \cdot \nabla Y_A\right) = \rho D_A \nabla^2 Y_A - 3 r_{TG} M_A \tag{7.55}$$

$$\rho\left(\frac{\partial Y_{GL}}{\partial t} + U \cdot \nabla Y_{GL}\right) = \rho D_{GL} \nabla^2 Y_{GL} + r_{TG} M_{GL} \tag{7.56}$$

$$\rho\left(\frac{\partial Y_E}{\partial t} + U \cdot \nabla Y_E\right) = \rho D_E \nabla^2 Y_E + 3 r_{TG} M_E \tag{7.57}$$

Also, an inert oil similar transport property of sunflower oil was added in system as a constraint to ensure the mass fraction restriction:

$$\sum_{i}^{n_i} Y_i = 1 \tag{7.58}$$

7.4.3.1 Numerical details for the micro and millidevices

For the numerical solution, high order discretization schemes were employed. A convergence criteria of RMS $= 1 \times 10^{-5}$ was defined for a solving iteration range of 500–5000. The oil and ethanol inlet were defined according to the ethanol/oil molar ratio and the residence time. Uniform velocities were set in each inlet, one for pure ethanol, other for pure vegetable oil. The outlet was considered to be a zero relative pressure and the fluid presented no-slip conditions at solid walls.

7.4.4 Results and discussion

The first step in a CFD analysis is to ensure an independence of the spatial grid discretization. The mesh discretion size and parameters were based on previous studies. After this, verification and validation of the model are required. For the BR, a validation was carried out using experimental results of Reyero et al. (2015) as shown in Fig. 7.12.

The CFD predictions showed good agreement with experimental data. The larger relative deviation (about 50%) was observed at 1 min; however, the relative deviations were below 5% for the remaining times. From CFD results were noticed that the lab-scale and industrial scale reactors presented similar results, with no significative differences on composition field. This behavior was expected due to the high agitation level considered, i.e., both stirred tank reactors presented ideal mixing behavior. A virtually complete conversion of the vegetable oil was achieved after 10 min of batch processing. After 15 min of processing, the saponification reaction starts to deprecate the biodiesel yield, as observed in Fig. 7.13. Also, the final soap content predicted by the CFD results was 0.141 g-soap/g-mixture, against the ≈0.165 g-soap/g-mixture reported by Reyero et al. (2015). Considering these results, the ideal batch time was considered to be 10 min. Further results of batch process are listed in Tables 7.2 and 7.3.

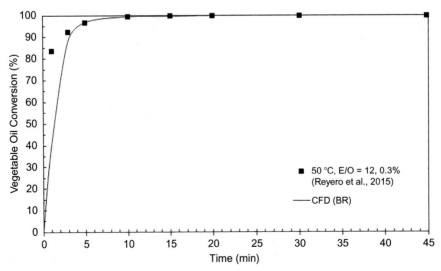

Fig. 7.12 Vegetable oil conversion from experimental data (Reyero et al., 2015) and CFD results in the BR. *(Credit: From authors.)*

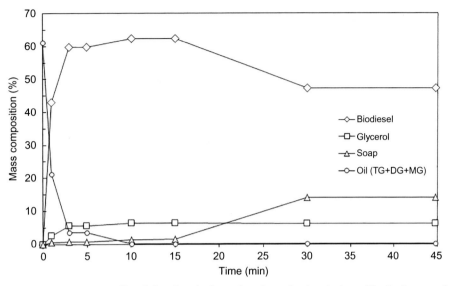

Fig. 7.13 Concentration profile of the chemical species along the batch time. *(Credit: From authors.)*

Table 7.2 Results of biodiesel production rate, number of equivalent microreactors, excess ethanol, and ERBPR.

Reactor	Biodiesel production rate (g/h)	n_{MR}[a]	MR units reduction factor[b]	Excess ethanol to recovery (g/h)	ERBPR (gh^{-1} EtOH/gh^{-1} BD)[c]
BR 1 L	3175.042	–	–	1476.423	0.465
BR 10 m^3	3.175×10^7	–	–	1.476×10^7	0.465
T-Shape	0.298	10,583.5	–	0.104	0.347
MTB micro	2.240	1417.4	7.5	0.706	0.315
MTB milli	18.990	167.2	63.3	5.840	0.308

[a]Number of equivalent micro/millidevices regarding the 1 L batch reactor, as given by Eqs. (7.20)–(7.22).
[b]Number of microreactor units reduction in comparison with the traditional T-shape design.
[c]ERBPR: ethanol recovery to biodiesel produced ratio, as given by Eq. (7.59).

For the microdevices, the mathematical model was already verified and validated from previous studies (Santana et al., 2016; Santana, Sanchez, & Taranto, 2017; Santana, Silva, & Taranto, 2015). Accordingly, the CFD results provided detailed information about the fluid flow field, allowing the comparison between the batch and the continuous biodiesel synthesis. In order to perform such comparison, three grouped parameters were proposed relating the biodiesel production rate, the unreacted ethanol flow rate, and the power consumption, as detailed following:

Table 7.3 Results of power requirement and consumption and MRPR and MRVR.

Reactor	$V_{reactor}$ (L)	W (W)[a]	P_W (Wh)[b]	MRPR (kg h^{-1}/Wh)[c]	MRVR (kg h^{-1}/L)[d]
BR 1 L	1	6.39	1.06	2.98	3.18
BR 10 m^3	10,000	4.59×10^5	7.64×10^4	0.04	0.32
T-Shape	1.17×10^{-5}	3.89×10^{-8}	3.89×10^{-8}	7.68×10^3	25.60
MTB micro	9.50×10^{-6}	6.33×10^{-8}	6.33×10^{-8}	3.54×10^4	235.82
MTB milli	7.28×10^{-5}	4.85×10^{-7}	4.85×10^{-7}	3.91×10^4	260.98
MTB milli (numbering-up)[e]	1.22×10^{-2}	4.28×10^{-3}	4.28×10^{-3}	745.66	260.98

[a]Power required for mechanical agitation in stirred batch reactors and pumping power for continuous micro/millidevices.
[b]The power consumption, P_W, for the batch time was considered equal to 10 min. The micro and millidevices were considered to operate continuously.
[c]MRPR: mass rate to power ratio, as given by Eq. (7.60).
[d]MRVR: mass rate to volume ratio, as given by Eq. (7.61).
[e]Proposed microplant considering a manifold with 4 millidevices, resulting in 42 modules of 4 millidevices each (a total of 168 millidevices to achieve the desired biodiesel throughput). The manifold pressure drop was considered to be 0.5 kPa, based on the device pressure drop, about 200 Pa each stream plus an extra 100 Pa pressure drop due to flow distributor and piping connections. These considerations were based on simulations results and previous study of Lopes et al. (2019) and Lopes, Santana, Andolphato, Silva, and Taranto (2019).

$$ERBPR = \frac{\dot{m}_{EtOH}}{\dot{m}_E} \text{ in } \frac{gh^{-1} \text{ of ethanol}}{gh^{-1} \text{ of biodiesel}} \tag{7.59}$$

The parameter *ERBPR* is the ethanol recovery to biodiesel produced ratio, given by the ratio of unreacted mass rate of ethanol requiring further recovery, to the unit of produced biodiesel mass rate.

$$MRPR = \frac{\dot{m}_E}{P_W} \text{ in } \frac{\text{kg h}^{-1} \text{ of biodiesel}}{\text{Wh of consumed energy}} \tag{7.60}$$

The parameter *MRPR* is the mass rate to power ratio, given by the ratio of produced biodiesel mass rate, in kg h^{-1}, to the unit of power consumption in Watts-hour.

$$MRVR = \frac{\dot{m}_E}{V_{reactor}} \text{ in } \frac{\text{kg h}^{-1} \text{ of biodiesel}}{\text{L of reactor volume}} \tag{7.61}$$

The parameter *MRVR* is the mass rate to volume ratio, given by the ratio of produced biodiesel mass rate, in kg h^{-1}, to the unit of reaction volume in L. Fig. 7.14 presents the oil conversion and the biodiesel production rate of the micro and millidevices in the residence time ranging from 2 to 60 s.

An increment on oil conversion with the residence time was observed in Fig. 7.14. This behavior was expected, once longer residence times results in time enough to complete the chemical reaction. The T-shape exhibited lower oil conversions than MTB

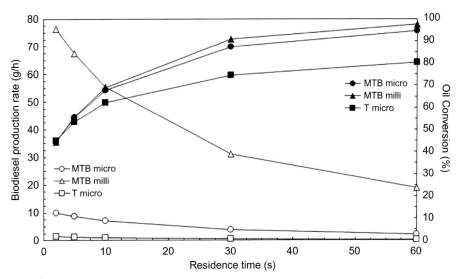

Fig. 7.14 Left Y-axis and blank markers: biodiesel production rate; right Y-axis and black markers: oil conversion. *(Credit: From authors.)*

devices. The higher oil conversion achieved in the T-shape was about 80% for $\tau = 60\,s$. The MTB micro and millidevice presented similar behavior, verifying the efficiency of the proposed micromixer design, i.e., the scale-up did no negatively affect fluid mixing and reaction performance. The superior oil conversion, about 97.7%, was obtained by MTB millidevice at $\tau = 60\,s$. Also, the biodiesel production rates in the MTB devices were larger than in the T-shape, for a factor of 7.5 to 63.3 for micro and millidevice, respectively. These results show that the design and the scale-up were effectively performed from the traditional T-shape to the MTB devices. The channel cross-section and longitudinal length increments regarding the T-shape were 4 and 16 times (channel cross section) and 2 and 4 (longitudinal length) for micro and millidevice, respectively. These increment factor results in an overall flow throughput increment of 8 and 64, respectively, due to flow rate increase and residence time adjustments. Accordingly, comparing the production rate increment factor, we can observe the efficient performance of the MTB to process the biodiesel synthesis.

Lower residence times resulted in larger biodiesel throughput since the reactants flow rates were larger. However, the oil conversions were lower, being below 70% for $\tau \leq 10\,s$ for all three devices. This is an undesired performance since one of the main goals of biodiesel synthesis from vegetable oil transesterification is to virtually convert all oil. Lower oil conversions will result in further processing, e.g., multiple pass reactors or even in additional operations for oil separation and recovery. In this context, the ideal residence times is about 60s or more. Further analysis were performed considering $\tau = 60\,s$.

From Table 7.2 was observed a production rate of 3.175 kg h^{-1} of biodiesel in the lab-scale batch reactor. As expected, the industrial scale, 10 m^3 reactor, presented an increase of 1×10^4 in biodiesel, due to the scale-up factor and the ideal behavior. The microreactors exhibited a very low production rate due to the flow rate limitations, especially the traditional T-shape design, with a 0.298 g/h of biodiesel production rate. The MTB micro, produced 2.240 g/h of biodiesel, while the MTB millidevice produced 18.990 g/h of biodiesel. The number of equivalent microreactor was estimated taking the lab-scale batch reactor as reference. From CFD predictions, more than 10,500 T-shape units are necessary to achieve the 1 L batch scale biodiesel throughput. The MTB microdevice reduced the number of units to 1418, while the MTB scale-up to the millidevice reduced even further the devices required to only 168. This result is due to the increment factor of 63.3 from the combination of cross-section and longitudinal length increments, testifying again the efficient design proposed.

Considering the global process, the unreacted ethanol flow rate is an important feature, since it requires separation and recycling to become the process economically feasible. The 1 L batch processing generated 1476.423 g/h of ethanol to be recovered. Due to the flow rate limitations, the excess ethanol flow rate in micro and millidevices are reduced, being only 0.104 g/h in the T-shape, 0.706 g/h in the MTB microreactor and 5.84 g/h in the MTB millidevice. However, the analysis of the ethanol recovery must be based on the ratio of excess ethanol to the biodiesel flow rate. In this context, the grouped parameter *ERBPR* was proposed. The examination of this parameter showed a lower relative amount of ethanol to be recovered in micro and millidevices regarding the batch reactors. The batch reactor exhibited the *ERBPR* of 0.465 gh^{-1} of ethanol/gh^{-1} of biodiesel against 0.308 gh^{-1} of ethanol/gh^{-1} of biodiesel of the MTB millidevice. This result means less energy amounts required in additional operations for ethanol recovery, usually carried out by evaporation, in the millidevice continuous processing. Santana, Tortola, Silva, and Taranto (2017) and Silva and Santana (2019) showed good performance of a continuous microevaporator applied to ethanol recovery from biodiesel synthesis. Experimental results showed a potential recovery above 80% of the ethanol for conditions of lower ethanol/biodiesel molar ratio and moderate flow rates and temperatures (100°C). The results highlight the potential application of microdevices to also perform other unit operations included in biodiesel synthesis process.

Other important features in PI analysis are related to energy consumption and size scales. Table 7.3 provided some parameters relating these parameters. The power consumption of mechanical agitation was estimated from CFD results, as well, the pumping power from the continuous operation in micro and millidevices. The energy consumption was estimated for a 10 min batch processing, according to the ideal batch time, and for a continuous operation of the micro and millireactors.

In order to achieve a biodiesel production throughput equal to the lab-scale batch reactor, a numbering-up procedure was estimated using MTB millidevices. Based on the CFD simulations and previous results from Lopes, Santana, Andolphato, Russo, et al. (2019) and Lopes, Santana, Andolphato, Silva, and Taranto (2019), it was required 168 units of MTB millireactor. Considering a manifold with 4 units, 42 modules were necessary. The manifold pressure drop was considered to be 0.5 kPa, based on the device pressure drop of about 200 Pa each stream plus an extra 100 Pa pressure drop due to flow distributor and piping connections, that can be controlled in the design and assembly of the microplant.

The absolute values of power requirements and power consumption could lead to misinterpretation of the results. In this context, the grouped parameter $MRPR$ was proposed, providing an insight of biodiesel produced to the unit of energy consumed. The examination of such parameter shows very low values for the microreactors, as expected due to the low flow rates. Considering the microplant estimated from the MTB millidevice parallelization, the power consumption was about 10^{-3} lower than the batch reactor. The MRPR was 745.66 kg h^{-1}/Wh against 2.98 kg h^{-1}/Wh of the 1 L batch reactor, highlighting the great potential of PI with continuous micro-plants. From Table 7.3, it was clear the inherent higher amount of power required to operate an industrial scale batch reactor. Despite the reactor performance can keep higher, as observed previously (Table 7.2), the scale-up results in lower energy efficiency in biodiesel production. The $MRPR$ of the lab-scale batch reactor was 2.98 kg h^{-1}/Wh against only 0.04 kg h^{-1}/Wh of the 10 m^3 stirred reactor. This result is related to the fifth-power dependence of the mechanical agitation power with the impeller diameter, resulting in very high power requirements in larger stirred tanks.

Another important feature in PI is the scale size of the equipment. The parameter $MRVR$ provides an estimation of the biodiesel produced per unit of the reaction volume required. The analysis of $MRVR$ shows that micro and millidevice provided the superior biodiesel yield per unit of reactor volume. The lab-scale batch reactor resulted in 3.18 kg h^{-1}/L against 0.32 kg h^{-1}/L of the industrial scale reactor. The T-shape presented a slightly superior $MRVR$ of 25.6 kg h^{-1}/L regarding the 1 L BR. The superior $MRVR$ were noticed for the MTB devices, ranging from 235.82 to 260.98 kg h^{-1}/L from the micro to the millidevice. Considering the aforementioned results, it is clear the potential of the use of Microfluidics concepts in PI.

7.5 Conclusions and future perspectives

Biomass-based fuels appear as interesting alternatives to fossil fuels. Biodiesel is a renewable fuel produced from vegetable oils or animal fats, presenting similar characteristics to the fossil diesel, consequently allowing its use without major modifications in the combustion engines. Recently, biodiesel synthesis has been explored in several research studies and acid, alkali, or enzymatic routes can be used to produce this biofuel. One of the most used routes is the transesterification reaction between triglycerides and short-chain alcohol in alkali media. Despite the advances, further development and optimization is still necessary. Difficulties inherent from physico-chemical properties, hinders the reactants mixing in continuous traditional reactors.

Micro and millidevices appears as interesting alternatives for an efficient continuous synthesis of biodiesel. Microfluidic devices can promote high mixing degree when using micromixer elements. High biodiesel yield and oil conversion were observed in short residence times in the order of few seconds to minutes. The micrometric dimensions also results in elevated pressure drops and limited flow rates. The use of micromixer elements in larger devices, e.g., in millimetric devices, proved to be reliable, allowing the design and operation of microchemical plants. The micromixer design is a fundamental key to achieve the desired efficiency and production rate.

In this context, the CFD appears as a powerful tool, allowing a rapid and low-cost analysis of several layouts before the manufacturing of the physical prototype. Considering these aspects, this chapter highlighted advances on biodiesel synthesis and the use of CFD. The proposed study case compared the performance of traditional macroscale batch reactors with micro and millidevices designed and optimized by CFD. The microdevice exhibited the superior performance considering aspects of volumetric production rate and energy efficiency; however, the limited throughput made unfeasible its usage in industrial demand. The estimations for scaled-up millidevices in a numbered-up microchemical plant can be noticed as an interesting alternative, providing a flow rate similar to a lab-scale batch reactor with a superior production rate to energy consumption. Also, the relative amount of excess reactant to the biodiesel produced was lower in micro and millidevices, proving its superior performance.

Despite these results, further advances are still required to become feasible the use of millidevices in modular microplants to achieve larger throughputs. For example, the biodiesel amount produced by the $10\,m^3$ stirred reactor requires about 1.7 million MTB millidevices, which is impracticable. In this context, further development must be achieved aiming larger devices with similar efficiency of Microfluidics devices. The development of efficient micromixer elements, micro and millidevices in general (reactors, separators, evaporators, among others), flow distributors, optimization of operating conditions, and even prediction tests of microplant module operation can be successfully assessed by CFD.

References

Anderson, J. D. J. (1995). *Computational fluid dynamics* (1st ed.). New York: McGraw-Hill.

ASTM. (2015). *ASTM D6751-15, Standard specification for biodiesel fuel blend stock (B100) for middle distillate fuels*. West Conshohocken, PA: ASTM International. https://www.astm.org.

Bakker, A. (2006). *Lecture 7—Meshing*. http://www.bakker.org/dartmouth06/engs150/07-mesh.pdf (Accessed 11.10.2020).

Bird, R. B., Stewart, W. E., & Lightfoot, E. N. (2006). *Transport phenomena* (2nd ed.). New Jersey: John Wiley & Sons Inc.

Bonechi, C., Consumi, M., Donati, A., Leone, G., Magnani, A., Tamasi, G., et al. (2017). Biomass: An overview. In F. Dalena, A. Basile, & C. Rossi (Eds.), *Bioenergy systems for the future* (pp. 3–42). Woodhead Publishing.

Chung, T. J. (2002). *Computational fluid dynamics*. Cambridge, United Kingdom: Cambridge University Press.

EIA (US Energy Information Administration). (2013). *International energy outlook 2013*. http://www.eia.gov/forecasts/archive/ieo13 (Accessed 07.04.2015).

ENEA. (2009). *The multiple reference frame model*. https://www.afs.enea.it/project/neptunius/docs/fluent/html/ug/node370.htm (Accessed 12.10.2020).

Fogler, H. S. (2006). *Elements of chemical reaction engineering*. Upper Saddle River, NJ: Prentice Hall PTR.

Freedman, B., Pryde, E. H., & Mounts, T. L. (1984). Variables affecting the yields of fatty esters from trans-esterified vegetable oils. *Journal of the American Oil Chemists' Society, 61*, 1638–1643.

Gent, S., Twedt, M., Gerometta, C., & Almberg, E. (2017). Introduction to feedstocks. In S. Gent, M. Twedt, C. Gerometta, & E. Almberg (Eds.), *Theoretical and applied aspects of biomass torrefaction* (pp. 17–39). Butterworth-Heinemann.

Ghasemzadeh, K., Ghahremani, M., Amiri, T. Y., & Basile, A. (2019). Performance evaluation of Pd-Ag membrane reactor in glycerol steam reforming process: Development of the CFD model. *International Journal of Hydrogen Energy, 44*, 1000–1009.

Han, W., Charoenwat, R., & Dennis, B. H. (2011). Numerical investigation of biodiesel production in capillary microreactor. In *ASME 2011 international design engineering technical conferences (IDETC) and computers and information in engineering conference (CIE); August, 28–31; Washington, DC, USA* (pp. 253–258).

Hirt, C. W., & Nichols, B. D. (1981). Volume of fluid (VOF) method for the dynamics of free boundaries. *Journal of Computational Physics, 39*(1), 201–225.

Jin, H., Wang, H., Wu, Z., Ren, Z., & Ou, Z. (2019). Numerical investigation on drag coefficient and flow characteristics of two biomass spherical particles in supercritical water. *Renewable Energy, 138*, 11–17.

Koyunoğlu, C., & Karaca, H. (2019). Application of industry 4.0 on biomass liquefaction study: A case study. *Procedia Computer Science, 158*, 401–406.

Laziz, A. M., KuShaari, K., Azeem, B., Yusup, S., Chin, J., & Denecke, J. (2020). Rapid production of biodiesel in a microchannel reactor at room temperature by enhancement of mixing behaviour in metanol phase using volume of fluid model. *Chemical Engineering Science, 219*, 115532.

Liu, B., Papadikis, K., Gu, S., Fidalgo, B., Longhurst, P., Li, Z., et al. (2017). CFD modelling of particle shrinkage in a fluidized bed for biomass fast pyrolysis with quadrature method of moment. *Fuel Processing Technology, 164*, 51–68.

Lopes, M. G. M., Santana, H. S., Andolphato, V. F., Russo, F. N., Silva, J. L., Jr., & Taranto, O. P. (2019). 3D printed micro-chemical plant for biodiesel synthesis in millireactors. *Energy Conversion and Management, 184*, 475–487.

Lopes, M. G. M., Santana, H. S., Andolphato, V. F., Silva, J. L., Jr., & Taranto, O. P. (2019). Flow uniformity data on 3D printed flow distributors. *Data in Brief, 23*, 103799.

López-Guajardo, E., Ortiz-Nadal, E., Montesinos-Castellanos, A., & Nigam, K. D. P. (2017). Process intensification of biodiesel production using a tubular micro-reactor (TMR): Experimental and numerical assessment. *Chemical Engineering Communications, 204*, 467–475.

Marjanovic, A. V., Stamenkovic, O. S., Todorovic, Z. B., Lazic, M. L., & Veljkovic, V. B. (2010). Kinetics of the base-catalyzed sunflower oil ethanolysis. *Fuel, 89*, 665–671.

McCabe, W. L., Smith, J. C., & Harriott, P. (1993). *Unit operations of chemical engineering* (5th ed.). New York; London: McGraw-Hill.

Mohadesi, M., Aghel, B., Maleki, M., & Ansari, A. (2020). Study of the transesterification of waste cooking oil for the production of biodiesel in a microreactor pilot: The effect of acetone as the co-solvent. *Fuel, 273*, 117736.

Moukalled, F., Mangani, L., & Darwish, M. (2016). *The finite volume method in computational fluid dynamics: An advanced introduction with OpenFOAM and matlab, fluid mechanics and its applications*. New York: Springer.

Pang, S. (2016). 9—Fuel flexible gas production: Biomass, coal and bio-solid wastes. In J. Oakey (Ed.), *Fuel flexible energy generation* (pp. 241–269). Woodhead Publishing.

Park, H. C., & Choi, H. S. (2019). Fast pyrolysis of biomas in a spouted bed reactor: Hydrodynamics, heat transfer and chemical reaction. *Renewable Energy, 146*, 1268–1284.

Pinheiro, D. S., Silva, R. R. O., Calvo, P. V. C., Silva, M. F., Converti, A., & Palma, M. S. A. (2018). Microreactor technology as a tool for the synthesis of a Glitazone drug intermediate. *Chemical Engineering and Technology, 41*, 1800–1807.

Pulliam, T. H., Lomax, H., & Zingg, W. D. (1999). *Fundamentals of computational fluid dynamics*. Toronto, Canada: University of Toronto Institute for Aerospace Studies.

Rahimi, M., Aghel, B., Alitabar, M., Sepahvand, A., & Ghasempour, H. R. (2014). Optimization of biodiesel production from soybean oil in a microreactor. *Energy Conversion and Management, 79*, 599–605.

Reyero, I., Arzamendi, G., Zabala, S., & Gandía, L. M. (2015). Kinetics of the NaOH-catalyzed transesterification of sunflower oil with ethanol to produce biodiesel. *Fuel Processing Technology, 129*, 147–155.

Ritchie, H., & Roser, M. (2017). *Fossil fuels*. Published online at OurWorldInData.org. https://ourworldindata.org/fossil-fuels (Accessed 01.03.2021).

Sabry, M. N., El-Emam, S. H., Mansour, M. H., & Shouman, M. A. (2018). Development of an efficient uniflow comb micromixer for biodiesel production at low Reynolds number. *Chemical Engineering & Processing: Process Intensification, 128*, 162–172.

Santana, H. S., da Silva, A. G. P., Lopes, M. G. M., Rodrigues, A. C., Taranto, O. P., & Silva, J. L., Jr. (2020). Computational methodology for the development of microdevices and microreactors with ANSYS CFX. *MethodsX, 7*, 82–103.

Santana, H. S., Sanchez, G. B., & Taranto, O. P. (2017). Evaporation of excess alcohol in biodiesel in a microchannel heat exchanger with Peltier module. *Chemical Engineering Research & Design, 124*, 20–28.

Santana, H. S., Silva, J. L., Jr., & Taranto, O. (2015). Numerical simulation of mixing and reaction of Jatropha curcas oil and ethanol for synthesis of biodiesel in micromixers. *Chemical Engineering Science, 132*, 159–168.

Santana, H. S., Silva, J. L., Jr., & Taranto, O. P. (2019). Development of microreactors applied on biodiesel synthesis: From experimental investigation to numerical approaches. *Journal of Industrial and Engineering Chemistry, 69*, 1–12.

Santana, H. S., Silva, J. L., & Taranto, O. P. (2015). Numerical simulations of biodiesel synthesis in microchannels with circular obstructions. *Chemical Engineering and Processing, 98*, 137–146.

Santana, H. S., Silva, J. L., & Taranto, O. P. (2019). Optimization of micromixer with triangular baffles for chemical process in millidevices. *Sensors and Actuators B: Chemical, 281*, 191–203.

Santana, H. S., Tortola, S., Reis, É. M., Silva, J. L., Jr., & Taranto, O. P. (2016). Transesterification reaction of sunflower oil and etanol for biodiesel synthesis in microchannel reactor: Experimental and simulation studies. *Chemical Engineering Journal, 302*, 752–762.

Santana, H. S., Tortola, D. S., Silva, J. L., & Taranto, O. P. (2017). Biodiesel synthesis in micromixer with static elements. *Energy Conversion and Management, 141*, 28–39.

SGB. (2016). *Top 10 uses for biofuel*. https://www.sgbiofuels.com/top-10-uses-for-biofuel/ (Accessed 15.10.2020).

Shah, I., Aziz, S., Soomro, A. M., Kim, K., Kim, S. W., & Choi, K. H. (2020). Numerical and experimental investigation of Y-shaped micromixers with mixing units based on cantor fractal structure for biodiesel applications. *Microsystem Technologies, 26*, 1783–1796.

Silva, J. L., Jr. (2019). An introduction to computational fluid dynamics and its application in microfluidics. In H. S. Santana, J. L. Silva Jr.,, & O. P. Taranto (Eds.), *Process analysis, design, and intensification in microfluidics and chemical engineering* (1st ed., pp. 50–78). Hershey PA: IGI Global, Engineering Science Reference.

Silva, J. L., Jr., Haddad, V. A., Taranto, O. P., & Santana, H. S. (2020). Design and analysis of new micromixers based on distillation column trays. *Chemical Engineering & Technology, 43*, 1–12.

Silva, J. L., Jr., & Santana, H. S. (2019). Experimental and numerical analyses of a micro-heat exchanger for ethanol excess recovery from biodiesel. In *Process analysis, design, and intensification in microfluidics and chemical engineering* (1st ed., pp. 167–194). Hershey PA: IGI Global, Engineering Science Reference.

Souza, M. R. P., Santana, H. S., & Taranto, O. P. (2020). Modeling and simulation using OpenFOAM of biodiesel synthesis in structured microreactor. *International Journal of Multiphase Flow, 132*, 103435.

Sun, J., Ju, J., Zhang, L., & Xu, N. (2008). Synthesis of biodiesel in capillary microreactors. *Industry Engineering Chemistry Research, 47*, 1398–1403.

Wen, Z., Yu, X., Tu, S.-T., Yan, J., & Dahlquist, E. (2009). Intensification of biodiesel synthesis using zigzag micro-channel reactors. *Bioresource Technology, 100*, 3054–3060.

Whitesides, G. M. (2006). The origins and the future of microfluidics. *Nature, 442*, 368–373.

Xie, T., Zhang, L., & Zu, N. (2012). Biodiesel synthesis in microreactors. *Green Processing and Synthesis, 1*, 61–70.

Zhang, X., Wiles, C., Painter, S., Watts, P., & Haswell, S. J. (2006). Microreactors as tools for chemical research. *Chemistry Today, 24*, 43–45.

CHAPTER 8

Control properties of intensified distillation processes: Biobutanol purification

Ernesto Flores-Cordero[a], Eduardo Sánchez-Ramírez[b], Gabriel Contreras-Zarazúa[b], César Ramírez-Márquez[b], and Juan Gabriel Segovia-Hernández[b]

[a]Biotechnology Engineering Department, University of Guanajuato, Campus Celaya-Salvatierra, Guanajuato, Gto., Mexico
[b]Chemical Engineering Department, University of Guanajuato, Guanajuato, Mexico

8.1 Introduction

Climate change, pollution, and the rising extraction costs of oil are some of the most important incentives to replace conventional fossil fuels. For this reason, there has been an increasing interest in the development and implementation of greener renewable fuels. In this sense, the biofuels produced from biomass, especially from lignocellulosic residues are a very attractive option for the researchers. The lignocellulosic residues are abundant and cheap because their costs are mainly associated with their recollection costs. Additionally, the lignocellulosic biomasses are not in competition with food crops avoiding in this way ethical dilemmas. Nowadays, the most produced biofuel from lignocellulosic residues is ethanol, however, in recent years the butanol has attracted special attention due to its physical properties such as currently higher energy density, lower steam pressure, less flammability, and hydrophobicity, making it ideal for use in the current conventional engines (Sánchez-Ramírez, Alcocer-García, et al., 2017; Sánchez-Ramírez, Quiroz-Ramírez, Segovia-Hernández, Hernández, & Ponce-Ortega, 2016). A proof of butanol potential is the road trip performed by Damey Ramey from Ohio to California and back in his unmodified car using 100% butanol (Dürre, 2007).

Nowadays, there are different routes for producing it, such as catalytic hydroformylation of propylene and hydrogenation also known as Oxo-Synthesis process. However, one of the most popular and sustainable ways to produce butanol is through the acetone-butanol-ethanol (ABE) fermentation route (Patraşcu, Bîldea, & Kiss, 2017). The ABE fermentation is not a new route, it has been used since World War I in order to satisfy the increasing demand during this period (Errico, Sanchez-Ramirez, Quiroz-Ramìrez, Segovia-Hernández, & Rong, 2016; Weizmann, 1919). Nowadays, ABE fermentation is carrying out using anaerobic bacterias such as *Clostridium acetobutylicum* or *Clostridium beijerinckii* (Patraşcu et al., 2017; Tashiro, Yoshida, Noguchi, & Sonomoto, 2013). Despite, the ABE fermentation had been used for decades for producing butanol, two important challenges must be overcome in order to increase the profitability of the process. The first one is the use of dilute sugar solutions, which generates dilute butanol solutions and large disposal loads, the second is the intensive use of energy required to purify the butanol. In this sense, the researchers have addressed these challenges by focusing on two main ways, the use of genetic engineering and processes engineering. The use of genetic engineering is focusing on the development of stronger microorganisms capable of tolerating higher sugar and butanol concentrations. On the other hand, the processes engineering is focusing on the design of new processes to remove the butanol from the fermentation vessel that a toxic butanol concentration inside the reactor is never reached as well as, the application and design of innovative processes separation capable of purifying the butanol with low energy consumptions. In this sense, this chapter is focusing on the study of control dynamics of novel separation processes for purifying butanol from the fermentation mixture.

The separation and purification of butanol from the ABE mixture is a complex procedure since it is obtained in a very diluted solution from the fermentation reactor, additionally, the presence of two azeotropes (Liu, Liu, & Feng, 2005), one homogeneous between ethanol-water, and another heterogeneous, between butanol-water (see Fig. 8.1), provokes that the butanol cannot be obtained with high purities by conventional processes (Errico et al., 2016). This implies that nonconventional processes such as azeotropic distillation or extractive distillation, liquid-liquid extraction, pervaporation, adsorption, vapor drag, must be used (Moreno & Cubillos Lobo, 2017). However, many of these methods are characterized by high energy consumptions, and more sophisticated process control structures, which can generate the worst control properties (Sánchez-Ramírez, Alcocer-García, et al., 2017).

Considering the different methods to purify butanol, one of the most promissory processes is the liquid -liquid extraction. In liquid-liquid extraction columns, an entrainer agent is added to the mixture coming from the fermentation broth, in this case, the entrainer must be immiscible in water and exhibit a high selectivity with respect to the compounds to be purified. A successful implementation of this process depends strongly on the appropriate selection of the extracting agent, which should have various desirable physicochemical properties, such as high surface tension with water to facilitate its separation, low viscosity, high stability (Sánchez-Ramírez, Alcocer-García, et al., 2017). According to a review of the literature, the hexyl acetate has proved that it can break the azeotropes of the ABE mixture obtaining in this way, important butanol recoveries, also it has physical properties the required physical properties, which makes it a potential solvent (Barton & Daugulis, 1992). Owing to the potential of hexyl acetate, many previous works have studied the potential of this solvent in a liquid-liquid extraction coupled to distillation processes, some examples are the works reported by

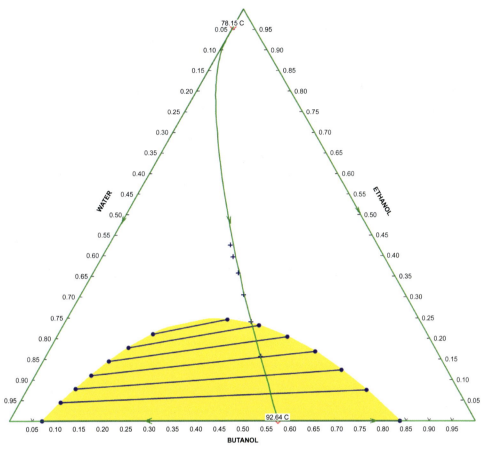

Fig. 8.1 Ternary diagram for the mixture butanol-ethanol-water, using the NRTL thermodynamic method.

Sánchez-Ramírez, Quiroz-Ramírez, Segovia-Hernández, Hernández, and Bonilla-Petriciolet (2015), Sánchez-Ramírez, Quiroz-Ramírez, Hernández, Segovia-Hernández, and Kiss (2017), Errico, Sánchez-Ramírez, Quiroz-Ramírez, Ben-Guang, and Segovia-Hernandez (2017), among others. All these previous works concluded that a liquid-liquid extraction process is possible; however, once the large amounts of water have been removed from the mixture, the energy required for purifying butanol up to the required purity using distillation is notably greater than the energy provided by butanol itself, for this reason, the reduction of energy consumption of the distillation stage is a key step. In this sense, the thermally coupled and dividing wall distillation arrangements have proved provided important energy reductions, these columns can save up to 30% in the capital invested and up to 40% in energy costs; (Isopescu, Woinaroschy, & Draghiciu, 2008). Despite, the intensified distillation processes such as the thermally coupled and the dividing wall columns can obtain important energy reductions, the greater complexity in the process structure means that its control properties may be affected. Controllability has a key role in the implementation of a process on an industrial scale, in order to always keep the product in the desired specifications of purity.

With the aforementioned in mind, in this chapter is presented the study of the control properties for different intensified distillation systems for the separation of ABE mixture, which is a poorly explored area. This work is based on the previous study reported by Errico et al. (2017), in which the synthesis, design and optimization of several intensified processes to purify butanol were performed. The total annual cost (TAC) and Eco-indicator 99 (EI99) were considered by Errico et al. (2017) as objective functions to evaluate the economic and the environmental impact, respectively, during the optimization procedure. It is important to mention, that Errico et al. (2017) used the differential evolution method with tabu list (DETL) to perform the optimization of the biobutanol separation processes; consequently, the distillation sequences used in this work are designs with minimum cost and environmental impact. Therefore, a control dynamic study of these distillation sequences for biobutanol purification is of utmost importance in order to determine the feasibility of these processes for potential industrial implementation. It is important to highlight that although there are already some previous works reported the control of distillation sequences applied to the ABE mixture (Kaymak, 2019; Patraşcu, Bîldea, & Kiss, 2019; Sánchez-Ramírez, Alcocer-García, et al., 2017; Sánchez-Ramírez et al., 2016), this is the first work that reports a control study for the sequences reported by Errico et al. (2017). The control study of this work was carried out using different control loops and generating disturbances in order to observe the capacity of the processes to respond to these disturbances. Finally, the IAE was used to evaluate the control performance.

8.2 Separation process for biobutanol

The dynamic analysis of the process is based on the study of control properties, which is applied to four systems in order to minimize the number of disturbances, specifically, when a change in the composition of the mixture is required. The four schemes studied in this work were reported by Errico et al. (2017) and they are based on hybrid extraction–distillation processes. The location of the extraction column is after the fermentation reactor, the entrainer is fed bottom of the extraction column. The raffinate phase leaving the bottom of the extractive column contains water and traces of acetone, butanol and entrainer. The extract phase obtained from the top of the extraction column is feed to a distillation sequence where butanol, acetone, and ethanol are recovered using conventional or intensified distillation processes. The butanol purification processes are shown in Fig. 8.2.

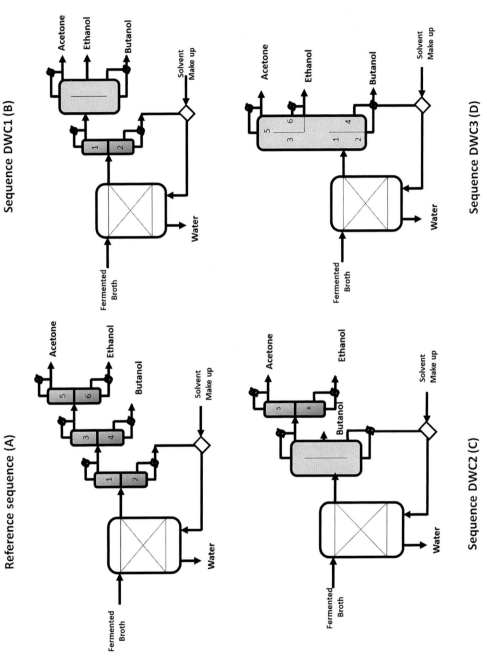

Fig. 8.2 Sequences studied. Reference sequence (A). Conventional sequence with L-L extraction and conventional distillation, and intensified sequences (DWC1-DWC7; B–D). First equipment, extraction column L-L, followed by conventional distillation columns or DWC type.

One conventional sequence and three intensified sequences are shown in Fig. 8.2. The three intensified sequences are the DWC1–3. The process intensification in the separation of ABE mixture is driven by the need for breakthrough changes in operations, focuses mainly on novel equipment. Under this idea, the process of separation of ABE mixture is subject to intensification in order to have the benefits in terms of the reduction in plant or equipment size, a dramatic increase in the production capacity, decreases in energy consumption per ton of product, or even a marked cut in wastes or by-products formation also qualify as process intensification (Ramshaw, 1995). However, when intensifying a process, an important point to study is the control. While the reduction in equipment size, energy savings, and minimal by-product waste can be achieved, this does not guarantee that the process can be carried out with a good operation. Therefore, the following section shows the control study of the sequences already pointed out.

8.3 Closed-loop analysis

The dynamic analysis of the process is based on the study of the control properties, which is applied to the system to minimize the number of disturbances. This is especially true when a change in the composition of the mixture is required or in order to maintain an initial composition in the face of a disturbance in the supply of the purification system, where it is necessary to use control systems that monitor and control these disturbances or changes. This is done with the purpose of knowing how stable they are or how operable they will be during the common changes that may emerge during its operation. It will also allow us to know if they are more stable than the reference sequence, which is an unenhanced sequence, with conventional separation operations (Fig. 8.2). This will lead to the conclusion of whether the intensified separation systems count on better control properties.

In this chapter, the study of the control properties was carried out in a closed loop. The control study carried out in the present work was through Aspen Dynamics®. The methodology that was implemented is based on what was reported by Segovia-Hernández and Bonilla-Petriciolet (2016), where, a common technique to evaluate the dynamics of a process is to generate a closed control loop using a PI-type controller and an IAE-type performance indicator to determine the optimal parameters of the controller. The choice of the PI as type of controller was due to its ability to compensate many practical industrial processes (O'Dwyer, 2000). This, in order to know its stability in the face of disturbances or changes in the system, will be reflected in decent operability within the process. Thus, the disturbances assessed in the control study were the purity of butanol and acetone streams (called SetPoint), and the perturbation in the flowrate of feed.

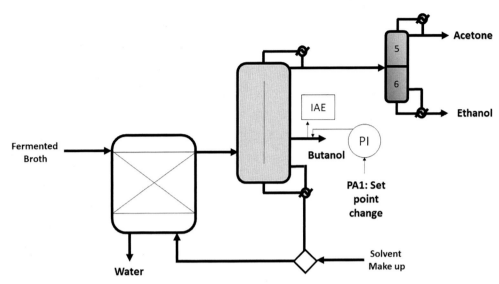

Fig. 8.3 Examples of the implementation of the closed control loops, PI controller and IAE, for the compounds of interest; acetone and butanol, in top and bottom outlets (H) and side (D). As well as the zone of the change of the Set-point (PA1) and the zone of the disturbance (PA2).

Therefore, this methodology has been implemented in this chapter with two different adjustment points (SetPoint, PA1; and feeding disturbance, PA2) using the instruments that are commonly utilized in the industry for a study of closed-loop control properties (Fig. 8.3). This dynamic analysis was performed for four sequences (models) of separation of the ABE mixture (Fig. 8.2), the reference sequence (A) and three intensified sequences (DWC1 to DWC 3; B–D). Some studies demonstrate that the control in intensified DWC schemes does not affect the control, and even in some cases, the control properties are improved concerning conventional schemes (Kiss, 2013; Serra, Espuna, & Puigjaner, 1999; Wang & Wong, 2007).

A classical closed-loop control system was implemented (in each sequence), individually and in an orderly manner in each global output stream (of the compounds of interest; acetone and butanol). The loop was first created. The controller was tuned. It was then simulated and the results of a loop were obtained. All the steps were repeated for the other output. This loop had a PI controller, Eq. (8.1), and an integration error criterion to evaluate the performance of the controller, for which the operating index of the absolute error integral was used; IAE, Eq. (8.2).

$$g_c(s) = K_c\left[1 + \frac{1}{\tau_i s}\tau_D s\right] \qquad (8.1)$$

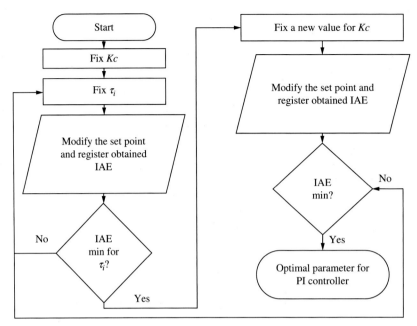

Fig. 8.4 Flowchart for tuning the optimal parameters under the IAE minimization criterion, of the PI controller. To obtain the optimal parameters K_c values between 0.5 and 250 and Ti between 0.5 and 150.

$$\text{IAE} = \int_0^\infty |\varepsilon(t)|dt \tag{8.2}$$

The purpose of the work is to tune the control parameters (PI) through the methodology shown in Fig. 8.4. It can be described as a search for the parameters K_c and τ_i where the value given by the IAE is optimized (minimized). The IAE is a standard optimization method applied to PI control loop design (Åström & Hägglund, 1995). In other words, the optimization of the IAE must be oriented to the disturbance rejection capacity of the control loop, that is, to minimize the absolute error $\varepsilon(t)$ integral caused by a step change of the disturbance.

This requires the exportation of files from Aspen Plus to Aspen Dynamics, where the control loops ascend naturally from the experience of the operation of conventional columns. The reflux flow rate was used for the upper product control (acetone), while the heat load was chosen for the lower product control (butanol). In the side stream, the product was directly controlled with the outflow. This is because a known structure is based on energy balance considerations, which gives rise to the so-called LV control structure in which the reflux flow rate L and the steam rate V (directly affected by the heat requirement supplied to the kettle) are used to control the compositions of distillates and lower outputs (García & Hernández, 2015). It is also worth mentioning that these control loops have been used with acceptable results in previous studies on thermally coupled systems, such as the DWC (Aguilera et al., 2017; García & Hernández, 2015). Takamatsu, Hashimoto, and Hashimoto (1987), show that the LV configuration is probably the best option for an industrial configuration since it has the closeness of the control variable, with respect to the manipulable variable and its implementation is simple, with basic instrumentation within a control scheme and with a fast response to system disturbances. The choice of the purity of the component of interest (acetone and butanol in this case) as the control variable is based on Ling and Luyben (2009). Ling and Luyben (2009), describe in their work that the purities of the components of interest are the main purpose of the control schemes. If adequate purity is not achieved, the cost of the final product declines and can make a process economically unviable.

After this, what changed was the set point in the tuning of the controller, since, in PA1, a change was made in the Set-point, that is, the change in the control variable, −1% in the purity (composition). According to Kapoor, McAvoy, and Marlin (1986), Skogestad and Morari (1987) and Khanam, Skogestad, and Shamsuzzoha (2017), the values of steady-state gains and time constant change drastically with small disturbances. Therefore, an acceptable closed-loop control study can be obtained based on the disturbance percentages close to 1% either positively or negatively. In PA2, the Set-point remained unchanged and a disturbance of −4% was made in the compounds of interest (butanol and acetone), and in the flow from the extraction column LL, without neglecting the balance. The percentage removed from a compound was added to another compound within the same flow, in this case, the percentage removed from acetone and butanol, was added to ethanol (Fig. 8.3). Feed disturbance is an exercise aimed at observing whether the control scheme is able to withstand changes in the source of the mixture composition. The 4% is a representative value without having a meaning of a daily variation, but a great significance in the robustness of the selected control system. All this was done to tune the optimal controller parameters (in the face of the two changes of the set-point) that will generate the best dynamic response, reflected in a lower number of disturbances to reach the set-point (Set-point; the purity of the compounds of interest at the output). The parameters of the controller that in its dynamic response will generate the lowest IAE were selected.

Once this was figured out, the elaboration of the closed control loop was begun for the compound that was analyzed at that time. It was then continued with the tuning of the PI controller, taking into account that all the loops of all the sequences for each compound of interest were done in an orderly manner, first the butanol, then the acetone of each sequence with PA1. It was all repeated with PA2. This procedure is described below, through the tuning of the optimal parameters of the controller under the criteria of minimization of the IAE. When tuning the controller, a new value of the IAE is obtained (dynamic Aspen has this criterion by default). Once this IAE value is obtained, it is compared with other tunings and the minimum value is obtained.

8.3.1 Results of the control analysis

The results of the study of the closed-loop control properties of the four sequences analyzed (Fig. 8.2) were obtained through the implementation and tuning of a classical control system in the global outputs (Figs. 8.3 and 8.4) for the compounds of interest (acetone and butanol). Analyzing exactly the dynamic behavior of the purification process, discarding the L-L extraction column and starting from the flow coming from it.

Sixteen control loops were implemented, of which, 8 belong to PA1, 4 to butanol and 4 to acetone; the other 8 belong to PA2, 4 to butanol and 4 to acetone, that is, pair of the compounds of interest for each sequence with two set-points, a total of 4 loops per sequence. We can make an emphasis in the fact that most of the sequences demonstrated good results in the attainment of excellent dynamic responses, by tuning the optimal parameters of the PI controller, as well as small numbers of the IAE.

The results obtained show the optimal parameters of the PI controller and the IAE values of the four separation models of the ABE mixture studied (the reference sequence and three intensified sequences), together with their dynamic response, according to the two points of adjustments made. Therefore, the results obtained according to the two adjustment points made will be displayed in an orderly manner below. Starting with the tables of the optimal parameters of the PI controller and their respective IAE (Tables 8.1 and 8.2). Then the individual dynamic responses of each implemented loop (Figs. 8.5–8.8). Since this is a tuning of the parameters K_c and τ_i, a range of these values is

Table 8.1 Optimal parameters of the PI controller and IAE values of the analyzed sequences, according to PA1.

	Acetone			Butanol		
Sequence	K_c (%%)	τ_i (min)	IAE	K_c (%%)	τ_i (min)	IAE
A	250	150	0.0291113	115	35	0.00595615
B	100	140	0.0233981	127	27	0.00451334
C	250	150	0.0319735	250	3	0.0137633
D	245	150	0.0292766	250	115	0.0829786

Table 8.2 Optimal PI controller parameters and IAE values, of the analyzed sequences, according to PA2.

	Acetone			Butanol		
Sequence	K_c (%%)	τ_i (min)	IAE	K_c (%%)	τ_i (min)	IAE
A	100	50	0.00016079	250	60	0.000197919
B	65	150	0.000188186	115	10	4.65626E-05
C	190	150	0.000166802	250	1	0.000782736
D	250	150	0.000315984	250	115	0.00348524

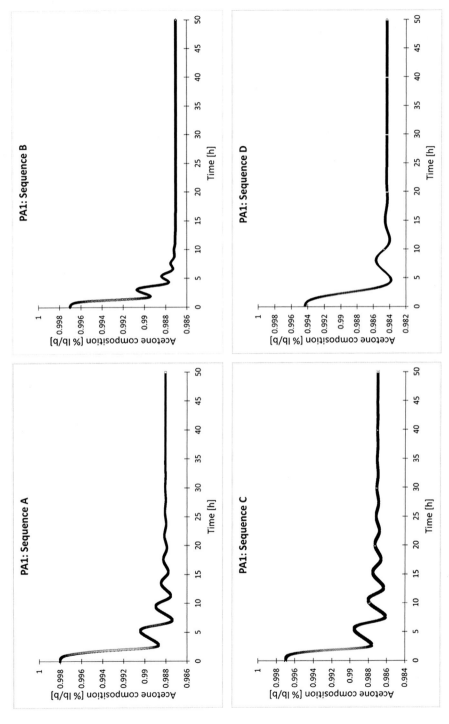

Fig. 8.5 Graphs of the closed-loop dynamic response of the analyzed sequences, for the acetone flow, according to the PA1. The units [% lb./lb.] represent the mass purity of the component.

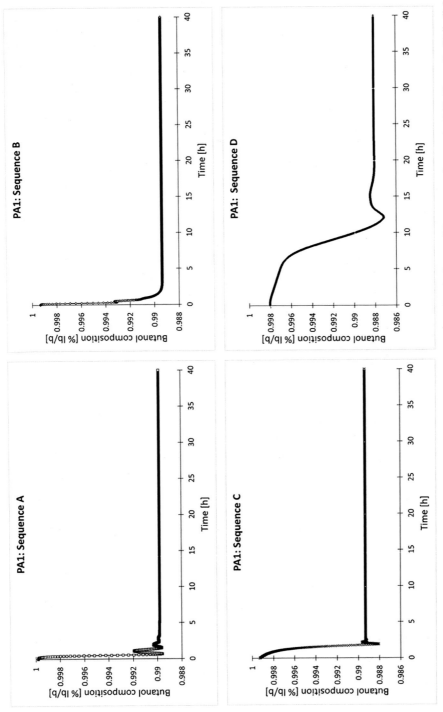

Fig. 8.6 Graphs of the closed-loop dynamic response of the analyzed sequences, for the butanol flow, according to the PA1. The units [% lb./lb.] represent the mass purity of the component.

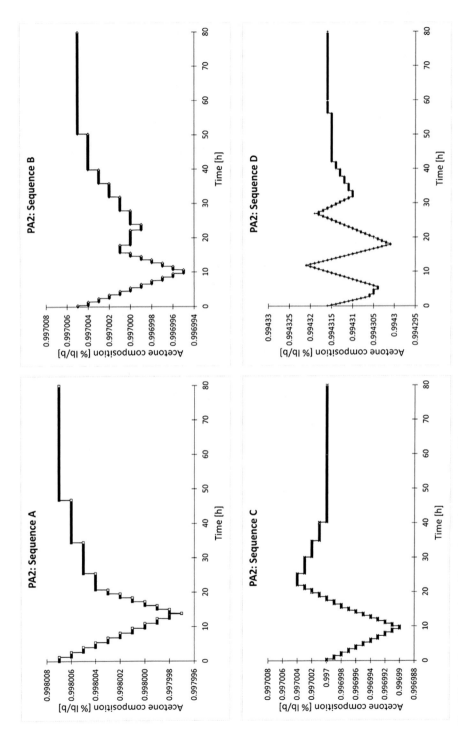

Fig. 8.7 Graphs of the closed-loop dynamic response of the analyzed sequences, for the flow of acetone, according to PA2. The units [% lb./lb.] represent the mass purity of the component.

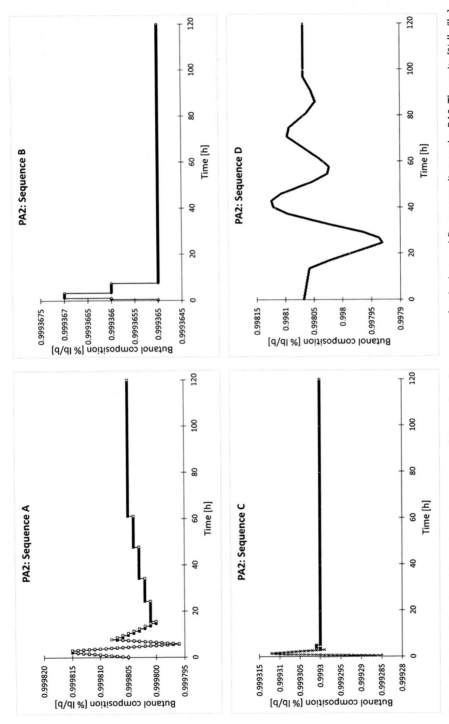

Fig. 8.8 Graphs of the closed-loop dynamic response of the analyzed sequences, for the butanol flow, according to the PA2. The units [% lb./lb.] represent the mass purity of the component.

obtained. The one with the lowest IAE value in each sequence is the optimal set of parameters in which the controller can have a better performance. That is, how well the system is prepared to withstand disturbances and how quickly the system can respond to stabilize the control variables according to the manipulable variables. It is also important to mention that the IAE criterion is favorable for comparing how a series of processes can behave in terms of control. Letting glimpse, which process is operationally favorable or if they are in equal circumstances operationally.

Figs. 8.7 and 8.8 show the expected behavior. The small disturbances shown in the Figures represent that even if there is a mismatch in the input composition, the control scheme and the tuned parameters are able to dampen the changes and maintain practically the system under the same purities. And even if the changes are almost imperceptible to the measuring equipment, in terms of control it represents that the system remained stable. In optimal control studies, this is what is expected of the systems studied.

8.3.1.1 Result analysis

To analyze and understand in detail the results obtained, the two adjustment points (PA1 and PA2) must be taken into account. As mentioned in the Tables and Figures above, each one has its results. Therefore, to perform this analysis, the results will be divided according to each set-point, until we reach the discussion of the results.

Set-point (PA1)

It was possible to obtain the optimal parameters of the controller (with its respective IAE) (Table 8.1), which allowed for the attainment of the best dynamic responses for each compound (Figs. 8.4 and 8.5) of all the analyzed sequences. Thus:

- In these results, we can see (according to Table 8.1) that for the acetone loop, the lowest IAE was that of sequence B (0.0233981) and the highest IAE was that of sequence C (0.0319735). For the butanol loop, the lowest IAE was that of sequence B (0.00451334) and the highest IAE was that of sequence D (0.0829786). Yet, it can be concluded that the best-analyzed sequence with this PA1 was B, since the IAE in acetone is lower, and in the same way it has the lowest IAE for butanol. Sequence D was the worst sequence.
- According to Fig. 8.4, the best dynamic behavior for the flow of acetone was that of sequence B, since its response was stabilized within 10 h. Sequence D was stabilized at 20 h. Sequence C was stabilized after 35 h, which showed the worst performance compared to the previous ones, although it is worth mentioning that it was not bad behavior in any sense. For Fig. 8.5, the best dynamic behaviors for the butanol flow were sequences A, B, C, since their response was stabilized within 5 h, being B the one that was stabilized the fastest. Sequence D was stabilized after 20 h. Yet, this last response does not reflect bad behavior.

- Making the comparison between the intensified sequences (B–D) and the reference sequence (A), with their respective IAE and dynamic responses it can be seen that they are all stable for these loops and PA1 results. After making the comparison, sequence B proved to be better than A; and all were similar in its majority after that, except for sequence D, which proved to be the most unstable.

Feeding disturbance (PA2)

It was possible to obtain the optimal controller parameters (with its respective IAE) (Table 8.2), which allowed us to obtain the best dynamic responses for each compound (Figs. 8.6 and 8.7) in most of the analyzed sequences; however, two of the analyzed loops were completely unstable. Thus:

- According to Table 8.2, for the acetone loop, the lowest IAE was that of sequence A (0.00016079) and the highest IAE was that of sequence D (0.000315984). For the butanol loop, the lowest IAE was that of sequence B (0.0000465626) and the highest IAE was that of sequence C (0.000782736); concluding that the best analyzed sequence with PA2 is B since the IAE in acetone is close to that of sequence A and has the lowest IAE for butanol.
- Based on Fig. 8.6, the best dynamic behavior for the acetone flow was that of sequence C. This is because its response was stabilized after 40 h. Most of the other sequences were stabilized between 40 and 50 h. Sequence D was stabilized after 60 h, which the worst performance compared to the previous ones, although it is worth mentioning that it was not a bad behavior. In Fig. 8.7, the best dynamic behaviors for the butanol flow were sequences B and C. Their response was stabilized before 20 h, being B, the one that was stabilized the fastest. Sequence A was stabilized between 40 and 60 h, and sequence D was stabilized after 100 h. Neither sequence A nor D were bad behaviors.
- Making the comparison between the intensified sequences (B–D) with the reference sequence (A), their respective IAE, and dynamic responses, we can first see that they are all stable for these loops and the PA2 results with the exception of the unstable ones. In this comparison, sequence B was better than sequence A, sequence C was almost similar to sequence A, and sequence D proved to be the most unstable.

8.3.1.2 Results discussion

Once the results have been analyzed, a discussion of each of the results can be interpreted, taking into account of course, that each of the proposed objectives were achieved. Out of all the analyzed sequences, most showed good closed-loop control properties with low IAE numbers and obtained the optimal controller parameters that generated the best dynamic response since stability was reached within a short period. It is important to mention that the study of tuning parameters K_c and τ_i and obtaining the minimum value of the IAE, help to identify in terms of control which system has the best control properties. That is to say, how fast the system returns to its nominal point and therefore which

process, if one is compared with the other, has better stability and better dynamic response. The evidence for this is that the IAE criterion is used to quantify behavior. Practically, the results based on the raised hypothesis will be discussed:
1. In the case of sequence B, the hypothesis was correct since it showed good control properties and the best stability. It also showed a better dynamic performance than the reference sequence A. Therefore, it can be concluded that sequence B proved to be the best-analyzed sequence.
2. Sequence C showed good control properties and good stability being almost as good as sequence A.
3. Sequence D showed good control properties with good stability, but it did not prove to be better than sequence A.

By mounting an LV-type control scheme, advantages are shown. The first one is to have a fast response capacity of the disturbance of the control variable with respect to the manipulated variable. This can be seen in Figs. 8.5 and 8.6. The responses turned out to be smooth, with slight overshooting and with effective stabilization to the required perturbation. The instrumentation for this type of control scheme would prove to be much simpler than if a crossover system were to be mounted. Furthermore, the K_c and τ_i parameters obtained are within the range of any commercial PI controller. Therefore, performing this type of control studies will undoubtedly impact the economy of the process, since the shorter the response time of the system, the less waste of raw material and energy.

From the results, it can be concluded that most of the intensified purification systems analyzed in this chapter demonstrated to have good control properties, surpassing, or approaching a conventional separation model, as mentioned in (Sánchez-Ramírez, Alcocer-García, et al., 2017; Sánchez-Ramírez et al., 2016), where good closed-loop control properties in intensified purification systems of the ABE mixture were also proved.

Taking into account that all the proposed objectives were carried out, it is worth mentioning that the results showed two behaviors that were not accounted for in this work and which directly influenced the control properties of the analyzed sequences.
1. The degree of intensification

The degree of intensification can be seen in Fig. 8.1 and in the obtained results since the sequences that showed the best control properties were those that are not completely intensified by having the combination of a conventional distillation column and a DWC. Here, sequence B was the best, and sequence D, which is almost completely intensified, was the worst with a DWC column and with the worst control properties out of all the sequences.
2. The position of the intensified team

Observing Fig. 8.1, and the obtained results, we can see that the purification sequences that began with the conventional distillation column and which left the DWC until the end, showed the best control properties, as in the case of sequence B. The sequences that started the purification process with the intensified equipment, the DWC column, showed the worse control properties, compared to the other sequences, similar to that of sequence D where it is only a single intensified equipment, a DWC column. In other words, the order of the equipment will provide stability to the process, according to the control properties that each equipment offer.

Consequently, for these two behaviors, it can be said that:
- The conventional column contributes greatly to the stability of the process, due to its simple design with good control properties, similar to the conventional sequence A. It compensates for the complexity of the intensified equipment design.
- The union of different tasks in a single equipment generates great complexity in all that is related to this. The more intensified equipment or a process is the poorer control properties it will have. It will be less controllable by showing great instability during its operation.

In conclusion, we can say that the process control properties are directly influenced by the degree of intensification and the position of the intensified equipment. Taking into account that the best results obtained occurred when processes not completely intensified were used and which began with conventional equipment. This is supported by the work provided by Lucero-Robles, Gómez-Castro, Ramírez-Márquez, and Segovia-Hernández (2016), who mentions that the degree of intensification and the position of the intensified equipment in the process, when thermally coupled columns are used such as DWC, can influence the control properties. Likewise, Torres-Ortega et al. (2018) showed that the degree of intensification affects the control properties of a process. All are presented in the results of this work.

8.4 Conclusions

The closed-loop control properties of four purification sequences of the ABE mixture (A, conventional and B–D, intensified) were studied and obtained. This was accomplished by obtaining most of the optimal parameters of the PI controller, as well as its dynamic response, along with its IAE, based on set points (PA1) and feeding disturbance (PA2), and making the comparison with a conventional purification model. In general, most of the sequences proved to have good control properties by having small IAE numbers and good dynamic performance, due to the tuning of the optimal parameters of the PI controller. Considering the four sequences, each one showed good control properties, with good dynamic performance; being sequence B the best of all sequences.

Also, two unexpected conclusions were reached. One is the degree of intensification and the second one the position of the intensified equipment for systems using thermally coupled columns such as DWCs, which directly influence the control properties of the process. These intensified purification processes proved to be more economical, with less environmental impact and with good control properties due to the good stability that they present during their operation. In a general way, we can say that these intensified purification systems showed good control properties, by showing a good dynamic performance during their simulator operation. This will surely be reflected in good operability when said systems are implemented at an industrial level proving to be stable and knowing that these control properties are directly influenced by the degree of intensification and by the position of the intensified equipment. The results obtained indicate the possibility to use intensified schemes in (bio)industrial applications and thus contribute to generate sustainable processes in the purification of (bio)products and to be able to introduce the (bio)industry in circular economy environments.

References

Aguilera, A. F., Alopaeus, V., Christensen, L. P., Contreras-Zarazúa, G., Errico, M., Feng, X., et al. (2017). *Process synthesis and process intensification: Methodological approaches*. Walter de Gruyter GmbH & Co KG.

Åström, K. J., & Hägglund, T. (1995). *PID controllers: Theory, design, and tuning. Vol. 2*. Research Triangle Park, NC: Instrument Society of America.

Barton, W. E., & Daugulis, A. J. (1992). Evaluation of solvents for extractive butanol fermentation with *Clostridium acetobutylicum* and the use of poly (propylene glycol) 1200. *Applied Microbiology and Biotechnology*, *36*(5), 632–639.

Dürre, P. (2007). Biobutanol: An attractive biofuel. *Biotechnology Journal*, *2*(12), 1525–1534.

Errico, M., Sánchez-Ramírez, E., Quiroz-Ramírez, J. J., Ben-Guang, R., & Segovia-Hernandez, J. G. (2017). Multiobjective optimal acetone–butanol–ethanol separation systems using liquid–liquid extraction-assisted divided wall columns. *Industrial & Engineering Chemistry Research*, 11575–11583.

Errico, M., Sanchez-Ramirez, E., Quiroz-Ramírez, J. J., Segovia-Hernández, J. G., & Rong, B. G. (2016). Synthesis and design of new hybrid configurations for biobutanol purification. *Computers & Chemical Engineering*, *84*, 482–492.

García, H. A., & Hernández, J. G. S. (2015). Desarrollo de un proceso de bajo costo de operación para la producción de biobutanol. *Jóvenes En La Ciencia*, *1*(2), 1613–1617.

Isopescu, R., Woinaroschy, A., & Draghiciu, L. (2008). Energy reduction in a divided wall distillation column. *Revista de Chimie*, *59*(1), 812–815.

Kapoor, N., McAvoy, T. J., & Marlin, T. E. (1986). Effect of recycle structure on distillation tower time constants. *AICHE Journal*, *32*(3), 411–418.

Kaymak, D. B. (2019). Design and control of an alternative process for biobutanol purification from ABE fermentation. *Industrial & Engineering Chemistry Research*, *58*(5), 1957–1965.

Khanam, A., Skogestad, S., & Shamsuzzoha, M. (2017). *Control structure selection of energy efficient divided wall column*. Lap Lambert: Academic Publishing.

Kiss, A. A. (2013). *Advanced distillation technologies: Design, control and applications*. John Wiley & Sons.

Ling, H., & Luyben, L. (2009). New control structure for divided-wall columns. *Industrial & Engineering Chemistry Research*, *48*(13), 6034–6049.

Liu, F., Liu, L., & Feng, X. (2005). Separation of acetone–butanol–ethanol (ABE) from dilute aqueous solutions by pervaporation. *Separation and Purification Technology*, *42*(3), 273–282.

Lucero-Robles, E., Gómez-Castro, F. I., Ramírez-Márquez, C., & Segovia-Hernández, J. G. (2016). Petlyuk columns in multicomponent distillation trains: Effect of their location on the separation of hydrocarbon mixtures. *Chemical Engineering Technology*, 2207–2216.

Moreno, J. A., & Cubillos Lobo, J. A. (2017). Biobutanol como combustible: Una alternativa sustentable. *Investigación Joven*, 4.

O'Dwyer, A. (2000). A summary of PI and PID controller tuning rules for processes with time delay. Part 1: PI controller tuning rules. *IFAC Proceedings*, *33*(4), 159–164.

Patrașcu, I., Bîldea, C. S., & Kiss, A. A. (2017). Eco-efficient butanol separation in the ABE fermentation process. *Separation and Purification Technology*, *177*, 49–61.

Patraşcu, I., Bîldea, C. S., & Kiss, A. A. (2019). Dynamics and control of a heat pump assisted azeotropic dividing-wall column for biobutanol purification. *Chemical Engineering Research and Design*, *146*, 416–426.

Ramshaw, C. (1995). *The incentive for process intensification, proceedings, 1st intl. conf. proc. intensif. for Chem. Ind.*, *18* (p. 1). London: BHR Group.

Sánchez-Ramírez, E., Alcocer-García, H., Quiroz-Ramírez, J. J., Ramírez-Márquez, C., Segovia-Hernández, J. G., Hernández, S., et al. (2017). Control properties of hybrid distillation processes for the separation of biobutanol. *Journal of Chemical Technology and Biotechnology*, *92*(5), 959–970.

Sánchez-Ramírez, E., Quiroz-Ramírez, J. J., Hernández, S., Segovia-Hernández, J. G., & Kiss, A. A. (2017). Optimal hybrid separations for intensified downstream processing of biobutanol. *Separation and Purification Technology*, *185*, 149–159.

Sánchez-Ramírez, E., Quiroz-Ramírez, J. J., Segovia-Hernández, J. G., Hernández, S., & Bonilla-Petriciolet, A. (2015). Process alternatives for biobutanol purification: Design and optimization. *Industrial & Engineering Chemistry Research*, *54*(1), 351–358.

Sánchez-Ramírez, E., Quiroz-Ramírez, J. J., Segovia-Hernández, J. G., Hernández, S., & Ponce-Ortega, J. M. (2016). Economic and environmental optimization of the biobutanol purification process. *Clean Technologies and Environmental Policy*, *18*(2), 395–411.

Segovia-Hernández, J. G., & Bonilla-Petriciolet, A. (2016). *Process intensification in chemical engineering*. New York: Springer.

Serra, M., Espuna, A., & Puigjaner, L. (1999). Control and optimization of the divided wall column. *Chemical Engineering and Processing: Process Intensification*, *38*(4–6), 549–562.

Skogestad, S., & Morari, M. (1987). Understanding the dynamic behavior of distillation columns. *Industrial & Engineering Chemistry Research*, *27*(10), 1848–1862.

Takamatsu, T., Hashimoto, I., & Hashimoto, Y. (1987). Selection of manipulated variables to minimize interaction in multivariate control of a distillation column. *International Chemical Engineering*, *27*, 669–677.

Tashiro, Y., Yoshida, T., Noguchi, T., & Sonomoto, K. (2013). Recent advances and future prospects for increased butanol production by acetone-butanol-ethanol fermentation. *Engineering in Life Sciences*, *13*(5), 432–445.

Torres-Ortega, C. E., Ramírez-Márquez, C., Sánchez-Ramírez, E., Quiroz-Ramírez, J. J., Segovia-Hernandez, J. G., & Rong, B. G. (2018). Effects of intensification on process features and control properties of lignocellulosic bioethanol separation and dehydration systems. *Chemical Engineering and Processing: Process Intensification*, *128*, 188–198.

Wang, S. J., & Wong, D. S. (2007). Controllability and energy efficiency of a high-purity divided wall column. *Chemical Engineering Science*, *62*(4), 1010–1025.

Weizmann, C. (1919). *Production of acetone and alcohol by bacteriological processes*. US Patent, 1(315) (p. 585).

CHAPTER 9

Assessment of modular biorefineries with economic, environmental, and safety considerations

Alexandra Barron[a,b], Natasha Chrisandina[a,b], Antioco López-Molina[c], Debalina Sengupta[b], Claire Shi[d], and Mahmoud M. El-Halwagi[a,b]

[a]Department of Chemical Engineering, Texas A&M University, College Station, TX, United States
[b]Gas and Fuels Research Center, Texas A&M Engineering Experiment Station, College Station, TX, United States
[c]Universidad Juárez Autónoma de Tabasco, Jalpa de Mendéz, Mexico
[d]Department of Chemistry, Rice University, Houston, TX, United States

9.1 Introduction to modular biorefineries

A biorefinery involves the conversion of biological feedstocks into value-added products such as chemicals or fuels (Sengupta & Pike, 2012; Stuart & El-Halwagi, 2013). These feedstocks may be composed of agricultural waste, municipal waste, and other biological materials suitable for conversion. Biorefineries have gained popularity with the increasing interest in sustainable development and the search for renewable sources of energy. Because of the distributed nature of biorefining feedstocks, *modular* units have been proposed as a promising option for biorefineries (LePree, 2015, 2016; O'Connor, O'Brien, & Choi, 2015, 2016; Roy, 2017). Although there are different ways of defining modules (e.g., Baldea, Edgar, Stanley, & Kiss, 2017), in this chapter, the term modular is used to designate a unit or a process which comprised pre-fabricated units that are assembled on-site or shipped from the manufacturer and is typically mounted on a steel structure. Modular units can also aid in achieving process intensification and resilience objectives (El-Halwagi et al., 2020; Sengupta & Yelvington, 2020). There are several comparisons to be drawn between modular processes and the more conventional "*stick-built*" (also known as built-on-site or "site-built") plants where the major elements of the plant are constructed and assembled on site. Most notable is the size distinction between the two; stick-built biorefineries typically have larger capacities than individual modules. Therefore, modular biorefineries must produce the same capacity by using multiple trains, each with a smaller capacity.

Modular biorefineries offer flexibility that stick-built refineries lack. This flexibility applies to capacity, product portfolio, mobility, and feedstock composition. In a modular plant, each module is distinct from the next and can be operated independently. This allows the plant to increase or decrease production capacity based on feedstock availability and market demand by taking modules off-line or starting up additional modules. This also allows maintenance within the plant to be streamlined because the module needing maintenance can simply be taken off-line as the feedstock to this module is diverted to a standby unit. Additionally, different modules can be operated in parallel to create a diverse product portfolio. In the same way that different modules can be used to create different products, they can be used to process different feedstocks. In this way, modular biorefineries are uniquely equipped to handle heterogeneous feedstocks such as municipal solid waste or diverse biomass. Finally, modular units can be moved from one area to another. This mobility means that individual modules can be removed and transported to other sites for reuse if necessary. Though not discussed here, mobile modular units may allow processing units to move with the seasonality of certain biomass feedstocks such as agricultural waste.

Modular biorefineries also benefit from low upfront costs, although they suffer from the lack of economy of scale enjoyed by the larger stick-built plants. Though multiple modules will be required for the plant to operate at maximum capacity, production can begin after installation of the first module. Instillation of additional modules can occur during normal plant operations. This is desirable for small municipalities that may not have the resources necessary to accrue the initial costs of a stick-built facility. As the first module begins turning profit, additional modules can be installed to increase production capacity. However, this "numbering up" approach (Weber & Snowden-Swan, 2019) does not benefit from the economy of scale. In this capacity, stick-built refineries are more economically scaled than modular refineries. While the six-tenths scaling rule can be applied when increasing the capacity of stick-built refineries, the cost of modular refineries scales linearly with increasing capacity. However, each additional module may cost less due to the learning effect (Lier & Grunewald, 2011). That is, each module should be fabricated at the same production facility; as each additional identical module is fabricated, the engineering and production costs will decrease, meaning that the capital cost of each additional module will decrease exponentially.

Due to the consistency of fabrication for each module, the autonomy, efficiency, and safety of each module can be greatly optimized. This reduces the operating, energy, and maintenance costs when compared to stick-built biorefineries. Due to automation, the staff necessary to operate each individual module is decreased. Instillation also requires fewer personnel than the construction of a stick-built biorefinery. The heat integration of each module can be well-optimized to improve energy efficiency throughout the module. Additionally, because modules can be fabricated and tested remotely at the production facility, safety can be improved in modular facilities. Because stick-built facilities must be fabricated and tested on site, increased operating staff are needed. Additionally, many safety concerns may not be realized until after operations have begun.

A significant advantage for modular units is the ability to support decentralized or distributed manufacturing. Implementing distributed modular processing facilities is already a common practice in the natural gas sector. It is often more economically advisable to process stranded natural gas on site and distribute the products than it is to extract the natural gas from the ground and ship it to a larger processing facility. Primus Green Energy has begun commercialization of modular methanol production from stranded natural gas throughout Texas, a trend that is becoming more popular as more stranded natural gas, is discovered throughout the United States. Several case studies and applications have been proposed for modular biorefineries. López-Molina et al. (2020) analyzed the performance of centralized biorefineries (where biomass is collected from various regions and transported to a large facility) versus decentralized/distributed biorefineries (where several biorefineries are installed each to treat biomass within a certain region) and derived an empirical expression for cost estimation. Tsagkari, Kokossis, and Dubois (2020) developed a modular method for capital cost estimation of biorefineries. Laibach, Müller, Pleissner, Raber, and Smetana (2021) described a case study for converting food waste to value-added chemicals. Infelise, Kazimierczak, Wietecha, and Kopania (2019) assessed a small-scale multipurpose modular biorefining technology for extracting high-quality fibers and chemicals. Dahmen, Lewandowski, Zibek, and Weidtmann (2019) assessed the concept of a modular lignocellulosic biorefinery and provided future perspectives. Gonzalez-Contreras, Lugo-Mendez, Sales-Cruz, and Lopez-Arenas (2021) evaluated an intensified lignocellulosic biorefineries-case study for the production of ethanol.

With the growing interest in modular biorefineries, it is important to have a systematic approach for comparison with stick-built biorefineries. This chapter provides a multicriteria approach for such comparison while highlighting the economic, safety, and environmental objectives.

9.2 Approach

In order to assess modular biorefineries versus stick-built biorefineries, the following procedure is proposed. First, the characteristics and availability of feedstocks are identified. Next, simulation and techno-economic analysis are carried out to identify a base case for both the modular option and the stick-built option. Life cycle analysis and risk assessment are used to quantify the environmental and safety objectives for each option. To conduct tradeoffs between the multiple objectives, the ε-constraint method is used. The economic objective (e.g., maximum profitability or minimum cost) is used as the objective function while the safety and environmental objectives are used as constraints whose bounds are successively altered to establish a tradeoff (Pareto curve). For the environmental objective, a life cycle assessment framework (e.g., Chouinard-Dussault, Bradt, Ponce-Ortega, and El-Halwagi (2011)) is utilized to track the emissions (especially the carbon footprint).

For the safety objective, the relative risk based on the Hazardous Process Stream Index (HPSI) (López-Molina et al., 2020) is employed. The HPSI is based on five indicators pertaining to pressure, density, molar flowrate, heat of combustion, and flash point as follows:

$$I_P = \frac{\text{pressure value of individual stream}}{\text{average pressure for all streams}} \quad (9.1)$$

$$I_\rho = \frac{\text{density value of individual stream}}{\text{average density for all streams}} \quad (9.2)$$

$$I_{MF} = \frac{\text{molar flow value of individual stream}}{\text{average molar flow for all streams}} \quad (9.3)$$

$$I_{\Delta Hc} = \frac{\text{heat of combustion of individual stream}}{\text{average heat of combustion for all streams}} \quad (9.4)$$

$$I_{Fp} = \frac{\text{flash point score of individual stream}}{\text{average flash point score for all streams}} \quad (9.5)$$

The result of HPSI is calculated through the following expressions.

$$HPSI = \left(\frac{I_P \cdot I_{MF} \cdot I_{\Delta Hc} \cdot I_{FP}}{I_\rho}\right) \cdot W \quad (9.6)$$

where W is defined as the ratio of the production capacity of the modular process with capacity i (CP_i) to the stick-built process capacity (CP_{base}).

$$W = \frac{CP_i}{CP_{base}} \quad (9.7)$$

The individual process stream risk (R_i) is estimated from the HPSI value using Eq. (9.8).

$$R_i = \frac{HPSI_i - HPSI_{min}}{HPSI_{max} - HPSI_{min}} \quad (9.8)$$

where $HPSI_i$ is the individual index value of the process streams, $HPSI_{min}$ is the smallest index value of the compared process streams, $HPSI_{max}$ is the maximum index value of the compared process streams.

Once the individual risk has been estimated, the total relative process risk (R_T) is estimated as follows:

$$R_T = \frac{\sum_{i}^{n} R_i}{PS} \quad (9.9)$$

where PS is the total number of process streams.

Next, the tradeoffs for the three objectives are considered for decision-making. Figure 9.1 shows the key steps in the proposed approach. In order to quantitatively

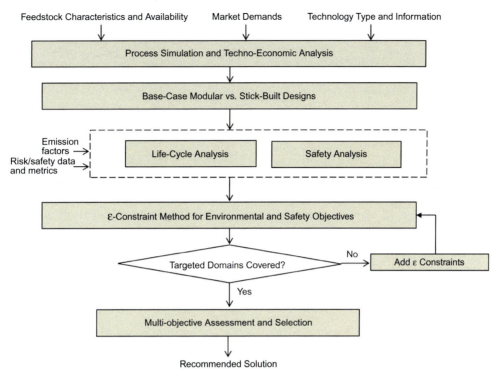

Fig. 9.1 Proposed approach for assessing modular vs stick-build biorefineries.

illustrate the characteristics of modular biorefineries and how they compare to stick-built biorefineries, the follow case study is used.

9.3 Case study

Methanol is an essential chemical with a global production capacity of about 110 million metric tons per annum "MTPA" (Methanol Institute, 2021). Various chemicals may be derived from methanol. Examples include olefins, dimethyl ether, formaldehyde, acetic acid, methyl tertiary-butyl ether (MTBE), gasoline, methyl methacrylate, methyl chloride, and methylamines (e.g., Al-Douri, Sengupta, & El-Halwagi, 2017). Methanol may be manufactured from different feedstocks and through several pathways. Common pathway in the US exploits the shale gas boom. Methane may be reformed into syngas which is subsequently converted into methanol (e.g., Julián-Durán, Ortiz-Espinoza, El-Halwagi, & Jiménez-Gutiérrez, 2014; Ehlinger, Gabriel, Noureldin, & El-Halwagi, 2014). The potential use of biomass offers various promising improvements especially for lowering the greenhouse gas (GHG) emissions. In this case study, it is desired to produce 7.000 t/yr of methanol via the conversion of refuse derived fuels

(RDFs) which is obtained from municipal solid waste (MSW). The RDF is gasified and processed to produce methanol. Two design options are to be considered: a single stick-built plant which produces 7000 t/yr or seven trains of modular plants each producing 1000 t/year.

This process consists of three main stages: gasification, purification, and synthesis. Gasification is the process that transforms liquid or solid substances into a gas phase. In the case of MSW, drying of the biomass is necessary to feed it to a fixed-bed reactor or fluidized bed subsequently. Inside the reactor, the pyrolysis of the biomass is carried out. For this, the biomass can be mixed with air or oxygen, the operating conditions for the gasification vary between 800°C and 1150°C at 1 atm of pressure (Salladini et al., 2018). The syngas purification section consists of removing H_2S and CO_2, which affect methanol production performance and, at the same time, generate operational concerns such as corrosion. This process is carried out at pressures between 15 and 32 bar and temperatures in a range of 30–180°C. There are several strategies for CO_2 removals, such as absorption, adsorption, membranes, or cryogenic distillation. However, absorption is the most used due to its high efficiency and low cost. The absorbent agents can be amines, carbonates, or new generation solvents such as Selexol (Salladini et al., 2018; Vakharia, Ramasubramanian, & Winston Ho, 2015). The synthesis of methanol from MSW is carried out inside a reactor that operates in a pressure range of 54–62 bar and a temperature between 240°C and 260°C (Salladini et al., 2018). The catalysts reported for this process are $Cu/ZnO/Al2O3$ (dos Santos, de Santos, & Prata, 2018) or syngas over a Cu-based catalyst (Park, Cho, Lee, Park, & Lee, 2019). The simplified flowsheet for the production of methanol from MSW is presented in Fig. 9.2.

Information reported in the bibliography about operating conditions, the density, and the molar compositions of all process streams were extracted from: Shehzad, Bashir, & Sethupathi, 2016; Lücking, 2017; Leo, 2018, which correspond to the gasification, purification, and synthesis sections, respectively.

The key technical and economic data are based on the studies of Iaquaniello et al. (2017), Salladini et al. (2018), and López-Molina et al. (2020). Table 9.1 summarizes key data. The fixed capital investment (FCI) is estimated based on the cost correlation derived by López-Molina et al. (2020):

$$FCI \text{ (in \$MM)} = 0.16 * N * (\text{Flowrate of biomass feed in 1000 tonnes/yr})^{0.84}$$

The working capital investment (WCI) is assumed to be 20% of FCI and the total capital investment (CAPEX) is the sum of the FCI and the WCI (El-Halwagi, 2017a, 2017b).

The results of the case study are shown by Table 9.2. If the economic objective alone is used for decision making, the stick built option is to be used because it provides a higher return on investment (ROI). But if there are limits on CO_2 emission (e.g., an emission

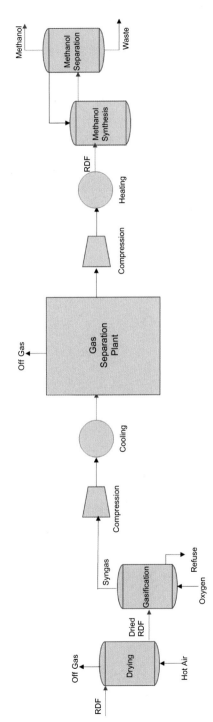

Fig. 9.2 Flowsheet of Methanol production from MSW gasification.

Table 9.1 Key data for the case study.

Cost of RDF ($/tonne)	Selling Price of methanol ($/tonne)	Yield of stick-built process (tonne methanol per tonne RDF)	Yield of modular process (tonne methanol per tonne RDF)	Price of electricity ($/kWh)
40	400	0.35	0.40	0.05

Table 9.2 Main results for the case study.

	CAPEX ($)	ROI (%/yr)	CO_2 emission (tonne/yr)	Total process risk	Risk level
Stick-built	14.3	10.5	14,167	0.22	Low
Modular	17.4	9.5	12,396	0.03	Very Low

upper bound of 13,000 t CO_2 per year) or risk level (e.g., very low), then the modular unit should be selected.

It is also worth noting that a profitability framework may be used to integrate the economic, environmental, and safety objectives through the use of a safety and sustainability weighted return on investment metric (SASWROIM) which extends the ROI metric and makes use of the augmented safety and sustainability objectives as well (El-Halwagi, 2017a, 2017b; Guillen-Cuevas et al., 2018). For the competition among multiple projects (e.g., modular, stick-built), the *Annual Safety and Sustainability Profit* "*ASSP*" for the pth project is defined as follows:

$$ASSP_p = AEP_p \left[1 + \sum_{i=1}^{N_{Indicators}} w_i \left(\frac{Indicator_i^{Base} - Indicator_{p,i}}{Indicator_i^{Base} - Indicator_i^{Target}} \right) \right] \quad (9.10)$$

where *AEP* is the annual economic profit of the pth option, *i* is an index for the different safety and sustainability indicators. The weighing factor w_i is a ratio representing the importance of the ith safety or sustainability indicator relative to the annual net economic profit. The term $Indicator_{p,i}$ represents the value of the ith sustainability indicator associated with the pth option (e.g., modular or stick-built), the term $Indicator_i^{Base}$ corresponds to the value of the ith sustainability indicator for the base case (stick-built), and the term $Indicator^{Target}$ corresponds to the target value of the ith sustainability indicator (which can be taken as the desired target by that biorefining sector is aspiring to achieve or a value that is slightly better performance than the better of two options: modular and stick-built). Consequently, the ratio $\left(\frac{Indicator_i^{Base} - Indicator_{p,i}}{Indicator_i^{Base} - Indicator_i^{Target}} \right)$ represents the fractional contribution of project p toward meeting the desired/targeted performance for the ith sustainability

Table 9.3 Data for the SASWROIM calculations.

Item	Value
Base-case ROI	10.5%/yr
Base-case GHG emission	14,167 Tonne CO_2/yr
Base-case total process risk	0.22
Target GHG emission	12,000 Tonne CO_2/yr
Target total process risk	0.02

metric and may be positive (if there is improvement in the indicator), zero (when the base-case is also the option), or negative (if there is worsening in the value of the indicator). The *Safety and Sustainability Weighted Return on Investment Metric* "*SASWROIM*" of option p (Guillen-Cuevas et al., 2018) is defined as:

$$SASWROIM_p = \frac{ASP_p}{TCI_p} \qquad (9.11)$$

For the case study, the ROI for the base-case design of stick-built biorefinery is 10.5%/yr. For the environmental indicator, the annual GHG emission is selected. For the safety indicator, the total process risk is chosen. Key data used in the SASWROIM calculations are given in Table 9.3.

Depending on the weights selected to value the and environmental and safety objectives relative to the economic objective, the value of the SASWROIM will change. For instance if w_{GHG} and w_{Risk} are selected to be 0.5, the SASWROIM for the modular biorefinery is calculated to be 17.9%/yr, rendering the modular system significantly more attractive than the stick-built biorefinery when safety and sustainability are included. It is interesting to note that the modular system has a higher SASWROIM than the stick-built plant as long as the weights w_{GHG} and w_{Risk} are higher than 0.06.

9.4 Conclusions

This chapter has presented an approach to assessing modular biorefineries compared to stick-built plants. Modular units offer more resilience and flexibility. They are also more appropriate for distributed manufacturing strategies. On the other hand, stick-built systems enjoy a more effective economy of scale for capital cost. The relative risk in modular biorefineries is typically lower than those for stick-built plants because the consequences of an accident in modular biorefineries are reduced through intensification. However, some safety aspects must be analyzed in more detail because the complexity of multitrain modular plants increases (more valves, pipes, instrumented systems, etc.). When modular units have higher levels of intensification that lead to higher performance, mass and energy efficiencies increase and GHG emissions typically decrease. Two approaches were proposed for the multi-criteria decision-making; "epsilon cuts" and a modified form of ROI. A case study on MSW-to methanol was solved to illustrate the proposed concepts and the multi-objective tradeoffs. The results tend to show that when only the economic factors are considered, stick-built plants have an advantage. On the other hand, when the environmental and safety considerations are properly accounted for, the modular units offer a superior performance.

References

Al-Douri, A., Sengupta, D., & El-Halwagi, M. M. (2017). Shale gas monetization—A review of downstream processing to chemicals and fuels. *Journal of Natural Gas Science and Engineering, 45*, 436–455.

Baldea, M., Edgar, T. F., Stanley, B. L., & Kiss, A. A. (2017). Modular manufacturing processes: Status, challenges, and opportunities. *AICHE Journal, 63*(10), 4262–4272.

Chouinard-Dussault, P., Bradt, L., Ponce-Ortega, J. M., & El-Halwagi, M. M. (2011). Incorporation of process integration into life cycle analysis for the production of biofuels. *Clean Technologies and Environmental Policy, 13*(5), 673–685. https://doi.org/10.1007/s10098-010-0339-8.

Dahmen, N., Lewandowski, I., Zibek, S., & Weidtmann, A. (2019). Integrated lignocellulosic value chains in a growing bioeconomy: Status quo and perspectives. *GCB Bioenergy, 11*(1), 107–117.

dos Santos, R. O., de Santos, L. S., & Prata, D. M. (2018). Simulation and optimization of a methanol synthesis process from different biogas sources. *Journal of Cleaner Production, 186*, 821–830. https://doi.org/10.1016/j.jclepro.2018.03.108.

Ehlinger, V. M., Gabriel, K. J., Noureldin, M. M., & El-Halwagi, M. M. (2014). Process design and integration of shale gas to methanol. *ACS Sustainable Chemistry & Engineering, 2*(1), 30–37.

El-Halwagi, M. M. (2017a). *Sustainable design through process integration: Fundamentals and applications to industrial pollution prevention, resource conservation, and profitability enhancement* (2nd ed.). IChemE/Elsevier.

El-Halwagi, M. M. (2017b). A return on investment metric for incorporating sustainability in process integration and improvement projects. *Clean Technologies and Environmental Policy, 19*(2), 611–617.

El-Halwagi, M. M., Sengupta, D., Pistikopoulos, E. N., Sammons, J., Eljack, F., & Kazi, M. K. (2020). Disaster-resilient design of manufacturing facilities through process integration: Principal strategies, perspectives, and research challenges, sustainable chemical process design. *Frontiers in Sustainability, 1*, 8. Open Access https://www.frontiersin.org/articles/10.3389/frsus.2020.595961/full. https://doi.org/10.3389/frsus.2020.595961.

Gonzalez-Contreras, M., Lugo-Mendez, H., Sales-Cruz, M., & Lopez-Arenas, T. (2021). Synthesis, design and evaluation of intensified lignocellulosic biorefineries-case study: Ethanol production. *Chemical Engineering and Processing Process Intensification, 159*, 108220.

Guillen-Cuevas, K., Ortiz-Espinoza, A. P., Ozinan, E., Jiménez-Gutiérrez, A., Kazantzis, N. K., & El-Halwagi, M. M. (2018). Incorporation of safety and sustainability in conceptual design via a return on investment metric. *ACS Sustainable Chemistry & Engineering, 6*(1), 1411–1416.

Iaquaniello, G., Centi, G., Salladini, A., Palo, E., Perathoner, S., & Spadaccini, L. (2017). Waste-to-methanol: Process and economics assessment. *Bioresource Technology, 243*, 611–619.

Infelise, L., Kazimierczak, J., Wietecha, J., & Kopania, E. (2019). GINEXTRA®: A small-scale multipurpose modular and integrated biorefinery technology. In *Biorefinery* (pp. 593–614). Cham: Springer.

Julián-Durán, L. M., Ortiz-Espinoza, A. P., El-Halwagi, M. M., & Jiménez-Gutiérrez, A. (2014). Techno-economic assessment and environmental impact of shale gas alternatives to methanol. *ACS Sustainable Chemistry & Engineering, 2*(10), 2338–2344.

Laibach, N., Müller, B., Pleissner, D., Raber, W., & Smetana, S. (2021). An integrated, modular biorefinery for the treatment of food waste in urban areas. *Case Studies in Chemical and Environmental Engineering*, 100118.

Leo, D. (2018). *Process modelling and simulation of a methanol synthesis plant using syngas streams obtained from biomass*. Master Politecnico di Milano. https://www.politesi.polimi.it/bitstream/10589/142748/3/2018_10_Leo.pdf.

LePree, J. (2015). Moving to modular. *Chemical Engineering*, *122*(1), 16.
LePree, J. (2016). Is modular right for your project? *Chemical Engineering*, *123*(1), 20.
Lier, S., & Grunewald, M. (2011). Net present value analysis of modular chemical production plants. *Chemical Engineering and Technology*, *34*(5), 809–816.
López-Molina, A., Sengupta, D., Shi, C., Aldamigh, E., Alandejani, M., & El-Halwagi, M. M. (2020). An integrated approach to the design of centralized and decentralized biorefineries with environmental, safety, and economic objectives. *Processes*, *8*, 1682. Open access www.mdpi.com/2227-9717/8/12/1682/pdf. https://doi.org/10.3390/pr8121682.
Lücking, L. E. (2017). *Methanol production from syngas: Process modelling and design utilising biomass gasification and integrating hydrogen supply*. Master Delft University of Technology. http://repository.tudelft.nl/.
Methanol Institute. (2021). https://www.methanol.org/the-methanol-industry/#:~:text=The%20methanol%20industry%20spans%20the%20entire%20globe%2C%20with,%28almost%2036.6%20billion%20gallons%20or%20138%20billion%20liters%29. Accessed on July 28, 2021.
O'Connor, J. T., O'Brien, W. J., & Choi, J. O. (2015). Standardization strategy for modular industrial plants. *Journal of Construction Engineering and Management*, *141*(9), 04015026.
O'Connor, J. T., O'Brien, W. J., & Choi, J. O. (2016). Industrial project execution planning: Modularization versus stick-built. *Practice Periodical on Structural Design and Construction*, *21*(1), 04015014.
Park, J., Cho, J., Lee, Y., Park, M.-J., & Lee, W. B. (2019). Practical microkinetic modeling approach for methanol synthesis from syngas over a Cu-based catalyst. *Industrial & Engineering Chemistry Research*, *58*(20), 8663–8673. https://doi.org/10.1021/acs.iecr.9b01254.
Roy, S. (2017). Consider modular plant design. *Chemical Engineering Progress*, *113*(5), 28–31.
Salladini, A., Agostini, E., Borgogna, A., Spadacini, L., Annesini, M. C., & Iaquaniello, G. (2018). Analysis on high temperature gasification for conversion of RDF into bio-methanol. In *Gasification for low-grade feedstock* (p. 143). IntechOpen.
Sengupta, D., & Pike, R. W. (2012). *Chemicals from biomass: Integrating bioprocesses into chemical production complexes for sustainable development*. CRC Press.
Sengupta, D., & Yelvington, P. (2020). Modular intensified processes promote resilient manufacturing. *Chemical Engineering Progress*, 25.
Shehzad, A., Bashir, M. J. K., & Sethupathi, S. (2016). System analysis for synthesis gas (syngas) production in Pakistan from municipal solid waste gasification using a circulating fluidized bed gasifier. *Renewable and Sustainable Energy Reviews*, *60*, 1302–1311. https://doi.org/10.1016/j.rser.2016.03.042.
Stuart, P., & El-Halwagi, M. M. (2013). *Integrated biorefineries: Design, analysis, and optimization*. Taylor and Francis/CRC.
Tsagkari, M., Kokossis, A., & Dubois, J. L. (2020). A method for quick capital cost estimation of biorefineries beyond the state of the art. *Biofuels, Bioproducts and Biorefining*, *14*(5), 1061–1088.
Vakharia, V., Ramasubramanian, K., & Winston Ho, W. S. (2015). An experimental and modeling study of CO2-selective membranes for IGCC syngas purification. *Journal of Membrane Science*, *488*, 56–66. https://doi.org/10.1016/j.memsci.2015.04.007.
Weber, R. S., & Snowden-Swan, L. J. (2019). The economics of numbering up a chemical process enterprise. *Journal of Advanced Manufacturing and Processing*, *1*(1–2), e10011.

CHAPTER 10

Production of biofuels and biobased chemicals in biorefineries and potential use of intensified technologies

Alvaro Orjuela[a] and Andrea del Pilar Orjuela[b]
[a]Department of Chemical and Environmental Engineering, Universidad Nacional de Colombia, Bogotá D.C., Colombia
[b]Process Solutions and Equipment SAS, Engineering Division, Bogotá D.C., Colombia

10.1 Introduction

Historically, the chemical industry has brought wellbeing to our society providing the required products to fulfill a variety of basic needs, including affordable food, drinking water, construction materials for shelter, personal care- and health-products, clothing materials, fuels, and more recently materials for electronics. Nevertheless, due to the environmental, social and economic challenges of modern times, this industry has been under strong internal and external pressures to achieve a rapid transition to a sustainable operation. As a result, some of the most important players of the global chemical market have set specific sustainability goals for the near future. For instance, the CO_2 emissions reductions targets recently established by some of the major chemical companies worldwide are summarized in Table 10.1.

The urgency to meet such goals has driven enormous transformations in the chemical sector, fostering the development, and implementation of technologies that are capable of enhancing productivity by several orders of magnitude, that incorporate biobased resources (virgin or recycled), and that employ renewable energies. At a large degree, this has been possible by exploiting the synergy of applying an integrated biorefinery production model together with the use of intensified processing technologies. Interestingly, these ideas are far from new, but have been only recently implemented at the industrial scale after a series of enabling technologies lately developed made it possible.

From one side, the use of biomass as raw material for the production of chemicals, materials, fuels, and energy have been carried out for centuries. Nevertheless, during recent years a dramatic revolution of the biology-biotechnology field has occurred. At some point, this revolution at the biological dimension might be considered as an additional domain of intensification to those commonly recognized (spatial, thermodynamic, functional, and temporal; Stankiewicz, Van Gerven, & Stefanidis, 2019; Van Gerven & Stankiewicz, 2009). A variety of advanced techniques such as next-generation

Table 10.1 Carbon-reduction targets of some of the leading chemical companies worldwide (Bettenhausen, 2021; Tullo, 2021).

Reduction plan	Company	Global sales ranking—2020	Reduction target	Reference date	Target date
MODEST	BASF	1	Maintain	2018	n/a
	ExxonMobil Chemical	11	Maintain	2018	n/a
	LG Chem	7	Maintain	2018	n/a
MODERATE	Air Liquide	12	30%	2015	2025
	Braskem	24	58%	2008	2030
	Covestro	21	50%	2005	2025
	Evonik Industries	19	50%	2008	2025
	Linde	9	35%	2018	2028
	Lotte Chemical	31	36%	2013	2028
	LyondellBasell Industries	10	15%	2015	2030
	Mitsui Chemicals	25	25%	2005	2030
	Sabic	5	25%	2010	2025
	Shin-Etsu Chemical	18	55%	1990	2025
	Solvay	28	26%	2018	2030
	Sumitomo Chemical	16	57%	2013	2050
	Toray Industries	17	30%	2013	2030
AMBITIOUS	Chemours	>50	Carbon negative		2050
	Dow	3	Carbon neutral		2050
	Dupont	14	Carbon neutral		2050
	Eastman Chemical	40	Carbon neutral		2050
	INEOS (at Antwerp)	4	Carbon neutral		2050
	Mitsubishi Chemical	8	Carbon neutral		2050
	PetroChina	13	Near carbon neutral		2050
	Reliance Industries	20	Carbon neutral		2035
	Sinopec	2	Carbon neutral		2030
	Yara	23	Carbon neutral		2050

sequencing (NGS), genome-wide association (GWA), gene- and genome-editing approaches involving clustered regularly interspaced short palindromic repeats (CRISPR), and development of induced pluripotent stem cells (iPSCs), have enabled a giant leap in life sciences (National Academies of the Sciences, Engineering, and Medicine, 2020). These, among other breakthroughs, enabled miniaturization and to carry out massively parallel genome sequencing reactions on a routine basis and at low costs. Also, they have permitted a rapid tuning of biochemical pathways for metabolic engineering, thus fostering the industrial biobased economy. For instance, these advancements made possible the development of several vaccines for SARS-CoV-2 in a 9–11 months period, which is an arduous process that typically used to take about 10 to 15 years on average to be completed.

On the other hand, rather than being incremental or connective, like in process improvement or process integration, process intensification represents a transformative paradigm for the chemical engineering discipline and the process industry. It is mostly based on ideas of disruptive integrative continuous processing and modular designs that have been possible as a result of a variety of technological breakthroughs. These ideas are not new, and as presented in Fig. 10.1, the first patents describing some of the intensified processes commercially used nowadays, date from almost a century ago. Few details of some of these early attempts for process intensification are here described:

- At the end of 1890s, the spiral heat exchangers were invented to enhance heat transfer efficiency, enabling better sterilization, and avoiding excessive fouling (Laurent, 1895).
- Continuous reactive distillation using homogeneous catalysts was designed in the early 1920s, as a way to produce esters for solvent applications, thus overcoming equilibrium limitations.
- Similarly, reactive absorption was invented almost in parallel, and counter-current columns were used to wash and react the effluent gases from coke ovens and blast furnaces, mainly in the extraction of ammonia (Gay, 1923).
- The effective use of membranes in the separation of helium and methane, using a slight pressure difference between both sides of the membrane, was early described by Lewis et al. (1924).
- The initial versions of the plate heat exchangers were designed as compact extended-surface devices for gas–gas applications, but claimed suitable for any type of fluids configuration (Lucke, 1931).
- The centrifugal mass transfer contactor (Higee) was invented in the early 1930s (Podbelnyak, 1932). It was envisioned as an enhanced method to reduce the size of the distillation columns for mixtures of close boiling point components, and to overcome the limitations of gravity as the single driving force of contact in counter-current gas-liquid columns.

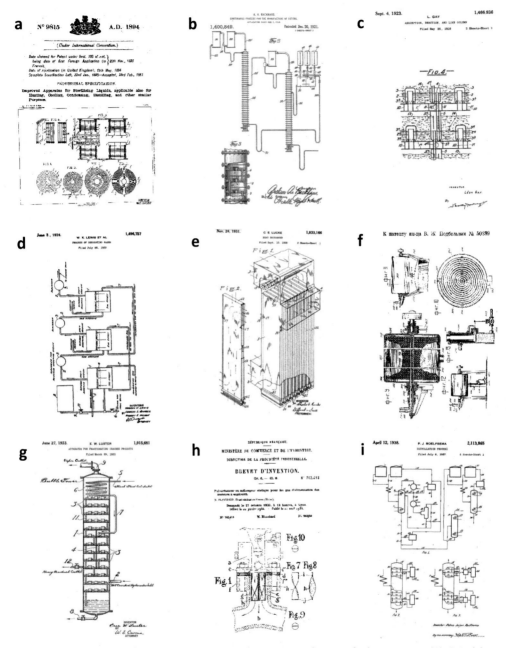

Fig. 10.1 Some of the first patents on currently recognized intensified processes. (A) Spiral heat exchanger, 1895 (Laurent, 1895). (B) Reactive distillation, 1921 (Backhaus, 1921). (C) Reactive adsorption, 1923 (Gay, 1923). (D) Membrane separation, 1924 (Lewis, Venarle, & Wilson, 1924). (E) Plate heat exchanger, 1931 (Lucke, 1931). (F) HiGee distillation, 1932 (Podbelnyak, 1932). (G) Dividing wall distillation column, 1933 (Luster, 1933). (H) Static mixer, 1934 (Blanchard, 1934). (I) Extractive and azeotropic distillation, 1938 (Roelfsema, 1937, 1938).

- The dividing wall distillation columns were initially invented for the separation of cracking products to avoid remixing problems, and the loss of valuable compounds contained in the feed within the heavy fractions in the bottoms (Luster, 1933).
- The static mixers were originally designed to enhance the performance of internal combustion engines by improving the gasoline-air mixing in the injection systems (Blanchard, 1934).
- Roelfsema (1937, 1938) reported the use of entrainers in distillations as early as 1927, in the separation of cyclohexane and benzene. However, he developed some of the firsts known continuous azeotropic and extractive distillation schemes for the separation of a variety of petrochemical mixtures.

Surprisingly, some of the aforementioned intensified processes were initially conceived and implemented within biorefineries (e.g., spiral heat exchangers for pulp processing (Laurent, 1895), reactive distillation for esterification of biobased alcohols (Backhaus, 1921)). Nevertheless, only few have been widely spread within the industry, and some of the less common, have just recently regained attention for their potential implementation in manufacturing facilities. This has resulted as a consequence of the rise of some enabling technologies including: devices miniaturization (e.g., vessels, mixers, reactors, pumps, sensors), modularization, rapid prototyping (e.g., 3D printings), synthesis of nanomaterials, thin films and microfibers (e.g., via vapor deposition or mechanochemistry), biomimicry (e.g., fractal structures), high-throughput experimentation, enhanced process control and optimization (e.g., real-time optimization algorithms), digitalization (e.g., digital twins, machine learning, internet of things, cloud computing, augmented reality), smart sensors (e.g., real-time chemical sensors, spectroscopic sensors), etc.

Despite the novel technologies accelerating the industrial implementation of process intensification in biorefineries, this has been achieved after centuries of scientific and technical developments. In this regard, this work aims to describe a past and current status of commercial scale biorefineries, and the implementation of some process intensification technologies in the production of different biobased derivatives. Some of the challenges and future directions in the adoption of intensified processes within biorefineries are also discussed.

10.2 Evolution and implementation of industrial biorefineries

The use of renewable resources for the production of chemicals, materials, fuels, and energy dates from ancient times of humanity when the "homo-chemicus" learned to transform matter (Meyer, 2011). Without a real understanding of the basic principles, early humans used vegetable- or animal-derived raw materials and minerals, applied physical, thermochemical or biological transformations, and obtained extracts, pigments, odorants, ointments, soaps, fuels, proto-medications, and many other valuable products. From that point, it took millennia and civilizations to achieve a large-scale exploitation of renewable resources for the production of biobased chemicals and materials; the days of the industrial revolution finally arrived. During this period, the production model of today's biorefineries was developed, and while it might appear as a recent topic, some of the current biobased chemicals market leaders originated from those early days. In particular, among many others, currently existing industrial biobased sectors such as naval stores, pulp and paper, oleochemicals, sucro-chemicals, and natural fragrance, flavors and colorants were born during that period.

10.2.1 Naval stores and lignocellulosic biorefineries

The boost of the naval stores, the resinous products from certain coniferous (e.g., tar, pitch, turpentine, rosin), is an example of the rise of the biobased chemical industry. Naval stores became paramount for the shipping business in Europe; mariners used tar to preserve ropes, applied pitch to the ships structures to make them watertight, and used turpentine as solvent for paints and coatings (Zinkel & Russel, 1989). In order to reduce dependence from Nordic manufacturers that were the main suppliers of naval stores in those days, the British developed a large-scale production in the colonies of New England and the Carolinas in the early eighteenth century. The resulting business, latter derived in the industrialization of pulp and paper, and made naval stores products available worldwide as one of the first biobased chemical commodities. This industry has reached our days in the form of pulp mills and wood-based biorefineries. Two of the oldest and currently operating lignocellulosic biorefineries are Borregaard in Sarpsborg, Norway, and Domsjö in Örnsköldsvik, Sweden (Mikkola, Sklavounos, King, & Virtanen, 2015). Also, Arizona Chemical Co. (currently part of Kraton Co.), that established in 1930 as a fusion of decades-old pulp mills, is still a world's leading biorefining company in the field of pine chemicals (e.g., terpenes, rosin acids, tall oil fatty acids, lignin-derived products, etc.) (Buisman & Lange, 2016).

10.2.2 Industrial natural flavors and fragrances

In parallel, another global biobased business was being born in Europe, the fragrance and flavor industry. The house of Antoine Chiris, founded in Grasse in 1768, is considered one of the first enterprises dedicated to the large-scale harvesting of natural materials to use them as feedstock of perfumery. That company expanded globally within the French and British colonies, having subsidiaries in every continent and exploiting the local biodiversity in each location. After years of flourishing, around 1930s the business contracted because of the rise of synthetic fragrances and aromas, and because American factories took the lead after the Second World War. Nevertheless, after numerous acquisitions and fusions, Chiris's company is present today being historically related with current

Givaudan (Givaudan, 2020). Despite the huge impact of lower-cost synthetic products and the developments of fossil-based fragrance and flavors, the segment of natural-based essential oils has a current market of around 18 billion dollars. Besides fragrances and flavors, some natural aroma chemicals used to be employed as feedstock for higher added-value derivatives. For instance, up to recent years natural citral isolated from *Litsea cubeba* or lemongrass, was typically used as raw material in the manufacture of vitamins.

Surprisingly, some of the current players of this highly valuable industry still rely on the use of ancient (but effective) processing technologies such as steam distillation extraction. This method is credited to Mary the Jewess (also Mary the Prophetess) from Alexandria, around the first century CE, and later documented by alchemist Jābir ibn Hayyan (known in Latin as Geber, 721–815, Hayyān, 1531). Fig. 10.2 depicts a typical steam distillation apparatus used in ancient times together with a modern distillery unit used in the extraction of rose damascene essential oil at the industrial scale.

10.2.3 Sugar and starchy biorefinery

Soon after Europeans arrived to America, the new continent became the main source of cane sugar, mostly produced under slave labor. Thousands of sugar mills were installed in Brazil, the Caribbean islands, and in the southern areas of North America, and cane sugar became one of the first globalized agricultural commodities. Besides sugar, different byproducts were also manufactured in these mills, including vinegar, alcohol, and molasses, thus becoming the model of today's sugar-based biorefinery. Interestingly, this extended sugar availability and supply enabled the origin of the candy and chocolate industry in Europe during the mid-1700s.

Around the same period, sugar in beetroot was discovered in Prussia, however, it was not exploited at that time because cane sugar price was still very low. Later in 1795, Achard, a Berliner chemist published his research on the complete exploitation of

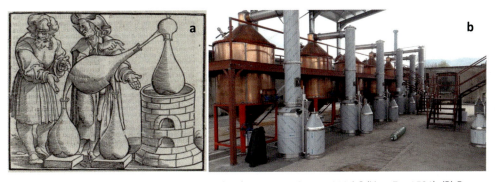

Fig. 10.2 Essential oils extraction units. (A) Jābir ibn Hayyan, Persia, 800 AC (Hayyān, 1531). (B) Recent construction of a rose damascene oil distillery plant, Slatina, Bulgary. *(Used with permission of Adlermech Co.)*

beetroot including sugar extraction, animal feed production, fermentation to ethanol and vinegar, manufacturing of other derivatives, and even manure harnessing (Marshall, 1862). Then, in the first decade of 1800s during Napoleonic wars, the shortage of cane sugar imports boosted the implementation of sugar beet extraction plants in France, based upon Achard's "biorefinery" ideas. This process rapidly spread through Europe and even in north America. The shortage of sugar also led to the development of the acid hydrolysis process of starches by Kirchoff, originating a huge industry based upon corn, wheat and potatoes (James, Hough, & Khan, 1989). As a consequence, sucrose and starches biorefineries fostered since, and some of the current top players use the crop-based biorefinery model, including Archer Daniels Midland (ADM), Royal DSM, Roquette, Cargill, Ingredion, Abengoa Bioenergy Corp., and Pacific Ethanol, among others.

10.2.4 Oleochemical biorefinery

In a later stage, in the early 1800s, the oleochemical industry was born in France. In 1813 Michel E. Chevreul elucidated fatty acids nature and together with Joseph L. Gay-Lussac developed a method for fatty acids separation and their use in candles manufacture. For this reason, 1825 is considered to be the beginning of oleochemical industry (GERLI, 2020). In the subsequent years, Procter and Gamble (P&G), and Thomas Emery & Sons (today's Emery Oleochemicals), two of the most important existing lipid-based companies were established. Initially using tallow for making soaps, candles and lamp oil, this industry turned into the current multibillion-dollar business that provides biobased commodities, fine chemicals, biofuels, and specialties for a large variety of applications. In addition to the above, some of the current largest biorefineries rely on the use of oleochemical feedstock, including some plants of Neste Oil OYJ, Renewable Energy Group Inc., and UOP LLC. These companies now use first (e.g., virgin) and second-generation (e.g., nonedible and waste) lipid feedstocks.

10.2.5 Starting the 1900s

After the first boost of the industrial revolution, commercial production of biobased chemicals from agricultural raw materials was fostered soon after First World War (Finlay, 2008). Despite the idea not being novel, the concept was recoined under the catching name "Farm Chemurgic" or "Chemurgy" (Hale, 1934). During this period, a large variety of applications were developed including oleochemical lubricants and fuels (i.e., biodiesel), sucrose and cellulose-based solvents (e.g., acetone, butanol, furfural), fiber and coatings from crop meals, olefins from alcohol, and even an alcohol-gasoline blended fuel known as "agrol". Surprisingly, one of the first legislations for a mandatory 10% blend of grain alcohol appeared in Iowa in 1933. A tipping point of the efforts to establish a renewable-based chemical industry came in 1942 during the Second World War. Due to the conflict in the Pacific, there was a shortage of natural rubber for automotive and airplanes, and the US government decided to develop a synthetic substitute. Despite ethanol conversion to butadiene enabled the production of biobased synthetic polybutadiene rubber in large quantities (i.e., nearly 360 kt in 1944), in the long-run petrochemical butadiene triumphed and even the industrial ethanol production was shifted from grains and molasses to the fossil feedstock (Finlay, 2008).

In the following years, the fossil-based feedstocks eventually took over the entire chemical sector, and postwar actions in the biobased field shifted from finding direct substitutes for petrochemicals in large markets, to obtaining valuable products from the available agricultural goods, but for smaller markets. An example of this model was the industrial development of soybean derivatives around mid-1900s, including lecithin, alkyd resins, proteins, epoxides, and many other valuable derivatives for niche applications. This also applied for the development of fermentation processes for certain carboxylic acids (e.g., citric, lactic, gluconic), antibiotics, hormones and enzymes to be used as ingredients in the food/feed and pharmaceutical industries. As a consequence of this transition, until recent decades most biobased chemicals and materials were obtained in single feedstock processing sites which were highly vulnerable to supply instabilities and market dynamics. This resulted in that only few of the biobased chemical businesses that appear during this period remained commercially viable and endured.

10.2.6 Current status

Nowadays, confronting the cascaded and cumulated environmental, economic and social impacts from the extended use of fossil resources, the chemical industry is urgently embracing sustainability as roadmap for future developments, and is coming back to the origins, to a broader use of biobased raw materials. Introducing more sustainable products into the market, dematerializing production, using renewable feedstocks, managing carbon footprint, reducing water use, and enhancing energy management constitutes a business opportunity and a social imperative for the current and future chemical companies (Dominguez de María, 2016). This has led to the implementation of integrated biorefineries, where diversified biobased feedstocks are transformed into a variety of products including food, feed, biofuels, biobased chemicals, materials, and energy. This complex model is certainly a challenge from a manufacturing standpoint but it helps stabilizing business performance, ensuring long-term sustainability, and becoming a blueprint of a future global biobased and circular economy. Figs. 10.3 and 10.4 present some of the current global leaders (as 2016) in the production of biofuels and biobased chemicals (Lokare, 2016).

The change toward the biorefinery model in the chemical industry has come with a variety of challenges that arise from working with biomass instead of fossil-based feedstocks. Besides the agricultural, environmental, social, logistics, and supply issues, there are many challenges from the processing standpoint. These include higher feedstock costs

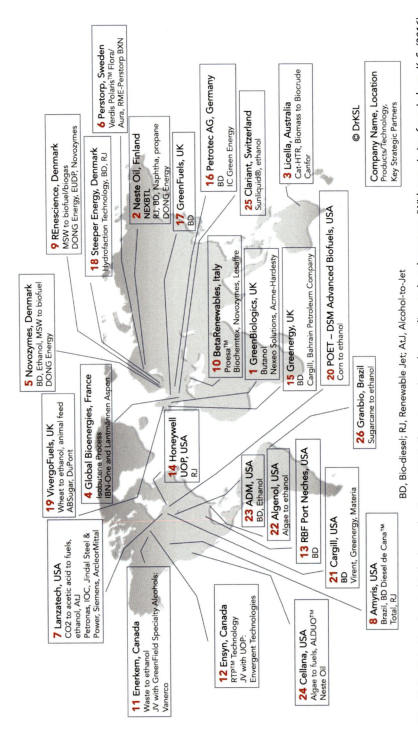

Fig. 10.3 Global leaders in biofuels production as 2016. Country assigned according to headquarters. (With permission from Lokare, K. S. (2016). 50 Hot biobased companies, 50 quick takes (Online). Available at: https://www.linkedin.com/pulse/50-hot-bio-based-companies-quick-takes-kapil-shyam-lokare-phd-1/?trk=hp-feed-article-title-publish (Accessed 04.03.2021).)

BD, Bio-diesel; RJ, Renewable Jet; AtJ, Alcohol-to-Jet

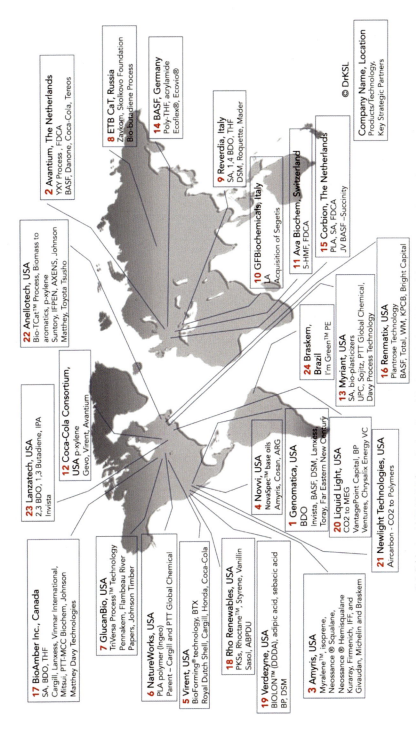

Fig. 10.4 Global leaders in biobased chemicals and biomaterials production as 2016. Country assigned according to headquarters. (With permission from Lokare, K. S. (2016). 50 Hot biobased companies, 50 quick takes (Online). Available at: https://www.linkedin.com/pulse/50-hot-bio-based-companies-quick-takes-kapil-shyam-lokare-phd-1/?trk=hp-feed-article-title-publish (Accessed 04.03.2021).)

PLA, Polylactic acid; FDCA, 2,5-furan dicarboxylic acid; SA, Succinic Acid; BDO, butanediol; BTX, Benzene, toluene, xylenes; 5-HMF, 5-Hydroxymethylfurfural; LA, Levulinic acid; IPA, Isopropyl alcohol; MEG, Ethylene glycol (per usable carbon atom), larger physicochemical heterogeneity, recalcitrance, higher chemical functionality, lower energy density, and high-water content, among many others. This forced the lead companies of the chemical business to develop more specialized knowledge and novel technologies for the biological, chemical, thermal and catalytic transformation of biomass, as well as for the upstream and downstream processing. This has been done by mean of internal research and development, but more often by undertaking joint-ventures, strategic partnerships, and alliances with existing or new players, in order to achieve leading positions in shorter times. In part, this explains the observed vertical disintegration of some key companies of the global chemical industry, and the increasing horizontal integration in specialized niche markets, as observed in Fig. 10.5. This is also a consequence of the need to focus on more profitable segments to avoid competing with growing Asian companies, and to disengage from a fossil-based past in terms of public perception.

10.3 Biorefineries and biobased chemicals market

As a result of the recent boost of the biobased chemical sector, a large number of biorefineries have been developed in the world, mostly in Europe (Fig. 10.6) and USA (Fig. 10.7) (Kostova, 2017; Parisi, 2020; U.S. Department of Agriculture, 2021). These have focused on the production of a variety of biofuels, chemicals, and materials, and by 2020, the global production of these derivatives was estimated in $400 billion (BIO, 2017). The biofuels share corresponds to nearly 35% of this market (GVR, 2020), and liquid fuels

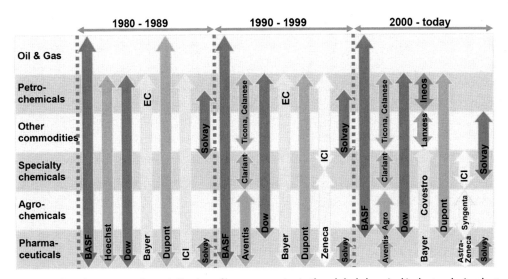

Fig. 10.5 Development and specialization of key companies in the global chemical industry during last decades. *(With permission from Fabri, J., Thünker, W. (2016). Innovative approach for the development of established refinery—Chemical cluster structures in Europe. OGEW/DGMK conference, Vienna.)*

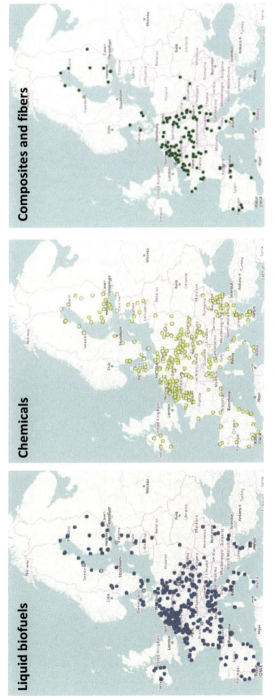

Fig. 10.6 Geographical distribution of current biorefineries (~1000 sites) producing biobased chemicals, liquid biofuels, and/or composites and fibers in Europe (Parisi, 2020).

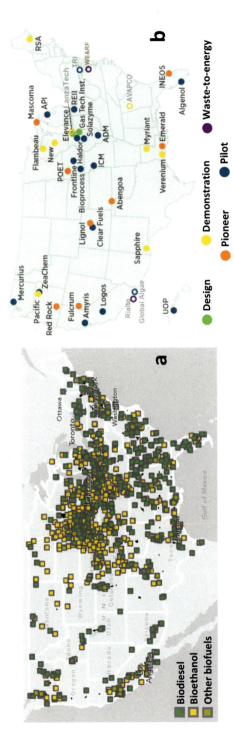

Fig. 10.7 Distribution of biorefineries in USA. (A) biofuels (~340 operating sites) (U.S. Department of Agriculture, 2021). (B) integrated biorefineries (~42 sites) (Kostova, 2017).

such as ethanol, biodiesel, hydrotreated vegetable oil and methyl esters (i.e., HVO and HFME), Fischer-Tropsch liquids, and others represent the larger fraction (74%). Biobased gases and solids represents 20% and 6% of the biofuels market, respectively. In the case of the biobased chemicals, this segment represents nearly 18% of the bioderived products market (MRF, 2019). The market is dominated by biorefineries and companies in Europe, Asia-pacific and USA, and the main product segments correspond to alcohols (41%), acids (23%), plastics (11%), lubricants (6%), surfactants (4%), solvents (4%), and others (11%) (MRF, 2019).

Remarkably, the biobased market is concentrated in a reduced number of molecules and materials. This considering the more than 400,000 kinds of molecules synthesized in living cells which could be used for different value-added products (National Academies of the Sciences, Engineering, and Medicine, 2020), and the numerous chemical routes for their further transformation. A recent work presented a comprehensive inventory and a visual map of currently recognized metabolic pathways to obtain thousands of valuable chemicals of current or potential industrial use (Lee et al., 2019). Nevertheless, the identification and selection of the biobased derivatives that are suitable for industrial exploitation from such a large inventory requires the consideration of economic, technical, market, and sustainability factors, among others (Reid, van Loon, Tollin, & Nieuwenhuizen, 2016).

The industrial biobased derivatives that have achieved current commercial stage are of different chemical nature and are used in a variety of applications. Fig. 10.8 presents a simplified map of main biomass sources and applications in current industrial biorefineries. As observed, cellulosic materials (i.e., wood, straw, waste) constitute the largest mass flows within the biobased industry. Lipids, starches, and sugars also represent a large fraction of the renewable materials exploited by the chemical industry. This feedstock distribution explains the current classification of the main three types of biorefineries: Lignocellulosic Biorefinery (LCB), Triglyceride Biorefinery (TGB), and Sugar and Starchy Biorefinery (SSB).

10.4 Key biobased fuels and chemicals for current industrial biorefineries

As above-mentioned, the production of biofuels has been carried for more than a century and products such as bioethanol, biomethanol, biogas, biochar, biodiesel, hydrocarbons from biomass gasification, liquefaction or hydrocracking, and even biomass itself, are already in a commercial stage. Typically, once technical and economic feasibility are assured, biofuels are assessed based upon heating values and the suitability for effective implementation in the existing infrastructures (e.g., combustion engines, fire boilers). However, in the case of the biobased chemicals, there is need to consider a variety of factors depending on the final applications. If the product is meant to be used as direct

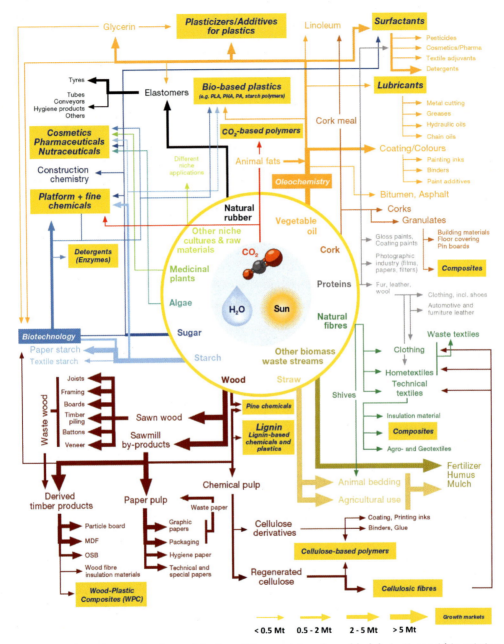

Fig. 10.8 Industrial routes for the exploitation of biomass into value-added derivatives within existing biorefineries. Arrow thickness represents relative market sizes. *(With permission from Nova Institute (2015). Industrial material use of biomass in Europe 2015 (Online). Available at: http://bio-based.eu/downloads/industrial-material-use-biomass-europe-2015-poster/ (Accessed 11.03.2021).)*

substitute, it must fulfill established technical specifications of the commercial product (including a certain price range). This might require the use of suitable processing technologies to reach such specifications including those related to the fulfillment of a certain sensory profile (e.g., color, appearance). On the other hand, if the product is meant to be used as a functional substitute, the chemical nature would not be a decisive factor, but instead an effective performance in the final application must be assured. In general, this turns the focus on obtaining a suitable compound with the reactivity, affinity, safety, solubility, or any other performance property that is required for the final application. In this case, the ability of fulfilling such requirements confers the biobased chemicals, a higher added value, generally becoming more profitable than biofuels. Examples of how biobased chemicals are employed at the industrial scale are described in Table 10.2 (Holladay, Bozell, White, & Johnson, 2007).

A systematic identification of key value-added derivatives to be produced in modern biorefineries started quite recently (Holladay et al., 2007; Werpy & Petersen, 2004). As a result, a reduced set of biobased chemicals were identified as potential platforms or building blocks for a variety of commercial products, thus enabling a similar production model to that used in petrochemical refineries. These were mainly selected based upon technological maturity and economic viability, and classified according to the biomass source and the functional groups in their molecular structures. Later in 2010, new key chemicals were shortlisted based upon additional criteria including industrial driving forces (Bozell & Petersen, 2010).

A subsequent comprehensive study enabled to identify a larger list of target biobased building blocks to be researched and industrialized, and they were classified according to the number of carbon atoms in their structure (de Jong, Higson, Walsh, & Wellisch, 2011). At this point, large chemical and pharmaceutical companies such as AkzoNobel, Dupont, Arkema, Evonik (Dominguez de María, 2016) and GSK (Alder et al., 2016) did a similar work, and based upon sustainability criteria under a life cycle approach, they

Table 10.2 Industrial applications of biobased chemicals.

Uses	Example
Direct substitute of existing fossil counterpart	Biobased butanol replacing that obtained from petrochemical propylene by hydroformylation (Morales, Frauenlob, Franke, & Börner, 2015)
Functional substitute	Natural polyols (e.g., sugars, fatty polyols) that can replace fossil-based polyols (e.g., polyglycols) in resins and polymers
Drop-in material	Fatty acid methyl esters for fuels and lubricants, triglycerides from vegetable oils as rubber and resins modifiers
Feedstock for chemical conversions	Acetic acid for synthesis of acetate solvents and fragrances, ethanol for ethers and biobased olefins, platform chemicals
Novel formulation or material	Natural fibers in composite materials, novel polyesters derived from 2,5-furandicarboxylic acid (FDCA) (Sousa et al., 2015)

identified their own target biobased products to be industrialized. Considering the evolution of markets and economy, recent works revised the current status of such key products, and tried to redefine additional targets (de Jong, Stichnothe, Bell, & Jørgensen, 2020; Rosales-Calderon & Arantes, 2019). Tables 10.3 and 10.4 summarize the historical evolution of the key biobased chemical platforms, biofuels, and biopolymers of industrial interest during the last two decades. As observed, the portfolio of biobased chemicals has grown extensively as well as the number of industrial players.

It is of particular interest the increasing participation of biobased monomers and polymers in the global plastics market (Fig. 10.9). Recent estimates indicate that the global production of biobased polymers already represent 1% of the total polymers market (Skoczinski et al., 2021). While this is still a small participation, it is growing at a faster pace than the fossil-based materials. Considering the large market of plastics (~380 Mt/yr), the increasing participation of biobased materials will have a huge impact in the future demand of biogenic products. Also, there will be need for developing novel and enhanced technologies to reduce costs of biomonomers and polymers in future biorefineries.

10.5 Potential for process intensification at the biorefineries

Similar to the traditional petrochemical refineries, and as observed in Fig. 10.10, the biorefineries consist of a complex network of processes of very different nature. Depending on the available feedstock and the target products, there is need for a variety of processing stages including: physical pretreatment (e.g., washing, grinding, sterilizing, swelling), extractions (e.g., mechanical, solvent, steam), chemical pretreatment (e.g., de-lignification), biopolymers break-up (e.g., cellulose/starch hydrolysis, lignin depolymerization), main and secondary transformations (i.e., thermal, chemical, catalytic, fermentative), a multiplicity of downstream purification and refining processes, and finally, a whole set of waste reclaiming and treatment processes. In this regard, biorefineries are suitable processing facilities for the implementation of intensified technologies. This is especially true because, in addition to the social, environmental, agricultural, political, and supply chain issues, the transformation of biomass involves a variety of technical challenges that jeopardizes economic feasibility and long-term sustainability. Table 10.5 summarizes some of the major technical challenges when using biobased feedstocks for chemical processes, and some potential intensification principles and applications to overcome them.

As observed in Fig. 10.10, there are numerous opportunities to implement process intensification technologies at the different stages of the biorefinery. Nevertheless, from a business standpoint and considering the limited financial resources, there is need to identify those stages where a more sustainable performance could be achieved. For instance, from a life cycle perspective, Akzonobel has recently observed that less than one-third of its environmental burdens are generated within their chemical

Table 10.3 Historical evolution of the key biobased platform chemicals for existing and future biorefineries (Alder et al., 2016; Biddy, Scarlata, & Kinchin, 2016; Bozell & Petersen, 2010; de Jong et al., 2011; de Jong et al., 2020; Holladay et al., 2007; Rosales-Calderon & Arantes, 2019; Werpy & Petersen, 2004).

# carbons	Compound	2004	2010	2011	2016	2019	2020	Industrial Potential as 2020	Capacity kton/yr	Companies
C1	Methane							Growth		Many
	Methanol							Growth	200	OCI (BioMCN), Sodra, Carbon Recycling International, W2C*
	Formaldehyde							Pipeline		BASF
	Syngas							Growth	760	Many
	Formic acid							Pipeline		Avantium
	Carbon dioxide							Pipeline		Climeworks
C2	Ethylene							Growth	200	Braskem
	Ethanol							Growth	86	Many
	Ethylene oxide							Growth	40	Croda, Biokim
	Ethylene glycol							Growth	175	India Glycols Ltd, HaldorTopsoe, UPM, Avantium, ENI/Versali
	Acetaldehyde							Growth		Jubilant
	Acetic acid							Growth	24.5	Sekab, Wacker, Godovari Biorefineries Ltd, Zeachem
C3	Glycolic acid							Pipeline		Metabolic Explorer (Metex)
	Oxalic acid							Pipeline		Avantium
	Propane							Growth	40	Neste/SHV
	Propylene							Pipeline		Braskem/Toyota Tsusho, Mitsubishi Chemical, Mitsui Chemicals
	Propanol							Pipeline		Braskem
	Isopropanol							Pipeline		Genomatica, Mitsui Chemicals
	Propylene glycol							Growth	120	ADM, Oleon, Avantium
	Acetone							Pipeline		Green Biologics, Celtic Renewables
	1,3-Propanediol							Growth	77	DuPont/Tate & Lyle, Glory Biomaterial, Shenghong Group
	Glycerol							Growth	1500	Many
	Epichlorohydrin							Growth	540	Yihai Kerry Group, Jiangsu Yangnong, Advance Biochemical Thailand
	Lactic acid							Growth	> 600	Corbion, NatureWorks, Galactic, Henan Jindan, BBCA
	Acrylic acid							Pipeline		Cargill/Novozymes, ADM/LC Chemicals, Perstorp, Arkema
	3-Hydroxy propionic acid							Pipeline		Cargill/Novozymes
	Malonic acid							Pipeline		Sirrus, Lygos
C4	Butanol							Pipeline	10	Green Biologics, Celtic Renewables
	Isobutanol							Growth	6-9	Butamax, Gevo
	Isobutene							Pipeline		Global Bioenergies
	1,4 Butanediol							Growth	30	Genomatica, Novamont, Dupont/Tate & Lyle, Godovari Biorefineries Ltd
	1,2 Butanediol									
	2,3 Butanediol							Pipeline		Intrexion
	Tetrahydrofuran							Growth		Novamont
	Ethyl acetate							Pipeline	36	Sekab(JRC), Zeachem, Greenyug
	Butyric acid							Pipeline		Metex, Kemin, Blue Marble Biomaterials
	Acetic anhydride							Growth		ZeaChem Inc.
	Crotonaldehyde							Growth		Godovari Biorefineries Ltd,
	Fumaric acid							Pipeline		Myriant
	Malic acid							Pipeline		Myriant
	Aspartic acid							Growth		Royal DSM, Ajinomoto Co. Inc.
	3-Hydroxybutyrolactone							Pipeline		Myriant
	Succinic acid							Growth	34	Myriant, Succinity (BASF /Corbion), Reverdia (Roquette)
C5	Isoprene/farnesene							Pipeline		Goodyear/ Genencor, GlycosBio, Amyris
	1,5-Pentanediamine							Growth	50	Cathay Industrial Biotech, CJ Cheiljedang
	Methyl methacrylate							Pipeline		Lucite/Mitsubishi Rayon, Evonik/Arkema
	Ethyl lactate							Growth		Corbion, Vertec BioSolvents
	Levulinic acid							Pipeline	2-3	Avantium, GFBiochemicals, Circa Group
	Xylitol							Growth	190	a.o. Danisco/Lenzing, Fortress
	Methylvinyl glycolate							Pipeline		Haldor Topsoe
	Furfural							Growth	360	Many
	Furfuryl alcohol							Growth		TransFurans Chemicals bvba
	Itaconic acid							Growth	90	Qingdao kehai, Zhejiang Guoguang, Jinan Huaming
	Glutamic acid							Growth		Global Biotech, Meihua, Fufeng, Juhua
C6	Caprolactam							Pipeline		Genomatica / Aquafil
	Lysine							Growth	1100	Global Biotech, Evonik/RusBiotech, BBCA, Draths, Ajinomoto
	Aniline							Pipeline		Covestro
	Sorbitol							Growth	1800	Roquette, Cargill, ADM, Ingredion
	Adipic acid							Pipeline		Genomatica
	Isosorbide							Growth	20	Roquette
	Cyrene							Pipeline		Circa Group
	2,5 Furan dicarboxylic acid							Pipeline		Avantium, ADM/Dupont, Corbion, Stora-Enso, Annikki
	Hydroxymethyl furfural							Growth	0.3	AVA Biochem
	Ethyl terbutyl ether							Growth		Global Bioenergies
	Glucaric acid							Pipeline		Rivertop renewables
	Citric acid							Growth	2000	Cargill, DSM, BBCA, Ensign, TTCA, RZBC
C7	Pentamethylene diisocyanate (PDI)							Growth		Covestro (70% biobased content), Mitsubishi Chemical
C8	p-Xylene							Pipeline		Annellotech, Origin Materials, BioBTX, Tesoro
	Terephthalic acid							Pipeline		UOP, Annellotech
C9	Pelargonic acid							Growth	25	Matrica (Novamont/Versalis JV)
	Azelaic acid							Growth	25	Matrica (Novamont/VersalisJV), Emery Oleochemicals
C10	Sebacic Acid							Growth	200	A.o. Arkema (Casda Biomaterials)
C11	Undecanedioic acid (UDDA)							Growth	24	Arkema
C12	12-Aminododecanoic acid							Growth	25	Evonik
	Dodecane-dioic Acid (DDDA)							Growth		Cathay Industrial Biotech
Cn	Biohydrocarbons							Pipeline		Amyris
	Terpenes							Pipeline		ADM, P2 Science

Table 10.4 Key industrialized biofuels and biopolymers (Brown et al., 2020; E4tech, RE-CORD, and WUR, 2015; IEA, 2020; IRENA, 2016, 2020; Skoczinski et al., 2021).

Biofuels	Capacity (Mton/yr)	Biopolymers	Capacity[a] (kton/yr)
Ethanol	86	Aliphatic polycarbonates (APC)	25.6
Fatty acid methyl esters (biodiesel)	37.8	Cellulose acetate (CA)	932.7
Hydrogenated vegetable oil (HVO)	7.85	Epoxy resins	1201.1
Hydrogenated fatty acid methyl esters (HFME)	–	Ethylene propylene diene monomer rubber (EPDM)	–
Methanol	0.2	Poly(butylene adipate-*co*-terephthalate) (PBAT)	268.3
Fischer-Tropsch liquids	–	Polyamides (PA)	236.4
Biogas[b]	104.6	Polybutylene succinate (PBS) and copolymers	89.4
Bio-oil (pyrolysis)	–	Polyethylene (PE)	217.2
Biobutanol	0.01	Polyethylene furanoate (PEF)	–
Hydrocarbons from alcohol	–	Polyethylene terephthalate (PET)	140.6
Dimethyl ether	–	Polyhydroxyalkanoates (PHA)	38.3
		Polylactic acid (PLA)	421.7
		Polypropylene (PP)	51.1
		Polytrimethylene terephthalate (PTT)	178.9
		Polyurethanes (PUR)	332.2
		Starch blends	396.1
		Casein polymers	12.8

[a] Estimated from plots (Skoczinski et al., 2021).
[b] Estimated with a lower heating value of 14 GJ/ton, and a current production of 35 Mtoe.

transformation facilities; instead, most of the impacts are embedded within supplies (Reid et al., 2016). In this case, working together with suppliers to improve their processes, or even changing them, would probably be a more cost-effective solution than implementing any intensified technology within the existing plants. Nevertheless, if process intensification is the selected approach, then the questions of when, where, how and to which extent to apply this method becomes a fundamental issue. This because the complexity arising from highly integrated, intensified, continuous or miniaturized processes only enables their implementation to very specific situations or within a very narrow window of operating conditions. Yet, when the implementation is possible, performance enhancement by intensification (i.e., conversion, throughput, footprints, safety, reliability, energy and material intensity, costs, etc.) could be groundbreaking.

Once process intensification is embraced, the main drivers for its industrial implementation may vary from one company to another. These drivers are not necessarily

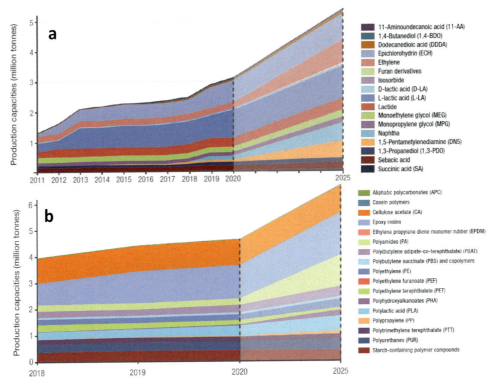

Fig. 10.9 Evolution of worldwide production capacity of (A) biomonomers and (B) biopolymers. *(With permission from Skoczinski, P., Carus, M., de Guzman, D., Käb, H., Chinthapalli, R., Ravenstijn, J., Baltus, W., & Raschka, A. (2021). Bio-based building blocks and polymers – Global capacities, production and trends 2020–2025. (Online) Available at: http://bio-based.eu/downloads/bio-based-building-blocks-and-polymers-global-capacities-production-and-trends-2020-2025/ (Accessed 11.03.2021).)*

related to the technical aspects of the process itself, but perhaps with a market or business context. For instance, from a business perspective, the improvements might be better focused at the logistics and supply chain optimization; in this case production in decentralized plants using modular devices might be the intensification pathway. Regarding the market driving forces, and as historical example, the boost of methyl terbutyl ether (MTBE) production via reactive distillation in the late 1970s and early 1980s, was a result of a large market demand after phasing out leaded additives for gasoline. In this case, the additional benefits of the intensification included reducing capital and operating costs of the new facilities that were supposed to be built anyways. Other driving forces might include enhancing environmental performance (e.g., carbon capture technologies, water purification), exploiting alternative feedstocks (e.g., gas-to-chemicals, changing the biobased feedstock), easy of applying, and even hazards reduction benefits (i.e., control of exothermic reactions, reduction of use of toxic chemicals).

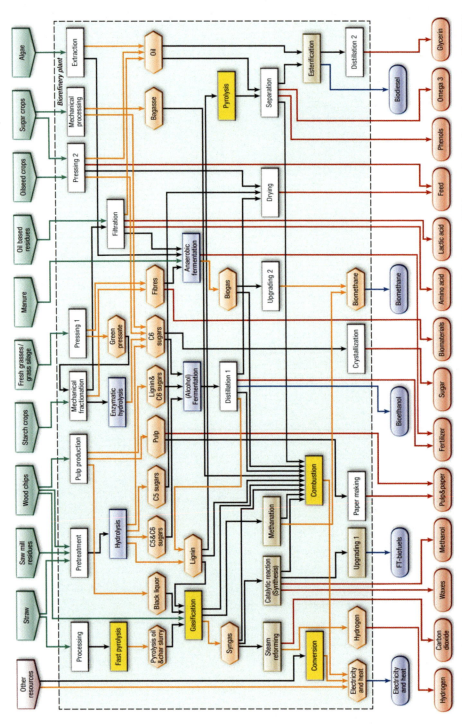

Fig. 10.10 Biorefinery networks for an industrial scale bioeconomy, some processes, and products. (With permission from Jungmeier, G. (2014). Approach for the integration of biorefineries in the existing industrial infrastructures. Workshop @i-SUP2014, Antwerp.)

Table 10.5 Some major technical challenges when using biomass as feedstock for biobased derivatives and some potential intensification approaches to overcome them.

Challenges	Potential impacts	Potential intensification approaches
Complex matrix of chemicals in feedstock and products	Side reactions, catalysts deactivation, microorganisms inhibition	More selective and resilient catalysts and microorganisms, selective separations
High chemical functionality	Thermal, chemical, and biological instability	Short residence time processing, improve molecular experiences
High O/C ratio	Low energy density and reduced yields	Preserve oxygen-rich compounds and enhance selectivity
High water content	High transport costs, energy intensive drying required, low energy density, waste generation	Wet processing, enhanced drying, heat integration, energy-efficient separations
Hydrophilic molecules	Commercial catalysts could be unsuitable	Enhanced accessibility materials, improve molecular experiences
Higher prices	Higher product prices than fossil-based counterpart	Maximize the effectiveness of intra- and intermolecular events, exploit synergies
Regional and seasonal supply instabilities	Sub utilized equipment, idle times	Resilient technologies for variable feedstocks and throughputs, modular equipment
Low bulk density	Lower productivity ($kg\,h^{-1}\,m^{-3}$) compared with fossil feedstock, high transport costs, energy intensity	Mechanical and thermal densification processes, decentralized modular equipment

In recent years, the RAPID Manufacturing Institute for Process Intensification of the AIChE has defined certain criteria and metrics to identify the potential for industrialization of a given intensified processes (AICHE, 2021; Bielenberg & Palou-Rivera, 2019). These are described in Table 10.6, and they overlap with the green chemistry and green engineering principles, and with the roadmap of process intensification established by the European Process Intensification Center (EUROPIC) in 2007 (Creative Energy, 2007). These criteria could be used to screen, rank and select among different intensification approaches for a certain type of process or product.

Table 10.6 Metrics and some proposed criteria to screen potential intensified processes with potential for industrial implementation (AICHE, 2021; Bielenberg & Palou-Rivera, 2019).

Metrics	Criteria[a]
Energy efficiency	Increase by 20%
Energy productivity (kg product/kJ)	double
Capital costs of intensified modules ($/yr)	Reduce 10×
Emissions/waste (kg/kg product)	Reduce by 20%
Manufacturing costs ($/kg product)	Reduce by 20%
Deployments costs ($)	Reduce by 50%

[a]With respect to commercial state of the art technology.

10.6 Implemented intensification technologies for bioderived chemicals

A recent comprehensive work described more than 100 intensification technologies that could be implemented in a variety of processes (Stankiewicz et al., 2019). These have been developed to exploit intensification principles within the spacial (Structure), functional (Synergy), thermodynamic (Energy) and temporal (Time) domains. From this group, nearly 50 technologies were initially recommended for exploration in the European Road Map of process intensification during the last decade (Creative Energy, 2007). As a result of recent studies and validations, nearly 30 of them are considered to have medium or high energy savings potential to be implemented at the productive scale (DOE, 2015), and they can provide a paramount contribution in the production of a variety of bioderived products. This is envisioned considering the increasing ecological burdens of exploiting renewable resources to match the growing demand for biobased chemicals, biofuels, and biopolymers. Also, there is need to deal with the food-water-energy security nexus involved in the use of biomass as industrial feedstock (Kamm, Gruber, & Kamm, 2010; Sanders et al., 2012).

Trying to do a comprehensive inventory of the different process intensification technologies that have been studied or implemented at the industrial scale for the production of biogenic fuels and chemicals would be very cumbersome. In addition to the large portfolio of intensification techniques and the variety of biobased commercial products, there is a large number of physical, thermal, chemical, and biomediated processes within the biorefineries where process intensification has been assessed. Also, the different intensified techniques could be combined to create an extended set of hybrid approaches (e.g., ultrasonic/microwave-assisted reactive distillation with enzyme-coated structured packings, Kiss, Jobson, & Gao, 2019). In this regard, far from being comprehensive, Table 10.7 presents a list of some of the intensification technologies that have been reported in the processing of different biobased materials, some of which have been implemented at the industrial scale. These include different synthesis routes and various stages of the corresponding processes, and rather than focusing only on specific biobased chemicals, they are classified according to the chemical functionality or final use.

Table 10.7 Some process intensification (PI) technologies used in the transformation and production of biobased fuels, chemicals, and polymers.

Type	Product	Stage or process	Process challenge	PI method	Status[a] Exp.	Status[a] Ind.	Remarks	Ref.
Biofuels	Ethanol	Saccharification	Multiple steps	Simultaneous reactions	X		Fluidized bed reactor	(Cardona & Sánchez, 2007)
		Fermentation	Multiple steps	Integrated extractions		X	Re-use of alkaline lignin solution possible	(Ranjan, Singh, Malani, & Moholkar, 2016; Antunes et al., 2018; Kumar, Ravikumar, Thenmoxhi, & Sankar, 2019; Yoon et al., 2019; Ayodele, Alsaffar, & Mustapa, 2020)
				Reactive separations	X		Solvent inhibition	
				Integrated reactions (hydrolysis-saccharification-fermentation)	X		Larger residence times	
		Separation	Azeotropic separation	Heat-integrated distillation		X	High energy savings	Sharma et al., 2020; Schubert, 2020; Gonzalez-Contreras, Lugo-Mendez, Sales-Cruz, & Lopez-Arenas, 2020)
				Extractive distillation		X	High purity ethanol	
				Swing adsorption		X	Complex control	
				Pervaporation		X	Lower energy consumption	
	Biodiesel	Transesterification/ esterification	Equilibrium reaction, catalyst deactivation, LL separation	Jet-stirred reactor (JSR)	X		>90% yield in first 5 min	(Bhatia et al., 2021; Kiss, 2019; Oh, Lau, Chen, Chong, & Choo, 2012; Tabatabaei et al., 2019; Yin et al., 2017)
				Oscillatory baffled reactor (OBR)	X		Very low methanol-to-oil ratio	
				Spinning reactors		X	Very low residence times (seconds)	
				Membrane reactor	X		Fouling still a problem	
				Reactive distillation		X	Esterification of fatty acids	
				Annular centrifugal contactor (ACC)		X	Extremely low residence time (up to 10s)	

Continued

Table 10.7 Some process intensification (PI) technologies used in the transformation and production of biobased fuels, chemicals, and polymers—cont'd

Type	Product	Stage or process	Process challenge	PI method	Status Exp.	Status Ind.	Remarks	Ref.
				Ultrasonic reactor		X	High yield but high energy consumption	
				Hydrodynamic cavitation reactor		X	Cost-effective and simple	
				Shockwave power reactor		X	Cost-effective and simple	
				Static mixer	X		Lower energy consumption	
				Microchannel reactor	X		Very small liquid capacity	
				Microwave reactor	X		High energy consumption	
				Dual-frequency counter-current pulsed ultrasound reactor	X		Dual-frequency better than single-frequency	
				Reactive absorption		X	Esterification of fatty acids	
				Supercritical reaction	X		High capital costs, no catalyst required	
				High-frequency magnetic impulse Cavitation reactor	X		Very low methanol-to-oil ratio and low catalyst loading	
	Biogas	Anaerobic biodigestion	Flocs formation	Ultrasound	X		24% more production	(Koniuszewska, Korzeniewska, Harnisz, & Czatzkowska, 2020)
			H_2S, CO_2, H_2O removal	Membrane		X	multiple steps for high recoveries	(Chen, Vinh-Thang, Avalos, Rodrigue, & Kaliaguine, 2015)
				Swing absorption		X	High capital and operating, complex control	
				Reactive adsorption		X	High capital and operating	

Table 10.7 Some process intensification (PI) technologies used in the transformation and production of biobased fuels, chemicals, and polymers—cont'd

Type	Product	Stage or process	Process challenge	PI method	Status Exp.	Status Ind.	Remarks	Ref.
	HVO	Hydrotreatment	Heat transfer issues	Microchannel reactor	X		Heavy and oligomerized hydrocarbons are the major (>95%) products	(Sinha & Rai, 2018)
	Methanol	Biobased syngas	Equilibrium reaction, hydrogen and CO_2 recovery	Periodic reactive adsorption	X		in situ water removal, 7% improvement in methanol yield	(Arora, Iyer, Bajaj, & Hasan, 2018)
				Membrane reactor	X		High yields, but far from being scaled up	(Dalena et al., 2018; Gallucci, 2018)
				Reactive adsorption		X	CO_2 amine absorption recovery	(Jaggai et al., 2020)
				Microchannel reactor	X		Equivalent conversions with packed reactor operating at GHSV seven times slower	(Hu et al., 2007)
				Partial oxidation gas turbine cycle (POGT)	X		12% gain in thermal efficiency, cost decrease of 7%	(Cornelissen, Tober, Kok, & van de Meer, 2006)
	FT Liquids	Fischer–Tropsch	Exothermic process, catalyst deactivation	Microchannel reactor	X		Best catalyst utilization and productivity	(Rauch, Kiennemann, & Sauciuc, 2013)
	BioOil	Fast pyrolysis	Heat transfer limitations	Autothermal/ Oxidative Pyrolysis	X		25% decrease in char	(Polin, Peterson, Whilmer, Smith, & Brown, 2019)
				Rotating cone reactor		X	No need of heat carrying gas	(Harmsen & Verkerk, 2020)

Continued

Table 10.7 Some process intensification (PI) technologies used in the transformation and production of biobased fuels, chemicals, and polymers—cont'd

Type	Product	Stage or process	Process challenge	PI method	Status Exp.	Status Ind.	Remarks	Ref.
	Butanol/ Isobutanol	Fermentation	Diluted broths, product inhibition, complex mixture	Heat-integrated distillation		X	High energy savings	(Gonzalez-Contreras et al., 2020; Ibrahim, Kim, & Abd-Aziz, 2018; Köhler, Rühl, Blank, & Schmid, 2014; Kushwaha, Srivastava, Mishra, Upadhyay, & Mishra, 2019; Morone & Pandey, 2014; Patrașcu, Bîldea, & Kiss, 2018)
				Simultaneous reactions	X		Resilient strains are required	
				Reactive separations	X		Difficult to implement	
				Continuous and integrated processes	X		Aseptic equipment issues	
				Oscillatory baffled reactors	X		>38% yields than stirred reactors	(Masngut & Harvey, 2012)
Chemicals	Carboxylic acids	Fermentation	Diluted broths, product inhibition, complex mixture	Simulated moving beds		X	10% of world citric acid is processed	(Beyerle & Diawara, 2017; Datta, Kumar, & Uslu, 2015; Inyang & Lokhat, 2020)
				Electrodialysis		X	High yield, high purity	
				Membranes		X	For clarification and concentration	
				Reactive extraction		X	Amine extraction	
				Reactive distillation		X	Established for lactic acid recovery	
		Hydrolysis of esters	High activation energy	Microwave	X		96.6% conversion in 4h below 200°C	(Nguyen, Lee, Su, Shih, & Chien, 2020)
		Oxidation	Highly exothermic	Microreactors	X		Higher selectivity and conversion	(Hommes, Heeres, & Yue, 2019)

Table 10.7 Some process intensification (PI) technologies used in the transformation and production of biobased fuels, chemicals, and polymers—cont'd

Type	Product	Stage or process	Process challenge	PI method	Status Exp.	Status Ind.	Remarks	Ref.
	Alcohols	Fermentation	Complex and diluted mixture	Heat-integrated distillation		X	High energy savings	(Ferreira, Meirelles, & Batista, 2013; Mendoza, Sánchez-Ramírez, Segovia-Hernández, Orjuela, & Hernández, 2021; Zhang et al., 2020)
				Structured packings		X		(Schubert, 2020)
		Guerbet condensation	Low yields	Combined reactions	X		High potential for existing industrial process	(Gabriëls, Hernández, Sels, Van Der Voort, & Verberckmoes, 2015)
			Water removal	Azeotropic distillation		X	Guerbet from ethanol or butanol not yet commercial process	
	Esters	Esterification/ Transesterification	Equilibrium reaction	Reactive distillation w/out combined processes		X	High energy savings	(Kiss, 2019; Kiss et al., 2019; Li, Duan, Fang, & Li, 2019; Santaella, Orjuela, & Narváez, 2015)
				Simulated moving beds		X	Mass transfer limitations	(Russo et al., 2019)
				Ultrasound		X	Enhanced enzymatic esterification	(Bansode & Rathod, 2017)
				Microwaves	X		Shorter reaction time in the esterification of a variety of starches	(Gilet et al., 2018)
				Membranes, Pervaporation	X		Polymeric catalytically active membranes, High selectivity for water removal	(Castro-Muñoz, Galiano, & Figoli, 2019; Qing et al., 2019)

Continued

Table 10.7 Some process intensification (PI) technologies used in the transformation and production of biobased fuels, chemicals, and polymers—cont'd

Type	Product	Stage or process	Process challenge	PI method	Status Exp.	Status Ind.	Remarks	Ref.
	Ethers	Alcohol etherification, or addition to olefin	Equilibrium reaction	Reactive distillation w/out combined processes (DWC, Membrane, pervaporation)		X	Completely biobased (e.g., DME, DEE) or partially biobased (e.g., MTBE, ETBE, TAME)	(Kiss & Bildea, 2018)
				Microwaves	X		Shorter reaction time in the modification of a variety of starches	(Gilet et al., 2018)
	Polyols, Glycols	Hydration	High pressure	Microreactors	X		Lower pressure operation	(Hommes et al., 2019)
						X	Biobased ethylene or propylene oxide ring-opening reaction	(SCI, 2019)
		Hydrogenation	Exothermic reaction, selectivity issues	Reactive distillation				
			Exothermic reaction, selectivity issues	Microreactors	X		1.5–2 times faster than stirred tank slurry reactor	(Hommes et al., 2019)
	BTX	Biomass FCC	Improve selectivity to aromatics	Shape selective zeolites in fluidized bed		X	>95% into aromatic chemicals and fuels	(Sudolsky, 2019)
	Aldehydes	Oxidation	Highly exothermic	Microreactors	X		Higher selectivity by effective heat removal	(Hommes et al., 2019)
	Acetals, ketals	Acetalization/ ketalization of biobased ethanol, glycerol	Equilibrium reaction	Simulated moving beds	X		Conversions >98%, purities >90%	(Faria, Graça, & Rodrigues, 2019)
				Reactive distillation	X		High purity of specific isomers	(Agirre et al., 2012; Hong et al., 2012; Hasabnis & Mahajani, 2014)

Table 10.7 Some process intensification (PI) technologies used in the transformation and production of biobased fuels, chemicals, and polymers—cont'd

Type	Product	Stage or process	Process challenge	PI method	Status Exp.	Status Ind.	Remarks	Ref.
	Furans	Saccharides dehydration	Selectivity	Microreactors	X		High conversion of less active saccharides	(Hommes et al., 2019)
	Epoxides	Triglycerides, fatty esters epoxidation	Highly exothermic, side ring opening	Microreactors	X		Higher oxirane number in very shorter time	(Hommes et al., 2019)
	Amines	Different routes	Selectivity control	Bifunctional catalysts		X	Improved selectivity	(Froidevaux, Negrell, Caillol, Pascault, & Boutevin, 2016)
	Peptides	Protein hydrolysis	Substrate and/or product inhibition,	Enzymatic membranes	X		Higher yields	(Satyawali, Vanbroekhoven, & Dejonghe, 2017)
	Biologics	Extraction	Mass transfer limitations	Static mixers		X	Higher yields	(Strube et al., 2018)
		Fermentation	Batch to continuous	Membrane reactor		X	Higher yields in aqueous two-phase extraction	
	Natural Products	Extraction	Complex mixtures, low volatility, thermolabile	Supercritical fluids		X	High yields at low temperature	(Chemat et al., 2020; Ingkaninan, Hazekamp, Hoek, Balconi, & Verpoorte, 2000)
				Twin-screw extruder		X	High extraction, high mixing	
				Mechanochemical devices		X	Reduced time and easier separation	
				Electric fields	X		Effective nonthermal techniques.	

Continued

Table 10.7 Some process intensification (PI) technologies used in the transformation and production of biobased fuels, chemicals, and polymers—cont'd

Type	Product	Stage or process	Process challenge	PI method	Status Exp.	Status Ind.	Remarks	Ref.
				Infrared	X		Reduced energy requirements	
				Ultraviolet	X		Reduced energy requirements	
				Solar steam distillation		X	Low energy cost	
				Acoustic cavitation	X		Very effective extraction Cells disruption	
				Microwaves	X		Commercial CBD/THC separation	
				Centrifugal partition chromatography		X		
Biopolymers and materials	Biobased polymers and materials	Polymerization, separation	Nanofibers required	Electrospinning		X	Unique functional nanostructures	(Stankiewicz, 2019; Stankiewicz & Yan, 2019)
			Surface modification	Plasma		X	Effective surface treatment	
			Initiation	Light; noncoherent		X	High rates at low temperature	
			Structured materials	ultrasound	X		Effective structure tuning	
			Activate specific functionalities	Magnetic fields	X		Separation with magnetic tags	
			Local heating	Microwaves	X		Shorter reaction time	

[a]Exploratory (Exp.) or industrially (Ind.) implemented.

As instance of the benefits of the incorporation of process intensification in the biorefineries, the following section describes the broad use of such technologies in the manufacture of a specific biobased chemical building block: succinic acid.

10.7 Process intensification in action—Succinic acid production

Succinic acid is a dicarboxylic acid also known as butanedioic acid. It is mainly used as raw material in the manufacture of chemical intermediates like tetrahydrofuran (THF), 1,4-butanediol and γ-butyrolactone. In the materials and polymers segment, succinic acid is also used as precursor of polyols, plasticizers, polyesters, and polyurethanes. It has been also employed in the production of food additives, cosmetics, fragrances, herbicides, detergents, pharmaceutical intermediates, and other commodity chemicals (Zeikus, Jain, & Elankovan, 1999). Some of its main properties are listed in Table 10.8 and some derivatives and applications are presented in Fig. 10.11.

The wide spectrum of industrial applications of succinic acid, as a building block for commodity chemicals, has led to an increased demand. The global market of biobased succinic acid reached nearly USD 200 million in 2019, and it is projected to grow up to USD 423 million by 2027 (Marketwatch, 2021). This compound can be obtained by petrochemical and biochemical pathways as presented in Fig. 10.12, and both are currently used by the main industrial producers.

The petrochemical route involves the catalytic hydrogenation of maleic acid or maleic anhydride, which are derived from n-butane and n-butenes (Mazière, Prinsen, Garcia, Luque, & Len, 2017). The catalytic hydrogenation reaction takes place in liquid phase under pressures ranging from 500 to 4000 kPa and temperatures between 120°C

Table 10.8 Properties of succinic acid (National Center for Biotechnology Information, 2021).

Molecular structure	Property		Value
HO—⟨structure⟩—OH	Molecular formula		$C_4H_6O_4$
	Molecular weight g/gmol		118.09
	Density g/cm³ (25°C) – Solid		1.572
	Melting point (°C)		185–187
	Boiling point at 101.3 kPa (°C)		235 (dehydration)
	Solubility g/100 g solvent	Water (0°C)	2.8
		Water (100°C)	121
		96% Ethanol (15°C)	10
		Ethyl ether (15°C)	1.2
	pK Water (25°C)	pK_1	4.16
		pK_2	5.61
	Flammability point (°C)		206

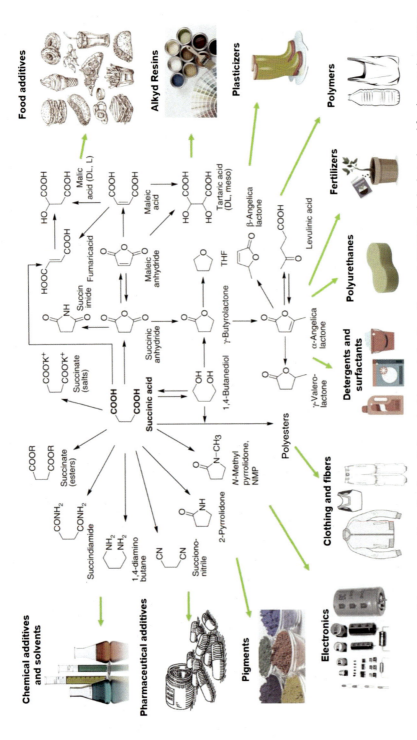

Fig. 10.11 Succinic acid derivatives and applications. (Adapted with permission from Kamm, B. (2009). Carbohydrate-based food processing wastes as biomass for biorefining of biofuels and chemicals. In: K. W. Waldron (Ed.), Handbook of waste management and co-product recovery in food processing. Boca Raton, FL: Woodhead Publishing Limited, p. 479–514. https://doi.org/10.1533/9781845697051.4.479).

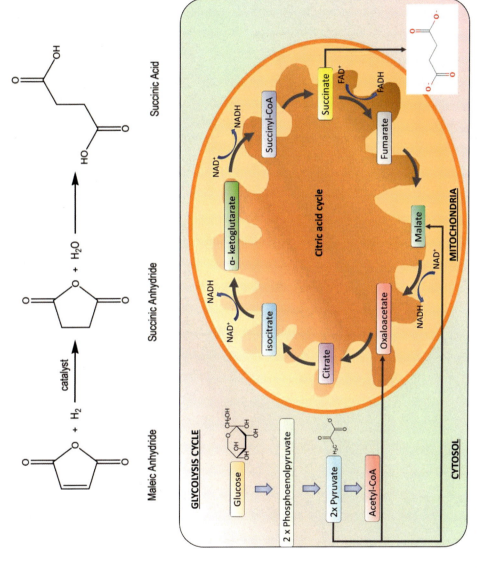

Fig. 10.12 Petrochemical and Biochemical pathways for succinic acid production.

and 180°C. During this step succinic anhydride is produced, and is later dissolved in hot water to produce succinic acid (Pinazo, Domine, Parvulescu, & Petru, 2015). In contrast, the biochemical production is carried out by mean of microbial fermentation of different renewable feedstocks. In this process succinate is generated within the mitochondria through the tricarboxylic acid cycle and acts as a metabolic intermediate to produce fumarate. Many strains have been identified for the industrial production of biobased succinic acid such as *Actinobacillus succinogenes*, *Anaerobiospirillum succiniciproducens*, *Mannheimia succiniciproducens*, *Basfia succiniciproducens*, *Bacteroides fragilis*, and recombinant *Escherichia coli* (Nghiem, Kleff, & Schwegmann, 2017). The fermentation process requires a sugar source (e.g., hydrolyzed starches, molasses, or hydrolyzed lignocellulosic materials), a nitrogen source (e.g., yeast extract, salts or amino acids), vitamins, and inorganic salts. The fermentation is carried out at temperatures between 30°C and 40°C, maintaining a pH between 5 and 7 and under a CO_2 atmosphere (Dai et al., 2019; Ferone, Raganati, Olivieri, & Marzocchella, 2019). The main succinic acid producing companies and their corresponding manufacturing pathways are summarized in Table 10.9.

Specifically in the production of biobased succinic acid, process intensification strategies have been explored and implemented at the different processing stages. Initially, in the fermentation stage, intensification has been applied by metabolic engineering of different microorganisms as well as on the bioreactor operation. Most of the research related to microbial engineering has focused on developing strains capable of consuming both C6 and C5 sugars (Desai & Rao, 2010; Tan, Chen, & Zhang, 2016) or capable of producing succinic acid under aerobic conditions (Lin, Bennet, & San, 2005; Vemuri, Eiteman, & Altman, 2002; Wu, Li, Zhou, & Ye, 2007). Other studies have centered on the overexpression of enzymes to improve the succinic acid yield in fermentation broths (Meng et al., 2016; Stols & Donelly, 1997). With regard to the bioreactor operation, research has focused on switching from batch operation approaches to continuous

Table 10.9 Main succinic acid producing companies (Sharma, Sarma, & Brar, 2020).

Route	Company	Country
Petrochemcial	Nippon Shokubai	Japan
	Gadiv Petrochemical Industries Ltd.	Israel
	Linyi Lixing Chemical Co.	China
	Anhui Sunsing Chemicals	China
	Kawasaki Kasei Chemicals	Japan
	Mitsubishi Chemical	Japan
Biobased	Succinity GmbH (Corbion Purac and BASF joint venture)	Germany
	Gc Innovation America (Formerly Myriant)	United States
	BioAmber[a] (joint venture with Mitsubishi)	United States
	Reverdia (Roquette-DSM Joint venture)	Netherlands

[a]BioAmber ceased operations on 2018 for insolvency.

operation schemes. This has been done through the implementation of immobilized cell reactors and membrane reactor systems that facilitate cell removal and recycling, while reducing nutrient depletion and product inhibition (Ferone et al., 2019; Lee, Lee, & Chang, 2008; Meynial-Salles, Dorotyn, & Soucaille, 2008; Urbance, Pometto 3rd, Dispirito, & Denli, 2004; van Heerden & Nicol, 2013).

Besides the myriad of intensification approaches used in the fermentation stage, multiple strategies are also applied at the downstream processing. The recovery of succinic acid from fermentation broths and the subsequent refining to meet the required commercial specifications is a complex and expensive endeavor. This is a consequence of the different nature of substrates, nutrients, microorganisms, and chemicals used during fermentation, and the resulting complex mixture in the final broth (i.e., diluted aqueous mixture of multiple chemicals). Because of these challenges, different processes have been developed and patented to recover and purify biobased succinic acid, many including innovative technologies that rely on process intensification approaches. Table 10.10 enumerates some of the technologies used by the main industrial producers of biobased succinic acid as well as some technologies currently under exploration.

10.8 Some challenges and future directions

In general, most of the intensification technologies and approaches listed in Tables 10.7 and 10.10 have been developed for specific processes and applications, and have resulted from recognizing particular limitations of each physical, chemical, or biochemical process. For instance, in the case of some etherification and esterification reactions, they are carried out in liquid phase, and near the boiling point of the reactive mixture, or at high enough temperatures to enable water removal. In this case, reactive distillation would be a desirable intensification approach to overcome chemical equilibrium limitations (Orjuela, Santaella, & Molano, 2016). Similarly, those techniques that would enable a selective product removal from the reactive mixture, would also be potential alternatives to explore (e.g., pervaporation, reactive membranes, azeotropic distillation, adsorption, microwaves heating, etc.). Then, once an intensified technology has been selected, the design approach can be systematically carried out using traditional heuristic process synthesis, a superstructure optimization, or by employing a combined methodology (Moncada, Aristizábal, & Cardona, 2016; Skiborowski, 2018). At this point, some of the main drawbacks of working with biobased chemicals appear.

Because the complex nature of biobased raw materials and the large number of derivatives, the isolation of pure components is generally challenging. Some of them are highly functionalized, reactive or they decompose easily during extraction and purification. Even if they can be effectively isolated, they can biodegrade or decompose during storage, for instance, due to heating or cooling, light exposure, contact with air or moisture, by changes in pH, or simply by biological means. Also, some of them have not been fully

Table 10.10 Process intensification (PI) technologies used in the recovery and purification of biobased succinic acid.

Stage or process	PI method	Remarks	Status Exp.	Status Ind.	Ref.
Recovery and purification	Cation exchange chromatography	Multistage evaporation to recover succinic acid salt followed by reaction with cation exchange resin. Succinic acid recovered through crystallization (BASF process)		X	(Schröder et al., 2015)
Recovery	Reactive pervaporation / Bipolar membrane electrodialysis	Diammonium succinate undergoes reactive evaporation to obtain monoammonium succinate (MAS). MAS crystals are recovered through evaporative crystallization. Succinic acid may be obtained through dilution of MAS in water and bipolar membrane electrodialysis or ion exchange chromatography. (BioAmber process)		X	(Dunuwila & Cockrem, 2012)
Recovery and purification	Microfiltration and nanofiltration	Evaporation and crystallization of recovery of crystals through centrifugation, microfiltration, and nanofiltration. Purification through activated carbon, anion, and cation exchange columns. (Reverdia process)		X	(Boit, Fiey, & Van De Graaf, 2015)

Table 10.10 Process intensification (PI) technologies used in the recovery and purification of biobased succinic acid—cont'd

Stage or process	PI method	Remarks	Status Exp.	Status Ind.	Ref.
Recovery	Crystallization	Crystallization at reduced pressure and pulverization of crystals. (Mitsubishi process)		X	(Mori, Takahashi, Suda, & Yoshida, 2015)
Recovery and purification	Vacuum evaporation Crystallization Simulated moving bed chromatography Nanofiltration	Broth is concentrated through vacuum evaporation and acidified. Succinic acid is crystallized and recovered through centrifugation. Mother liquor separated through simulated moving bed chromatography to obtain ammonium sulfate and remaining succinic acid which is purified through nanofiltration. (Gc Innovation America process)		X	(Tosukhowong, 2015)
Recovery and purification	Ion exchange chromatography Bipolar electrodialysis	Magnesium or calcium succinate is recovered through filtration and purified through ion exchange and subsequent bipolar electrodialysis. (Succinity Gmbh Process)		X	(Krieken & Breugel, 2009)
Recovery	Reactive extraction	Reactive extraction of organic acids with high molecular weight amines, followed by temperature swing and pH swing to recover succinic acid.	X		(Alexandri et al., 2019)

Continued

Table 10.10 Process intensification (PI) technologies used in the recovery and purification of biobased succinic acid—cont'd

Stage or process	PI method	Remarks	Status Exp.	Status Ind.	Ref.
Recovery	Reactive extraction	Removal of succinic acid from fermentation broths with mixture of trialkylphosphine oxides.	X		(Krzyżkowska & Regel-Rosocka, 2020)
Recovery	Nanofiltration	Membrane nanofiltration of fermentation broths to recover succinic acid.	X		(Borda, de Araujo, & Habert, 2020)
Recovery	Bipolar membrane electrodialysis	Fermentation broth was preclarified through ultrafiltration and concentrated through nanofiltration. Inorganic salts were removed with an ion exchange resin. The pretreated broth was fed to a large-scale EDBM process and the succinic acid recovery was evaluated.	X		(Szczygiełda, Antczak, & Prochaska, 2017) (Prochaska, Antczak, Regel-Rosocka, & Szczygiełda, 2018)
Recovery	Membrane supported reactive extraction	In situ recovery of succinic acid from an aqueous stream using a PTFE-based membrane and reaction with tri-n-octylamine in n-decanol.	X		(Gössi et al., 2020)
Recovery	Supercritical CO_2 extraction	Reactive extraction using supercritical CO_2 and tri-n-octylamine.	X		(Henczka & Djas, 2018)
Recovery	Adsorption in two phase partitioning bioreactor	Shifting the pH of the broth with CO_2 in order to use amorphous polymers to sequester the succinic acid.	X		(Hepburn & Daugulis, 2011)

Table 10.10 Process intensification (PI) technologies used in the recovery and purification of biobased succinic acid—cont'd

Stage or process	PI method	Remarks	Status Exp.	Status Ind.	Ref.
Recovery	Adsorption	A variety of resins and polymers were evaluated to remove succinic acid from fermentation broths through adsorption.	X		(Davison, Nghiem, & Richardson, 2004; De Wever & Dennewald, 2017)
Recovery	Ionic liquid extraction	Phosphonium-based ILs were used to extract succinic acid from fermentation broths at different ratios. The ILs were recovered through reduced pressure distillation and pH variation.	X		(Oliveira, Araújo, Ferreira, Rebelo, & Marrucho, 2012)
Recovery	Nanofiltration Vacuum distillation Reactive extraction Supported liquid membrane	Fermentation broth was filtered through ceramic membrane and passed through activated carbon. The solution was vacuum distilled to remove volatile fatty acids and ethanol and concentrate the succinic acid. tri-n-octylamine (TOA) dissolved in octanol was used for liquid-liquid extraction. A hydrophobic membrane was used to study the removal of succinic acid from the fermentation broth to a NaOH solution.	X		(Omwene, Yagcioglu, Sarihana, Karagunduz, & Keskinler, 2020)

Continued

Table 10.10 Process intensification (PI) technologies used in the recovery and purification of biobased succinic acid—cont'd

Stage or process	PI method	Remarks	Status Exp.	Status Ind.	Ref.
Recovery and purification	Microfiltration and nanofiltration-assisted crystallization	Spiral wound microfiltration and nanofiltration membranes were used to remove bacterial cells, proteins, and multivalent ions. The clarified broth was concentrated and crystallized to recover succinic acid.	X		(Thuy & Boontawan, 2017)
Recovery and purification	Reactive distillation	Succinic acid is recovered as diethyl succinate. Succinate salt is precipitated, dried, simultaneously acidified and partially esterified in ethanol, and completely esterified in a reactive distillation column to recover diethyl succinate.	X		(Orjuela, Orjuela, Lira, & Miller, 2013)
Recovery and purification	Forward osmosis Crystallization	Fermentation broth was pretreated with activated carbon and concentrated using forward osmosis membranes. The concentrated broth was crystallized to recover succinic acid crystals.	X		(Law et al., 2019; Law & Mohammad, 2018)
Recovery and purification	Aqueous two phase extraction and esterification	Extraction of succinic acid salts with ethanol and purification through esterification.	X		(Matsumoto & Tatsumi, 2018)

Table 10.10 Process intensification (PI) technologies used in the recovery and purification of biobased succinic acid—cont'd

Stage or process	PI method	Remarks	Status Exp.	Status Ind.	Ref.
Recovery and purification	Reactive extraction and crystallization	Reactive extraction of succinic acid from fermentation supernatants using trihexylamine in 1-octanl and diisooctylamine and dihexylamine in a mixture of 1-hexanol and 1-octanol. Back extraction with trimethylamine and crystallization of succinic acid in a rotating evaporator.	X		(Kurzrock, Schallinger, & Weuster-Botz, 2011;Kurzrock & Weuster-Botz, 2011)

studied yet. All these factors limit the possibility for evaluation of physicochemical properties, phase equilibria, kinetics, reaction mechanisms, stability, health and safety data, and environmental burdens. As a result, there is lack of models able to describe the behavior of this type of compounds, and most of the existing predictive methods (e.g., group contribution) are simply not accurate enough. The high content of polar groups and heteroatoms that can induce ionic and/or hydrogen bonding interactions, limits the predictive capabilities of existing methods that were mostly developed for nonpolar hydrocarbons. Despite a variety of methods have been increasingly developed to improve predictive and adjusting capabilities for biocompounds, they are mostly specific (e.g., for saccharides, glycols, amines, triglycerides) and can only be applied for a limited range of operating conditions. Thereby, high uncertainty is expected in the results obtained from the computational design of intensified processes when working with biobased chemicals. This is a major concern from a process engineering perspective and for an effective technology derisking.

In addition to this, there is still an incomplete understanding of some of the complex phenomena occurring in a variety of intensified processes, and the modeling capabilities are limited. For instance, clogging, swelling, fouling, and changes of ionic environments on the surface, are some of the typical phenomena occurring when putting membranes in contact with multicomponent mixtures. However, a complete and combined mathematical description of such factors, and their effects on activity, permeability or selectivity, are yet to be determined. Thereby, a highly resource-consuming experimental approach is generally required to develop empirical models for this type of processes. But even if they are developed, then the limiting factor would become the computational capacity to solve such models. For instance, an accurate description of a membrane-integrated fermentation reactor would require the simultaneous solution of highly nonlinear dynamic models including biomediated kinetics, lumped mass-energy-momentum interactions in the microscale, and hydrodynamic analogies and/or rate-based models for the macro scale (Helmi & Gallucci, 2020). But then, it is important to recall that the membrane reactor is just one single part of the entire process, which at the end must be globally optimized considering multiple feedstocks and products, and several processes.

Another factor to consider is that not all the intensified configurations generate a benefit for a certain process. For instance, the use of combined reaction-separation processes requires that kinetics and mass transfer at the desired operating conditions, are both fast enough to achieve cost-effectiveness. This has been referred as the operating window, and in some cases is too narrow for a suitable implementation at the industrial scale. In some cases, the operating window is not known beforehand, and the promising intensified technologies end up being affected by negative synergies or by controllability issues. In this sense, there is need to develop systematic methods capable of a rapid screening and identification of suitable intensified technologies at the early stages of the conceptual design. Recently, some attempts to develop and implement such methodologies for process intensifications in the biobased field have been reported (Anantasarn, Suriyapraphadilok, & Babi, 2017; Lutze, Gani, & Woodley, 2010; Lutze, Román-Martinez, Woodley, & Gani, 2012).

Besides the abovementioned, most of the current biorefinery facilities are relatively new, and they have not yet benefited from extensive process optimization as the petrochemical refineries have during last century. Therefore, the previously validated models for the intensified technologies must be implemented within simulation tools for the optimization of the integrated facilities (Sadhukhan, Ng, & Martinez, 2014; Stuart & El-Halwagi, 2012). But, considering the different issues of using biobased materials as industrial feedstocks, this optimization has to be done not only in economic terms, but instead considering the different sustainability dimensions (i.e., social, environmental, economic) and also within circular economy approaches (Demirel, Li, El-Halwagi, & Hasan, 2020; Mohan, Dahiya, Amulya, Katakojwala, & Vanitha, 2019). Besides, as biomass supply chain is paramount, the models must incorporate a life cycle perspective for the entire portfolio of processes and products. Here again the lack of information, not only from the process but from the life cycle, hinders the screening and identification of the most suited technology for a certain application.

Finally, the need for decarbonization of the chemical industry will require the use of nonconventional methods relying on sustainable electrical energy. In this case, electricity-driven intensified process such as microwave- or ultrasound-assisted reactions,

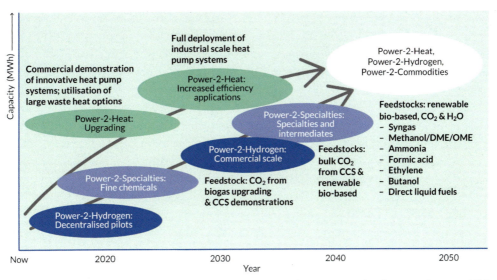

Fig. 10.13 Development road map for the electrification of the chemical industry in Europe. *(With permission from VoltaChem/TNO (2016). Empowering the chemical industry—Opportunities for electrification (Online). Available at: https://www.voltachem.com/publications/empowering-the-chemical-industry (Accessed 14.03.2021).)*

ohmic heating, electrodialysis, electrophoresis, and electrosynthesis, among others, will have a major role in the future. In this context, Fig. 10.13 presents a road map recently proposed for the electrification of the European chemical industry in the coming decades (VoltaChem/TNO, 2016). Because of the low installed capacity based upon sustainable sources (i.e., solar, eolic, tidal, geothermal), current attempts for electrification of chemical processes have been focused on the production of high value added and low demand biobased compounds. For instance, the synthesis of furan dicarboxylic (FDCA), glycolic, maleic and tartaric acid have been already carried out under high yields via electrosynthesis methods (VoltaChem/TNO, 2016). As the capacity expands, it is expected that electrification could reach biobased commodities and biofuels production, also including captured CO_2 as feedstock. This will create a much larger degree of complexity, as the biorefinery facilities will become key nodes within the global electric grid.

10.9 Concluding remarks

Current biorefineries evolved from the early days of the industrial revolution. After surviving troubling waters during most of the twentieth century when fossil-based resources took over the chemical industry, biorefineries have reached an effervescent present. In part, the recent boost of harnessing biobased resources as feedstock for chemicals, fuels and materials was a result of the urgent need for more sustainable production models. In this regard, it is expected that such models will rely on advanced processing technologies including the exploitation of integrated biobased feedstocks, for a tunable and diversified portfolio of products, and using intensified technologies.

While the biology-biotechnology revolution enabled a more effective production and exploitation of different biogenetic resources, intensification breakthroughs have made possible disruptive integrative continuous processing and modular designs within the biorefineries. As presented, examples of these achievements are present in the current industrial production of a variety of bioderived fuels, chemicals, polymers, and materials. As a result of this, current markets of biobased products are growing at a faster pace than those of the traditional fossil-based products, calling the attention of additional industrial players. Nevertheless, a broader implementation of intensified technologies within biorefineries will be confronted with important challenges from the processing standpoint, including potential negative synergies, highly complex interactions, reduced degrees of freedom (i.e., more constrains), and consequently, reduced operating windows. These factors can derive in a series of issues that can limit industrial implementation, such as stiffness of the integrated systems, and difficult controllability.

As a consequence, future advancements and research efforts in the implementation of intensified technologies for biorefineries must be directed to a better understanding of the fundamentals of the technologies, to a better description and modeling of the physicochemical interaction occurring when handling biobased molecules, to develop systematic methods for technologies conception, design, assessment, and selection, and finally to develop comprehensive methods with a systemic view for process design using sustainability criteria. In particular, these future developments and efforts must be aligned with the electrification trends of the different production sectors worldwide. Also, process intensification and biorefining fundamentals must be incorporated in the contents of academic curricula. Thus, future professional practitioners, scientists and technicians will consider intensification as a natural processing pathway rather than an exotic one, and will be able to use a systemic view, and enhanced complex thinking abilities to deal with the forthcoming challenges of integrated biorefineries.

Acknowledgments

This work has been partially funded by the Royal Academy of Engineering under the grant IAPP18-19\65, and the project entitled: *Valorization of Urban Used Cooking Oils by transformation into value-added oleochemicals. Study Case for Bogota, Colombia.* Also, by the British Council and MINCIENCIAS through the Institutional Links call, and the project entitled "*Development of technologies for the production of valuable biobased fatty esters from agroindustrial residual streams*" Code 526061819, and Contract 584-2020.

References

Agirre, I., Güemez, M. B., Motelica, A., van Veen, H. M., Vente, J. F., & Arias, P. L. (2012). A techno-economic comparison of various process options for the production of 1, 1-diethoxy butane. *Journal of Chemical Technology & Biotechnology*, *87*(7), 943–954. https://doi.org/10.1002/jctb.3704.

AICHE. (2021). *About RAPID: Institute Roadmap (Online)*. Available at: https://www.aiche.org/rapid/about (Accessed 11.03.2021).

Alder, C. M., Hayler, J. D., Henderson, R. K., Redman, A. M., Shukla, L., Shuster, L. E., et al. (2016). Updating and further expanding GSK's solvent sustainability guide. *Green Chemistry*, *18*(13), 3879–3890. https://doi.org/10.1039/C6GC00611F.

Alexandri, M., Vlysidis, A., Papapostolou, H., Tverezovskaya, O., Tverezovskiy, V., Kookos, I. K., et al. (2019). Downstream separation and purification of succinic acid from fermentation broths using spent sulphite liquor as feedstock. *Separation and Purification Technology*, *209*, 666–675. https://doi.org/10.1016/j.seppur.2018.08.061.

Anantasarn, N., Suriyapraphadilok, U., & Babi, D. K. (2017). A computer-aided approach for achieving sustainable process design by process intensification. *Computers & Chemical Engineering*, *105*, 56–73. https://doi.org/10.1016/j.compchemeng.2017.02.025.

Antunes, F., Chandel, A. K., Brumano, L. P., Terán, R., Peres, G. F. D., Ayabe, L. E. S., et al. (2018). A novel process intensification strategy for second-generation ethanol production from sugarcane bagasse in fluidized bed reactor. *Renewable Energy*, *124*, 189–196. https://doi.org/10.1016/j.renene.2017.06.004.

Arora, A., Iyer, S. S., Bajaj, I., & Hasan, M. M. F. (2018). Optimal methanol production via sorption-enhanced reaction process. *Industrial & Engineering Chemistry Research*, *57*(42), 14143–14161. https://doi.org/10.1021/acs.iecr.8b02543.

Ayodele, B. V., Alsaffar, M. A., & Mustapa, S. I. (2020). An overview of integration opportunities for sustainable bioethanol production from first- and second-generation sugar-based feedstocks. *Journal of Cleaner Production*, *245*, 118857. https://doi.org/10.1016/j.jclepro.2019.118857.

Backhaus, A. A. (1921). *Continuous process for the manufacture of esters*. United States Patent US1400849.

Bansode, S. R., & Rathod, V. K. (2017). An investigation of lipase catalysed sonochemical synthesis: A review. *Ultrasonics Sonochemistry*, *38*, 503–529. https://doi.org/10.1016/j.ultsonch.2017.02.028.

Bettenhausen, C. (2021). Chemical firms will coalesce around more ambitious climate goals. *Chemical & Engineering News*, *99*(2), 33. https://doi.org/10.1021/cen-09902-cover2.

Beyerle, M., & Diawara, M. C. (2017). Industrial purification of biobased chemicals—Meeting the challenge of efficient desalting. *Industrial Biotechnology*, *13*(1), 23–27. https://doi.org/10.1089/ind.2017.29070.mbe.

Bhatia, S. K., Bathia, R. K., Jeon, J., Pugazhendhi, A., Awasthi, M. K., Kumar, D., et al. (2021). An overview on advancements in biobased transesterification methods for biodiesel production: Oil resources, extraction, biocatalysts, and process intensification technologies. *Fuel*, *185*, 119117. https://doi.org/10.1016/j.fuel.2020.119117.

Biddy, M. J., Scarlata, C., & Kinchin, C. (2016). *Chemicals from biomass: Market assessment of bioproducts with near-term potential*. Technical Report NREL/TP-5100-65509 Denver, USA: National Renewable Energy Laboratory.

Bielenberg, J., & Palou-Rivera, I. (2019). The RAPID Manufacturing Institute—Reenergizing US efforts in process intensification and modular chemical processing. *Chemical Engineering and Processing*, *138*, 49–54. https://doi.org/10.1016/j.cep.2019.02.008.

BIO. (2017). *The biobased economy: Measuring growth and impacts*. (Online). Available at: https://www.bio.org/sites/default/files/legacy/bioorg/docs/Biobased_Economy_Measuring_Impact.pdf (Accessed 11.03.2021).

Blanchard, H. (1934). *Static sprayer or mixer for gas supply to internal combustion engines*. France, Patent No. FR762413.

Boit, B., Fiey, G., & Van De Graaf, M. (2015). *Process for manufacturing succinic acid from a fermentation broth using nano filtration to purify recycled mother liquor*. United States Patent US10189767B2.

Borda, P. A., de Araujo, F., & Habert, A. C. (2020). Effect of feed conditions and added solutes on the performance of membrane nanofiltration of succinic acid solutions. *Brazilian Journal of Chemical Engineering*, *37*, 283–295. https://doi.org/10.1007/s43153-020-00029-7.

Bozell, J. J., & Petersen, G. R. (2010). Technology development for the production of biobased products from biorefinery carbohydrates—The US Department of Energy's "Top 10" revisited. *Green Chemistry*, *12*(4), 539–554. https://doi.org/10.1039/B922014C.

Brown, A., Walsheim, L., Landälv, I., Saddler, J., Ebadian, M., McMillan, J. D., et al. (2020). *Advanced biofuels—Potential for cost reduction*. IEA Bioenergy. IEA Bioenergy.

Buisman, G. J., & Lange, J. H. (2016). Arizona chemical: Refining and upgrading of bio-based and renewable feedstocks. In P. D. de Maria (Ed.), *Industrial biorenewables: A practical viewpoint* (pp. 21–57). New York: John Wiley & Sons.

Cardona, C. A., & Sánchez, Ó. J. (2007). Fuel ethanol production: Process design trends and integration opportunities. *Bioresource Technology*, *98*(12), 2415–2457. https://doi.org/10.1016/j.biortech.2007.01.002.

Castro-Muñoz, R., Galiano, F., & Figoli, A. (2019). Chemical and bio-chemical reactions assisted by pervaporation technology. *Critical Reviews in Biotechnology*, *39*(7), 884–903. https://doi.org/10.1080/07388551.2019.1631248.

Chemat, F., Vian, M. A., Fabiano-tixier, A. S., Nutrizio, M., Jambrak, A. R., Munekata, P. E. S., et al. (2020). A review of sustainable and intensified techniques for extraction of food and natural products. *Green Chemistry*, *22*(8), 2325. https://doi.org/10.1039/C9GC03878G.

Chen, X. Y., Vinh-Thang, H., Avalos, A., Rodrigue, D., & Kaliaguine, S. (2015). Membrane gas separation technologies for biogas upgrading. *RSC Advances*, *5*(31), 24399–24448. https://doi.org/10.1039/C5RA00666J.

Cornelissen, R., Tober, E., Kok, J., & van de Meer, T. (2006). Generation of synthesis gas by partial oxidation of natural gas in a gas turbine. *Energy*, *31*(15), 3199–3207. https://doi.org/10.1016/j.energy.2006.03.028.

Creative Energy. (2007). *European roadmap for process intensification*. (Online). Available at: https://efce.info/efce_media/-p-531-EGOTEC-34kdll9fiqnru8376q05juj2d2.pdf?rewrite_engine=id (Accessed 11.03.2021).

Dai, Z., Guo, F., Zhang, S., Zhang, W., Yang, Q., Dong, W., et al. (2019). Bio-based succinic acid: An overview of strain development, substrate utilization, and downstream purification. *Biofuels, Bioproducts and Biorefining*, *14*(5), 965–985. https://doi.org/10.1002/bbb.2063.

Dalena, F., Senatore, A., Basile, M., Knani, S., Basile, A., & Lulianelli, A. (2018). Advances in methanol production and utilization, with particular emphasis toward hydrogen generation via membrane reactor technology. *Membranes (Basel)*, *8*(4), 98. https://doi.org/10.3390/membranes8040098.

Datta, D., Kumar, S., & Uslu, H. (2015). Status of the reactive extraction as a method of separation. *Journal of Chemistry*, 853789. 1–16 https://doi.org/10.1155/2015/853789.

Davison, B. H., Nghiem, N. P., & Richardson, G. L. (2004). Succinic acid adsorption from fermentation broth and regeneration. *Applied Biochemistry and Biotechnology*, *113*, 653–669. https://doi.org/10.1385/ABAB:114:1-3:653.

de Jong, E., Higson, A., Walsh, P., & Wellisch, M. (2011). *Bio-based chemicals—Value added products from biorefineries*. IEA Bioenergy.

de Jong, E., Stichnothe, H., Bell, G., & Jørgensen, H. (2020). *Bio-based chemicals—A 2020 update*. IEA Bioenergy.

De Wever, H., & Dennewald, D. (2017). Screening of sorbents for recovery of succinic and itaconic acid from fermentation broths. *Journal of Chemical Technology & Biotechnology*, *93*(2), 385–391. https://doi.org/10.1002/jctb.5366.

Demirel, S. E., Li, J., El-Halwagi, M., & Hasan, M. M. F. (2020). Sustainable process intensification using building blocks. *ACS Sustainable Chemistry & Engineering*, *8*(48), 17664–17679. https://doi.org/10.1021/acssuschemeng.0c04590.

Desai, T. A., & Rao, C. V. (2010). Regulation of arabinose and xylose metabolism in *Escherichia coli*. *Applied and Environmental Microbiology*, *76*(5), 1524–1532. https://doi.org/10.1128/AEM.01970-09.

DOE. (2015). Quadrennial technology review: An assessment of energy technologies and research opportunities. In *Annex 6J: Process intensification: Technology assessmentes* Department of Energy.

Dominguez de María, P. (Ed.). (2016). *Industrial biorenewables: A practical viewpoint* (1st ed.). New York: John Wiley & Sons. https://doi.org/10.1002/9781118843796.

Dunuwila, D., & Cockrem, M. (2012). *Methods and systems of producing dicarboxylic acids*. United States Patent US20120259138A1.

E4tech, RE-CORD and WUR. (2015). *From the sugar platform to biofuels and biochemicals. Final report for the European Comission Directorate-General Energy, s.l.* European Commission. No. ENER/C2/423-2012/SI2.673791.

Faria, R., Graça, N., & Rodrigues, A. (2019). Green fuels and fuel additives production in simulated moving bed reactors. In A. Górak, & A. Stankiewicz (Eds.), *Intensification of biobased processes* (pp. 145–165). Royal Society of Chemistry. https://doi.org/10.1039/9781788010320-00145.

Ferone, M., Raganati, F., Olivieri, G., & Marzocchella, A. (2019). Bioreactors for succinic acid production processes. *Critical Reviews in Biotechnology*, *39*(4), 571–586. https://doi.org/10.1080/07388551.2019.1592105.

Ferreira, M. C., Meirelles, A. J. A., & Batista, E. A. C. (2013). Study of the fusel oil distillation process. *Industrial & Engineering Chemistry Research*, *52*(6), 2336–2351. https://doi.org/10.1021/ie300665z.

Finlay, M. R. (2008). Old efforts at new uses: A brief history of chemurgy and the American search for biobased materials. *Journal of Industrial Ecology*, *7*(3–4), 33–46. https://doi.org/10.1162/108819803323059389.

Froidevaux, V., Negrell, C., Caillol, S., Pascault, J., & Boutevin, B. (2016). Biobased amines: From synthesis to polymers; present and future. *Chemical Reviews*, *116*(22), 14181–14224. https://doi.org/10.1021/acs.chemrev.6b00486.

Gabriëls, D., Hernández, W. Y., Sels, B., Van Der Voort, P., & Verberckmoes, A. (2015). Review of catalytic systems and thermodynamics for the Guerbet condensation reaction and challenges for biomass valorization. *Catalysis Science & Technology*, *5*, 3876–3902. https://doi.org/10.1039/C5CY00359H.

Gallucci, F. (2018). Inorganic membrane reactors for methanol synthesis. In A. Basile, & F. Dalena (Eds.), *Methanol science and engineering* (pp. 493–518). Elsevier. https://doi.org/10.1016/B978-0-444-63903-5.00018-2.

Gay, L. (1923). *Absorption, reaction and like column*. United States Patent US1466936.

GERLI. (2020). *Cyberlipid history*. (Online). Available at: http://cyberlipid.gerli.com/history/ (Accessed 11.03.2021).

Gilet, A., Quettier, C., Wiatz, V., Bricout, H., Ferreira, M., Rousseau, C., et al. (2018). Unconventional media and technologies for starch etherification and esterification. *Green Chemistry*, *20*(6), 1152–1168. https://doi.org/10.1039/C7GC03135A.

Givaudan. (2020). *Timeline 1768-2020*. (Online). Available at: https://www.givaudan.com/our-company/rich-heritage/timeline (Accessed 11.03.2021).

Gonzalez-Contreras, M., Lugo-Mendez, H., Sales-Cruz, M., & Lopez-Arenas, T. (2020). Synthesis, design and evaluation of intensified lignocellulosic biorefineries—Case study: Ethanol production. *Chemical Engineering and Processing: Process Intensification*, *159*, 108220. https://doi.org/10.1016/j.cep.2020.108220.

Gössi, A., Burgener, F., Kohler, D., Urso, A., Kolvenbach, B. A., Riedl, W., et al. (2020). In-situ recovery of carboxylic acids from fermentation broths through membrane supported reactive extraction using membrane modules with improved stability. *Separation and Purification Technology*, *241*, 116694. https://doi.org/10.1016/j.seppur.2020.116694.

GVR. (2020). *Biofuels market size, share & trends analysis report by form (solid, liquid, gaseous), by region (North America, Europe, Asia Pacific, Central & South America, Middle East & Africa), and segment forecasts, 2020-2027*. (Online). Available at: https://www.grandviewresearch.com/industry-analysis/biofuels-market (Accessed 11.03.2021).

Hale, W. J. (1934). *The farm chemurgic: Farmward the star of destiny lights our way*. Boston, Mass: The Stratford Company.

Harmsen, J., & Verkerk, M. (2020). *Process intensification—Breakthrough in design, industrial innovation practices, and education* (pp. 142–151). De Gruyter Textbook. https://doi.org/10.1515/9783110657357.

Hasabnis, A., & Mahajani, S. (2014). Acetalization of glycerol with formaldehyde by reactive distillation. *Industrial & Engineering Chemistry Research*, *53*, 12279–12287. https://doi.org/10.1021/ie501577q.

Hayyān, J. I. (1531). *The three books on alchemy by Geber, the great philosopher and alchemist*. Strasbourg: Grüninger, J.

Helmi, A., & Gallucci, F. (2020). Latest developments in membrane (bio)reactors. *Processes*, *8*(10), 1239. https://doi.org/10.3390/pr8101239.

Henczka, M., & Djas, M. (2018). Reactive extraction of succinic acid using supercritical carbon dioxide. *Separation Science and Technology*, *53*(4), 655–661. https://doi.org/10.1080/01496395.2017.1405033.

Hepburn, A. J., & Daugulis, A. J. (2011). The use of CO2 for reversible pH shifting, and the removal of succinic acid in a polymer-based two-phase partitioning bioreactor. *Journal of Chemical Technology & Biotechnology, 87*(1), 42–50. https://doi.org/10.1002/jctb.2763.

Holladay, J. E., Bozell, J. J., White, J. F., & Johnson, D. (2007). *Top value-added chemicals from biomass volume II—Results of screening for potential candidates from biorefinery lignin*. Richland, USA: Pacific Northwest National Laboratory.

Hommes, A., Heeres, H. J., & Yue, J. (2019). Catalytic transformation of biomass derivatives to value-added chemicals and fuels in continuous flow microreactors. *ChemCatChem, 11*(19), 4671–4708. https://doi.org/10.1002/cctc.201900807.

Hong, X., McGiveron, O., Lira, C. T., Orjuela, A., Peereboom, L., & Miller, D. J. (2012). A reactive distillation process to produce 5-hydroxy-2-methyl-1,3-dioxane from mixed glycerol acetal isomers. *Organic Process Research & Development, 16*(5), 1141–1145. https://doi.org/10.1021/op200072x.

Hu, J., Wang, Y., Cao, C., Elliot, D. C., Stevens, D. J., & White, J. F. (2007). Conversion of biomass-derived syngas to alcohols and C2 oxygenates using supported Rh catalysts in a microchannel reactor. *Catalysis Today, 120*(1), 90–95. https://doi.org/10.1016/j.cattod.2006.07.006.

Ibrahim, M. F., Kim, S. W., & Abd-Aziz, S. (2018). Advanced bioprocessing strategies for biobutanol production from biomass. *Renewable and Sustainable Energy Reviews, 91*, 1192–1204. https://doi.org/10.1016/j.rser.2018.04.060.

IEA. (2020). *Renewables 2020 Analysis and forecast to 2025*. (Online). Available at: https://www.iea.org/reports/renewables-2020 (Accessed 04.03.2021).

Ingkaninan, K., Hazekamp, A., Hoek, A. C., Balconi, S., & Verpoorte, R. (2000). Application of centrifugal partition chromatography in a general separation and dereplication procedure for plant extracts. *Journal of Liquid Chromatography & Related Technologies, 23*(14), 2195–2208. https://doi.org/10.1081/JLC-100100481.

Inyang, V. M., & Lokhat, D. (2020). Separation of carboxylic acids: Conventional and intensified processes and effects of process engineering parameters. In M. Daramola, & A. Ayeni (Eds.), *Valorization of biomass to value-added commodities* (pp. 469–505). Cham: Springer. https://doi.org/10.1007/978-3-030-38032-8_22.

IRENA. (2016). *Innovation outlook: Advanced liquid biofuels*. Abu Dhabi: International Renewable Energy Agency.

IRENA. (2020). *Reaching zero with renewables: Eliminating CO2 emissions from industry and transport in line with the 1.5 °C climate goal*. Abu Dhabi: International Renewable Energy Agency.

Jaggai, C., Imkaraaz, Z., Samm, K., Pounder, A., Koylass, N., Chakrabarti, D. P., et al. (2020). Towards greater sustainable development within current mega-methanol (MM) production. *Green Chemistry, 22*, 4279–4294. https://doi.org/10.1039/D0GC01185A.

James, C. E., Hough, L., & Khan, R. (1989). Sucrose and Its derivatives. In W. Herz, H. Grisebach, G. Kirby, & C. Tamm (Eds.), *Progress in the chemistry of organic natural products* (pp. 117–184). Vienna: Springer-Verlag. https://doi.org/10.1007/978-3-7091-9002-9_4.

Kamm, B., Gruber, P. R., & Kamm, M. (2010). *Biorefineries—Industrial processes and products: Status quo and future directions*. Weinheim: Wiley-VCH Verlag GmbH & Co. https://doi.org/10.1002/9783527619849.

Kiss, A. A. (2019). Novel catalytic reactive distillation processes for a sustainable chemical industry. *Topics in Catalysis, 62*, 1132–1148. https://doi.org/10.1007/s11244-018-1052-9.

Kiss, A. A., & Bîldea, C. S. (2018). Intensified downstream processing in biofuels production. In A. Gorak, & A. Stankiewicz (Eds.), *Catalyst-free organic synthesis* (pp. 62–85). Royal Society of Chemistry. https://doi.org/10.1039/9781788010320-00062.

Kiss, A. A., Jobson, M., & Gao, X. (2019). Reactive distillation: Stepping up to the next level of process intensification. *Industrial and Engineering Chemistry Research, 58*(15), 5909–5918. https://doi.org/10.1021/acs.iecr.8b05450.

Köhler, K., Rühl, J., Blank, L. M., & Schmid, A. (2014). Integration of biocatalyst and process engineering for sustainable and efficient n-butanol production. *Engineering in Life Sciences, 15*(1), 4–19. https://doi.org/10.1002/elsc.201400041.

Koniuszewska, I., Korzeniewska, E., Harnisz, M., & Czatzkowska, M. (2020). Intensification of biogas production using various technologies: A review. *International Journal of Energy Research, 44*(8), 6240–6258. https://doi.org/10.1002/er.5338.

Kostova, B. (2017). *Current status of biorefininf in USA*. (Online). Available at: https://nachhaltigwirtschaften.at/resources/iea_pdf/events/20171023_bioenergy-task-42/kostova_biorefinerystatus-usa.pdf?m=1512152154 (Accessed 11.03.2021).

Krieken, J. V., & Breugel, J. V. (2009). *Process for the preparation of a monovalent succinate salt*. United States Patent US 8,865.438 B2.

Krzyżkowska, A., & Regel-Rosocka, M. (2020). The effect of fermentation broth composition on removal of carboxylic acids by reactive extraction with Cyanex 923. *Separation and Purification Technology, 236*, 116289. https://doi.org/10.1016/j.seppur.2019.116289.

Kumar, M. R., Ravikumar, R., Thenmoxhi, S., & Sankar, M. K. (2019). Choice of pretreatment technology for sustainable production of bioethanol from lignocellulosic biomass: Bottle necks and recommendations. *Waste and Biomass Valorization, 10*, 1693–1709. https://doi.org/10.1007/s12649-017-0177-6.

Kurzrock, T., & Weuster-Botz, D. (2011). New reactive extraction systems for separation of bio-succinic acid. *Bioprocess and Biosystems Engineering, 34*, 779–787. https://doi.org/10.1007/s00449-011-0526-y.

Kurzrock, T., Schallinger, S., & Weuster-Botz, D. (2011). Integrated separation process for isolation and purification of biosuccinic acid. *Biotechnology Progress, 27*(6), 1623–1628. https://doi.org/10.1002/btpr.673.

Kushwaha, D., Srivastava, N., Mishra, I., Upadhyay, S. N., & Mishra, P. K. (2019). Recent trends in biobutanol production. *Reviews in Chemical Engineering, 35*(7), 475–504. https://doi.org/10.1515/revce-2017-0041.

Laurent, P. M. F. (1895). *Improved apparatus for sterilizing liquids applicable for heating, cooling, condensing, distilling, and other similar purposes*. Great Britain Patent GB189409815A.

Law, J. Y., & Mohammad, A. W. (2018). Osmotic concentration of succinic acid by forward osmosis: Influence of feed solution pH and evaluation of seawater as draw solution. *Chinese Journal of Chemical Engineering, 26*(5), 976–983. https://doi.org/10.1016/j.cjche.2017.10.003.

Law, J. Y., Mohammad, A. W., Tee, Z. K., Zaman, N. K., Jahim, J. M., Santanaraj, J., et al. (2019). Recovery of succinic acid from fermentation broth by forward osmosis assisted crystallization process. *Journal of Membrane Science, 583*, 139–151. https://doi.org/10.1016/j.memsci.2019.04.036.

Lee, S. Y., Kim, H. U., Chae, T. U., Cho, J. S., Kim, J. W., Shin, J. H., et al. (2019). A comprehensive metabolic map for production of bio-based chemicals. *Nature Catalysis, 2*(1), 18–33. https://doi.org/10.1038/s41929-018-0212-4.

Lee, P. C., Lee, S. Y., & Chang, H. N. (2008). Cell recycled culture of succinic acid-producing *Anaerobiospirillum succiniciproducens* using an internal membrane filtration system. *Journal of Microbiology and Biotechnology, 18*(7), 1252–1256.

Lewis, W. K., Venarle, C. S., & Wilson, R. E. (1924). *Process for separating gases*. United States Patent US1496757.

Li, C., Duan, C., Fang, J., & Li H. (2019). Process intensification and energy saving of reactive distillation for production of ester compounds. *Chinese Journal of Chemical Engineering, 27*(6), 1307–1323. https://doi.org/10.1016/j.cjche.2018.10.007.

Lin, H., Bennet, G. N., & San, K.-Y. (2005). Metabolic engineering of aerobic succinate production systems in *Escherichia coli* to improve process productivity and achieve the maximum theoretical succinate yield. *Metabolic Engineering, 7*(2), 116–127. https://doi.org/10.1016/j.ymben.2004.10.003.

Lokare, K. S. (2016). *50 Hot bio-based companies, 50 quick takes*. (Online). Available at: https://www.linkedin.com/pulse/50-hot-bio-based-companies-quick-takes-kapil-shyam-lokare-phd-1/?trk=hp-feed-article-title-publish (Accessed 04.03.2021).

Lucke, C. E. (1931). *Heat exchanger*. United States Patent US1833166.

Luster, E. W. (1933). *Apparatus for fractionating cracked products*. United States Patent US1915681.

Lutze, P., Gani, R., & Woodley, J. M. (2010). Process intensification: A perspective on process synthesis. *Chemical Engineering and Processing: Process Intensification, 49*(6), 547–558. https://doi.org/10.1016/j.cep.2010.05.002.

Lutze, P., Román-Martinez, A., Woodley, J. M., & Gani, R. (2012). A systematic synthesis and design methodology to achieve process intensification in (bio) chemical processes. *Computers & Chemical Engineering, 36*, 189–207. https://doi.org/10.1016/j.compchemeng.2011.08.005.

Marketwatch. (2021). *Global bio-based succinic acid market research report 2021*. (Online). Available at: https://www.marketwatch.com/press-release/bio-based-succinic-acid-market-2021—market-share-top-manufacturers-globally-market-size-and-forecast-to-2027-with-top-growth-companies-2021-03-22 (Accessed 22.03.2021).

Marshall, F. (1862). *Population and trade in France in 1861-62*. London: Chapman and Hall.

Masngut, N., & Harvey, A. P. (2012). Intensification of biobutanol production in batch oscillatory baffled bioreactor. *Procedia Engineering, 42*, 1079–1087. https://doi.org/10.1016/j.proeng.2012.07.499.

Matsumoto, M., & Tatsumi, M. (2018). Extraction and esterification of succinic acid using aqueous two-phase systems composed of ethanol and salts. *Solvent Extraction Research and Development, 25*(2), 101–107. https://doi.org/10.15261/serdj.25.101.

Mazière, A., Prinsen, P., Garcia, A., Luque, R., & Len, C. (2017). A review of progress in (bio)catalytic routes from/to renewable succinic acid. *Biofuels, Bioproducts and Biorefining, 11*(5), 908–931. https://doi.org/10.1002/bbb.1785.

Mendoza, J. D. J., Sánchez-Ramírez, E., Segovia-Hernández, J. G., Orjuela, A., & Hernández, S. (2021). Recovery of alcohol industry wastes: Revaluation of fusel oil through intensified processes. *Chemical Engineering and Processing Process Intensification, 163*, 108329. https://doi.org/10.1016/j.cep.2021.108329.

Meng, J., Wang, B., Liu, D., Chen, T., Wang, Z., & Zhao, X. (2016). High-yield anaerobic succinate production by strategically regulating multiple metabolic pathways based on stoichiometric maximum in Escherichia coli. *Microbial Cell Factories, 15*(1), 141. https://doi.org/10.1186/s12934-016-0536-1.

Meyer, M. (2011). Chemistry: How it all started. *The UNESCO Courier, 16*, 11–13.

Meynial-Salles, I., Dorotyn, S., & Soucaille, P. (2008). A new process for the continuous production of succinic acid from glucose at high yield, titer, and productivity. *Biotechnology and Bioengineering, 99*(1), 129–135. https://doi.org/10.1002/bit.21521.

Mikkola, J.-P., Sklavounos, E., King, A. W., & Virtanen, P. (2015). The biorefinery and green chemistry. In R. Bogel-Lukasik (Ed.), *Ionic liquids in the biorefinery concept: Challenges and perspectives* (pp. 1–37). Cambridge: Royal Society of Chemistry. https://doi.org/10.1039/9781782622598-00001.

Mohan, S. V., Dahiya, S., Amulya, K., Katakojwala, R., & Vanitha, T. K. (2019). Can circular bioeconomy be fueled by waste biorefineries—A closer look. *Bioresource Technology Reports, 7*, 100277. https://doi.org/10.1016/j.biteb.2019.100277.

Moncada, J., Aristizábal, V., & Cardona, C. A. (2016). Design strategies for sustainable biorefineries. *Biochemical Engineering Journal, 116*(15), 122–134. https://doi.org/10.1016/j.bej.2016.06.009.

Morales, G., Frauenlob, R., Franke, R., & Börner, A. (2015). Production of alcohols via hydroformylation. *Catalysis Science & Technology, 5*(1), 34–54. https://doi.org/10.1039/C4CY01131G.

Mori, Y., Takahashi, G., Suda, H., & Yoshida, S. (2015). *Processes for producing succinic acid*. United States Patent US9035095B2.

Morone, A., & Pandey, R. A. (2014). Lignocellulosic biobutanol production: Gridlocks and potential remedies. *Renewable and Sustainable Energy Reviews, 37*, 21–35. https://doi.org/10.1016/j.rser.2014.05.009.

MRF. (2019). *Bio-based chemicals market: Information by type (bioplastics, bio-lubricants, bio-solvents, bio-alcohols, bio-based acids, bio-surfactants), by application (food & beverage, pharmaceuticals, agriculture, packaging)—Global forecast till 2023*. (Online). Available at: https://www.marketresearchfuture.com/reports/bio-based-chemicals-market-5706 (Accessed 11.03.2021).

National Center for Biotechnology Information. (2021). *PubChem compound summary for CID 1110, succinic acid*. (Online). Available at: https://pubchem.ncbi.nlm.nih.gov/compound/Succinic-acid (Accessed 22.03.2021).

National Academies of the Sciences, Engineering, and Medicine. (2020). *Safeguarding the bioeconomy*. Washington DC: The National Academies Press. https://doi.org/10.17226/25525.

Nghiem, N. P., Kleff, S., & Schwegmann, S. (2017). Succinic acid: Technology development and commercialization. *Fermentation, 3*(2), 26–40. https://doi.org/10.3390/fermentation3020026.

Nguyen, H. C., Lee, H., Su, C. H., Shih, W., & Chien, C. (2020). Green process for fatty acid production from soybean oil through microwave-mediated autocatalytic synthesis. *Chemical Engineering and Processing Process Intensification*, *147*, 107782. https://doi.org/10.1016/j.cep.2019.107782.

Oh, P. P., Lau, H. L. N., Chen, J., Chong, M. F., & Choo, Y. M. (2012). A review on conventional technologies and emerging process intensification (PI) methods for biodiesel production. *Renewable and Sustainable Energy Reviews*, *16*(7), 5131–5145. https://doi.org/10.1016/j.rser.2012.05.014.

Oliveira, F. S., Araújo, J. M. M., Ferreira, R., Rebelo, L. P. N., & Marrucho, I. M. (2012). Extraction of L-lactic, L-malic, and succinic acids using phosphonium-based ionic liquids. *Separation and Purification Technology*, 137–146. https://doi.org/10.1016/j.seppur.2011.10.002.

Omwene, P. I., Yagcioglu, M., Sarihana, Z. B. O., Karagunduz, A., & Keskinler, B. (2020). Recovery of succinic acid from whey fermentation broth by reactive extraction coupled with multistage processes. *Journal of Environmental Chemical Engineering*, *8*(5), 104216. https://doi.org/10.1016/j.jece.2020.104216.

Orjuela, A., Orjuela, A., Lira, C. T., & Miller, D. J. (2013). A novel process for recovery of fermentation-derived succinic acid: Process design and economic analysis. *Bioresource Technology*, *139*, 235–241. https://doi.org/10.1016/j.biortech.2013.03.174.

Orjuela, A., Santaella, M. A., & Molano, P. A. (2016). Process intensification by reactive distillation. In J. G. Segovia-Hernández, & A. Bonilla-Petriciolet (Eds.), *Process intensification in chemical engineering. s.l.* (pp. 131–181). Springer International Publishing. https://doi.org/10.1007/978-3-319-28392-0_6.

Parisi, C. (2020). *Distribution of the bio-based industry in the EU*. Luxembourg: JRC Scientific information systems and databases report. https://doi.org/10.2760/745867.

Patrașcu, I., Bîldea, C. S., & Kiss, A. A. (2018). Eco-efficient downstream processing of biobutanol by enhanced process intensification and integration. *ACS Sustainable Chemistry & Engineering*, *6*(4), 5452–5461. https://doi.org/10.1021/acssuschemeng.8b00320.

Pinazo, J. M., Domine, M. E., Parvulescu, V., & Petru, F. (2015). Sustainability metrics for succinic acid production: A comparison between biomass-based and petrochemical routes. *Catalysis Today*, *239*, 17–24. https://doi.org/10.1016/j.cattod.2014.05.035.

Podbelnyak, V. Z. (1932). *Absorption and distillation apparatus*. USSR Patent SU50239A1.

Polin, J. P., Peterson, C. A., Whilmer, L. E., Smith, R. G., & Brown, R. C. (2019). Process intensification of biomass fast pyrolysis through autothermal operation of a fluidized bed reactor. *Applied Energy*, *249*, 276–285. https://doi.org/10.1016/j.apenergy.2019.04.154.

Prochaska, K., Antczak, J., Regel-Rosocka, M., & Szczygiełda, M. (2018). Removal of succinic acid from fermentation broth by multistage process (membrane separation and reactive extraction). *Separation and Purification Technology*, *192*, 360–368. https://doi.org/10.1016/j.seppur.2017.10.043.

Qing, W., Li, X., Shao, S., Shi, X., Wang, J., Feng, Y., et al. (2019). Polymeric catalytically active membranes for reaction-separation coupling: A review. *Journal of Membrane Science*, *583*, 118–138. https://doi.org/10.1016/j.memsci.2019.04.053.

Ranjan, A., Singh, S., Malani, R. S., & Moholkar, V. S. (2016). Ultrasound-assisted bioalcohol synthesis: Review and analysis. *RSC Advances*, *6*(70), 65541–65562. https://doi.org/10.1039/C6RA11580B.

Rauch, R., Kiennemann, A., & Sauciuc, A. (2013). Fischer-Tropsch synthesis to biofuels (BtL Process). In K. S. Triantafyllidis, A. A. Lappas, & M. Stöcker (Eds.), *The role of catalysis for the sustainable production of biofuels and bio-chemicals* (pp. 397–443). Elsevier B.V. https://doi.org/10.1016/B978-0-444-56330-9.00012-7.

Reid, A., van Loon, M., Tollin, S., & Nieuwenhuizen, P. (2016). Akzonobel: Biobased raw materials. In P. Dominguez de Maria (Ed.), *Industrial biorenewables: A practical viewpoint* (pp. 1–20). New York: John Wiley & Sons. https://doi.org/10.1002/9781118843796.ch1.

Roelfsema, P. J. (1937). *Distillation process*. United States Patent US2085546A.

Roelfsema, P. J. (1938). *Distillation process*. United States Patent US2113965.

Rosales-Calderon, O., & Arantes, V. (2019). A review on commercial-scale high-value products that can be produced alongside cellulosic ethanol. *Biotechnology for Biofuels*, *12*(240), 1–58. https://doi.org/10.1186/s13068-019-1529-1.

Russo, V., Tesser, R., Rossano, C., Vitello, R., Turco, R., Salmi, T., et al. (2019). Chromatographic reactor modelling. *Chemical Engineering Journal*, *377*, 119692. https://doi.org/10.1016/j.cej.2018.08.078.

Sadhukhan, J., Ng, K. S., & Martinez, E. (2014). *Biorefineries and chemical processes: Design, integration and sustainability analysis.* New York: John Wiley & Sons, Ltd. https://doi.org/10.1002/9781118698129.

Sanders, J., Clark, J. H., Harmsen, G. J., Heeres, H. J., Heijnen, J. J., Kersten, S. R. A., et al. (2012). Process intensification in the future production of base chemicals from biomass. *Chemical Engineering and Processing: Process Intensification, 51*, 117–136. https://doi.org/10.1016/j.cep.2011.08.007.

Santaella, M. A., Orjuela, A., & Narváez, P. C. (2015). Comparison of different reactive distillation schemes for ethyl acetate production using sustainability indicators. *Chemical Engineering and Processing: Process Intensification, 96*, 1–13. https://doi.org/10.1016/j.cep.2015.07.027.

Satyawali, Y., Vanbroekhoven, K., & Dejonghe, W. (2017). Process intensification: The future for enzymatic processes? *Biochemical Engineering Journal, 121*, 196–223. https://doi.org/10.1016/j.bej.2017.01.016.

Schröder, H., Haefner, S., Von Abendroth, G., Hollmann, R., Raddatz, A., Ernst, H., et al. (2015). *Microbial succinic acid producers and purification of succinic acid.* United States Patent US 9023632 B2.

Schubert, T. (2020). Production routes of advanced renewable C1 to C4 alcohols as biofuel components—A review. *Biofuels, Bioproducts and Biorefining, 14*(4), 845–878. https://doi.org/10.1002/bbb.2109.

SCI. (2019). *Renewable ethylene, ethylene oxide & ethylene glycol.* (Online). Available at: https://www.scidesign.com/products/technology-license/renewable-ethylene-ethylene-oxide-ethylene-glycol/ (Accessed 14.03.2021).

Sharma, B., Larroche, C., & Dussap, C.-G. (2020). Comprehensive assessment of 2G bioethanol production. *Bioresource Technology, 313*, 123630. https://doi.org/10.1016/j.biortech.2020.123630.

Sharma, S., Sarma, S. J., & Brar, S. K. (2020). Bio-succinic acid: An environment-friendly platform chemical. *International Journal of Environment and Health Sciences (IJEHS), 2*(2), 69–80. https://doi.org/10.47062/1190.0202.01.

Sinha, A. K., & Rai, A. (2018). Process intensification for hydroprocessing of vegetable oil. In A. Górak, & A. Stankiewicz (Eds.), *Intensification of biobased processes* (pp. 188–209). The Royal Society of Chemistry. https://doi.org/10.1039/9781788010320-00188.

Skiborowski, M. (2018). Process synthesis and design methods for process intensification. *Current Opinion in Chemical Engineering, 22*, 216–225. https://doi.org/10.1016/j.coche.2018.11.004.

Skoczinski, P., Carus, M., de Guzman, D., Käb, H., Chinthapalli, R., Ravenstijn, J., Baltus, W., & Raschka, A. (2021). *Bio-based building blocks and polymers – Global capacities, production and trends 2020-2025.* (Online) Available at: http://bio-based.eu/downloads/bio-based-building-blocks-and-polymers-global-capacities-production-and-trends-2020-2025/. (Accessed 3 November 2021).

Sousa, A. F., Vilela, C., Fonseca, A. C., Matos, M., Freire, C. S. R., Gruter, G. M., et al. (2015). Biobased polyesters and other polymers from 2,5-furandicarboxylic acid: A tribute to furan excellency. *Polymer Chemistry, 6*(33), 5961–5983. https://doi.org/10.1039/C5PY00686D.

Stankiewicz, A. I. (2019). Alternative energy forms in manufacturing, processing and applications of biopolymers and biomaterials. In A. Górak, & A. Stankiewicz (Eds.), *Intensification of biobased processes* (pp. 488–506). Royal Society of Chemistry. https://doi.org/10.1039/9781788010320-00488.

Stankiewicz, A. I., & Yan, P. (2019). 110th anniversary: The missing link unearthed: materials and process intensification. *Industrial & Engineering Chemistry Research, 58*(22), 9212–9222. https://doi.org/10.1021/acs.iecr.9b01479.

Stankiewicz, A., Van Gerven, T., & Stefanidis, G. (2019). *The fundamentals of process intensification.* Weinheim, Germany: Wiley-VCH.

Stols, L., & Donelly, M. (1997). Production of succinic acid through overexpression of NAD(+)-dependent malic enzyme in an *Escherichia coli* mutant. *Applied and Environmental Microbiology, 63*(7), 2695–2701. https://doi.org/10.1128/AEM.63.7.2695-2701.1997.

Strube, J., Ditz, R., Kornecki, M., Huter, M., Schmidt, A., Thiess, H., et al. (2018). Process intensification in biologics manufacturing. *Chemical Engineering and Processing Process Intensification, 133*, 278–293. https://doi.org/10.1016/j.cep.2018.09.022.

Stuart, P. R., & El-Halwagi, M. M. (2012). *Integrated biorefineries: Design, analysis, and optimization (green chemistry and chemical engineering).* Boca Raton, Florida: CRC Press. https://doi.org/10.1201/b13048.

Sudolsky, D. (2019). Commercializing renewable aromatics for biofuels, biobased chemicals and plastics chemical recycling. *Industrial Biotechnology, 15*(6), 330–333. https://doi.org/10.1089/ind.2019.29192.dsu.

Szczygiełda, M., Antczak, J., & Prochaska, K. (2017). Separation and concentration of succinic acid from post-fermentation broth by bipolar membrane electrodialysis (EDBM). *Separation and Purification Technology, 181*, 53–59. https://doi.org/10.1016/j.seppur.2017.03.018.

Tabatabaei, M., Aghbashlo, M., Dehhaghi, M., Panahi, H. K. S., Mollahosseini, A., Hosseini, M., et al. (2019). Reactor technologies for biodiesel production and processing: A review. *Progress in Energy and Combustion Science, 74*, 239–303. https://doi.org/10.1016/j.pecs.2019.06.001.

Tan, Z., Chen, J., & Zhang, X. (2016). Systematic engineering of pentose phosphate pathway improves Escherichia coli succinate production. *Biotechnology for Biofuels, 9*(1), 262. https://doi.org/10.1186/s13068-016-0675-y.

Thuy, N. T. H., & Boontawan, A. (2017). Production of very-high purity succinic acid from fermentation broth using microfiltration and nanofiltration-assisted crystallization. *Journal of Membrane Science, 524*, 470–481. https://doi.org/10.1016/j.memsci.2016.11.073.

Tosukhowong, T. (2015). *A process for preparing succinic acid and succinate ester*. Worldwide Patent application WO2015085198A1.

Tullo, A. H. (2021). C&EN's global top 50 for 2021. *Chemical & Engineering News, 99*(27), 27–33. https://doi.org/10.1021/cen-09927-cover.

U.S. Department of Agriculture. (2021). *Biorefinery tool map gallery—Biorefineries*. (Online). Available at: https://www.usda.gov/energy/maps/maps/SingleMap.htm?Open&Competitors (Accessed 11.03.2021).

Urbance, S. E., Pometto, A. L., 3rd, Dispirito, A. A., & Denli, Y. (2004). Evaluation of succinic acid continuous and repeat-batch biofilm fermentation by *Actinobacillus succinogenes* using plastic composite support bioreactors. *Applied Microbiology and Biotechnology, 65*(6), 664–670. https://doi.org/10.1007/s00253-004-1634-2.

Van Gerven, T., & Stankiewicz, A. (2009). Structure, energy, synergy, time—The fundamentals of process intensification. *Industrial & Engineering Chemistry Research, 48*(5), 2465–2474. https://doi.org/10.1021/ie801501y.

van Heerden, C., & Nicol, W. (2013). Continuous succinic acid fermentation by *Actinobacillus succinogenes*. *Biochemical Engineering Journal, 73*(5), 5–11. https://doi.org/10.1016/j.bej.2013.01.015.

Vemuri, G. N., Eiteman, M. A., & Altman, E. (2002). Succinate production in dual-phase *Escherichia coli* fermentations depends on the time of transition from aerobic to anaerobic conditions. *Journal of Industrial Microbiology and Biotechnology, 28*(6), 325–332. https://doi.org/10.1038/sj/jim/7000250.

VoltaChem/TNO. (2016). *Empowering the chemical industry—Opportunities for electrification*. (Online). Available at: https://www.voltachem.com/publications/empowering-the-chemical-industry (Accessed 14.03.2021).

Werpy, T., & Petersen, G. (2004). *Top value added chemicals from biomass volume I—Results of screening for potential candidates from sugars and synthesis gas*. Golden, USA: National Renewable Energy Laboratory. https://doi.org/10.2172/15008859.

Wu, H., Li, Z.-M., Zhou, L., & Ye, Q. (2007). Improved succinic acid production in the anaerobic culture of an Escherichia coli pflB ldhA double mutant as a result of enhanced anaplerotic activities in the preceding aerobic culture. *Applied and Environmental Microbiology, 73*(24), 7837–7843. https://doi.org/10.1128/AEM.01546-07.

Yin, X., Zhang, X., Wan, M., Duan, X., You, Q., Zhang, J., et al. (2017). Intensification of biodiesel production using dual-frequency counter-current pulsed ultrasound. *Ultrasonics Sonochemistry, 37*, 136–143. https://doi.org/10.1016/j.ultsonch.2016.12.036.

Yoon, L. W., Ngoh, G. C., Chua, A. S. M., Fazly, M., Patah, A., & Teoh, W. H. (2019). Process intensification of cellulase and bioethanol production from sugarcane bagasse via an integrated saccharification and fermentation process. *Chemical Engineering and Processing Process Intensification, 142*, 107528. https://doi.org/10.1016/j.cep.2019.107528.

Zeikus, J., Jain, M., & Elankovan, P. (1999). Biotechnology of succinic acid production and markets for derived industrial products. *Applied Microbiology and Biotechnology, 51*(5), 545–552. https://doi.org/10.1007/s002530051431.

Zhang, H., Ye, Q., Chen, L., Wang, N., Xu, Y., & Li, Y. (2020). Purification of isopropanol-butanol-ethanol (IBE) from fermentation broth: Process intensification and evaluation. *Chemical Engineering and Processing Process Intensification, 158*, 108182. https://doi.org/10.1016/j.cep.2020.108182.

Zinkel, D. F., & Russel, J. (1989). *Naval stores: Production, chemistry, utilization*. New York: Pulp Chemicals Association.

CHAPTER 11

Modeling and optimization of supply chains: Applications to conventional and intensified biorefineries

Fernando Israel Gómez-Castro[a], Yulissa Mercedes Espinoza-Vázquez[a], and José María Ponce-Ortega[b]

[a]Departamento de Ingeniería Química, División de Ciencias Naturales y Exactas, Universidad de Guanajuato, Guanajuato, Guanajuato, Mexico
[b]Facultad de Ingeniería Química, División de Estudios de Posgrado, Universidad Michoacana de San Nicolás de Hidalgo, Morelia, Michoacán, Mexico

Nomenclature

$CP(m)$ [kUSD/kt]	unitary production cost for the product "m"
$CTP(m,k,n)$ [kUSD/y]	cost due to the transportation of the product "m" from the facility "k" to the facility "n".
$Demand(m,n)$ [kt/y]	demand of the product "m" in the market "n"
$D_1(i,j,k)$ [km]	distance from the source "j" of biomass "i" to the facility "k"
$D_2(k,n)$ [km]	distance from the facility "k" to the market "n"
$FP(m,k,n)$ [kt/y]	mass production of the product "m" in the facility "k" to the market "n"
i	index for possible raw material: barley, corn, sorghum, wheat (see Fig. 11.3)
j	index for possible source of raw material: AGS, BC, ..., ZAC (see Fig. 11.3)
k	index for possible facilities: BC, COAH, ..., VER (See Fig. 11.3)
LC [kUSD/m^2]	land cost per square meter
$LS(m,k)$ [m^2/L]	land surface required per liter of product "m"
m	index for possible products: BE or LA (see Fig. 11.3)
M_{11} [ ]	10,000,000
n	index for markets: QRO, NL, ..., HGO (see Fig. 11.3)
$PD(m,n)$ [kL/y]	total demand of the product "m" in the market "n"
$Profit$ [kUSD/y]	difference between the income and the total cost
$RMA(i,j)$ [kt/y]	availability of raw material "i" in the source "j"
$RMAF(i,j,k,m)$ [kt/y]	mass of raw material "i" from the source "j" entering to the facility "k" to produce "m"
$RMUC(i)$ [kUSD/kt]	unitary cost for the raw material "i"
$SP(m)$ [kUSD/y]	selling price for the product "m"
$S(m,k)$ [ ]	binary variable, existence of the product "m" in the facility "k"
$TCL(k)$ [kUSD/y]	cost due to land acquisition to build the facility "k"
$TCSB$ [kUSD/km kt]	unitary transportation cost of solid lignocellulosic biomass
$TCLP$ [kUSD/km kt]	unitary transportation cost of the liquid product
$TCP(k)$ [kUSD/y]	total production cost for the biorefinery "k"
$TCRM(k)$ [kUSD/y]	cost of the raw material to be processed in the facility "k"

$TSP(i,j,k)$ [kUSD/y]	cost due to the transportation of biomass "i" from the source "j" to the facility "k"
$Udemand(m)$ [kt/y]	unsatisfied demand of the product "m"
$Yld(m)$ [ ]	yield of the raw material to the product "m"
$Y(k)$ [ ]	binary variable, existence of the facility "k"
$\rho(m)$ [kt/kL]	density of the product "m".

11.1 Introduction

Biofuels are part of the strategies to replace the use of fossil fuels, therefore reducing the environmental impact associated with the use of petroleum derivatives, particularly in terms of greenhouse gas emissions. Moreover, biofuels could be key in the transition to a production scheme based on the circular economy principles. Additionally, a proper planning on biofuels production could be fundamental to achieve various of the sustainable development goals. Biofuels can be obtained from materials containing triglycerides, or from materials containing sugars. An example of this is biodiesel, which is a mixture of fatty acid alkyl esters (FAAEs), derived from triglyceride molecules. Triglycerides and alcohol are converted to alkyl esters (biodiesel) via a catalyzed transesterification reaction (Gaurav, Leite, Ng, & Rempel, 2013). Another example of biofuel is biojet fuel, which is produced from nonedible vegetable oils, with the hydrotreating process being the most commonly used (Romero, Gómez, Gutiérrez, & Hernández, 2016). On the other hand, bioethanol and biobutanol can be obtained from sugar-containing materials. Bioethanol is used in blends of 10 vol% (Khuong et al., 2016). Another important biofuel is biobutanol, which is blended with gasoline in proportions of around 16 vol% (Zhang, Jing, Yongmei, Yangdong, & Yong, 2016). Of the two biofuels, biobutanol is particularly interesting because it is less corrosive to fuel system component materials compared to bioethanol. For the production of biofuels, there is a wide range of options available, stands up lignocellulosic biomass, which has great potential to provide raw material for manufacturing a wide variety of chemicals and materials. The biomass comes mainly from residues of corn, sugarcane, sorghum, wheat, and barley. Of these residues, 20% is burned, such biomass burning is recognized as a significant source of greenhouse gas emissions (Damián-Huato et al., 2013). Therefore, there is a major interest in assessing the energy potential of biomass.

Unfortunately, the industrial production of biofuels is not economically competitive (Xue, Zhao, Chen, Yang, & Bai, 2017). By integrating the production of high value-added products into biofuel production, overall profitability will be improved, making the biorefinery scheme more attractive to biobased companies. Analyzing the chemicals with the greatest potential, from both an economic and technological point of view, 30 potential candidates have been identified (Werpy & Petersen, 2004). Those candidates are listed in Table 11.1.

Table 11.1 Top 30 high value-added products.

Carbon number	Potential top 30 candidates
1	Carbon monoxide and hydrogen
3	Glycerol, 3 hydroxypropionic acid, lactic acid, malonic acid, propionic acid, serine
4	Acetoin, aspartic acid, fumaric acid, 3-hydroxybutyrolactone, malic acid, succinic acid, threonine
5	Arabinitol, furfural, glutamic acid, itaconic acid, levulinic acid, proline, xylitol, xylonic acid
6	Aconitic acid, citric acid, 2,5 furan dicarboxylic acid, glucaric acid, lysine, levoglucosan, sorbitol

A recent report by Biddy, Scarlata, and Kinchin (2016) has described some of the biomass derivatives with the greatest potential in the market. Those products and their uses are shown in Table 11.2.

The development of biorefineries for the production of biofuels has several aspects that must be included for production and distribution planning. From the selection of raw materials to the markets in which it must be distributed. Additionally, since biofuel production is not economically sustainable, the production of high value-added products must be considered. All these aspects must be taken into account because, even if processes using modern technologies are considered (as do intensified processes), if the

Table 11.2 Top 12 high added value products.

Bioproduct	Uses
Butadiene (1,3-)	Production of polybutadiene and styrene-butadiene rubbers
Butanediol (1,4-)	Production of polymers, solvents, and specialty chemicals
Ethyl lactate	To replace volatile organic petroleum-derived compounds. As a solvent
Fatty alcohols	Production of anionic a nonionic surfactant for household cleaners and industrial application
Furfural	Production of foundry resins, plastics, pharmaceuticals, agro-chemical products, and nonpetroleum-derived chemicals
Glycerin	Humectant in food and personal care products, and more than 1500 uses
Isoprene	Production of polyisoprene rubber, styrene co-polymers, and butyl rubber
Lactic acid	For food, pharmaceuticals, personal care products, industrial uses, and polymers
Propanediol (1,3-)	Polymers, personal care products, solvents, and lubricants
Propylene glycol	Production of consumer products such as antiperspirants, etc., and for industrial use for the production of unsaturated polyester resins
Succinic acid	Production of commodity chemicals, polymers, surfactants, and solvent
Xylene (para)	Production of terephthalic acid (TA) and dimethyl terephthalate (DMT)

supply chain is not properly planned, the biorefinery will not be economically feasible or environmentally friendly. Therefore, the detailed study of the supply chain and its optimization through mathematical techniques is essential for the selection of the best distribution alternative as well as the selection of the best location for the plants. Such a supply chain must find the best trade-off between the production scale and transportation cost (de Jong et al., 2017). This chapter is focused on the supply chain for the production of biofuels and high value-added production in a biorefinery scheme. First, the most important objective functions to be considered for the optimization of supply chains for biorefining are mentioned. The elements involved in supply chain analysis and some insights on the generation of this information are also mentioned. Then, the effect of intensified processes on the global supply chain analysis is mentioned. The methodology for modeling and optimizing the supply chain for the production of biofuels and high value-added products is presented. Finally, a case study is developed and discussed.

11.2 Objective functions for supply chain optimization

Biorefineries appear to be a viable solution to replace traditional fossil fuel refineries, but their implementation requires exploration of several aspects, such as raw material selection and processing routes, products, collection sites, processing and markets, as well as many other sustainability criteria. Therefore, the best choice is not easily visible, so the use of a mathematical optimization models are valuable tools, since they allow well-informed decision making and analysis of scenarios. For optimization purposes, one or more objective functions are required, linked to a set of constraints given by a system of equations. The selection of a given objective function will drive the selection of a given supply chain, thus special care must be taken when defining such function. Some of the most common objective functions used to plan the supply chain for a biobased production are described in this section.

A considerable number of works on supply chain optimization focus on economic objectives, such as profit maximization or net present value. For example, Kim, Realff, and Lee (2011) proposed a model for the optimal design of biomass supply chain networks under conditions of uncertainty due to varying market prices, product demand, and processing technologies. This project was based on the southeastern region of the United States for the production of gasoline and diesel, whose objective function is to maximize profit. As a result, a model capable of making decisions for biofuel conversion processing infrastructure, including processing locations, volumes, supply networks, and transportation logistics from forest resources to conversion and from conversion to final markets was obtained. On the other hand, Marvin, Schmidt, and Daoutidis (2013) proposed a mixed integer linear program to determine the location and capacity of biomass processing facilities to evaluate the Midwestern U.S.

biofuel supply chain. Eight types of biomass, seven processing technologies will be considered to produce biofuels. The objective function was taken as the net present value, satisfying market demand.

Avami (2012) proposed the development of a supply chain for biodiesel production in Iran, having TAC minimization as a target function. The final model proposes the centralization of the network designed with low costs, concluding that algae are feasible for biodiesel production in certain areas of the country. Rendon-Sagardi, Sánchez-Ramírez, Cortes-Robles, Alor-Hernández, and Cedillo-Campos (2014) developed a supply chain for the production of bioethanol and fuel from sugarcane and sorghum grains, blending at 10% volume. Several aspects were concluded from this analysis. Firstly, Mexico is not selfsufficient in the production of fuel; it requires imports from other countries to meet the demand for it. On the other hand, if gasoline is blended with bioethanol, only 0.8% of the internal demand would be covered; however, the amount of CO_2 would be reduced in 1282.95 million tons between 2014 and 2030.

Most of the work mentioned above optimizes the economic aspect of the supply chain as a single objective. In the design of a supply chain, not only one objective should be considered (Beamon, 1998). An example of this is the work by Murillo-Alvarado, Ponce-Ortega, Serna-González, Castro-Montoya, and El-Halwagi (2013) who developed a generalized disjunctive programming model that takes into account the simultaneous selection of products, raw materials, and processing steps. The optimal solution can consist of multiproduct and multifeedstock biorefineries. Taking into account two objectives, profit maximization and minimization of greenhouse gas emissions, while considering the number of processing steps. The main products are bioethanol, biodiesel, and hydrogen, while sugarcane, jatropha, and microalgae appear as raw materials in the optimal pathways. In other work, Santibañez-Aguilar, González-Campos, Ponce-Ortega, Serna-González, and El-Halwagi (2014) developed a generalized disjunctive programming model that takes into account the simultaneous selection of products, raw materials, and processing steps. The optimal solution can consist of biorefineries of multiple products and raw materials. Taking into account three objectives, profit maximization, minimization of greenhouse gas emissions, and maximization of the social impact, which is modeled in terms of the number of generated employments. On the other hand, Geng and Sun (2021), propose the development of a supply chain for the production of biodiesel from kitchen waste for Jiangsu Province. Taking into account economic, environmental, and social objectives, using a heuristic algorithm. The design results of the supply chain network show that the pretreatment facilities are located in the places with large population and large supply of urban kitchen waste, so as to reduce the transportation cost of kitchen waste. Similarly, it is important to note that the production of biofuels requires a large amount of water (Gerbens-Leenes, Hoekstra, & van der Meer, 2009). However, this objective function has been neglected and ignored by considering only the environmental impact.

11.3 Elements of the supply chain for a biorefinery

There are several elements involved in the supply chain for the production of biofuels and bioproducts. First, there is the type of raw material to be used, it is important to know what kind of materials can be used to produce biofuels. It is necessary to know what amounts of this biomass are available and its unitary cost. This may represent a challenge, since such cost is not easily available or even determined, as occurs with some residues. In case there is information about its variation with respect to the seasonal availability, it can be of useful for the planning of the supply chain. All the data required to determine a proper supply chain for the production of biofuels and/or bioproducts can be obtained from different sources. Information about the availability of crops is usually reported by government agencies, as the Department of Energy in the United States, or the Service of Agrifood and Fisheries Information (Servicio de Información Agroalimentaria y Pesquera, SIAP) and the Ministry of Agriculture, Livestock, Rural Development, Fisheries and Food (Secretaría de Agricultura, Ganadería, Desarrollo Rural, Pesca y Alimentación, SAGARPA) in Mexico. Moreover, scientific papers can be found with reports on the availability of sources of vegetables oils (e.g., Rosillo-Calle, Pelkmans, & Walter, 2009), or lignocellulosic biomass (e.g., Espinoza-Vázquez, Gómez-Castro, & Ponce-Ortega, 2021). Additionally, it is necessary to define if the available raw material is enough to satisfy the demand of the considered biofuels and bioproducts, or if a set of them must be preferred to make the production profitable. Once the potential raw materials, their location, and availability are determined, the next required information is about how the materials will be transported to the facility (or facilities) where they will be processed into biofuels and/or bioproducts. It is necessary to know the distances between the biomass sources and the potential locations of the facilities. This information can be acquired from different applications, such as Google Maps. Likewise, it is necessary to know the type of transportation needed, i.e., in the case of solid transportation, 40-ton trucks are generally used, but they vary depending on the restrictions of each country. In addition to this information, the cost of fuel per kilometer per ton must be known in order to be able to relate the profit appropriately. Likewise, it must be considered how much greenhouse gases are released per kilometer traveled, since it is also desired that the supply chain has a low environmental impact. Such information can be found reported in literature for a number of cases (e.g., Espinoza-Vázquez et al., 2021; Méndez-Vázquez et al., 2016).

Another important aspect to consider in the supply chain is the location of the facilities. First of all, it is necessary to analyze which locations have the infrastructure to be able to build a biorefinery, since if the area is very urban it will be impossible to find an available area. Some information can be useful to eliminate alternatives from a first insight, reducing the complexity of the problem. Additionally, since the supply chain is expected not only to be economically feasible, but also to have a low water footprint, the locations must have enough water for industrial use, regions with water scarcity can be removed from the set of feasible locations. On the other hand, information associated with the biomass processing is the technology employed to obtain the biofuels and/or bioproducts. There must be knowledge about the mean yields that each technology can achieve, their cost per unit of biofuel produced, and the utility requirements. Moreover, it is necessary to have information about the costs for utilities, and even the cost of the industrial land for each potential location. In the case of the information related to the production process, estimations for the yields from biomass to the desired product, together with approximated production costs, can be found in literature (e.g., Sassner, Galbe, & Zacchi, 2008; You et al., 2008).

Likewise, it is important to know the final sales price per ton of product. Moreover, to evaluate the social impact, it would be desirable to have information about the potential jobs generated by the processing of each product with each type of raw material, such information can be obtained from sources such as IMPLAN (Santibañez-Aguilar et al., 2014). Finally, it is important to transport the facilities to the markets. It is important to consider that, for a given raw material, the fuels demand may not be met, so strategic markets must be selected for supply chain planning. In addition, it is important to select the type of transportation. In the case of liquids, a 28-ton transport could be selected, although this varies depending on the country Adding the above considerations about the transportation of raw material to the facilities.

11.4 Process intensification and supply chain

Within the planning of a supply chain, production costs associated with the equipment used for the transformation of raw materials into products are required. In this area, two types of processes can be considered: conventional or intensified. Conventional processes, as their name indicates, are processes based on widely studied technologies. Intensified processes, on the other hand, seek to significantly enhance these conventional processes. The main objective of the intensified processes is to achieve technological implementations, all this while occupying less space as well as being more efficient in energy management, i.e., more environmentally friendly (Reay, Ramshaw, & Harvey, 2013). There is a wide variety of specialized books and research papers that approach the various issues related to process intensification (Kiss, 2014; Segovia-Hernández & Bonilla-Petriciolet, 2016; Tian, Demirel, Hasan, & Pistikopoulos, 2018). Research on process intensification has focused on decreasing energy demand and processing costs, although the use of intensified technologies may also have a positive effect on the process operation and safety. Decreasing energy requirements has a direct effect on reducing CO_2 emissions, but process intensification has further effects on environmental impact, as the need of smaller equipment, which is translated into lower use of resources and lower land needs.

Concerning supply chains, there are currently very few studies that consider the application of process intensification. In this aspect, Villicaña-García, Ramírez-Márquez, Segovia-Hernández, and Ponce-Ortega (2021), developed a mathematical programming model for planning the production and distribution of solar grade silicon used for the construction of solar panels to satisfy the demand for electricity in the residential sector in Mexico. In this study, process intensification was used to improve the different ways of obtaining solar grade silicon. The results showed that the application of intensified processes allowed obtaining better production costs as well as a higher silicon production rate. On the other hand, Monsiváis-Alonso, Mansouri, and Román-Martínez (2020) focused on the production of omega-3 polyunsaturated fatty acid concentrates from the residual oil of a tuna processing plant in Mexico. They proposed a framework for integrating economic, environmental and social aspects with process design decisions by formulating and solving a multiobjective optimization problem that includes options for intensified processing in a superstructure. The results obtained, from an optimization approach, demonstrated that process intensification has the potential to improve processing unit operations, as well as to overcome challenges with respect to lipid processing. It is important to mention that, for the development of supply chains, the use of intensified processes does not have a considerable effect on design. This is because supply chains are mainly based on the logistics of transporting both raw materials and products. However, the use of intensified processes can reduce production costs and emissions of greenhouse gases due to the process, making the chain more sustainable.

11.5 Modeling and optimization of the supply chain

Optimization and mathematical modeling tools can be used to select the best combination of biofuels and high added value products. The development of a proper model to achieve this purpose implies relating the interactions between the different elements of the supply chain, which have been mentioned in Section 11.3. The model must be able to choose between types of raw materials, select the source of the raw material, transport the raw material to a certain plant, and so on. To help the modeler to visualize all the potential interactions, a superstructure must be developed. An example of a superstructure for the supply chain of a biofuel is shown in Fig. 11.1, taking into account the basic elements.

In Fig. 11.1, raw material 1 can be produced by supplier 1, supplier 2, and so on, or may be produced only by one supplier. Raw materials can be processed in one or more facilities, and either a single product or a combination of products can be obtained. These products may be transported to all or to a given market. For a given case, some of the interactions between the elements on the superstructure may not exist, or other elements can be included.

Modeling and optimization of supply chains 369

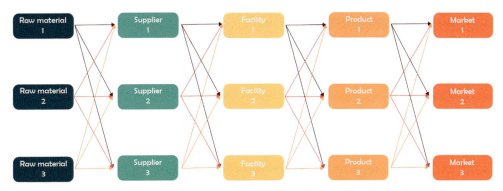

Fig. 11.1 General superstructure for the supply chain of biofuels and high added value products.

11.5.1 Modeling discrete decisions

Optimization involves a decision-making process, where one decision has characteristics that make it better than the other options. Likewise, a mathematical optimization model consists of one or more objective functions and a set of constraints in the form of a system of equations, which in turn shape the feasible region. Both the objective function and the constraints are in terms of a vector of decision variables, on whose values each possible solution will depend. The objective of optimization is to find the best solution of decision models, where the optimization model can be linear (LP), nonlinear (NLP), mixed integer linear (MILP) or mixed integer nonlinear (MINLP); depending on the type of objective function, constraints, and/or decision variables. Therefore, the optimization model can be represented only with continuous variables or with discrete variables, or with a combination of both types of variables. The different types of optimization models are shown in Table 11.3.

In the case of supply chains, decision-making involves the use of binary variables in conjunction with continuous variables, which implies that these types of problems are modeled as MILP or MINLP.

11.5.2 The generalized disjunctive programming

Once the superstructure has been defined, the mathematical model representing it must be developed. The different potential decision involved in a supply chain optimization

Table 11.3 Optimization models.

Constraints	Variable	Model
Linear	Continuous	LP
	Continuous and discrete	MILP
Nonlinear	Continuous	NLP
	Continuous and discrete	MINLP

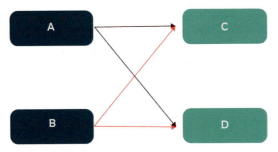

Fig. 11.2 Example of supply chain structure for decision-making.

statement can be represented through a generalized disjunctive programming (GDP) model. In the GDP, a set of constraint depending on the selection of a given route are represented through disjunctions, while other constrains which are not dependent of the selected supply chain can be added outside of the disjunctions. Raman and Grossmann (1994), reported integer programming techniques to optimize logical decisions expressed in terms of disjunctions. The GDP model allows a combination of algebraic and logical equations, which facilitates the representation of discrete decisions. Türkay and Grossmann (1996) proposed a logic-based outer approximation algorithm to solve nonlinear GDP problems for process networks involving two terms in each disjunction. In general, the GDP model includes Boolean variables, disjunctions, and logical propositions, as shown below. Fig. 11.2 shows an example of superstructure for a supply chain, where two kinds of raw material (A and B) can be selected to produce two possible products (C and D).

In the case of the raw material, a logic proposition Y_A would indicate that the raw material A will be selected for this process (Y_A is true), while the logic proposition Y_B would indicate that the raw material B is chosen (Y_B is true). Similarly, Y_C and Y_D represent the potential selection of products C and D, respectively. The relationship between elements in the superstructure can be stated in terms of logical propositions, as shown in Table 11.4.

Table 11.4 Relationship between elements in the superstructure.

Name	Connector	Logical proposition	Binary equivalent
NOT	$\neg$	$\neg Y_i$	$1 - y_i$
OR	$\vee$	$Y_i \vee Y_j$	$y_i + y_j \geq 1$
EXCLUSIVE OR	$\underline{\vee}$	$Y_i \underline{\vee} Y_j$	$y_i + y_j = 1$
IF THEN	$\Rightarrow$	$Y_i \Rightarrow Y_j$	$y_i - y_j \leq 0$
AND	$\wedge$	$Y_i \wedge Y_j$	$y_i \geq 1, y_j \geq 1$
IFF THEN	$\Leftrightarrow$	$Y_i \Leftrightarrow Y_j$	$y_i = y_j$

As an example, having $\neg Y_A$, would indicate that raw material A is not selected. The proposition $Y_A \vee Y_B$ would imply that either A, B, or both raw materials are used simultaneously, i.e., A or B are true, or both are simultaneously true. On the other hand, if the proposition $Y_A \underline{\vee} Y_B$ is used, it implies that only a single raw material can be selected. Likewise, the proposition $Y_A \Rightarrow Y_C$ is an implication, which means that if A is selected then C will be produced. However, if the raw material A is not selected, there is still a possibility that C will be produced from B. The proposition $Y_A \wedge Y_B$ would imply that both raw materials are used, i.e., Y_A and Y_B are true. Finally, the double implication $Y_A \Leftrightarrow Y_C$ means that the raw material A and the product C are dependent of each other. If A is selected as raw material, C would be produced. If A is not selected, C is not produced.

Implications are used to represent the interactions between the components of the superstructure, and more complex implications can be formulated and converted into constraints in terms of binary variables. To do this, it is necessary to modify the implications in a form known as a conjunctive normal form, where the composed propositions are linked by AND operators, as follows:

$$(Y_i \vee Y_m) \wedge (Y_j \vee Y_m) \wedge (Y_k \vee Y_m) \quad (11.1)$$

Once all the constraints and relationships between the elements of the superstructure have been established, the model for each element must be developed and the objective function of the optimization problem must be determined. A general representation of the GDP statement is given by:

$$\min Z = \sum_{k \in K} c_k + f(\overline{x}) \quad (11.2)$$

s.t.

$$g(\overline{x}) \leq 0 \quad (11.3)$$

$$\bigvee_{j \in J_k} \begin{bmatrix} Y_{jk} \\ h_{jk}(\overline{x}) \leq 0 \\ c_k = \gamma_{jk} \end{bmatrix}, k \in K \quad (11.4)$$

$$\Omega(Y) = True \quad (11.5)$$

$$x_i \geq 0, c_k \geq 0, Y_{jk} \in \{true, false\} \quad (11.6)$$

where $x \in R^n$ is the vector of continuous variables and Y_{jk} are the Boolean variables. c_k are continuous variables associated to the costs and γ_{jk} are fixed values for the cost of a given alternative; $f: R^n \to R^1$ is the term for continuous variables x in the objective function and $g: R^n \to R^q$ are sets of common constraints that hold independently of discrete decisions. Finally, the term $\Omega(Y) = True$ corresponds to the logical proposition in terms of Boolean variables. Once the GDP model has been developed, it can be transformed into a mixed-integer linear programming or mixed-integer nonlinear programming model through relaxation techniques, which are presented in the next subsection.

11.5.3 Relaxation of a generalized disjunctive programming

To solve the optimization problem stated above, the model must be considered in terms of discrete variables instead of logical variables. To achieve this, there are two relaxation methods that can be used: the Big-M approach and the convex hull approach. These methods are described below.

11.5.3.1 The Big-M approach
Consider the follow disjunction:

$$\vee_{i \in D} \begin{bmatrix} Y_i \\ h_i(\bar{x}) \leq 0 \end{bmatrix} \tag{11.7}$$

where $h_i(\bar{x})$ is a nonlinear function. Equation 11.7 can be relaxed through the Big "M" method as:

$$h_i(\bar{x}) \leq M_i(1 - y_i) \tag{11.8}$$

where the right-side term in the inequality is replaced by $M_i(1 - y_i)$. The parameter M_i must be big enough to exceed the region where $h_i(\bar{x})$ is defined. Thus, if Y_i is true, it would imply that $y_i = 1$, and the relaxed constraint would remain $h_i(\bar{x}) \leq 0$, satisfying the condition given by the logical proposition. On the other hand, if Y_i is false, $y_i = 0$, and the constraint would be $h_i(\bar{x}) \leq M_i$. Therefore, if M_i is big enough, a feasible region is defined, where the solution would be defined by other logical propositions. For the case of having an equality in the proposition, as shown below:

$$\vee_{i \in D} \begin{bmatrix} Y_i \\ g_i(\bar{x}) = 0 \end{bmatrix} \tag{11.9}$$

the equality must be converted to an equivalent set of inequalities, as shown below:

$$\vee_{i \in D} \begin{bmatrix} Y_i \\ g_i(\bar{x}) \leq 0 \\ -g_i(\bar{x}) \geq 0 \end{bmatrix} \tag{11.10}$$

Thus, there would be two constraints:

$$g_i(\bar{x}) \leq M_i(1 - y_i) \tag{11.11}$$

$$g_i(\bar{x}) \geq -M_i(1 - y_i) \quad (11.12)$$

For this example, if $y_i = 1$, the constraints $g_i(\bar{x}) \leq 0$ and $g_i(\bar{x}) \geq 0$. To accomplish these constraints $g_i(\bar{x}) = 0$, which is equal to the original constraint. On the other hand, if the constraints are $g_i(\bar{x}) \leq M_i$ and $g_i(\bar{x}) \geq M_i$, which, if M_i is big enough, would generate a feasible region, implying that the solution must be given by other logical propositions.

11.5.3.2 The convex hull approach

For this approach, each variable x_j in the vector of variables $\bar{x}$ is represented as a sum of variables (disaggregated variables), as shown below:

$$x_j = \sum_{i \in I} v_{ji} \quad (11.13)$$

Each variable v_{ji} can be related to a logical variable Y_i. For this case, the equivalent disjunction is given by:

$$y_i h_i \left(\frac{\bar{v}_i}{y_i} \right) \leq 0 \quad (11.14)$$

$$v_{ji} \geq 0 \quad (11.15)$$

$$v_{ji} \leq v_{ji}^{up} y_i \quad (11.16)$$

where $\bar{v}_i$ is the vector of disaggregated variables, v_{ji}^{up} is the upper bound of the variable, this limit can be determined in terms of the upper bound of the variable x_j. On the other hand, if the constraint is an equality, as in Eq. (11.9), the following constraint is obtained:

$$y_i g_i \left(\frac{\bar{v}_i}{y_i} \right) = 0 \quad (11.17)$$

Eqs. (11.15) and (11.16) must be added to Eq. (11.17), to complete the convex hull equivalent of Eq. (11.9).

Both methods can be used to relax the disjunctions shown in the GDP model, however the Big-M approach is easier to apply than the convex hull. On the other hand, the Big-M approach is used if the constraints within the disjunction are linear, while the convex hull is more recommended for nonlinear situations.

11.6 Case study: Optimization of the supply chain for the production of bioethanol and high value-added products in Mexico

In this section, a case of study related to the determination of the optimal supply chain for the production of bioethanol and levulinic acid in Mexico is presented. Mathematical optimization will be used, developing a GDP model to represent the supply chain. For this case of study, four types of residues from crops have been considered as raw material. Those crops are corn, sorghum, wheat, and barley, distributed along the 32 states in Mexico. Five potential locations for the biorefineries are proposed: Baja California, Coahuila, Querétaro, Guanajuato, and Veracruz. The states are established as feasible locations since they have the highest industrial infrastructure in the country. Each facility can produce either one or both products. The products obtained on each biorefinery are distributed to the markets, which are selected in terms of their highest demand for gasoline (for bioethanol) (Sistema de Información Energética, n.d.) and for levulinic acid for its use as solvent (Petróleos Mexicanos, 2018).

The selection of the best supply chain to produce bioethanol and levulinic acid is carried out aiming to maximizing the profit. The economic objective represents the difference between the income (i.e., sales of products) and the total cost. The total cost involves the cost of production, raw material cost, land cost, and transportation cost. Table 11.5 presents the assumed production costs and the reported yields of the products, as well, the density of the products. In Fig. 11.3, the problem statement is presented.

The profit is estimated as shown in Eq. (11.18):

$$Profit = \sum_m \sum_k \sum_n FP(m,k,n) \cdot SP(m) - \sum_k TCP(k) - \sum_k TCRM(k) \\ - \sum_k TCL(k) - \sum_i \sum_j \sum_k TSP(i,j,k) - \sum_m \sum_k \sum_n CTP(m,k,n)$$

(11.18)

In Eq. (11.18), $FP(m,k,n)$ is the mass production of each product m in the facility k and $SP(m)$ is the selling price for the product m. In the case of bioethanol, sales price has been taken as 1902.57 kUSD/kt (El Cronista, 2020) and for levulinic acid is taken as 8280 kUSD/kt (eBioChem, n.d.).

Subject to:

$$Y(BC) + Y(COAH) + Y(GTO) + Y(QRO) + Y(VER) \geq 1 \quad (11.19)$$

Table 11.5 Production costs, density, and yields.

Product	Production cost (kUSD/kt)	Density (kt/kL)	Yield (t product/t biomass)	Source
Bioethanol	832.23	0.00081	0.1741	Sassner et al. (2008)
Levulinic acid	2148.23	0.00114	0.1940	Gozan, Ryan, and Krisnandi (2017)

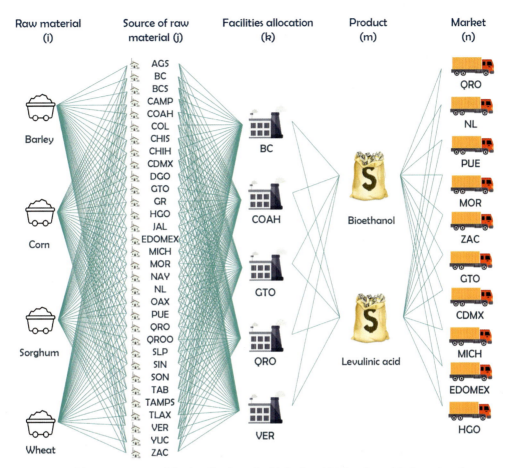

Fig. 11.3 Problem statement of the localization of a biofuel and high value-added product factory.

$$S(BE, BC) + S(LA, BC) + S(BE, COAH) + S(LA, COAH) + S(BE, GTO)$$
$$+ S(LA, GTO) + S(BE, QRO) + S(LA, QRO) + S(BE, VER) + S(LA, VER) \geq 1$$
(11.20)

Eq. (11.19) implies that at least one of the biorefineries k must exist, while Eq. (11.20) sets that at least one of the products m must be obtained in any of the facilities k. Moreover, each facility k can either exist or not. A logical variable $Y(k)$ is assigned to each facility. Furthermore, each biorefinery may produce any of the desired products or both. A logical variable $S(m,k)$ is thus assigned to each product. These considerations can be represented in terms of a nested disjunction for each facility, as follows:

$$\begin{bmatrix} Y(BC) \\ TCP(BC) = \sum_m \sum_n FP(m, BC, n) \cdot CP(m) \\ TCRM(BC) = \sum_m \sum_n \frac{FP(m, BC, n)}{Yld(m)} \cdot RMUC(i) \\ TCL(BC) = \sum_m \sum_n \frac{FP(m, BC, n)}{\rho(m)} \cdot LS(m, BC) \cdot LC \\ S(BE, BC) \\ \sum_n FP(BE, BC, n) = \sum_i \sum_j RMAF(i, j, BC, BE) \cdot Yld(BE) \\ S(LA, BC) \\ \sum_n FP(LA, BC, n) = \sum_i \sum_j RMAF(i, j, BC, LA) \cdot Yld(LA) \end{bmatrix} \vee \begin{bmatrix} -S(BE, BC) \\ \sum_n FP(BE, BC, n) = 0 \end{bmatrix} \vee \begin{bmatrix} -S(LA, BC) \\ \sum_n FP(LA, BC, n) = 0 \end{bmatrix} \vee \begin{bmatrix} -Y(BC) \\ TCP(BC) = 0 \\ TCRM(BC) = 0 \\ TCL(BC) = 0 \end{bmatrix}$$

(11.21)

$$Y(COAH)$$

$$TCP(COAH) = \sum_m \sum_n FP(m, COAH, n) \cdot CP(m)$$

$$TCRM(COAH) = \sum_m \sum_n \frac{FP(m, COAH, n)}{Yld(m)} \cdot RMUC(i)$$

$$TCL(COAH) = \sum_m \sum_n \frac{FP(m, COAH, n)}{\rho(m)} \cdot LS(m, COAH) \cdot LC$$

$$\left[\begin{array}{c} S(BE, COAH) \\ \sum_n FP(BE, COAH, n) = \sum_i \sum_j RMAF(i, j, COAH, BE) \cdot Yld(BE) \end{array} \right] \vee \left[\begin{array}{c} \neg S(BE, COAH) \\ \sum_n FP(BE, COAH, n) = 0 \end{array} \right]$$

$$\left[\begin{array}{c} S(LA, COAH) \\ \sum_n FP(LA, COAH, n) = \sum_i \sum_j RMAF(i, j, COAH, LA) \cdot Yld(LA) \end{array} \right] \vee \left[\begin{array}{c} \neg S(LA, COAH) \\ \sum_n FP(LA, COAH, n) = 0 \end{array} \right]$$

$$\vee \left[\begin{array}{c} \neg Y(COAH) \\ TCP(COAH) = 0 \\ TCRM(COAH) = 0 \\ TCL(COAH) = 0 \end{array} \right]$$

(11.22)

$$\begin{bmatrix} Y(GTO) \\ TCP(GTO) = \sum_m \sum_n FP(m, GTO, n) \cdot CP(m) \\ TCRM(GTO) = \sum_m \sum_n \frac{FP(m, GTO, n)}{Yld(m)} \cdot RMUC(i) \\ TCL(GTO) = \sum_m \sum_n \frac{FP(m, GTO, n)}{\rho(m)} \cdot LS(m, GTO) \cdot LC \\ \begin{bmatrix} S(BE, GTO) \\ \sum_n FP(BE, GTO, n) = \sum_i \sum_j RMAF(i,j, GTO, BE) \cdot Yld(BE) \end{bmatrix} \vee \begin{bmatrix} \neg S(BE, GTO) \\ \sum_n FP(BE, GTO, n) = 0 \end{bmatrix} \\ \begin{bmatrix} S(LA, GTO) \\ \sum_n FP(LA, GTO, n) = \sum_i \sum_j RMAF(i,j, GTO, LA) \cdot Yld(LA) \end{bmatrix} \vee \begin{bmatrix} \neg S(LA, GTO) \\ \sum_n FP(LA, GTO, n) = 0 \end{bmatrix} \end{bmatrix} \vee \begin{bmatrix} \neg Y(GTO) \\ TCP(GTO) = 0 \\ TCRM(GTO) = 0 \\ TCL(GTO) = 0 \end{bmatrix}$$

(11.23)

$$Y(QRO)$$

$$TCP(QRO) = \sum_m \sum_n FP(m, QRO, n) \cdot CP(m)$$

$$TCRM(QRO) = \sum_m \sum_n \frac{FP(m, QRO, n)}{Yld(m)} \cdot RMUC(i)$$

$$TCL(QRO) = \sum_m \sum_n \frac{FP(m, QRO, n)}{\rho(m)} \cdot LS(m, QRO) \cdot LC$$

$$\begin{bmatrix} S(BE, QRO) \\ \sum_n FP(BE, QRO, n) = \sum_i \sum_j RMAF(i,j, QRO, BE) \cdot Yld(BE) \end{bmatrix} \vee \begin{bmatrix} \neg S(BE, QRO) \\ \sum_n FP(BE, QRO, n) = 0 \end{bmatrix}$$

$$\begin{bmatrix} S(LA, QRO) \\ \sum_n FP(LA, QRO, n) = \sum_i \sum_j RMAF(i,j, QRO, LA) \cdot Yld(LA) \end{bmatrix} \vee \begin{bmatrix} \neg S(LA, QRO) \\ \sum_n FP(LA, QRO, n) = 0 \end{bmatrix}$$

$$\vee \begin{bmatrix} \neg Y(QRO) \\ TCP(QRO) = 0 \\ TCRM(QRO) = 0 \\ TCL(QRO) = 0 \end{bmatrix}$$

(11.24)

$$Y(VER)$$

$$TCP(VER) = \sum_m \sum_n FP(m, VER, n) \cdot CP(m)$$

$$TCRM(VER) = \sum_m \sum_n \frac{FP(m, VER, n)}{Yld(m)} \cdot RMUC(i)$$

$$TCL(VER) = \sum_m \sum_n \frac{FP(m, VER, n)}{\rho(m)} \cdot LS(m, VER) \cdot LC$$

$$S(BE, VER)$$

$$\sum_n FP(BE, VER, n) = \sum_i \sum_j RMAF(i, j, VER, BE) \cdot Yld(BE)$$

$$S(LA, VER)$$

$$\sum_n FP(LA, VER, n) = \sum_i \sum_j RMAF(i, j, VER, LA) \cdot Yld(LA)$$

$$\left\{ \left[\sum_n FP(BE, VER, n) = 0 \right] \vee \left[\sum_n FP(LA, VER, n) = 0 \right] \right\} \vee \begin{bmatrix} \neg Y(VER) \\ TCP(VER) = 0 \\ TCRM(VER) = 0 \\ TCL(VER) = 0 \end{bmatrix} \quad (11.25)$$

with $\neg S(BE, VER)$ and $\neg S(LA, VER)$ labels over the respective bracketed terms.

TCP(k) is the total production cost for the biorefinery k, CP(m) is the unitary production cost for the product m. TCRM(k) is the cost due to the raw material entering to the facility k, Yld(m) is the yield to the product m. The values of CP(m) and Yld(m) are shown in Table 11.5. On the other hand, RMUC(i) is the unitary cost for the raw material i and has been taken as 87.5 kUSD/kt, which is an estimated selling price for agricultural residues for animal feeding (Mercado Libre, n.d.). TCL(k) is the land acquisition cost for the facility k, LS(m,k) is the land surface required per liter of product, defined as 0.66 m^2/L. Finally, LC is the land cost per square meter and is taken as 0.4048 kUSD/m^2. Values for LC and LS(m,k) are estimated from the production, land surface, and land cost reported for the refinery "Francisco I. Madero", located in Ciudad Madero, Tamaulipas, Mexico (Informador, 2019).

The model requires some specifications, the total demand Demand(m,n) of a product m in a market n (kt/y). It is given by the demand of the product m in the market n, PD (m, n) by the density of the product m, ρ(m).

$$Demand(BE, n) = PD(BE, n) \cdot \rho(BE), \quad \forall n \quad (11.26)$$

$$Demand(LA, n) = PD(LA, n) \cdot \rho(BE), \quad \forall n \quad (11.27)$$

Additionally, the cost to transport biomass i from the source j to the facility k, TSP(i,j,k), is given by:

$$TSP(i, j, k) = \sum_m RMAF(i, j, k, m) \cdot D_1(i, j, k) \cdot TCSB, \forall i, j, k \quad (11.28)$$

where RMAF(i,j,k,m) is the quantity of raw material i obtained from the source j entering to the biorefinery k, D_1(i,j,k) represents the distance from the source j of biomass i to biorefinery k. TCSB is the transportation cost of the solid lignocellulosic biomass, taken as 0.0087 kUSD/km kt. The cost to transport the product m from the facility k to the market n, CTP(m,k,n), is given by:

$$CTP(m, k, n) = FP(m, k, n) \cdot D_2(k, n) \cdot TCLP, \forall m, k, n \quad (11.29)$$

In Eq. 11.29, FP(m,k,n) is the mass production of the product m in the facility k to the market n, D_2(k,n) is the distance from the facility k to the market n and TCLP is the transportation cost or the liquid product, which value is taken as 0.0131 kUSD/km kt. TCSB and TCLP are computed in terms of the cost for diesel (GlobalPetroPrices, 2020) and the estimated diesel requirements for the transportation of solid (Beetrack, 2019) and liquid (OAS, n.d.). The unsatisfied demand for each product m, UDemand(m), is given by:

$$UDemand(BE) = \sum_n Demand(BE, n) - \sum_k \sum_n FP(BE, k, n), \forall n, k \quad (11.30)$$

$$UDemand(LA) = \sum_n Demand(LA, n) - \sum_k \sum_n FP(LA, k, n), \forall n, k \quad (11.31)$$

The constraint for the raw material is as follows. $RMAF(i,j,k,m)$ cannot be higher than the availability of raw material i in the location j, $RMA(i,j)$. This is represented as follows:

$$\sum_k \sum_m RMAF(i, j, k, m) \leq RMA(i,j), \forall i, j \quad (11.32)$$

The constraints for the product are as follows. $FP(m,k,n)$, must not be higher than the total demand for that product:

$$FP(BE, BC, n) \leq PD(BE, n), \forall n \quad (11.33)$$

$$FP(LA, BC, n) \leq PD(LA, n), \forall n \quad (11.34)$$

To solve this optimization problem, the big "M" strategy is used to relax the following constraints for all facility "k", for example for "BC":

$$TCP(BC) - \sum_n FP(BE, BC, n) \cdot CP(BE) - \sum_n FP(LA, BC, n) \cdot CP(LA)$$
$$\leq M_{11} \cdot (1 - Y(BC)) \quad (11.35)$$

$$TCP(BC) - \sum_n FP(BE, BC, n) \cdot CP(BE) - \sum_n FP(LA, BC, n) \cdot CP(LA)$$
$$\geq -M_{11} \cdot (1 - Y(BC)) \quad (11.36)$$

$$TCP(BC) \leq M_{11} \cdot Y(BC) \quad (11.37)$$

$$TCP(BC) \geq -M_{11} \cdot Y(BC) \quad (11.38)$$

$$TCL(BC) - \sum_n \frac{FP(BE, BC, n)}{\rho(BE)} \cdot LS(BE, BC) \cdot LC$$
$$- \sum_n \frac{FP(LA, BC, n)}{\rho(LA)} \cdot LS(LA, BC) \cdot LC \leq M_{11} \cdot (1 - Y(BC)) \quad (11.39)$$

$$TCL(BC) - \sum_n \frac{FP(BE, BC, n)}{\rho(BE)} \cdot LS(BE, BC) \cdot LC$$
$$- \sum_n \frac{FP(LA, BC, n)}{\rho(LA)} \cdot LS(LA, BC) \cdot LC \geq -M_{11} \cdot (1 - Y(BC)) \quad (11.40)$$

$$TCL(BC) \leq M_{11} \cdot (1 - Y(BC)) \quad (11.41)$$

$$TCL(BC) \geq -M_{11} \cdot (1 - Y(BC)) \quad (11.42)$$

$$TCRM(BC) - \sum_n \frac{FP(BE,BC,n)}{Yld(BE)} \cdot RMUC(i) - \sum_n \frac{FP(LA,BC,n)}{Yld(LA)} \cdot RMUC(i)$$
$$\leq M_{11} \cdot (1 - Y(BC)) \tag{11.43}$$

$$TCRM(BC) - \sum_n \frac{FP(BE,BC,n)}{Yld(BE)} \cdot RMUC(i) - \sum_n \frac{FP(LA,BC,n)}{Yld(LA)} \cdot RMUC(i)$$
$$\geq -M_{11} \cdot (1 - Y(BC)) \tag{11.44}$$

$$TCRM(BC) \leq M_{11} \cdot (1 - Y(BC)) \tag{11.45}$$

$$TCRM(BC) \geq -M_{11} \cdot (1 - Y(BC)) \tag{11.46}$$

$$\sum_n FP(BE,BC,n) - \sum_i \sum_j RMAF(i,j,BC,BE) \cdot Yld(BE) \leq M_{11} \cdot (1 - Y(BC)) \tag{11.47}$$

$$\sum_n FP(BE,BC,n) - \sum_i \sum_j RMAF(i,j,BC,BE) \cdot Yld(BE) \geq -M_{11} \cdot (1 - Y(BC)) \tag{11.48}$$

$$\sum_n FP(BE,BC,n) \leq M_{11} \cdot (1 - Y(BC)) \tag{11.49}$$

$$\sum_n FP(BE,BC,n) \geq -M_{11} \cdot (1 - Y(BC)) \tag{11.50}$$

$$\sum_n FP(LA,BC,n) - \sum_i \sum_j RMAF(i,j,BC,LA) \cdot Yld(LA) \leq M_{11} \cdot (1 - Y(BC)) \tag{11.51}$$

$$\sum_n FP(LA,BC,n) - \sum_i \sum_j RMAF(i,j,BC,LA) \cdot Yld(LA) \geq -M_{11} \cdot (1 - Y(BC)) \tag{11.52}$$

$$\sum_n FP(LA,BC,n) \leq M_{11} \cdot (1 - Y(BC)) \tag{11.53}$$

$$\sum_n FP(LA,BC,n) \geq -M_{11} \cdot (1 - Y(BC)) \tag{11.54}$$

The remaining constraints are not relaxed because they are not logical constraints. This problem is solved in GAMS software. The resulting MILP problem consists of 886 equalities and 456 inequalities. The optimization problem is solved with BARON, in an HP equipment with an Intel Core i5-8300H CPU, 12.00 GB of RAM.

According to the results, the maximum profit is 670781.47 kUSD per year, which implies installing a biorefinery on each of the proposed locations. Also, facilities should produce the amounts of levulinic acid and bioethanol presented in Table 11.6. Table 11.7 shows the amounts of bioethanol received by each market and Table 11.8 presents such information for levulinic acid. All the levulinic acid demand considered on the mathematical model is satisfied, while 93.17% of the ethanol demand is satisfied.

Fig. 11.4A shows the quantities of raw material used by the supply chain for each type of stubble. It can be seen that the states with the greatest contribution are Durango, Jalisco, and Sinaloa, mainly with corn residues. On the other hand, Fig. 11.4B shows the inputs from each state to the plants.

Table 11.6 Results: Quantities of product for each facility (kt/y).

Facility product	BC	COAH	GTO	QRO	VER
Bioethanol	20.276	188.619	254.574	139.594	–
Levulinic acid	–	15.979	30.085	15.556	37.963

Table 11.7 Demand satisfied of bioethanol for each facility.

Facility market	BC	COAH	GTO	QRO	VER
CDMX	–	16.382	–	35.510	–
GTO	–	–	64.173	–	–
HGO	–	–	30.115	29.912	–
EDOMEX	–	–	103.358	–	–
MICH	–	–	56.928	–	–
MOR	–	–	–	–	–
NL	–	157.301	–	–	–
PUE	–	–	–	–	–
QRO	–	–	–	74.172	–
ZAC	20.276	14.936	–	–	–

Table 11.8 Demand satisfied of levulinic acid for each facility.

Facility market	BC	COAH	GTO	QRO	VER
CDMX	–	–	–	–	5.094
GTO	–	–	20.095	–	–
HGO	–	–	–	1.389	–
EDOMEX	–	–	–	2.979	8.141
MICH	–	–	9.99	–	–
MOR	–	–	–	–	4.663
NL	–	14.119	–	–	–
PUE	–	–	–	–	20.065
QRO	–	–	–	11.188	–
ZAC	–	1.86	–	–	–

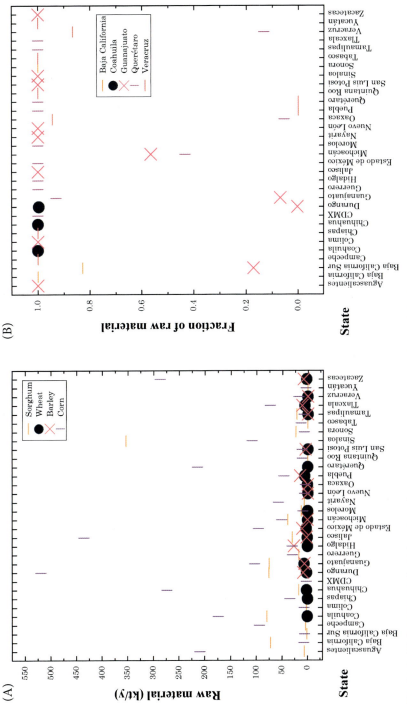

Fig. 11.4 (A) Raw material used and (B) fraction of raw material used for each facility.

It can be seen that of all the possibilities considered in the superstructure presented in the Fig. 11.3, the best solution consists in the existence of all the plants but with a specified distribution, this allows us to satisfy the demand for levulinic acid in the ten markets considered and almost all the demand for bioethanol in all the markets except Morelos and Puebla. Additionally, it is possible to add objective functions and employ multiobjective optimization techniques to ensure supply chain sustainability. It is important to mention that the presented model can be further developed for other case studies by modifying the parameters and constraints involved. The mathematical model developed can be improved by selecting other feedstocks and other high value-added products to increase the profit of the biorefinery. The environmental effects should be evaluated by means of multicriteria decision-making. Similarly, other objective functions must be considered, including social aspects, in order to have a sustainability scheme.

11.7 Conclusion

When developing a production scheme for biofuels and high value-added products, it is important to consider not only the production process itself but also all aspects involved in the supply chain, from the selection of raw materials to the final use of the products. The realization of such an evaluation will give a global vision of the whole production scheme, allowing assessing the total cost for the complete supply chain. Considering intensified processes may allow lower parameters in terms of greenhouse gas emissions and production costs, however, the modeling of the supply chain will not be severely modified in comparison with the case were considering conventional processes. Nevertheless, including intensified processes would reduce the costs and environmental impact associated with the processing of the raw materials. A strategy to determine the best supply chain is to develop logic-based mathematical models, with the selection of appropriate objective functions; this tool will help us find the best configuration of the elements involved in the supply chain to minimize or maximize objective functions. To ensure a sustainable production, not only economic objectives must be evaluated but also social, environmental and water aspects must be included to ensure the sustainability of the supply chain.

References

Avami, A. (2012). A model for biodiesel supply chain: A case study in Iran. *Renewable and Sustainable Energy Reviews, 16*(6), 4196–4203. https://doi.org/10.1016/j.rser.2012.03.023.

Beamon, B. M. (1998). Supply chain design and analysis: Models and methods. *International Journal of Production Economics, 55*, 281–294. https://doi.org/10.1016/S0925-5273(98)00079-6.

Beetrack. (2019). *Costos de transporte terrestre en México, ¿cómo se calculan?*. https://www.beetrack.com/es/blog/costos-de-transporte-terrestre-en-mexico (Accessed 24 October 2020) (Spanish).

Biddy, M. J., Scarlata, C., & Kinchin, C. (2016). *Chemicals from biomass a market assessment of bioproducts with near-term potential*. Golden, CO (United States): National Renewable Energy Lab. (NREL).

Damián-Huato, M. A., Cruz-León, A., Ramírez-Valverde, B., Romero-Arenas, O., Moreno-Limón, S., & Reyes-Muro, L. (2013). Maíz, Alimentación Y Productividad: Modelo Tecnológico Para Productores De Temporal De México. *Agricultura Sociedad y Desarrollo, 10*(2), 157–176.

de Jong, S., Hoefnagels, R., Wetterlund, E., Petterson, K., Faaij, A., & Junginger, M. (2017). Cost optimization of biofuel production—The impact of scale, integration, transport and supply chain configurations. *Applied Energy, 195*, 1055–1070. https://doi.org/10.1016/j.apenergy.2017.03.109.

eBioChem, n.d. Wholesale levulinic acid. http://www.ebiochem.com/product/levulinic-acid-15676 (Accessed 24 October 2020).

El Cronista. (2020). *El Gobierno subió los precios del bioetanol y del biodiesel y suma presión al valor de las naftas.* https://www.cronista.com/economiapolitica/El-Gobierno-subio-los-precios-del-bioetanol-y-del-biodiesel-y-suma-presion-al-valor-de-las-naftas-20201014-0005.html (Accessed 24 October 2020) (Spanish).

Espinoza-Vázquez, Y. M., Gómez-Castro, F. I., & Ponce-Ortega, J. M. (2021). Optimization of the supply chain for the production of biomass-based fuels and high-added value products in México. *Computers & Chemical Engineering, 107181.* https://doi.org/10.1016/j.compchemeng.2020.107181.

Gaurav, A., Leite, M. L., Ng, F. T. T., & Rempel, G. L. (2013). Transesterification of triglyceride to fatty acid alkyl esters (biodiesel): Comparison of utility requirements and capital costs between reaction separation and catalytic distillation configurations. *Energy & Fuels, 27*, 6847–6857. https://doi.org/10.1021/ef401790n.

Geng, N., & Sun, Y. (2021). Multiobjective optimization of sustainable WCO for biodiesel supply chain network design. *Discrete Dynamics in Nature and Society.* https://doi.org/10.1155/2021/6640358, 1026-0226.

Gerbens-Leenes, W., Hoekstra, A. Y., & van der Meer, T. H. (2009). The water footprint of bioenergy. *Proceedings of the National Academy of Sciences, 106*(25), 10219–10223. https://doi.org/10.1073/pnas.0812619106.

GlobalPetroPrices. (2020). *México precios del diesel, 19-Oct-2020.* https://es.globalpetrolprices.com/Mexico/diesel_prices/#:~:text=M%C3%A9xico%20-%20precios%20del%20diesel%3A%20Mostramos%20los%20precios,un%20m%C3%A1ximo%20de%2019.85%20%28Mexican%20Peso%29%20a%2027-jul-2020 (Accessed 24 October 2020) (Spanish).

Gozan, M., Ryan, B., & Krisnandi, Y. (2017). Techno-economic assessment of levulinic acid plant from Sorghum Bicolor in Indonesia. *IOP Conference Series: Materials Science and Engineering, 345*, 012012.

Informador, E. (2019). *Serán siete las refinerías de México en 2022.* https://www.informador.mx/Seran-siete-las-refinerias-de-Mexico-en-2022-t201905090003.html (Accessed 27 May 2020). (Spanish).

Khuong, L. S., Zulkifli, N. W. M., Masjuki, H. H., Mohamad, E. N., Arslan, A., Mosarof, M. H., et al. (2016). A review on the effect of bioethanol dilution on the properties and performance of automotive lubricants in gasoline engines. *RSC Advances, 6*(71), 66847–66869. https://doi.org/10.1039/C6RA10003A.

Kim, J., Realff, M. J., & Lee, J. H. (2011). Optimal design and global sensitivity analysis of biomass supply chain networks for biofuels under uncertainty. *Computers and Chemical Engineering, 35*, 1738–1751. https://doi.org/10.1016/j.compchemeng.2011.02.008.

Kiss, A. A. (2014). *Process intensification technologies for biodiesel production: Reactive separation processes.* Springer Science & Business Media.

Marvin, W. A., Schmidt, L. D., & Daoutidis, P. (2013). Biorefinery location and technology selection through supply chain optimization. *Industrial and Engineering Chemistry Research, 52*, 3192–3208. https://doi.org/10.1021/ie3010463.

Méndez-Vázquez, M. A., Gómez-Castro, F. I., Ponce-Ortega, J. M., Serafin-Muñoz, A. H., Santibañez-Aguilar, J. E., & El-Halwagi, M. M. (2016). Mathematical optimization of the production of fuel pellets from residual biomass. *Computer Aided Chemical Engineering, 38*, 133–138. https://doi.org/10.1016/B978-0-444-63428-3.50027-8.

Mercado Libre, n.d. 3000 pacas de rastrojo de sorgo y de maíz. https://articulo.mercadolibre.com.mx/MLM-776767869-3000-pacas-de-rastrojo-de-sorgo-y-de-maiz-_JM#position=1&type=item&tracking_id=373cabcd-850e-4e4d-97e3-e39d2a33c3f1 (Accessed 09 June 2020) (Spanish).

Monsiváis-Alonso, R., Mansouri, S. S., & Román-Martínez, A. (2020). Life cycle assessment of intensified processes towards circular economy: Omega-3 production from waste fish oil. *Chemical Engineering and Processing Process Intensification, 158.* https://doi.org/10.1016/j.cep.2020.108171, 108171.

Murillo-Alvarado, P. E., Ponce-Ortega, J. M., Serna-González, M., Castro-Montoya, A. J., & El-Halwagi, M. M. (2013). Optimization of pathways for biorefineries involving the selection of feedstocks, products, and processing steps. *Industrial and Engineering Chemistry Research, 52*, 5177–5190. https://doi.org/10.1021/ie303428v.

OAS, n.d. Los costos del transporte de cargas entre los países del MERCOSUR. https://www.oas.org/usde/publications/Unit/oea75s/ch08.htm#2.%20An%C3%A1lisis%20de%20costos%20para%20Uruguay. (Accessed 24 October 2020) (Spanish).

Petróleos Mexicanos. (2018). *Anuario estadístico.* https://www.pemex.com/ri/Publicaciones/Anuario%20Estadistico%20Archivos/anuario-estadistico_2018.pdf (Accessed 07 July 2020) (Spanish).

Raman, R., & Grossmann, I. E. (1994). Modelling and computational techniques for logic based integer programming. *Computers & Chemical Engineering, 18*(7), 563–578. https://doi.org/10.1016/0098-1354(93)E0010-7.

Reay, D., Ramshaw, C., & Harvey, A. (2013). *Process intensification: Engineering for efficiency, sustainability and flexibility.* Butterworth-Heinemann.

Rendon-Sagardi, M. A., Sánchez-Ramírez, C., Cortes-Robles, G., Alor-Hernández, G., & Cedillo-Campos, M. G. (2014). Dynamic analysis of feasibility in ethanol supply chain for biofuel production in México. *Applied Energy, 123*, 358–367. https://doi.org/10.1016/j.apenergy.2014.01.023.

Romero, A., Gómez, F. I., Gutiérrez, C., & Hernández, S. (2016). Localización óptima de una planta para la producción de turbosina renovable. *Revista de Simulación y Laboratorio, 3*, 8–15 (Spanish).

Rosillo-Calle, F., Pelkmans, L., & Walter, A. (2009). *A global overview of vegetable oils, with reference to biodiesel. A report for the IEA Bioenergy Task* (p. 40).

Santibañez-Aguilar, J. E., González-Campos, J. B., Ponce-Ortega, J. M., Serna-González, M., & El-Halwagi, M. M. (2014). Optimal planning and site selection for distributed multiproduct biorefineries involving economic, environmental and social objectives. *Journal of Cleaner Production, 65*, 270–294. https://doi.org/10.1016/j.jclepro.2013.08.004.

Sassner, P., Galbe, M., & Zacchi, G. (2008). Techno-economic evaluation of bioethanol production from three different lignocellulosic materials. *Biomass and Bioenergy, 32*(5), 422–430. https://doi.org/10.1016/j.biombioe.2007.10.014.

Segovia-Hernández, J. G., & Bonilla-Petriciolet, A. (2016). *Process intensification in chemical engineering.* Springer.

Sistema de Información Energética, n.d.. Volumen de ventas internas de petrolíferos por entidad federativa. http://sie.energia.gob.mx/movil.do?action=cuadro&cvecua=PMXE2C03 (Accessed 07 July 2020) (Spanish).

Tian, Y., Demirel, S. E., Hasan, M. F., & Pistikopoulos, E. N. (2018). An overview of process systems engineering approaches for process intensification: State of the art. *Chemical Engineering and Processing Process Intensification, 133*, 160–210. https://doi.org/10.1016/j.cep.2018.07.014.

Türkay, M., & Grossmann, I. E. (1996). Logic-based MINLP algorithms for the optimal synthesis of process networks. *Computers & Chemical Engineering, 20*(8), 959–978. https://doi.org/10.1016/0098-1354(95)00219-7.

Villicaña-García, E., Ramírez-Márquez, C., Segovia-Hernández, J. G., & Ponce-Ortega, J. M. (2021). Planning of intensified production of solar grade silicon to yield solar panels involving behavior of population. *Chemical Engineering and Processing Process Intensification, 161*. https://doi.org/10.1016/j.cep.2020.108241, 108241.

Werpy, T., & Petersen, G. (2004). *Top value added chemicals from biomass, Volume I: Results of screening for potential candidates from sugars an synthesis gas.* Golden, CO (US): National Renewable Energy Lab.

Xue, C., Zhao, J., Chen, L., Yang, S.-T., & Bai, F. (2017). Recent advances and state-of-the-art strategies in strain and process engineering for biobutanol production by *Clostridium acetobutylicum. Biotechnology Advances, 35*, 310–322. https://doi.org/10.1016/j.biotechadv.2017.01.007.

You, Y. D., Shie, J. L., Chang, C. Y., Huang, S. H., Pai, C. Y., Yu, Y. H., et al. (2008). Economic cost analysis of biodiesel production: Case in soybean oil. *Energy & Fuels, 22*(1), 182–189. https://doi.org/10.1021/ef700295c.

Zhang, Q., Jing, D., Yongmei, L., Yangdong, W., & Yong, C. (2016). Towards a green bulk-scale biobutanol from bioethanol upgrading. *Journal of Energy Chemistry, 25*(6), 907–910. https://doi.org/10.1016/j.jechem.2016.08.010.

CHAPTER 12

Life cycle approach for the sustainability assessment of intensified biorefineries

M. Collotta[a], P. Champagne[b], G. Tomasoni[a], and W. Mabee[c]

[a]DIMI, Department of Mechanical and Industrial Engineering, University of Brescia, Brescia, Italy
[b]Institut national de la recherche scientifique, Québec City, QC, Canada
[c]Queen's University, Department of Geography and Planning, Mackintosh-Corry Hall, Kingston, ON, Canada

12.1 Introduction

A growing interest in biorefineries—facilities dedicated to the processing of biomass for the production of bio-based fuels and other bioproducts, including materials and chemicals—has been observed in the recent years, with increasing attention paid to these opportunities by decision-makers and increasing awareness in the general public. Biofuels in particular have taken on a prominent role in public discourse, as their role as an alternative to fossil-based transport fuels has focused attention on these products (Demırbas, 2017).

For this reason, the scientific community has endeavored to assess the impacts of biorefineries, and particularly their potential to reduce greenhouse gas (GHG) emissions and improve overall sustainability of the transportation fuel mix. The use of life cycle assessment (LCA) has highlighted a series of issues, including environmental challenges as well as the social impact generated by existing biorefinery technologies. Information developed through LCA has in turn promoted the development of new types of biorefineries, including new "intensive" facilities capable of delivering so-called third generation biofuel products, such as algal biofuels or bioplastics.

As a consequence, in addition to the environmental and economic aspects, issues relating to potential social impacts related to biorefineries are emerging more clearly. For these reasons, it is important that the potential impacts of emerging technologies be accurately assessed at the early stage of process development. This assessment is particularly complex as it must be based on data that is both difficult to access and uncertain; however, new tools that can help mitigate these effects, especially with regard to the evaluation of environmental aspects. In this chapter, we present the main tools that can be used to assess the environmental and social impacts associated with intensive biorefineries, as well as key trends in the scientific literature regarding the sustainability of these facilities. Finally, a case study is presented to illustrate how the assessment of environmental sustainability carried out in the early stages of process development can guide the development itself toward more sustainable solutions.

12.2 LCA methodology

According to ISO 14040, the LCA is a methodology to study the environmental aspects and potential impacts throughout a product's life cycle (i.e., cradle-to-grave) from raw materials acquisition through production, use, and disposal (ISO, 2006).

The LCA dates back to the 1980s and has been the reference for the birth and development of similar methodologies dedicated to the assessment of economic (Life Cycle Costing, or LCC) and social (Social Life Cycle Assessment, or SLCA) impacts (Finnveden et al., 2009; Muralikrishna & Manickam, 2017).

LCA can be used for various purposes, such as product eco-labeling or public decision-making support, but the most important is eco-design. In fact, thanks to this methodology it is possible to identify the critical environmental aspects within the life cycle of a product, of a material and/or of a production process. These critical aspects can then be taken into consideration when making choices between alternatives related to, for example, the components of a product, the location of a production facility or, the production technology to be adopted, in order to choose the alternatives that present a lower environmental impact.

According to ISO 14040 and ISO 14044, LCA involves four steps (ISO, 2006):

(1) The definitions of the goal and scope of the analysis, which need to be clearly defined and be consistent with the intended application. The definition of the scope, in particular, includes crucial aspects such as the products in the system to be analyzed, including their boundaries, the functional unit considered, and the allocation procedures to be adopted.

(2) The generation of a Life Cycle Inventory (LCI), where relevant mass and energy inputs and outputs are identified and quantified. This can be done using either primary data, i.e., data gathered from the field, or secondary data, i.e., data gathered from literature or databases.

(3) The environmental Life Cycle Impact Assessment (LCIA), where impact categories, category indicators and characterization models are selected, mass and energy flows identified and quantified in the inventory are sorted and assigned to specific impact categories, and one or more impact assessment methodology are then applied to quantify the environmental impacts.

(4) The interpretation, where the results of the previous phases are evaluated in relation to the goal of the analysis and proper conclusions are drawn.

Fig. 12.1 shows the relationships that exist between the various phases of the LCA methodology. In particular, although at first glance it appears to have a linear structure, in

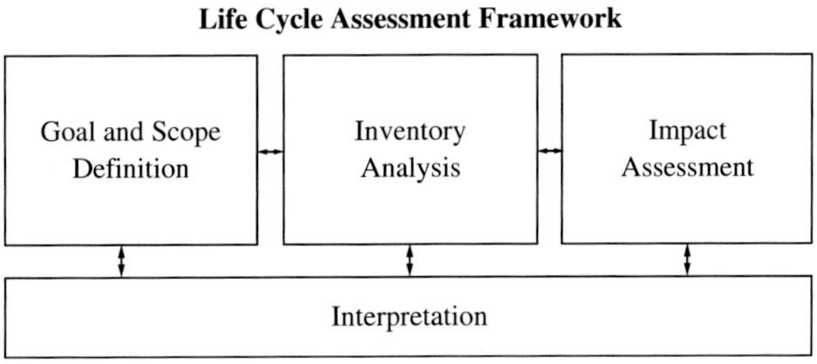

Life Cycle Assessment Framework

Fig. 12.1 LCA methodology. *(Adapted from ISO, 2006. ISO 14040 international standard. Environmental management-life cycle assessment—Principles and framework. International Organization for Standardization.)*

reality it is a recursive process in which the various phases are gradually refined until a result is obtained that is consistent with the initial purpose and objectives.

One of the most relevant aspects of the LCA methodology is that the different impact assessment methodologies applicable can consider several environmental impact categories at the same time: climate change, resource consumption, soil acidification, freshwater eutrophication, particulate matter formation, etc. This enables the acquisition of a much more complete and accurate knowledge of the different environmental implications related to the life cycle of the technology, product, or material being assessed. In fact, it is not uncommon to observe how the introduction of a technological innovation, including those for the intensification of biorefineries, produces benefits with respect to a specific environmental impact category, or a set of impact categories, in the face, however, of a worsening of performance observed in other environmental impact categories.

12.3 LCA tools

The practical application of the LCA methodology can be particularly laborious and time-consuming, both because of the need to access specific databases for the collection of secondary data for the life cycle inventory phase, and because of the complexity of the calculations required for the application of the impact assessment methods which, as highlighted above, can take into account numerous categories of environmental impact.

For this reason, several software tools have been developed to support practitioners in the application of LCA methodology over the years. Although it is possible to carry out an LCA analysis without the use of any specific software, nowadays both within the academic environment and within the industry, the adoption of these tools is quite common. For this reason, it is worth presenting the main alternatives available on the market today (Bach, Mohtashami, & Hildebrand, 2018; Lüdemann & Feig, 2014).

In particular, a first distinction within these tools can be made between those that are designed to be applied within a specific sector or context, and the more generic ones that can be applied in any type of sector. For example, the PackageSmart software, developed by EarthShift LLC Global (Papadaskalopoulou et al., 2019), allows to carry out a detailed study to determine the "most sustainable" combination between design and type of material employed for the packaging of a product; while OneClickLCA, created by Bionova LTD, has been developed specifically for the construction sector.

The most widely adopted tools that can be applied in any type of industry and sector are SimaPro, Gabi, Umberto, and OpenLCA. All of these software can provide more than adequate support for evaluating environmental impacts, where each has its strengths and weaknesses, summarized in Table 12.1, that differ ranging from costs, user interfaces, accessible databases, data processing approaches, and support provided for the interpretation of results.

With reference to the costs, OpenLCA is freeware, which is not very common among its competitors, and provides all the necessary means for the development of an LCA study.

In particular, this tool is more intuitive than SimaPro, Gabi, and Umberto, for which the user needs to understand fundamental elements before the tool can be used effectively, such as the difference between product and elementary flows, or the difference between project and system products.

Another disadvantage is the absence of preinstalled databases and predefined calculation methods that force the user to import them from external sources that have to be supplied separately. For this reason, GreenDelta has recently launched an excellent project with the design of Nexus (Cirot, Di Noi, Burhan, & Srocka, 2019): the development of a platform with a vast collection of downloadable and importable databases, which at the same time presents a nonnegligible problem, where a user with little experience in the sector is faced with a large amount of information. The evaluation and choice of databases suited to a particular study requires time and experience, and some of these databases can be subject to a relatively high fee.

SimaPro, Gabi and Umberto are generally considered to be more intuitive than OpenLCA. Unlike the latter offers a "guided tour" that shows the most important features of the software and a LCA Wizard, a tool that semiautomates the creation of the study model. Two other advantages of these LCA tools are the existence of predefined databases and the side window of the screen dedicated to the four phases of the LCA, making it easy to navigate from one phase to another. In summary, the user interface intuitively represents the theoretical structure of the LCA methodology.

Regionalization is another important functionality in LCA tools, that is the consideration of aspects related to the geographical location of the different process units that constitute the system to be analyzed. In particular, most software adopts a flow-based regionalization. In Simapro, Gabi, Umberto, and OpenLCA, the elementary flows are

Table 12.1 Pros and cons of the main LCA Tools.

LCA Tool	Developer	License	User-friendliness	Available inventory databases	Flexibility (data entry and manipulation of datasets)	Support for the interpretation phase
Gabi	Sphera Solutions GmbH	Commercial	−	+	+	++
Open-LCA	GreenDelta GmbH	Open source	+	−	−	+
Simapro	PRé Sustainability B.V.	Commercial	−	+	++	++
Umberto	ifu Institut für Umweltinformatik Hamburg GmbH	Commercial	−	+	+	+

extended with the addition of regional information and the location of the process is defined in the regional elementary flows. LCIA is also calculated combining preaggregated geographically explicit characterization factors and regional LCI data (Frischknecht et al., 2019).

In terms of default setting, Simapro, Gabi, and Umberto have more editable settings compared to OpenLCA and allow for greater manipulation of the dataset. Only Simapro presents an automated computational iteration to support the optimization of the analysis. Although most software offers the possibility to choose between different inventory databases. Simapro also provides a geometric based entry format, while Gabi, Umberto, and OpenLCA are limited to a spreadsheet format (Bach et al., 2018).

In conclusion, despite their advantages and disadvantages, while all the cited software are a valid alternative for LCA, one can argue that OpenLCA is more suitable for inexperienced users, tools like Simapro, Gabi, and Umberto are more devoted to expert users and they are more appropriate for conducting more complex analyses.

12.4 State of the art of LCA in the biorefineries sector

Although the interest in biorefineries has been present within the scientific community for a long period of time, it is only in the last years that there has been a very marked increase in the number of studies based on the application of LCA to biorefineries (Balasubramanian et al., 2011; Bishop, Styles, & Lens, 2021; Collotta et al., 2019; Collotta, Champagne, Mabee, Tomasoni, & Alberti, 2019; Li, Wang, & Yan, 2018; Liu et al., 2021). The goals and scopes of these LCA studies are very diverse, ranging from the comparison between a biorefinery and a conventional production process (Bishop et al., 2021; Blanco, Iglesias, Morales, Melero, & Moreno, 2020; Garcia Gonzalez, Levi, & Turri, 2017; Papadaskalopoulou et al., 2019), to the comparison between different technological alternatives for one of the phases of the production process of a biorefinery (Collotta et al., 2017; Collotta & Tomasoni, 2017), to the choice of the location of the biorefinery (Buchspies, Kaltschmitt, & Neuling, 2020; Collotta et al., 2016; Collotta et al., 2016; Kim & Dale, 2009; Lopes et al., 2019). From the point of view of the products produced, most of the studies have focused on biofuels, but there are also many studies addressing bioplastics (Bishop et al., 2021; Devadas et al., 2021; Ita-Nagy, Vázquez-Rowe, Kahhat, Chinga-Carrasco, & Quispe, 2020; Walker & Rothman, 2020) or other chemical products (Bello, Salim, Feijoo, & Moreira, 2021; Moretti et al., 2021).

The main focus of these studies is often related to environmental impact, particularly the impact category "climate change," which is determined by the emission of greenhouse gases, primarily CO_2, N_2O, and CH_4. This is not surprising given the importance of addressing GHG emissions and the need to develop robust industrial solutions that can achieve the goal of reducing these emissions (Medeiros, Sales, & Kiperstok, 2015). However, it is necessary to take into account that simply looking only at climate change indicators may not provide a sufficiently complete view of the environmental implications of biorefineries. In fact, it is not uncommon that an improvement in a specific impact category, such as climate change, or in a set of impact categories, produces a concomitant worsening in other impact categories. However, this problem can be easily overcome by LCA practitioners, given the fact that most impact assessment methods available in literature and implemented within LCA software easily allow the calculation of the impacts for a broad set of impact categories.

While on the one hand, the wide use of LCA methodology is a positive factor, since it is the most robust methodology available today, considering more critically the manner in which LCA studies are conducted, it is necessary to note that there remain some fairly common shortcomings that have also been highlighted in a number of reviews, as summarized in more detail below.

In some cases, a relevant weakness reported is the unclear definition of system boundaries and the functional unit (the elementary quantification of the system function against which all inventory data are reported) (Collotta, Busi, et al., 2016; Collotta, Champagne, et al., 2016; Liu et al., 2021). Looking at the process stages and system boundaries, all studies generally consider the material acquisition stage, while some downstream phases are more frequently neglected in the analysis. It is important to underline the importance of the definition of the system boundaries correlated to the final results in order to have a clear understanding of the analysis (Tillman, Ekvall, Baumann, & Rydberg, 1994).

One of the main critical issues encountered in LCAs is the scarcity of primary inventory data (Cirot et al., 2019). Primary data is information collected directly in the field and is distinct from secondary data, which is drawn from the literature or commercially available databases. While the use of secondary data reduces the amount of time and cost required to conduct the analysis, primary data generally ensure a higher quality of the results obtained.

The lack of primary data is mainly due to the limited number of operating, large-scale, intensified biorefineries available for data gathering. Many of the candidate processes for implementation at this scale are still at lower Technology Readiness Levels. The lack of empirical, observed data in operating facilities increases the uncertainty of the input data to LCAs, and thus increases the potential uncertainty in results, which some studies try to overcome through a sensitivity analysis.

Another important issue that often arises in LCA studies applied to biorefineries is the relevant impact of the regional location, the impact of local climate on biological growth rates, and the energy mix (i.e., the group of different primary energy sources from which secondary energy—such as electricity—is produced) available in the specific location (Itten, Frischknecht, Stucki, Scherrer, & Psi, 2014). The location where a biorefinery is placed may also allow some synergies when industrial symbiosis practices are pursued and the use of co-products between different facilities can be strategically employed to reduce overall impacts.

A related issue is water availability, which is dependent on geographical location, and is often one of the most critical factors affecting the feasibility of the processes within a biorefinery, particularly in the case of biofuels. For example, the use of fresh water in the process provides a reduction in term of operational cost and water treatment process (evaporation) compared to the use of salt water (Gendy & El-Temtamy, 2013). In addition, locations that have higher sunlight and temperature potentials presents a better productivity in terms of growing rate in the cultivation of microalgae (Medeiros et al., 2015). Moreover, water and land use are two important factors to consider in the analysis; these factors can change over time and must be monitored regularly, as they can significantly affect GHG emissions.

12.5 Social LCA methodology

In addition to the environmental impacts, issues related to potential social impacts associated with the production and consumption of biofuels and other bioproducts are becoming more evident. This chapter also aims to review the scientific literature in the field of assessment of social impacts with respect to biorefineries and their products.

To date, there is still no clear methodology that has reached a sufficient level of consensus within the scientific community to establish itself as a reference approach to assess social impacts (Ben Ruben, Menon, & Sreedharan, 2018; Mattioda, Tavares, Casela, & Junior, 2020). Furthermore, a clear and unique definition of social impact is far to come; on the contrary, there are almost as many definitions of social impact as the number of methods used to measure it.

However, there are some elements that are frequently present in the assessment methodologies, including: the definition of the objectives, the identification, and involvement of the stakeholders, the definition and measurement of a set of indicators, the multi-objective approach, and the use of questionnaires administered to stakeholders. Among the methodologies most frequently adopted, the social life cycle assessment (Social-LCA), which has also spread thanks to the publication of a specific manual by UNEP (UNEP, 2009, 2011; Valdivia et al., 2013) is noteworthy.

S-LCA is defined as a methodological approach to evaluate the positive and negative social impacts of a product or a service throughout its life cycle (Jørgensen, Le Bocq, Nazarkina, & Hauschild, 2008). The scope in conducting an S-LCA is to promote social conditions and the overall socioeconomic performance of a product for all its stakeholders, leading to a holistic approach, where different stakeholders, with diverse interests are involved.

Table 12.2 shows an example of a set of reference indicators for assessing social impacts in relation to each of the relevant stakeholders (Schenker et al., 2020). Each of the indicators is designed to be assessed either quantitatively or qualitatively. However,

Table 12.2 Example of indicators for the assessment of social impacts (Schenker et al., 2020).

Workers	Users	Local communities	Small-scale entrepreneurs
• Occupational health and safety • Remuneration • Child labor • Forced labor • Discrimination • Freedom of association and collective bargaining • Work-life balance	• Health and safety • Responsible communication • Privacy • Accessibility • Affordability • Effectiveness and comfort	• Health and safety • Access to tangible resources • Community engagement • Skill development • Contribution to economic development	• Meeting basic needs • Access to services and inputs • Women's empowerment • Child labor • Health and safety • Land rights • Trading relationship

both the set of stakeholders and the set of indicators need to be adapted to each individual case where the methodology is applied, taking into account what the objectives of the evaluation are and what data and information are accessible.

At the production process design stage, stakeholder knowledge, and engagement in the system are very limited, and data are not readily available to assess social aspects. Therefore, S-LCA applied to the early stage of production process design is a highly complex task, but at the same time, it is a top priority to ensure appropriate social considerations when the process is implemented. However, there remains a lack of systematic and quantitative methodologies to guide practitioners in S-LCA application.

The utilization of S-LCA can be designed to not only evaluate the social consequences of the product, but also to improve the social conditions of the stakeholders involved and to provide information that can be used to support informed decisions. Furthermore, there is a need and a growing interest to have a social impacts assessment in the biorefineries sector. It is clear that there are considerable challenges associated with conducting social analyses and to identify relevant data and documents related to this topic. These challenges are due to the number and variety of factors that can have social impacts that are often difficult to measure; conversely, environmental impacts or variations in the economic sphere are more readily identifiable and are also influenced by a decidedly lower number of factors than a social phenomenon. Further, to date, there are no detailed background databases that could be used to create a representative inventory for social LCA studies (Mattioda et al., 2020) and this requires to utilize a variety of literature sources and expert inputs to explore these issues.

12.6 LCA at the early stage of development—Case study

The importance of the third generation of biofuels (microalgae) and how this feedstock could overcome some of the challenges associated to the previous generation first and second generation biofuels is the main focus of this case study (adapted from (Collotta, Champagne, Mabee, & Tomasoni, 2018)). Particular advantages, as an example, are that microalgae can be cultivated on non-arable land, resulting in a lower water footprint, and have been demonstrated to successfully assimilate organics and nutrients from wastewater, thereby reducing their environmental impact (Clarens, Resurreccion, White, & Colosi, 2010; Hwang, Church, Lee, Park, & Lee, 2016; Usher, Ross, Camargo-Valero, Tomlin, & Gale, 2014). Several wastewater treatment processes employing microalgae have been investigated in literature for municipal (Sturm & Lamer, 2011; Wang et al., 2010) and industrial wastewater (Balasubramanian et al., 2011; Chinnasamy, Bhatnagar, Hunt, & Das, 2010) as well animal manure (Wang et al., 2010; Woertz, Feffer, Lundquist, & Nelson, 2009).

In combination, the use of wastewater and flue gas (CO_2 – the carbon source for microalgal growth) from other industrial facilities (i.e., cement plants) could positively affect the entire environmental footprint of the production process (Clarens et al., 2010). For this reason, it has been suggested that the location and the distance to the biodiesel plant should be relatively close to these other sources of nutrients and CO_2 (Collotta, Busi, et al., 2016; Collotta, Champagne, et al., 2016).

To understand the benefits of early-stage LCA, we consider the potential environmental benefits from the combined used of sources such as wastewater and CO_2 in the microalgae cultivation phase (lipid extraction and transesterification excluded) for biodiesel production. A comparative LCA is presented based on a hypothetical microalgae production facilities (open pond microalgae production technologies) located and operated in the Canadian context (Kingston, ON). The functional unit is the cultivation of 1 kg of *Chlorella vulgaris*, with a concentration of 0.1% in the cultivation medium, on a dry mass basis. *C. vulgaris* has an annual average biomass productivity of 24.75 (g/m2/d) and a growth rate of 0.99 (day^{-1}). Six different scenarios for a large-scale industrial system have been proposed with an inventory of mass and energy balance. Three alternatives were considered for procuring carbon dioxide (CO_2): use of CO_2 supplied from a conventional production process (CO_2 is a co-product of the ammonia (NH_3) production process); CO_2 recovered from the flue gas of a cement plant; and finally, the direct injection of flue gas from a cement plant into the microalgal cultivation pond. In parallel, fresh water taken from Lake Ontario with the addition of chemical nutrients (i.e., fertilizers) was considered, as was the use wastewater from a municipal wastewater treatment plant (WWTP).

The system boundaries of all scenarios include CO_2 production or recovery, with the exception of cases where direct flue gas injection is adopted (flue gas is considered a waste or by-product from another process). Each scenario ends after the microalgal cultivation phase.

Harvesting and drying phases, as well lipid extraction and transesterification have been excluded from this analysis and are considered beyond the scope of this study.

In terms of life cycle inventory, all secondary data have been extracted from sources such as the Ecoinvent database (Frischknecht et al., 2005), literature, (Clarens et al., 2010; Clarens, Nassau, Resurreccion, White, & Colosi, 2011; Collet et al., 2014) and expert inputs. As the case study refers to a hypothetical cultivation plant, the inventory data refer to the functional unit (the cultivation of 1 Kg of *C. vulgaris*, with a concentration of 0.1% in the cultivation medium, on a dry mass basis) and it was not possible to take into account their possible dependence on the actual capacity of the plant. A specific energy mix for the North American context has been selected for the pumping of fresh/wastewater, injection of CO_2/flue gas, and for paddle wheel handling. The water loss has been estimated from the cultivation pond and the removal of nutrients from wastewater, where wastewater was used, have been considered as an environmental credit in the Scenarios 2, 4, and 6. Moreover, the impacts related to the construction and maintenance of the production facility were not included within the system boundary.

Scenario 1—CO_2 produced and fresh water

The first scenario represents the standard microalgae production process, using fresh water and nutrient inputs along with CO_2 sourced from a conventional monoethanolamine (MEA) based process, i.e., the most widely used CO_2 production process, based on the absorption by monoethanolamine of waste CO_2 emitted in ammonia production. CO_2 is transported in cylinders in liquid form, with transport distance assumed to be 100 km from the supplier location to the facility plant production.

Scenario 2—CO_2 produced and wastewater

The second scenario uses CO_2 from a conventional MEA process, transported via cylinder for 100 km; wastewater replaces fresh water and nutrient inputs.

Scenario 3—Flue gas injection and fresh water

In this scenario, cement plant flue gas was used as a CO_2 source; this gas had a CO_2 concentration of 22%, and did not need to be transported. Fresh water and nutrients were externally supplied (as in Scenario 1).

Scenario 4—Flue gas injection and wastewater

In this scenario, cement plant flue gas was used as a CO_2 source (as in Scenario 3); wastewater replaced freshwater and nutrient inputs.

Scenario 5—CO_2 recovery and fresh water

In this scenario, CO_2 was recovered from cement plant flue gas using the MEA process. The recovered CO_2 was then injected to the microalgal pond with no need for transport. Fresh water and nutrients were externally supplied.

Scenario 6—CO_2 recovery and wastewater

In this scenario, CO_2 was recovered from cement plant flue gas (as in Scenario 5); fresh water and nutrient inputs were replaced with wastewater.

All the calculations were made with Simapro software and the Recipe impact assessment method was used, which include the following impact categories: ozone depletion, climate change human health, climate change ecosystem, human toxicity, photochemical oxidant format, particulate matter formation, ionizing radiation, terrestrial acidifications and ecotoxicity, freshwater eutrophication, and ecotoxicity, marine ecotoxicity, urban and agricultural land occupation, natural land transformation, metal, and fossil depletion.

Fig. 12.2 shows the results of the impact assessment for the six scenarios considered against the different impact categories of the Recipe method. The impact categories shown are in the midpoint version (midpoint categories look at the impact earlier along the cause-effect chain, before the endpoint is reached).

The negative bars represent the avoided impacts in the categories freshwater eutrophication, marine eutrophication, terrestrial ecotoxicity, and freshwater ecotoxicity. These avoided impacts relate to the environmental credits attributed to Scenarios 2, 4, and 6 for the removal of nutrients from wastewater (which would otherwise be considered environmental pollutants).

As can be seen, Scenario 1 (i.e., the scenario using conventionally produced CO_2 and freshwater with the addition of industrially produced fertilizers) shows the highest environmental impact in all categories, while Scenario 4 (involving the direct injection of cement plant flue gas and the use of wastewater to supply nutrients to microalgae) shows the lowest. The most pronounced differences between the scenarios with the lowest and highest environmental impact are found in the categories influenced by the

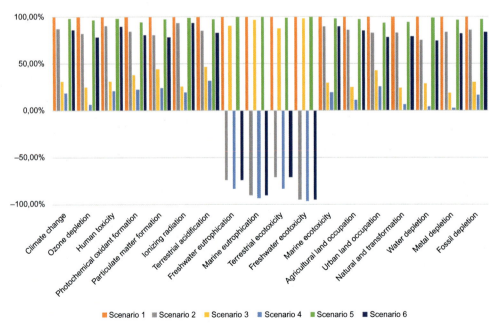

Fig. 12.2 Environmental impact of the six scenarios compared with respect to ReCiPe midpoint impact categories.

environmental credits attributed to Scenarios 2, 4, and 6, while among the other categories, significant differences are found for ozone depletion, natural land transformation, water depletion, and metal depletion. These categories are mainly influenced by differences in nutrient input patterns, which therefore seems to give a higher contribution in the reduction of the environmental impact that the use of the cement plant flue gas.

12.7 Conclusions

Development of sustainable biorefineries, capable to delivering new bioproducts (such as biodiesel) and meeting diverse criteria of environmental, economic, and social sustainability, is essential to meet policy goals (Živković et al., 2017).

Although there is still little evidence emerging from the literature, it is nevertheless clear that the effectiveness of public policies aimed at improving the sustainability on the various stakeholders cannot be taken for granted. Moreover, from this derives the fact that it is very important to reinforce the efforts that are today going in different directions to develop and standardize methods for the assessment sustainability beyond the environmental impacts and primarily including the social impacts; this, with the aim of guiding both the technological development and the public incentive policies in the correct direction for a sustainable bio-economy.

This comprehensive approach need to considers all the impact categories that contribute to the aspects of natural environment, human health resources, and society (Elkington, 1998).

LCA and S-LCA need to become a standard framework when selecting and developing intensification paths for biorefineries, even at an early stage of their development.

As illustrated through the case study relating the cultivation of bioalgae for biofuels, through these assessments it is possible to draw useful indications to support choices related to the configuration of the system that will be able to reduce impacts on the environment in an effective way. From this, it can also be deduced the importance of these methodologies as tools to support the design, not only of bio-products or biorefineries, but also of larger systems such as supply chains or company clusters in which biorefineries can be integrated through practices of industrial symbiosis.

References

Bach, R., Mohtashami, N., & Hildebrand, L. (2018). Comparative overview on LCA software programs for application in the Façade design process. *Journal of Facade Design and Engineering*, 7. https://doi.org/10.7480/jfde.2019.1.2657. No 1 Spec. Issue Powerskin 2019.

Balasubramanian, L., Subramanian, G., Nazeer, T. T., Simpson, H. S., Rahuman, S. T., & Raju, P. (2011). Cyanobacteria cultivation in industrial wastewaters and biodiesel production from their biomass: A review. *Biotechnology and Applied Biochemistry*, *58*, 220–225. https://doi.org/10.1002/bab.31.

Bello, S., Salim, I., Feijoo, G., & Moreira, M. T. (2021). Inventory review and environmental evaluation of first- and second-generation sugars through life cycle assessment. *Environmental Science and Pollution Research*. https://doi.org/10.1007/s11356-021-12405-y.

Ben Ruben, R., Menon, P., & Sreedharan, R. (2018). Development of a social life cycle assessment framework for manufacturing organizations. In *2018 International conference on production and operations management society (POMS)* (pp. 1–6). https://doi.org/10.1109/POMS.2018.8629496.

Bishop, G., Styles, D., & Lens, P. N. L. (2021). Environmental performance comparison of bioplastics and petrochemical plastics: A review of life cycle assessment (LCA) methodological decisions. *Resources, Conservation and Recycling, 168*. https://doi.org/10.1016/j.resconrec.2021.105451.

Blanco, J., Iglesias, J., Morales, G., Melero, J. A., & Moreno, J. (2020). Comparative life cycle assessment of glucose production from maize starch and woody biomass residues as a feedstock. *Applied Sciences, 10*. https://doi.org/10.3390/APP10082946.

Buchspies, B., Kaltschmitt, M., & Neuling, U. (2020). Potential changes in GHG emissions arising from the introduction of biorefineries combining biofuel and electrofuel production within the European Union—A location specific assessment. *Renewable and Sustainable Energy Reviews, 134*. https://doi.org/10.1016/j.rser.2020.110395.

Chinnasamy, S., Bhatnagar, A., Hunt, R. W., & Das, K. C. (2010). Microalgae cultivation in a wastewater dominated by carpet mill effluents for biofuel applications. *Bioresource Technology, 101*, 3097–3105. https://doi.org/10.1016/j.biortech.2009.12.026.

Cirot, A., Di Noi, C., Burhan, S. S., & Srocka, M. (2019). LCA database creation: Current challenges and the way forward. *Indonesian Journal of Life Cycle Assessment and Sustainability, 3*, 41–51.

Clarens, A. F., Nassau, H., Resurreccion, E. P., White, M. A., & Colosi, L. M. (2011). Environmental impacts of algae-derived biodiesel and bioelectricity for transportation. *Environmental Science & Technology, 45*, 7554–7560.

Clarens, A. F., Resurreccion, E. P., White, M. A., & Colosi, L. M. (2010). Environmental life cycle comparison of algae to other bioenergy feedstocks. *Environmental Science & Technology, 44*, 1813–1819.

Collet, P., Lardon, L., Hélias, A., Bricout, S., Lombaert-Valot, I., Perrier, B., et al. (2014). Biodiesel from microalgae—Life cycle assessment and recommendations for potential improvements. *Renewable Energy, 71*, 525–533. https://doi.org/10.1016/j.renene.2014.06.009.

Collotta, M., Busi, L., Champagne, P., Mabee, W., Tomasoni, G., & Alberti, M. (2016). Evaluating microalgae-to-energy -systems: Different approaches to life cycle assessment (LCA) studies. *Biofuels, Bioproducts and Biorefining, 10*. https://doi.org/10.1002/bbb.1713.

Collotta, M., Champagne, P., Mabee, W., & Tomasoni, G. (2018). Wastewater and waste CO2 for sustainable biofuels from microalgae. *Algal Research, 29*. https://doi.org/10.1016/j.algal.2017.11.013.

Collotta, M., Champagne, P., Mabee, W., Tomasoni, G., & Alberti, M. (2019). Life cycle analysis of the production of biodiesel from microalgae. *Green Energy and Technology*. https://doi.org/10.1007/978-3-319-93740-3_10.

Collotta, M., Champagne, P., Mabee, W., Tomasoni, G., Alberti, M., Busi, L., et al. (2016). Environmental assessment of co-location alternatives for a microalgae cultivation plant: A case study in the City of Kingston (Canada). *Energy Procedia*, 29–36. https://doi.org/10.1016/j.egypro.2016.09.007.

Collotta, M., Champagne, P., Mabee, W., Tomasoni, G., Leite, G. B., Busi, L., et al. (2017). Comparative LCA of flocculation for the harvesting of microalgae for biofuels production. *Procedia CIRP*, 756–760. https://doi.org/10.1016/j.procir.2016.11.146.

Collotta, M., Champagne, P., Tomasoni, G., Alberti, M., Busi, L., & Mabee, W. (2019). Critical indicators of sustainability for biofuels: An analysis through a life cycle sustainabilty assessment perspective. *Renewable and Sustainable Energy Reviews, 115*. https://doi.org/10.1016/j.rser.2019.109358.

Collotta, M., & Tomasoni, G. (2017). The environmental sustainability of biogas production with small sized plant. *Energy Procedia*.

Demırbas, A. (2017). The social, economic, and environmental importance of biofuels in the future. *Energy Sources, Part B: Economics, Planning, and Policy, 12*, 47–55. https://doi.org/10.1080/15567249.2014.966926.

Devadas, V. V., Khoo, K. S., Chia, W. Y., Chew, K. W., Munawaroh, H. S. H., Lam, M.-K., et al. (2021). Algae biopolymer towards sustainable circular economy. *Bioresource Technology*. https://doi.org/10.1016/j.biortech.2021.124702.

Elkington, J. (1998). Partnerships from cannibals with forks: The triple bottom line of 21st-century business. *Environmental Quality Management, 8*, 37–51.

Finnveden, G., Hauschild, M. Z., Ekvall, T., Guinée, J., Heijungs, R., Hellweg, S., et al. (2009). Recent developments in life cycle assessment. *Journal of Environmental Management, 91*, 1–21. https://doi.org/10.1016/j.jenvman.2009.06.018.

Frischknecht, R., Jungbluth, N., Althaus, H. J., Doka, G., Dones, R., Heck, T., et al. (2005). The ecoinvent database: Overview and methodological framework. *International Journal of Life Cycle Assessment, 10*, 3–9.

Frischknecht, R., Pfister, S., Bunsen, J., Haas, A., Känzig, J., Kilga, M., et al. (2019). Regionalization in LCA: Current status in concepts, software and databases—69th LCA forum, Swiss Federal Institute of Technology, Zurich, 13 September, 2018. *International Journal of Life Cycle Assessment, 24*, 364–369. https://doi.org/10.1007/s11367-018-1559-0.

Garcia Gonzalez, M. N., Levi, M., & Turri, S. (2017). Development of polyester binders for the production of sustainable polyurethane coatings: Technological characterization and life cycle assessment. *Journal of Cleaner Production, 164*, 171–178. https://doi.org/10.1016/j.jclepro.2017.06.190.

Gendy, T. S., & El-Temtamy, S. A. (2013). Commercialization potential aspects of microalgae for biofuel production: An overview. *Egyptian Journal of Petroleum, 22*, 43–51. https://doi.org/10.1016/j.ejpe.2012.07.001.

Hwang, J.-H., Church, J., Lee, S.-J., Park, J., & Lee, W. H. (2016). Use of microalgae for advanced wastewater treatment and sustainable bioenergy generation. *Environmental Engineering Science, 33*. https://doi.org/10.1089/ees.2016.0132.

ISO. (2006). *ISO 14040 international standard. Environmental management-life cycle assessment—Principles and framework*. International Organization for Standardization.

Ita-Nagy, D., Vázquez-Rowe, I., Kahhat, R., Chinga-Carrasco, G., & Quispe, I. (2020). Reviewing environmental life cycle impacts of biobased polymers: Current trends and methodological challenges. *International Journal of Life Cycle Assessment, 25*, 2169–2189. https://doi.org/10.1007/s11367-020-01829-2.

Itten, R., Frischknecht, R., Stucki, M., Scherrer, P., & Psi, I. (2014). *Life cycle inventories of electricity mixes and grid* (pp. 1–229). Paul Scherrer Inst.

Jørgensen, A., Le Bocq, A., Nazarkina, L., & Hauschild, M. (2008). Methodologies for social life cycle assessment. *International Journal of Life Cycle Assessment, 13*, 96–103.

Kim, S., & Dale, B. E. (2009). Regional variations in greenhouse gas emissions of biobased products in the United States-corn-based ethanol and soybean oil. *International Journal of Life Cycle Assessment, 14*, 540–546. https://doi.org/10.1007/s11367-009-0106-4.

Li, J., Wang, Y., & Yan, B. (2018). The hotspots of life cycle assessment for bioenergy: A review by social network analysis. *Science of the Total Environment, 625*, 1301–1308. https://doi.org/10.1016/j.scitotenv.2018.01.030.

Liu, Y., Lyu, Y., Tian, J., Zhao, J., Ye, N., Zhang, Y., et al. (2021). Review of waste biorefinery development towards a circular economy: From the perspective of a life cycle assessment. *Renewable and Sustainable Energy Reviews, 139*. https://doi.org/10.1016/j.rser.2021.110716.

Lopes, T. F., Carvalheiro, F., Duarte, L. C., Gírio, F., Quintero, J. A., & Aroca, G. (2019). Techno-economic and life-cycle assessments of small-scale biorefineries for isobutene and xylo-oligosaccharides production: A comparative study in Portugal and Chile. *Biofuels, Bioproducts and Biorefining, 13*, 1321–1332. https://doi.org/10.1002/bbb.2036.

Lüdemann, L., & Feig, K. (2014). Comparison of software solutions for life cycle assessment (LCA)—A software ergonomic analysis. *Logistics Journal*. https://doi.org/10.2195/lj_NotRev_luedemann_de_201409_01.

Mattioda, R. A., Tavares, D. R., Casela, J. L., & Junior, O. C. (2020). Chapter 9—Social life cycle assessment of biofuel production. In J. Ren, A. Scipioni, A. Manzardo, & H. Liang (Eds.), *Biofuels for a more sustainable future* (pp. 255–271). Elsevier. https://doi.org/10.1016/B978-0-12-815581-3.00009-9.

Medeiros, D. L., Sales, E. A., & Kiperstok, A. (2015). Energy production from microalgae biomass: Carbon footprint and energy balance. *Journal of Cleaner Production, 96*, 493–500. https://doi.org/10.1016/j.jclepro.2014.07.038.

Moretti, C., Corona, B., Hoefnagels, R., Vural-Gürsel, I., Gosselink, R., & Junginger, M. (2021). Review of life cycle assessments of lignin and derived products: Lessons learned. *Science of the Total Environment, 770*. https://doi.org/10.1016/j.scitotenv.2020.144656.

Muralikrishna, I. V., & Manickam, V. (2017). Chapter five—Life cycle assessment. In I. V. Muralikrishna, & V. Manickam (Eds.), *Environmental management* (pp. 57–75). Butterworth-Heinemann. https://doi.org/10.1016/B978-0-12-811989-1.00005-1.

Papadaskalopoulou, C., Sotiropoulos, A., Novacovic, J., Barabouti, E., Mai, S., Malamis, D., et al. (2019). Comparative life cycle assessment of a waste to ethanol biorefinery system versus conventional waste management methods. *Resources, Conservation and Recycling, 149*, 130–139. https://doi.org/10.1016/j.resconrec.2019.05.006.

Schenker, U., Head, M., Collotta, M., Andro, T., Viot, J., Whatelet, A., et al. (2020). *Handbook for product social impact assessment*. https://product-social-impact-assessment.com/wp-content/uploads/2021/04/20-01-Handbook2020.pdf. (Accessed 13 April 2022).

Sturm, B. S. M., & Lamer, S. L. (2011). An energy evaluation of coupling nutrient removal from wastewater with algal biomass production. *Applied Energy, 88*, 3499–3506. https://doi.org/10.1016/j.apenergy.2010.12.056.

Tillman, A.-M., Ekvall, T., Baumann, H., & Rydberg, T. (1994). Choice of system boundaries in life cycle assessment. *Journal of Cleaner Production, 2*, 21–29. https://doi.org/10.1016/0959-6526(94)90021-3.

UNEP. (2009). *Guidelines for social LCA of products*. (WWW Document).

UNEP. (2011). *Towards a life cycle sustainability assessment. Making informed choices on products*. (WWW Document).

Usher, P. K., Ross, A. B., Camargo-Valero, M. A., Tomlin, A. S., & Gale, W. F. (2014). An overview of the potential environmental impacts of large-scale microalgae cultivation. *Biofuels, 5*, 331–349. https://doi.org/10.1080/17597269.2014.913925.

Valdivia, S., Ugaya, C. M. L., Hildenbrand, J., Traverso, M., Mazijn, B., & Sonnemann, G. (2013). A UNEP/SETAC approach towards a life cycle sustainability assessment—Our contribution to Rio +20. *International Journal of Life Cycle Assessment, 18*, 1673–1685. https://doi.org/10.1007/s11367-012-0529-1.

Walker, S., & Rothman, R. (2020). Life cycle assessment of bio-based and fossil-based plastic: A review. *Journal of Cleaner Production, 261*. https://doi.org/10.1016/j.jclepro.2020.121158.

Wang, L., Li, Y., Chen, P., Min, M., Chen, Y., Zhu, J., et al. (2010). Anaerobic digested dairy manure as a nutrient supplement for cultivation of oil-rich green microalgae Chlorella sp. *Bioresource Technology, 101*, 2623–2628. https://doi.org/10.1016/j.biortech.2009.10.062.

Wang, L., Min, M., Li, Y., Chen, P., Chen, Y., Liu, Y., et al. (2010). Cultivation of green algae Chlorella sp. in different wastewaters from municipal wastewater treatment plant. *Applied Biochemistry and Biotechnology, 162*, 1174–1186. https://doi.org/10.1007/s12010-009-8866-7.

Woertz, I., Feffer, A., Lundquist, T., & Nelson, Y. (2009). Algae grown on dairy and municipal wastewater for simultaneous nutrient removal and lipid production for biofuel feedstock. *Journal of Environmental Engineering, 135*, 1115–1122. https://doi.org/10.1061/(ASCE)EE.1943-7870.0000129.

Živković, S. B., Veljković, M. V., Banković-Ilić, I. B., Krstić, I. M., Konstantinović, S. S., Ilić, S. B., et al. (2017). Technological, technical, economic, environmental, social, human health risk, toxicological and policy considerations of biodiesel production and use. *Renewable and Sustainable Energy Reviews, 79*, 222–247. https://doi.org/10.1016/j.rser.2017.05.048.

CHAPTER 13

Social impact assessment in designing supply chains for biorefineries

Sergio Iván Martínez-Guido[a], Juan Fernando García-Trejo[a], and José María Ponce-Ortega[b]

[a]Facultad de Ingeniería, Universidad Autónoma de Querétaro, Amazcala, Querétaro, Mexico
[b]Facultad de Ingeniería Química, División de Estudios de Posgrado, Universidad Michoacana de San Nicolás de Hidalgo, Morelia, Michoacán, Mexico

13.1 Introduction

Supply chain optimization has been a topic with high attention over 20 years ago, in terms of Hassini, Surti, and Searcy (2012), all the information, resources, and funds need to be management to achieve profit maximization in a supply chain, without leaving out social well-being maximization and environmental impact reduction. In this way, social impact assessment has attracted growing attention, from different investigations. Santibañez-Aguilar, González-Campos, Ponce-Ortega, Serna-González, and El-Halwagi (2014) proposed a mathematical supply chain optimization of an ethanol and biodiesel biorefinery, integrating job creation as social impact tool, identifying that job generation has a beneficial impact in social terms (due to the per-capita income), but negatively in environmental aspects due to the relation more jobs-more anthropogenic activities. Similarly, Yue, Slivinsky, Sumpter, and You (2014) quantified the total number jobs by construction and operation phase of a bioelectricity supply chain network, as social objective, obtaining that it is possible to create 160,000 jobs (direct and indirect jobs) whit a maximum plant capacity. Mota, Gomes, Carvalho, and Barbosa-Povoa (2015) included in their mathematical model a regional factor, which was based on the regional unemployment percent, in this way, higher impact is received by jobs creation in locations with lower employ rates, the results showed that social benefit and the total cost for the supply chains increase proportionally. In the same way, Cambero and Sowlati (2016) integrated the hours of work (job) as impact variable in a forest-based biorefinery supply chain, having that social benefit increases with the implementation of a bigger number of installed plants. Chazara, Negny, and Montastruc (2017) found the relation between job creation (social benefit) and environmental impact reduction, the reason in this positive relation is because the job source is a biorefinery, with the goal of CO_2 fixation. Fathollahi-Fard, Hajiaghaei-Keshteli, and Mirjalili (2018) incorporated job opportunities based on manufactured products demand using the scenario probability. Popovic, Barbosa-Póvoa, Kraslawki, and Carvalho (2018) suggested quantitative indicators for social sustainability assessment of supply chains, including labor practices, decent work and human rights, stabling setting up that all these variables are possible to measure only in supply chains already established, because of the nature of the involved factors in the quantification (resigned employees, working hours, etc.). On the other hand, Martínez-Guido et al. (2019a, b) proposed a new way to evaluate the social impact in a food network optimization, using the human development index (HDI) as analysis tool, designing mathematic equations, which involve how the economic benefit impact in health, income, and education sectors.

13.2 Current perspective of biorefinery systems: Conventional fuel replacement by bioenergy sources

Since 1965, biofuels have received more interest from academic, industrial, and political sectors, due to their versatility in sub-product generation, efficiency, and cost effectiveness. Biofuels are characterized by being derivate from biological sources: as plants, animals, and microorganism, including to the biodegradable, nontoxic and environmentally friendly aspects (Anto et al., 2020). World Bioenergy Association (2017) reported renewable energy participation globally percent (see Fig. 13.1), being Asia the continent with the highest exa-Joules (EJ) obtained from bio-resources (35 EJ), and Oceania with the lowest value (0.91 EJ). Global demands are satisfied by energy sources as Fig. 13.2 shows, where 61% is supplied by traditional biofuels (Smil, 2017). Most of this demand is satisfied

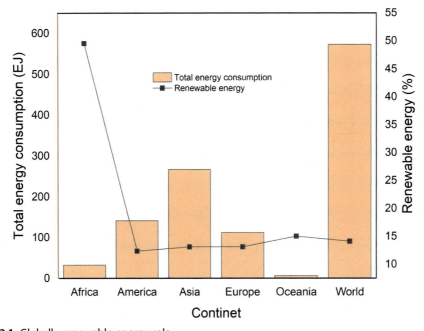

Fig. 13.1 Globally renewable energy role.

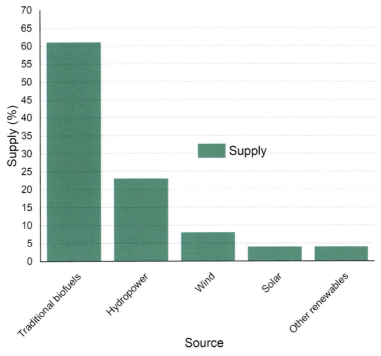

Fig. 13.2 Renewable energy supply by source.

by biomass consumption, which includes crops, energy crops, aquatic plants, forestry wastes, and lastly, agro-residues (Stameković et al., 2020), with the capacity to produce bioethanol, biodiesel (as first-generation fuels) and advanced biofuels like biogas, bio-oil, biochar, hydrogen, syngas, pellets, and bio-jet fuel (Gutierrez-Antonio, Gómez-Castro, de Lira-Flores, & Hernández, 2017). Li, Xu, Xie, and Wang (2018) indicated that only 2% of global transport fuel is obtained from biofuel sources, with the goal to achieve 27% by 2050, blends of ethanol gasoline (E10 and E85), and biodiesel 20% (B20) are the most common fuel sources used (Office of Energy Efficiency and Renewable Energy (BETO), 2020). In 2019, developed countries reached a biofuel production of 160.9 mnL-L, while in developing countries only 66.3 mnL-L were produced (Subramaniam, Masron, & Nik-Azman, 2020). On the other hand, Schwerz et al. (2020) obtained 284 ha of woody energy crop plantation to satisfy the electric energy demand for one year of Federico Westphalen, Brasil (1884 kWh). In addition, woody energy has been used as heat and power source in residential uses, obtaining 20 kW/tonne of burned pellet in Stirling engines working with efficiencies of 72% (Cardozo, Erlich, Malmquist, & Alejo, 2014). A study performed by Anderson and Toffolo (2013) provided an economic analysis of wood pellet production, reporting the pellet production cost by USD$34/MWh,

USD$92/MWh by electric energy generation and USD$34/MWh by heat power. Therefore, Yun, Clift, and Bi (2020) raised the scenario of 20% greenhouse reduction in Europe, with wood pellet use; also, China, India, Malaysia, and Indonesia are expected to obtain 2%–6% from solid biofuels. Similarly, Martínez-Guido et al. (2019a, b) performed a strategic planning of agro-residue pellets used in Mexican power plants, having that it is possible to satisfy 50% of the national demand with a CO_2 emissions reduction of 25%. In this way, a study in USA for the electric energy consumption was carried out by Kim, Choi, and Seok (2020), having that only 11% of this demand is satisfied by renewable energy, 45% from biomass, 4% from biomass waste, 21% from biofuel, and 9% from woody energy crops. Sagastume-Gutiérrez, Cabello-Eras, Hens, and Vandecasteele (2020) analyzed potential electricity production from available biomass (agriculture, livestock, and agro-industrial wastes) in Colombia, obtaining that is possible to yield 100,000 GWh from direct combustion, and 63,000 GWh from anaerobic processes. Mortensen et al. (2020) identified that each GJ of hydrogen has the capacity to replace 2.5 GJ of biomass in terms of volume; however, in terms of production cost more aspects should be considered.

Biomass energy has been linked to land degradation, land competition, resource depletion, deforestation, biodiversity loss, and food security problems (Wang, Bui, Zhang, & Pham, 2020). In this way, it is necessary to evaluate all the alternatives in all the possible ways, including: biomass harvest, transportation, intermediate storage, processing operation, final product storage, and expedition, which are the important points to achieve an integral supply chain (Sulaiman, Abdul-Rahim, & Ofozor, 2020). In which there are added economic, environmental, and social aspects having as result the big picture of the problem. In this way, Nunes, Causer, and Ciolkosz (2020) suggested that supply chain modeling results an attractive tool to achieve goals as: predict supply chain performance, select the most efficient supply chain configuration, optimize sizing of supply chain components to minimize cost/environmental impact, increase the social benefit, and even optimize scheduling of supply chain operations.

13.3 Biorefineries evaluation using optimization tools

As was described in previous section, there is a huge market for biofuel consumption under the biorefinery concept, due to the growing energy demand and the scarcity of conventional fuels. However, it is necessary to take into consideration the big picture and all the details of each system production. Developing biofuel production new technologies, creating new policy's actions (governmental incentives, environmental improvement), economic analysis (including financial risk), biomass supply security (Díaz-Trujillo, Tovar-Facio, Nápoles-Rivera, & Ponce-Ortega, 2019), strategic decisions in logistic and operational aspects, production capacity, and interactions with surrounding environment are some of the constraints that have higher impact in the decision-making process (López-Díaz et al., 2018). In this way, process system engineering tools (mathematical optimization) (El-Halwagi, 1997) provide scenarios in which is possible to maximize yields and social benefit, cost, and environmental impact, integrating all the variables involved in biofuel supply chains.

Case studies as the one formulated by Hernández-Pérez, Sánchez-Tuirán, Ojeda, El-Halwagi, and Ponce-Ortega (2019) showed the application of mathematic and computational techniques to the biodiesel biorefinery production process using microalgae as raw material, 47.23% of CO_2 reduction/y was the optimal objective solution with an economic value of 51.69 M$/y by the integration of this supply chain as energy alternative.

13.4 Supply chain optimization

Fig. 13.3 shows the big picture of using pellets in power plants and how HDI could have a benefit by the biorefinery integration. The problem resides in that the electricity demand ($E_{j,t}^{Max-DE}$) is currently satisfied from conventional sources mainly ($F_{j,f,t}^{CFuel}$); however, with the constant population growth, overexploitation, and even the extinction of certain resources it is put at risk. Hence, there are agro-residue fluxes ($FC_{r,i,t}^{Residues}$) from some crops planted to satisfy nutritional human demand, which can be introduced in a pellet

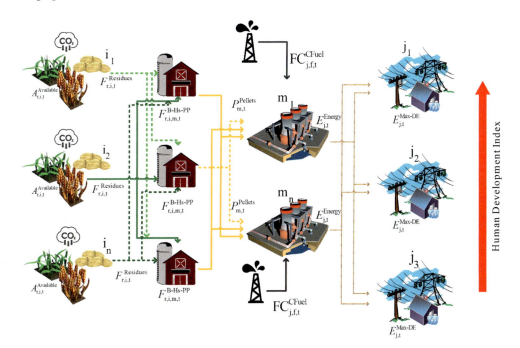

Fig. 13.3 Big picture of integrated pellet systems.

($P_{m,t}^{Pellets}$) processing plant (m_n), due to its physiochemical characteristics, and use it as energy source in some types of electric generation plant (J_n). If this alternative is carried out, it is possible to obtain some beneficial aspects as: agro-residues revalorization, CO_2 emissions reduction (natural fixation), and job generation.

However, it is necessary to take into consideration distribution pathways, supply chain cost, electric energy balances, environmental impact measurements, biomass scheduling, and design and planning facility installation. Particularly, as was described in Section 13.1, social impact has becoming crucial to achieve sustainable process; hence, human development index is used as reference tool to evaluate this impact.

Therefore, due the huge number of variables included, it is necessary to carry out a big picture analysis using optimization process engineering tools, guaranteeing the evaluation of the process and its resilience in different scenarios. In next section is described the proposed mathematic model, which describes all the supply chain activities.

13.5 Problem statement

The addressed problem can be stated as follows: Given the different crops produced in the Michoacán state and the available agro-residues in the region, it is possible to take into consideration these biomass fluxes and characteristics for bio-pellets production which can be used as electric energy source, performing an environmental energy generation, and reducing oil's dependence. However, due to the variability for agro-residues availability (type, amount, and localization and even chemical characteristics), then the problem consists in determining the optimal pellets supply chain configuration. Nevertheless, it is necessary to take into consideration that agro-residues are not the only variables, which affect the optimal configuration, also aspects as biomass transport, pellets processing, biofuel distribution, plant installation and localization, and even the calorific power should be included. In this work, environmental variables such as CO_2 fixation (by crops production), and generation were also included. In addition, human development index was used to calculate social benefit generation using local agro-residues. In this way, an optimization model described in Section 13.5 was proposed, where there are included all the equations that represent the big picture of the supply chain, afterwards the model was codified in GAMS® platform, and solved as it is described in Section 13.6 using parameters from the current situation of the Michoacán state.

13.6 Mathematical model

Table 13.1 shows the main set used for the mathematic model formulation, also in Section 13.9 is described all the nomenclature used on the equations.

Eq. (13.1) shows the flux in tonnes of agro-residues (r) collected from the selected planted crops ($F_{r,i,t}^{Residues}$), this flux depends on the area ($A_{r,i,t}^{Available}$) used for harvesting,

Table 13.1 Sets used in the mathematical model.

Set	Values	Description
r	1–11	Type of agro-residues collected and used as pellet production source
i	1–113	Michoacán municipalities, which can be harvesting sites
t	1–52	Period of time (weeks)
m	1–113	Michoacán municipalities, which can be processing sites
j	1–113	Michoacán municipalities, which can be energy consumption sites

and the yield of residual biomass linked to each crop ($\beta_{r,i,t}^{yield}$). Harvesting used area ($A_{r,i,t}^{Available}$) is linked with two factors, current area ($A_{r,i,t}^{current}$) used for each vegetable production, and new area ($A_{r,i,t}^{new}$) with the possibility to transform it into cropping activity (taking into consideration the lower environmental impact by soil use change), as Eq. (13.2) describes. To achieve the lower possible environmental impact, the used area ($A_{r,i,t}^{Available}$) is constrained by the maximum capacity ($AT_{r,i,t}$) for each crop (see Eq. 13.3). However, it is necessary to take into consideration distribution pathways analysis, supply chain cost, electric energy balances, environmental impact measurements, biomass scheduling scrutiny, and design and planning facility installation. Particularly, as it is described in Section 13.1, social impact has becoming crucial to achieve sustainable processes; hence, the human development index is used as reference tool to evaluate this impact.

Therefore, due to the huge number of variables included, it is necessary to carry out a big picture analysis using optimization process engineering tools, guaranteeing the evaluation of the process and its resilience in different scenarios. In next section is described the proposed mathematic model, which describes all the supply chain activities.

Eq. (13.1) shows the flux in tonnes of agro-residues (r) collected from the selected planted crops ($F_{r,i,t}^{Residues}$), this flux depends on the area ($A_{r,i,t}^{Available}$) used for harvesting, and the yield of residual biomass linked to each crop ($\beta_{r,i,t}^{yield}$). Harvesting used area ($A_{r,i,t}^{Available}$) is linked with two factors, current area ($A_{r,i,t}^{current}$) used for each vegetable production, and new area ($A_{r,i,t}^{new}$) with the possibility to transform it into cropping activity (taking into consideration the lower environmental impact by soil use change), as Eq. (13.2 describes. To achieve the lower possible environmental impact, the used area ($A_{r,i,t}^{Available}$) is constrained by the maximum capacity ($AT_{r,i,t}$) for each crop (see Eq. 13.3).

$$F_{r,i,t}^{Residues} = A_{r,i,t}^{Available} \cdot \beta_{r,i,t}^{yield}, \forall r \in R, i \in I, t \in T \quad (13.1)$$

$$A_{r,i,t}^{Available} = A_{r,i,t}^{current} + A_{r,i,t}^{new}, \forall r \in R, i \in I, t \in T \quad (13.2)$$

$$A_{r,i,t}^{Available} \leq AT_{r,i,t}, \forall r \in R, i \in I, t \in T \quad (13.3)$$

After harvesting, agro-residues ($F_{r,i,t}^{Residues}$) are collected and sent to the pellet processing ($F_{r,i,m,t}^{B-Hs-PP}$) plants, the sent flux ($F_{r,i,m,t}^{B-Hs-PP}$) is equal to the received flux ($F_{r,m,t}^{B-PP}$),

Eqs. (13.4) and (13.5) represent these balances of biomass distribution. Eq. (13.6) is the constrain for biomass distribution, in this way, the sent flux from harvesting ($F_{r,i,m,t}^{B-Hs-PP}$) must be equal or lower than the one of the plant hub capacity ($HC_{m,t}$).

$$\sum_m F_{r,i,m,t}^{B-Hs-PP} = F_{r,i,t}^{Residues}, \forall r \in R, i \in I, t \in T \quad (13.4)$$

$$\sum_i F_{r,i,m,t}^{B-Hs-PP} = F_{r,m,t}^{B-PP}, \forall r \in R, m \in M, t \in T \quad (13.5)$$

$$\sum_r F_{r,m,t}^{B-PP} = HC_{m,t}, \forall m \in M, t \in T \quad (13.6)$$

All the residues received ($F_{r,m,t}^{B-PP}$) at processing plants are used for pellet manufacturing, each kind of residue has linked to its own yield factor [(ϕ_r^{yield}) tonnes of pellets per tonnes of processed residue]; hence, Eq. (13.7) presents the pellet production fluxes ($P_{m,t}^{Pellets}$) in each installed plant. Afterward, pellet fluxes ($P_{m,t}^{Pellets}$) are distributed to the thermoelectric power plants ($P_{m,j,t}^{P-Distribu}$), in this way, Eqs. (13.8) and (13.9) represent these flux distributions, tonnes of pellets sent from processing plants ($P_{m,j,t}^{P-Distribu}$) are equal to the received fluxes at power plants ($FP_{j,t}^{P-Ther}$).

In Eq. (13.10), the electric energy ($E_{j,t}^{Energy}$) balance is defined, there are two main sources considered for cogeneration, conventional fuels ($FC_{j,f,t}^{CFuel}$), and pellets ($FP_{j,t}^{P-Ther}$), in this way, each used flux (depending on the demand and pellets availability), are multiplied by their respective conversion factor [(η^{PYE}) for pellets and (μ_f^{FYE}) for conventional fuel]. Electric energy generation ($E_{j,t}^{Energy}$) is constrained by Eq. (13.11), accordingly with the maximum electricity demand in each place ($E_{j,t}^{Max-DE}$).

$$P_{m,t}^{Pellets} = \sum_r F_{r,m,t}^{B-PP} \cdot \phi_r^{yield}, \forall t \in T \quad (13.7)$$

$$\sum_j P_{m,j,t}^{P-Distribu} = P_{m,t}^{Pellets}, \forall m \in M, t \in T \quad (13.8)$$

$$FP_{j,t}^{P-Ther} = \sum_m P_{m,j,t}^{P-Distribu}, \forall j \in J, t \in T \quad (13.9)$$

$$E_{j,t}^{Energy} = FP_{j,t}^{P-Ther} \cdot \eta^{PYE} + FC_{j,f,t}^{CFuel} \cdot \mu_f^{FYE}, \forall j \in J, t \in T \quad (13.10)$$

$$E_{j,t}^{Energy} \geq E_{j,t}^{Max-DE}, \forall j \in J, t \in T \quad (13.11)$$

Pellet facility installation capital costs ($CPP_m^{capital}$) are evaluated by Eq. (13.12), fixed costs (V^{fixed}) are linked with the plant existence represented by binary variable (Y_m^{Var}), while variable costs (V^{cost}) are liked with agro-residues processed flux ($HC_{m,t}$) in each plant. Eq. (13.13) shows the processing capacity ($HC_{m,t}$) of each plant constrain ($HC_{m,t}^{max}$) given that the volume accepted for transformation. In the other hand, transportation cost ($TC^{transpo-cost}$) are defined by residues ($F_{r,i,m,t}^{B-Hs-PP}$) and pellets ($P_{m,j,t}^{P-Distrbu}$)

distribution in Eq. (13.14), multiplying transported fluxes by the respectively unitary cost ($RT_{i,m}^{rm-trans}$, $PT_{m,j}^{pel-trans}$).

If the electric energy is obtained from conventional fuels, Eq. (13.15) is activated given the fuel cost ($FC_{j,t}^{fuel-cost}$), using the combusted flux ($FC_{j,f,t}^{CFuel}$) by unitary cost ($UF_{f}^{cost-Fuel}$). Addition of capital cost ($CPP_{m}^{capital}$), raw material and pellets distribution costs ($TC^{transpo-cost}$) and conventional fuel use costs ($FC_{j,t}^{fuel-cost}$) give the electric energy total cost (TOT), which is shown by Eq. (13.16).

$$CPP_{m}^{capital} = V^{fixed} \cdot Y_{m}^{Var} + V^{cost} \cdot \sum_{t} HC_{m,t}, \forall m \in M \quad (13.12)$$

$$HC_{m,t} \cdot Y_{m}^{Var} \geq HC_{m,t}^{max}, \forall m \in M, t \in T \quad (13.13)$$

$$TC^{transpo-cost} = \sum_{r,i,m,t} F_{r,i,m,t}^{B-Hs-PP} \cdot RT_{i,m}^{rm-trans} + \sum_{m,j,t} P_{m,j,t}^{P-Distribu} \cdot PT_{m,j}^{pel-trans} \quad (13.14)$$

$$FC_{j,t}^{fuel-cost} = \sum_{f} FC_{j,f,t}^{CFuel} \cdot UF_{f}^{cost-Fuel}, \forall j \in J, t \in T \quad (13.15)$$

$$TOT = \sum_{m} CPP_{m}^{capital} + \sum_{j,t} FC_{j,t}^{fuel-cost} + TC^{transpo-cost} \quad (13.16)$$

Job generation is a high important variable, due to its relationship with the social impact assessment. The way to obtain the created jobs number was using the flux of money assigned to this activity (CPE_m). (P_m^{People}), which represents the percent of money from facility capital cost ($CPP_{m}^{capital}$) that is used for salaries, (P_r^{rm}) represents the percent of money used to pay the agro-residues use flux ($F_{r,m,t}^{B-PP}$) as technique to add a value to the main crops (see Eq. 13.17). Job number (P_m^{P-jobs}) is calculated with Eq. (13.18), dividing money flux (CPE_m) assigned to workers by the respective salary per worker (S_m^{salary}).

$$CPE_m = P_m^{People} \cdot CPP_{m}^{capital} + \sum_{r,t} F_{r,m,t}^{B-PP} \cdot P_r^{rm}, \forall m \in M \quad (13.17)$$

$$P_m^{P-jobs} \frac{CPE_m}{S_m^{salary}}, \forall m \in M \quad (13.18)$$

Afterwards, for social impact assessment, it is necessary the taxes calculation. Tax recovery was taken into to account based on its destination use (human health, education, and income dimensions). Eqs. (13.19) and (13.20) show how the taxes used for human health dimension are calculated. First, there is a factor taxes (F_m^{HH-F}) charged to the facility [from capital cost ($CPP_m^{capital}$)] given the industrial contribution (T_m^{HH-tax}) to the human health (Eq. 13.19), for each worker the industry must pay a factor contribution (TB_m) also charged to the capital cost ($CPP_m^{capital}$), plus a factor contribution (TA_m) charged directly to each worker (CPE_m) salary (Eq. 13.20).

$$T_m^{HH-tax} = CPP_m^{capital} \cdot F_m^{HH-F}, \forall m \in M \quad (13.19)$$

$$T_m^{HH} = CPE_m \cdot TA_m + CPP_m^{capital} \cdot TB_m + T_m^{HH-tax}, \forall m \in M \quad (13.20)$$

Therefore, taxes collected by human health, when pellets are integrated as biofuel (T_m^{HH}), are compared with the current flux assigned by the government (CVHH$_m$), obtaining the percent increase contribution in this dimension (I_m^{HH}), as Eq. (13.21) describes.

$$I_m^{HH} = \frac{T_m^{HH}}{CVHH_m}, \forall m \in M \quad (13.21)$$

For education dimension, there are included the taxes charged to workers salary (T_m^{Edu}), designated by a respective factor (TE$_m$) multiplied by the flux of money used for jobs salary (CPE_m) as Eq. (13.22) shows. Similarly, as in the human health dimension, the collected new education taxes (I_m^{Edu}) are compared with the current situation (CVE$_m$) (see Eq. 13.23). Under the same concept, taxes assigned to the income dimension (T_m^i) are calculated by Eq. (13.24) [a factor (TI$_m$) multiplied to the salary (CPE_m)], and compared with the current income situation (CVI$_m$), as Eq. (13.25) describes.

$$T_m^{Edu} = CPE_m \cdot TE_m, \forall m \in M \quad (13.22)$$

$$I_m^E = \frac{T_m^{Edu}}{CVE_m}, \forall m \in M \quad (13.23)$$

$$T_m^i = CPE_m \cdot TI_m, \forall m \in M \quad (13.24)$$

$$I_m^I = \frac{T_m^i}{CVI_m}, \forall m \in M \quad (13.25)$$

All the taxes collected for human health (I_m^{HH}), education (I_m^E), and income (I_m^I) dimensions are divided by the 113 municipalities that conform Michoacán State, to obtain the HDI benefit expect (HDI^{Exp}) determined by Eq. (13.26). The result obtained in Eq. (13.26) is compare with the HDI current value (HDI^{CURR}) to obtain the comparison ($HDI^{Tot-Benefit}$) between the new HDI value reached and the current situation (see Eq. 13.27).

$$HDI^{Exp} = \frac{\sum_m I_m^{HH}}{113} + \frac{\sum_m I_m^E}{113} + \frac{\sum_m I_m^I}{113} \quad (13.26)$$

$$HDI^{Tot-Benefit} = \left(\frac{HDI^{Exp}}{HDI^{CURR}}\right) \cdot 100 \quad (13.27)$$

In terms of environmental emissions generated by the pellet integration (E^{PP}), it was calculated by Eq. (13.28). Emission by the biomass ($EM_{i,m}^{Bio-trans}$) and pellets ($EM_{m,j}^{Pellet-trans}$) distribution are multiplied by the respective transported flux ($F_{r,i,m,t}^{B-Hs-PP}$, $P_{m,j,t}^{P-Distribu}$), plus the emissions generated (EM) by the pellets production flux ($P_{m,t}^{Pellets}$), and the emissions ($EM^{Pellet-use}$) generated by pellets flux combusted ($FP_{j,t}^{P-Ther}$) in power plants.

$$E^{PP} = \sum_{m,t} P^{\text{Pellets}}_{m,t} \cdot \text{EM} + \sum_{r,i,m,t} F^{\text{B-Hs-PP}}_{r,i,m,t} \cdot \text{EM}^{\text{Bio-trans}}_{i,m} + \sum_{m,j,t} P^{\text{P-Distribu}}_{m,j,t} \cdot \text{EM}^{\text{Pellet-trans}}_{m,j}$$
$$+ \sum_{j,t} FP^{\text{P-Ther}}_{j,t} \cdot \text{EM}^{\text{Pellet-use}}$$
(13.28)

If the electric energy demand is satisfied with conventional fuels, generated emissions are calculated ($E^{\text{CC-use}}$) by Eq. (13.29), multiplying the used fluxes ($FC^{\text{CFuel}}_{j,f,t}$) by the respective factor (FE^{CC}_{f}).

$$E^{\text{CC-use}} = \sum_{j,f,t} FC^{\text{CFuel}}_{j,f,t} \cdot FE^{\text{CC}}_{f} \qquad (13.29)$$

All the biomass considered as agro-residue source as part of natural plant respiration carry out the CO_2 absorption ($E^{\text{abs-co2}}$), in this way in Eq. (13.30), there are multiplied the agro-residue flux ($F^{\text{Residues}}_{r,i,t}$) used by the fraction absorbed ($F^{\text{CO2-fixed}}_{r}$) by each tonne of biomass.

$$E^{\text{abs-co2}} = \sum_{r,i,t} F^{\text{Residues}}_{r,i,t} \cdot F^{\text{CO2-fixed}}_{r} \qquad (13.30)$$

($E^{\text{Total-Emiss}}$) represents the total CO_2 tonnes emitted by all the electric energy supply chain (Eq. 13.31), including the addition of the pellet emissions (E^{PP}) and the conventional fuel emissions ($E^{\text{CC-use}}$) minus the natural absorption ($E^{\text{abs-co2}}$).

$$E^{\text{Total-Emiss}} = E^{PP} + E^{\text{CC-use}} - E^{\text{abs-co2}} \qquad (13.31)$$

As objective function, there are the total emissions ($E^{\text{Total-Emiss}}$) and total cost (TOT) minimization, plus the HDI maximization ($HDI^{\text{Tot-Benefit}}$) as it is described in Eq. (13.32).

$$F \cdot O \cdot = \{\text{Minimizing} E^{\text{Total-Emiss}}, TOT \cdot \text{Maximizing} HDI^{\text{Tot-Benefit}}\} \qquad (13.32)$$

13.7 Mathematical modeling solution

The proposed mathematical model was used to calculate each involved variable for the pellet production supply chain optimization. The formulated equations can be fed with parameters from any other case study from other region, or even with parameters from other country. In this way, the multiobjective problem was solved using ε constrain method. The basic strategy of this method is transformed the multiobjective problem into a series of single objective optimization problems (Eq. 13.32). Hence, keeping in mind to satisfy the Michoacán electric energy demand using the bio-pellets production as energy source, emissions minimization was used as single objective, while human development variable is turned into an inequality constrain, obtaining the point A in Fig. 13.5. Similarly, to obtain point E in Pareto's diagram (Fig. 13.5), the problem was solved using the human development index as single objective, turning emissions goal into an inequality constrain which take values between $\pm\infty$. Afterward, to obtain scenarios B, C, and D with emissions goal turned into an inequality constrain, which was fixed to a specific value, while HDI objective is leaved as single free objectives; all the resulting analyzed points are shown in Fig. 13.5, which represents the tradeoff between emissions minimization and human development increase. However, to obtain these final emissions, human development and even economic values for each point shown in Pareto's diagram is necessary to calculate all the variables show in Fig. 13.3. For example, variables such as raw material yields, transportation cost, processing pellets cost, distribution to the electric plants. In addition, the optimal configuration for each point is finding using the proposed mathematical model.

13.8 Case study

Michoacán state is one of 32 states, which conform the Mexican republic, located in western Mexico, and it has a stretch of coastline on the Pacific Ocean to the southwest. It is border by the states of Colima and Jalisco to the west and northwest, Guanajuato to the north, Querétaro to the northeast, the State of México to the east, and Guerrero to the southeast (see Fig. 13.4). This state is divided in 113 municipalities, being Morelia the capital.

The main economic activity in this region is the tourism; however, there are municipalities with high area of km^2 cultivated by different crops, as banana, barley, broad beans, chickpea, coffee bean, kidney beans, lentil, peanut, rice, sesame, sorghum, sugarcane, and wheat. Being this part of the crops with capacity to pellets production due to its higher calorific power (Ríos-Badrán, Luzardo-Ocampo, García-Trejo, Santos-Cruz, & Gutiérrez-Antonio, 2020). From these agro-industrial activities, it is possible to take into account the solid biofuel production for energy integration as green alternative. In this way, Fig. 13.5 shows the cost-emissions-HDI trade-off generated by pellets integration in electricity network in Michoacan State, Mexico. A, B, C, D and E represent scenarios analyzed by the constraint method, in which multiobjective analysis is transformed into single optimization problems (maximization or minimization). Therefore, scenario A represents emissions maximum reduction solution, where HDI and total cost take values of $\pm\infty$, while scenario E has minimum pellets integration cost HDI and emissions take values of $\pm\infty$. B, C, and D show intermediate points over these limits using total cost as inequality constrain. Hence, HDI is linked as response variable of money distribution, as salaries and taxes charged to processing facilities installed.

From Fig. 13.5 is possible to notice that the solution with the lower emission implied a higher economic inversion, and the HDI received a beneficial impact by the

Social impact assessment in designing supply chains 417

Fig. 13.4 Michoacán localization.

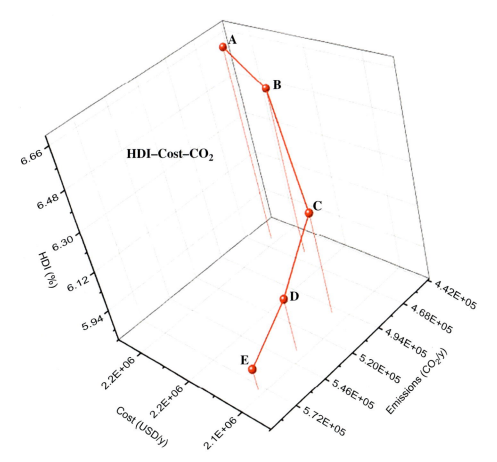

Fig. 13.5 Pareto's diagram solution.

introduction of a new industry (scenario A). Similarly, the solution with the minimum inversion in electricity production network has linked the higher emissions, and the lower HDI increase (scenario E). Noticed that even when solution E shows the minimum cost invested, pellet integration as energy source is considered in approximately 80% of energy demand, this means that pellet production is cheaper in comparison with conventional fuel uses.

From 13 agro-residues selected as biomass source for production of pellets (see Fig. 13.6), only 11 were selected by the optimization mathematical model. To satisfy 723,600 GJ of energy demand, wheat residues contribute with the biomass used for 13.21% of produced pellets, while lentil contributes with only 0.27%, being wheat agro-residue the highest used biomass and lentil the lowest used in scenario A. Similarly, in scenario E, wheat and lentil are the biomass residues most and less used in pellet production, however in this solution lentil contributed with 0.61% and wheat with 18.28%. In this scenario, there are produced 20,985 pellet tonnes, 11.03% lower in comparison with scenario A. The reason of why wheat and lentil are the most or less biomass used is due to its flux availability.

If a conventional fuel is replaced using pellets, hence it is necessary the facility installation to carry out this new fuel production, creating new jobs in all the supply chain areas. For biomass transportation, production (including maintenance and administrative work), and pellet distribution. Therefore, the integration of this alternative energy source, brings economic benefits to the population, which are measured in three main ways, education, human health, and income of the new money inputs.

Hence, Fig. 13.7 shows how if a higher flux of pellets is used as energy resource, the educational (A), human health (B), and income (C) benefit percentages increase

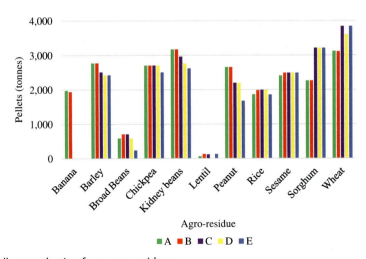

Fig. 13.6 Pellets production from agro-residues.

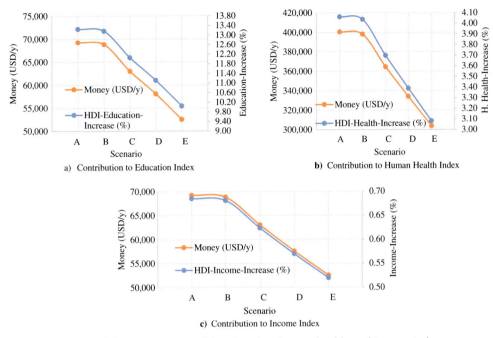

Fig. 13.7 Relation of the money invested in education, human health, and income index.

respectively, in this way it is possible to noticed, based on blue and orange lines behavior, that to reach a HDI benefit increase it is necessary the introduction of new job sources and the improvement of the current options.

All the invested money amounts and increased benefit percentage in each aspect (educational, human health and per-capita) are considered additional to the current situation in Michoacán (current governmental supports in each index).

Notice in Fig. 13.7 that if these three graphs (A), (B), and (C) are compared, Education is the parameter with the highest benefit, and income is the aspect with the lowest benefit, this is due to currently education is the sector with the lowest governmental support if it is compared with the human health and income aspects, hence USD$62,242/y invested by the pellets integration in scenario A [graph (A)], reached 13.26% of increase in education index, in the other hand, in scenario E [graph (A)] only 10% is reached by the USD$52,500/y from generated taxes of new economic activity: solid fuel production; this is the lowest money flux. Similarly, injection of USD$65,000/y in income index obtains only 0.69% of increase benefit [graph (C), Fig. 13.7], in comparison with the current situation, however, a profit of this percent in terms of per-capita income, represents more benefits in a good way in social development.

Mexico is scored 76th from 189 places at international level (UNDP, 2019), particularly at national level, Michoacán State is in 27th place (CONAPO, 2000), from the

32 states that conform Mexico, only five places from the worst situation, in the other hand Mexico City is scored in 1st place. Taking into account the information about Michoacán state and Mexico City, Fig. 13.8 shows the comparison of current situation of both places, noticed that Michoacán invests 51.53% less that Mexico City in education. However, if a new economic activity, as pellet production, is introduced in Michoacán, therefore it is possible to reduce from 0.66% (scenario E) or 0.87% (in scenario A), the difference of 51.53%, resulting of 50.87% and 50.66%, respectively, for each scenario, as Fig. 13.8 describes.

Similarly, in Fig. 13.9 are shown the human health current situation comparisons between Mexico City and Michoacán, if the new solid fuel is adopted as industry, it is possible that Michoacán become closer in 2.27% (scenario E) or 3.00% (scenario A) from Mexico City HDI in terms of Human Health factor, depending on the chosen scenario. In Income index of Michoacán (see Fig. 13.10), it is possible to be near Mexico City current situation in 0.17% (scenario E), and 0.22% (scenario A).

Also, if scenario A is adopted in Michoacán, the introduction of new pellet facilities for the electric energy production, Michoacán takes the 26th place at HDI at national level. Noticed in Fig. 13.11 that introducing a new alternative for electric energy production in Michoacán State, also has environmental benefits, for each scenario are shown the fraction of emissions in percent of the current situation, to satisfy 723,600 GJ of energy demand with the pellet combustion. In this way, if 723,600 GJ of electric energy demands are satisfied with the solid biofuel (scenario A), then it is possible to reduce 22.16% of CO_2 tonnes per year. In the other hand, if 578,880 GJ are satisfied with pellet combustion (scenario E), only 2.68% of CO_2 tonnes per year are reduced. Almost 90% of CO_2 tonnes generated by electric energy production from pellets are linked to the biomass and pellet distribution, this is due to the trucks used for transportation consuming conventional fuels.

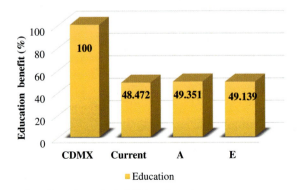

Fig. 13.8 Comparison of Michoacán with CDMX in education.

Social impact assessment in designing supply chains 421

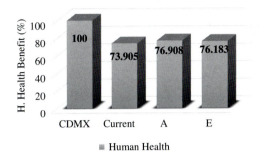

Fig. 13.9 Comparison of Michoacán with CDMX in Human health.

Fig. 13.10 Comparison of Michoacán with CDMX in Income.

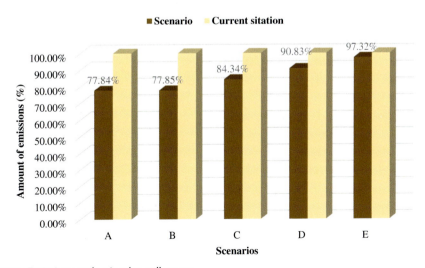

Fig. 13.11 Emissions reduction by pellets use.

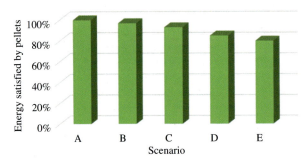

Fig. 13.12 Percent of energy satisfied by pellet combustion.

Fig. 13.12 shows the percent of energy from 723,600 GJ satisfied by solid biofuel combustion, only in scenario A, the only energy source for electricity production is pellets. In scenario E only 144,720 GJ are satisfied with conventional sources while 578,880 GJ comes from pellet integration.

As was possible to notice in the evaluated case study, biomass (particularly agro-residues) used as energy source results in attractive solution to reduce oil's dependence, achieving lower emissions values, and higher human development parameters. Also, it is possible to deduce that local supply chains optimization helps to improve supply electric energy security, fostering a sovereign culture. However, still being questionable parts of this particular supply chain, as new public politics, which contributes to carry out environmental proposal implementation in real life.

13.9 Conclusions

With the introduction of solid biofuels as alternative for conventional fuels, it is possible to obtain benefits in different ways as economic, environmental, and social aspects. The use of mathematical optimization gives the overview of how the facility location for pellets production is a feasible alternative in terms of HDI. In addition, pellets used as fuel can be integrated in the Mexican electric network, which shows that new economic activity integration is linked with education, human health, and income aspects, given the possibility to achieve a better scenario in terms of social impact for the Michoacán population, without leaving aside the environmental impact reduction.

Acknowledgments

Financial support provided by SEP, through grant PRODEP-UAQ/332/19 - for the postdoctoral stay of S.I. Martínez-Guido is gratefully acknowledged.

Nomenclature

Variables

$(A_{r,i,t}^{Available})$	available area for the agro-residues production (km²)
$(A_{r,i,t}^{new})$	new area used for harvesting (km²)
(CPE_m)	assigned money flux for salary workers (USD)
$(CPP_m^{capital})$	capital cost generated by pellets production (USD/y)
$(E_{j,t}^{Energy})$	energy flux produced at power plants (MJ)
(E^{PP})	emissions generated by pellets integration as biofuel for energy production (CO$_2$/y)
(E^{CC-use})	emissions generated by conventional fuel for energy production (CO$_2$/y)
$(E^{abs-co2})$	fixed emissions (CO$_2$/y)
$(E^{Total-Emiss})$	total emissions (CO$_2$/y)
$(F_{r,i,t}^{Residues})$	agro-residues collected for the pellets production (tonnes)
$(F_{r,i,m,t}^{B-Hs-PP})$	agro-residues flux sent it from harvesting sites to pellets processing plants (tonnes)
$(F_{r,m,t}^{B-PP})$	agro-residues flux received at pellets processing plants (tonnes)
$(FC_{j,t}^{fuel-cost})$	fuel transportation cost (USD/y)
$(FC_{j,f,t}^{CFuel})$	conventional fuel flux used as energy source in power plants (tonne)
$(FP_{j,t}^{P-Ther})$	pellets flux received at power plants (tonnes)
$(HC_{m,t})$	flux of biomass processed at pellets processing plants (tonnes).
(HDI^{Exp})	expected HDI benefit
$(HDI^{Tot-Benefit})$	total HDI obtained benefit
(I_m^{Edu})	education obtained benefit
(I_m^{HH})	human health obtained benefit
(I_m^{I})	income obtained benefit
$(P_{m,t}^{Pellets})$	pellets flux produced in each facility (tonnes)
$(P_{m,j,t}^{P-Distribu})$	pellets flux sent it from processing plants (tonnes)
(P_m^{P-jobs})	created jobs number (number)
$(TC^{transpo-cost})$	transportation cost (USD/y)
(TOT)	total cost (USD/y)
(T_m^{HH})	collected taxes used in human health dimension (USD)
(T_m^{Edu})	collected taxes used in education dimension (USD)
(T_m^{i})	collected taxes used in income dimension (USD)

Binary variable

(Y_m^{Var})	binary variable for pellets facility existence

Parameters

$(A_{r,i,t}^{current})$	current area platted with the crops used as residues (km²)
$(AT_{r,i,t})$	maximum limit for considered area designed for cropping activities (km²)
(CVE_m)	current education invested money (USD)
$(CVHH_m)$	current human health invested money (USD)
(CVI_m)	current income invested money (USD)
$(E_{j,t}^{Max-DE})$	maximum electric energy demand (MJ)

$(EM_{i,m}^{Bio-trans})$	emission factor by biomass distribution (CO_2/tonne km^2)
$(EM_{m,j}^{Pellet-trans})$	emission factor by pellets distribution (CO_2/tonne km^2)
(EM)	emission factor by pellets production (CO_2/tone)
$(EM^{Pellet-use})$	emission factor by pellets combustion (CO_2/tone)
(FE_f^{CC})	emission factor by conventional fuel combustion (CO_2/tone)
$(F_r^{CO2-fixed})$	fixed emission factor by each crop tonne (CO_2/tone)
(F_m^{HH-F})	charged taxes factor to the pellets industry (%)
$(HC_{m,t}^{max})$	biomass maximum capacity to process in each installed facility (tonnes)
(HDI^{CURR})	current HDI situation
$(PT_{m,j}^{pel-trans})$	pellets transportation unitary cost (USD/tonne km^2)
(P_m^{People})	salary factor used to workers payment (%)
(P_r^{rm})	unitary cost of agro-residues (USD/tonne)
$(RT_{i,m}^{rm-trans})$	biomass transportation unitary cost (USD/tonne km^2)
(S_m^{salary})	salary payment by hired people (USD/people)
(TA_m)	charged taxes factor to the worker's salary (%)
(TB_m)	charged taxes factor by each hired worker to the facility (%)
(TE_m)	charged taxes factor to the worker's salary (%)
(TI_m)	charged taxes factor to the worker's salary (%)
$(UF_f^{cost-Fuel})$	fuel transportation unitary cost (USD/tonne km^2)
(V^{fixed})	fixed cost by pellets facility installation (USD)
(V^{cost})	variable cost by pellets production process (USD/tonne)
$(\beta_{r,i,t}^{yield})$	agro-residues yield factor (tonnes/km^2)
(η^{PYE})	energy generation yield by pellets use (MJ/tonne of pellets)
(ϕ_r^{yield})	pellets production yield linked to each selected crop (tonne of pellets per residue tonne)
(μ_f^{FYE})	energy generation yield by conventional fuel use (MJ/tonne of fuel)

References

Anderson, J. O., & Toffolo, A. (2013). Improving energy efficiency of sawmill industrial sites by integration with pellet and CHP plants. *Applied Energy*, *111*, 791–800.

Anto, S., Mukherjee, S. S., Muthappa, R., Mathimani, T., Deviram, G., Kumar, S., et al. (2020). Algae as green energy reserve: Technological outlook on biofuel production. *Chemosphere*, *242*, 125079.

Cambero, C., & Sowlati, T. (2016). Incorporating social benefits in multi-objective optimization of forest based bioenergy and biofuel supply chains. *Applied Energy*, *178*, 721–735.

Cardozo, E., Erlich, C., Malmquist, A., & Alejo, L. (2014). Integration of a wood pellet burner and a Stirling engine to produce residential heat and power. *Applied Thermal Engineering*, *73*, 671–680.

Chazara, P., Negny, P., & Montastruc, L. (2017). Quantitative method to assess the number of jobs created by production systems: Application to the multi-criteria decision analysis for sustainable biomass supply chain. *Sustainable Production and Consumption*, *12*, 134–154.

CONAPO "Population National Council". (2000). *National human development index*. http://www.conapo.gob.mx/work/models/CONAPO/Resource/211/1/images/desarrollo_humano.pdf. Accessed 15-09-20.

Díaz-Trujillo, L. A., Tovar-Facio, J., Nápoles-Rivera, F., & Ponce-Ortega, J. M. (2019). Effective use of carbon pricing on climate change mitigation projects: Analysis of the biogas supply chain to substitute liquefied-petroleum gas in Mexico. *Processes*, *7*, 668.

El-Halwagi, M. M. (1997). *Pollution prevention through process integration: Systematic design tools* (1st ed.). San Diego California: Academic Press.

Fathollahi-Fard, A. M., Hajiaghaei-Keshteli, M., & Mirjalili, S. (2018). Multi-objective stochastic closed-loop supply chain network design with social considerations. *Applied Soft Computing*, *71*, 505–525.

Gutierrez-Antonio, C., Gómez-Castro, F. I., de Lira-Flores, J. A., & Hernández, S. (2017). A review on the production processes of renewable jet fuel. *Renewable and Sustainable Energy Reviews*, *79*, 709–729.

Hassini, E., Surti, C., & Searcy, C. (2012). A literature review and a case study of sustainable supply chains with a focus on metrics. *International Journal of Production Economics*, *140*(1), 69–82.

Hernández-Pérez, L. G., Sánchez-Tuirán, E., Ojeda, K. A., El-Halwagi, M. M., & Ponce-Ortega, J. M. (2019). Optimization of microalgae-to-biodiesel production process using a metaheuristic technique. *ACS Sustainable Chemistry & Engineering*, 7, 8490–8498.

Kim, G. S., Choi, S. K., & Seok, J. H. (2020). Does biomass energy consumption reduce total energy CO_2 emissions in the US? *Journal of Policy Modelling*, 42(5), 953–967.

Li, M., Xu, J., Xie, H., & Wang, Y. (2018). Transport biofuels technological paradigm based conversion approaches towards a bio-electric energy framework. *Energy Conversion and Management, 172*, 554–566.

López-Díaz, D. C., Lira-Barragán, L. F., Rubio-Castro, E., Serna-González, M., El-Halwagi, M. M., & Ponce-Ortega, J. M. (2018). Optimization of biofuels production via water-energy-food nexus framework. *Clean Technologies and Environmental Policy*, 20, 1443–1466.

Martínez-Guido, S. I., González-Campos, J. B., & Ponce-Ortega, J. M. (2019). Strategic planning to improve the human development index in disenfranchised communities through satisfying food, water and energy needs. *Food and Bioproducts Processing*, 117, 14–29.

Martínez-Guido, S. I., Ríos-Badrán, I. M., Gutiérrez-Antonio, C., & Ponce-Ortega, J. M. (2019). Strategic planning for the use of waste biomass pellets in Mexican power plants. *Renewable Energy*, 130, 622–632.

Mortensen, A. W., Mathiesen, B. V., Hansen, A. B., Pedersen, S. L., Grandal, R. D., & Wenzel, H. (2020). The role of electrification and hydrogen in breaking the biomass bottleneck of the renewable energy system—A study on the Danish energy system. *Applied Energy*, 275, 115331.

Mota, B., Gomes, M. I., Carvalho, A., & Barbosa-Povoa, A. P. (2015). Towards supply chain sustainability: Economic, environmental and social design and planning. *Journal of Cleaner Production*, 105, 14–27.

Nunes, L. J. R., Causer, T. P., & Ciolkosz, D. (2020). Biomass for energy: A review on supply chain management models. *Renewable and Sustainable Energy Reviews*, 120, 109658.

Office of Energy Efficiency & Renewable Energy (BETO). (2020). *Bioenergy: Biofuels basics*. https://www.energy.gov/eere/bioenergy/biofuels-basics. Accessed 30-08-20.

Popovic, T., Barbosa-Póvoa, A., Kraslawki, A., & Carvalho, A. (2018). Quantitative indicators for social sustainability assessment of supply chain. *Journal of Cleaner Production*, 180, 748–768.

Ríos-Badrán, I. M., Luzardo-Ocampo, I., García-Trejo, J. F., Santos-Cruz, J., & Gutiérrez-Antonio, C. (2020). Production and characterization of fuel pellets from rice husk and wheat straw. *Renewable Energy*, 145, 500–507.

Sagastume-Gutiérrez, A., Cabello-Eras, J. J., Hens, L., & Vandecasteele, C. (2020). The energy potential of agriculture, agroindustrial, livestock, and slaughterhouse biomass wastes through direct combustion and anaerobic digestion. The case of Colombia. *Journal of Cleaner Production*, 269, 122317.

Santibañez-Aguilar, J. E., González-Campos, J. B., Ponce-Ortega, J. M., Serna-González, M., & El-Halwagi, M. M. (2014). Optimal planning and site selection for distributed multiproduct biorefineries involving economic, environmental and social objectives. *Journal of Cleaner Production*, 65, 270–294.

Schwerz, F., Neto, D. D., Caron, B. O., Nardini, C., Sgarbossa, J., Eloy, E., et al. (2020). Biomass and potential energy yield of perennial Woody energy crops under reduced planting spacing. *Renewable Energy*, 153, 1238–1250.

Smil, V. (2017). *Energy transitions: Global and national perspectives* (2nd ed.). California: PRAEGER.

Stameković, O. S., Siliveru, K., Veljković, V. B., Banković-Ilić, I. B., Tasić, M. B., Ciampitti, I. A., et al. (2020). Production of biofuels from sorghum. *Renewable and Sustainable Energy Reviews*, 124, 109769.

Subramaniam, Y., Masron, T. A., & Nik-Azman, N. H. (2020). Biofuels, environmental sustainability, and food security: A review of 51 countries. *Energy Research and Social Science*, 68, 101464.

Sulaiman, C., Abdul-Rahim, A. S., & Ofozor, C. A. (2020). Does wood biomass energy use reduce CO2 emissions in European Union member countries? Evidence from 27 members. *Journal of Cleaner Production*, 253, 119996.

UNDP "United Nations Development Programme". (2019). *Human Development Report 2019: Inequalities in human development in the 21st century*. http://hdr.undp.org/sites/all/themes/hdr_theme/country-notes/MEX.pdf. Accessed 01-09-20.

Wang, Z., Bui, Q., Zhang, B., & Pham, T. L. A. (2020). Biomass energy production and its impacts on the ecological footprint: An investigation of the G7 countries. *Science of the Total Environment*, 743, 140741.

World Bioenergy Association. (2017). *WBA global bioenergy statistics 2017*. https://worldbioenergy.org/uploads/WBA%20GBS%202017_hq.pdf. Accessed 13-08-20.

Yue, D., Slivinsky, M., Sumpter, J., & You, F. (2014). Sustainable design and operation of cellulosic bio-electricity supply chain networks with life cycle economic, environmental, and social optimization. *Industrial and Engineering Chemistry Research, 53*(10), 4008–4029.

Yun, H., Clift, R., & Bi, X. (2020). Process simulation, techno-economic evaluation and market analysis of a supply chains for torrefied Wood pellets from British Columbia: Impacts of plant configuration and distance to market. *Renewable and Sustainable Energy Reviews, 127*, 109745.

Index

Note: Page numbers followed by *f* indicate figures and *t* indicate tables.

A

Absorption, 19–20, 298
Acetone–butanol–ethanol (ABE) fermentation, 14–15, 273–274
   conventional separation sequences, 60–61, 60*f*
   intensified separation schemes, 218–219
Acid pretreatment, for lignocellulosic biomass, 50–51
AD. *See* Azeotropic distillation (AD)
Agrol, 312–313
Agro-residues, 410–412, 418
AIRCOMP module, 177
Alcohol to jet synthetic paraffinic kerosene (ATJ-SPK), 119, 121
Alkaline pretreatment, for lignocellulosic biomass, 51, 51*t*
Anaerobic digestion, 161–163
Anaerobic fermentation, 60–61
Annual safety and sustainability profit (ASSP), 300–301
Aquaporin-based biomimetic FO membranes, 73
Aromatic compounds, 119
Aspen Dynamics simulator, 76
Aspen Plus process simulator, 130–131
Aspen Process Economic Analyzer, 185–188
Autohydrolysis. *See* Steam explosion
Azeotropic distillation (AD), 58, 59*f*, 72, 207–208, 274

B

Barium oxide (BaO), 90
Batch reactors (BRs), 233, 255–260, 259*t*, 262*f*
2,3-BDO. *See* 2,3-Butanediol (2,3-BDO)
Bean straw, proximate and ultimate analysis for, 175, 176*t*
Bernoulli's principle, 102
Big-M approach, 372–373
Bioalcohols, production of
   block flow diagram, 44–45, 45*f*
   distillation-based separation process, 55–61
   membrane-assisted reactive distillation, 72–78
   membrane-based separation process, 61–68
   pretreatment technologies for lignocellulosic biomass, 44–55
   process control, 68–72
Biobased chemicals
   industrial applications, 321*t*
   market, 316–319
Biobased feedstocks for chemical process, 322, 325*f*
Biobutanediol, 217–222
Biobutanol, 198, 217–222
   butanol purification, 276, 277*f*
   closed-loop analysis, 278–290, 279*f*
      degree of intensification, 289
      DWC column, 289–290
      feeding disturbance, 279, 281, 288
      integral of absolute error (IAE), 280, 280*f*, 282–287, 282*t*
      LV-type control scheme, 280–281, 289
      PI controller, 279–281, 280*f*, 282*t*
      reflux flow rate, 280–281
      set-point, 279, 281, 287–288
   conventional production process, 13–14, 15*f*
   hybrid extraction–distillation process, 276
   separation process for, 60–61, 276–278
Bioderived chemicals, implemented intensification technologies for, 328–337, 329–336*t*
Biodiesel, 197, 209–214, 232
   definition, 232–233
   dividing wall columns (DWC), 211–212
   dry washing of, 210–211
   ethanol recovery, 264
   heating value of, 209–210
   production
      conventional process, 15–17, 16*f*
      hydrodynamic cavitation, 98–106, 104–105*t*, 107*f*
      intensification technologies, 88–111
      microwave, 88–92, 91–92*t*, 93*f*
      reactive distillation, 106–108, 109*t*, 110*f*
      ultrasonic cavitation, 93–97, 98*f*, 99–101*t*
   purification, 210–211, 211*f*
   reactive DWC, 212–214, 213*f*
   separation, 210–211, 211*f*

Biodiesel *(Continued)*
   sulfur content, 87–88
   transesterification of glycerides, 210
Bioethanol
   conventional production process, 13–14, 14f
   distillation, 55–60
   process intensification, 23–27
   purification, 23–24
   reactive zones, 24
   separation by membrane-assisted reactive distillation, 72–78, 73f
Biofuels, 362
   classification, 41–42
   definition, 2, 41
   first generation, 41–42
   fourth generation, 41–42
   primary, 41–42
   production
      conventional processes, 13–22
      process intensification, 5–13, 22–29
   research articles, 158, 159f
   second-generation, 41–42
   solid, liquid, or gaseous, 42
   third generation, 41–42
Biogas
   composition, 19, 161–162
   conventional production process, 19, 19f, 161–163
   process intensification, 170–171
   solar-assisted bioreactor, 28
Biogas-lift bioreactor, 170–171
Biogasoline, conventional production process, 15, 16f
Biohydrogen. *See* Renewable hydrogen
Biojet fuel. *See also* Renewable aviation biofuel
   certified processes for production of, 119, 120f
   conventional processing of bio-oil from pyrolysis, 134–136, 135f
   process intensification, 130–132
      economic assessment, 145–147
      environmental assessment, 147–148
      hydrotreatment reactor, 140–142, 143t
      industrial scale implementation, 132–134
      intensified distillation columns, 126–130, 127–128f
      intensified reactors, 121–126
      one-step hydroprocessing, 122, 123t
      OTC configurations, 144–145, 144t
      problem statement, 134–136
      process simulation, 139–145
      reactive stage modeling, 136–138, 137–138t
      SC configurations, 144, 144t
      separation zone modeling, 138–139
      stabilizer reactor, 136–137, 137t, 140–142, 142t
      TES configurations, 145, 145t
   purification, 27
Biomass, 408
   classification, 2, 2–3f
   combustion process, 185, 186t
   conversion processes for biofuel production, 3–4, 4f
      biochemical, 3–4
      biological, 3–4
      liquid, gaseous, or solid state, 3
      thermochemical, 4
   defined, 2, 158
   energy extraction process from, 231, 232f
   gasification, 189
   lignocellulosic, 3
   particle size distribution, 184–185, 184t
   starch, 3
   sugar, 3
   triglyceride, 3
Bio-monomers, 322, 325f
Bio-oil
   conventional process from pyrolysis into biojet fuel, 134–136, 135f
   from *Jatropha curcas*, 134, 135t
   mixture, 138, 139f
Bio-polymers, 322, 325f
Biorefinery, 197–198
   and biobased chemicals market, 316–319
   characteristics, 198–199
   conversion process, 199
   current, 316–319, 317f, 406–408
   distillation process challenges in, 199–201
   evaluation using optimization tools, 408–409
   industrial, 223–224
      current status, 313–316
      evolution and implementation, 309–316
      key bio-based fuels and chemicals for, 319–322, 321t, 323–324t
      natural flavors and fragrances, 310–311
      naval stores and lignocellulosic biorefineries, 310

oleochemical biorefinery, 312
  starting 1900s, 312–313
  sugar and starchy biorefinery, 311–312
intensified and hybrid distillation technologies for, 201–206
  dividing wall columns (DWC), 201, 202f
  heat pump-assisted distillation (HPD), 203–204, 204f
  for high value-added processes from biomass, 207–224
  hybrid extraction-distillation (ED), 204–206, 205f
  hybrid membrane-distillation (MD), 206f, 207–209
  reactive distillation (RD), 201–203, 203f
life cycle assessment (LCA) in, 394–396
modular, 293–295
  annual safety and sustainability profit (ASSP), 300–301
  assessment, 295–297
  case study, 297–301, 300t
  costs, 294
  Hazardous Process Stream Index (HPSI), 295–296
  safety and sustainability weighted return on investment metric (SASWROIM), 300–301, 301t
  vs. stick-built biorefineries, 295, 297f
process intensification, 322–327, 326f, 327–328t
supply chain elements for, 366–367
in USA, 316–319, 318f
Biowastes, 43
Bourdon reaction, 29
Briquettes, 168
  conventional production process, 21–22, 22f
  defined, 21
  minimum net calorific power, 21
BRs. See Batch reactors (BRs)
Butanedioic acid. See Succinic acid
2,3-Butanediol (2,3-BDO)
  hybrid-evaporator assisted vapor recompression distillation scheme, 219–220, 220f
  hybrid extraction-distillation of butanol-based solvent, 221–222, 221f
  industrial applications, 217–218
  process intensification schemes, 198, 219–222
Butanol-ethanol-water, ternary diagram for, 274, 275f

Butanol purification, 276, 277f
Butyl chloride, 216–217, 217f

C
Calcium chloride ($CaCl_2$), 207–208
Carbon footprint, 88
Carbon-reduction targets, 305, 306t
Carbon to nitrogen ratio (C/N), 161
Catalytic hydrothermolysis synthesized kerosene (CH-SK), 119, 121
Cauchy boundary condition, 241–242
Cavitation, 93
  hydrodynamic, 98–106, 104–105t, 107f
  ultrasonic, 93–97, 98f, 99–101t
Cavitation number, 102–106
CE. See Cellulosic ethanol (CE)
Cells, 244
Cellulose, 44, 52
Cellulosic ethanol (CE), 197, 207–209
  dividing wall columns (DWC), 208–209
  fermentation, 207–208
  pretreatment, 207–208
  production from actual fermentation broth, 208–209, 208–209f
  separation, 207–208
Ceramic PV membranes, 63
CFD. See Computational fluid dynamics (CFD)
Chemical kinetics
  global oil transesterification mechanism, 261–262
  vegetable oil transesterification with parallel saponification reactions, 258–260, 259t
Chicken fat, 123t
Choline hydroxide (ChOH), 89–90
Closed-loop analysis, 278–290, 279f
  degree of intensification, 289
  DWC column, 289–290
  feeding disturbance, 279, 281, 288
  integral of absolute error (IAE), 280, 280f, 282–287, 282t
  LV-type control scheme, 280–281, 289
  PI controller, 279–281, 280f, 282t
  reflux flow rate, 280–281
  set-point, 279, 281, 287–288
Clostridium acetobutylicum, 273–274
Clostridium beijerinckii, 273–274
Clustered regularly interspaced short palindromic repeats (CRISPR), 305–307
CO_2 emissions, biojet fuel production, 147, 149t

Combined heat and power production (CHP), 174–175
Compact heat exchanger, 12
Computational fluid dynamics (CFD), 12, 124, 235–236
  batch reactor modeling, 258–260
  case study, 254–267
  continuous micro and millireactors modeling, 261–262
  in macroscale reactors, 247–249
  mathematical model, 257–258
    discretization, 242–243
    fluid flow, 238–242
  mesh generation, 236–237, 237f, 243–246, 244f
  in microscale reactors, 250–253
  model basic steps, 236, 236f
  partial differential equations (PDEs), discretization of, 236–237
  in processes analysis, 246–247
  transforming physical process geometry into, 237, 237f
Computational particle fluid dynamics (CPFD), 248–249
COMSOL Multiphysics 5.3 software, 249
Continuous closed-circulating fermentation (CCCF) systems, 62, 63f
Continuous micro and millireactors modeling, 261–262
Continuous surface force (CSF) model, 253
Continuum species transfer (CST), 253
Controllability analysis, 75–78, 77f
Conventional azeotropic distillation, 58–60, 59f
Conventional distillation columns, for biojet fuel production, 128–129
Conventional fuel replacement, by bioenergy sources, 406–408
Conventional liquid-liquid assisted distillation, 61, 61f
Conventional pellet production process, 175–179, 181f
Conventional production process
  biobutanol, 13–14, 15f
  biodiesel, 15–17, 16f
  bioethanol, 13–14, 14f
  biogas, 19, 19f, 161–163
  biogasoline, 15, 16f
  briquettes, 21–22, 22f
  fuel pellets, 22, 22f, 168–170

  green diesel, 17–18, 17f
  renewable aviation biofuel, 18, 18f
  renewable hydrogen, 163–167, 164f
  syngas, 20, 20f
Conventional reactors, 6, 7f
Convex hull approach, 372–373
COVID-19, 157–158

D

Dark fermentation, of carbohydrate, 164–165
Decision-making, supply chain for, 369–370, 370f
Differential evolution method with tabu list (DETL), 276
Dirichlet boundary condition, 241
Distillation, 8–9, 9f
  for biobutanol separation, 60–61
  for bioethanol separation, 55–60
  columns, 9
  extractive, 55, 56–57f
  membrane, 66
  process challenges in biorefineries, 199–201
Distillation-pervaporation hybrid system, 64–65, 66f
Distillation-pressure swing adsorption, 23–24
Dividing wall columns (DWC), 56–58, 58f
  biodiesel, 211–212
  biorefinery, 201
  cellulosic ethanol (CE), 208–209
  furfural, 215–216, 216f
  industrial, 201, 202f
  intensified ABE separation schemes, 218–219
  levulinic acid (LA), 198, 223
  types, 199f, 201
Dry reforming, 29
Dry washing, of biodiesel, 210–211
DWC. See Dividing wall columns (DWC)

E

Edges, 244
Effective mass yield (EMY), 75–76
Electric energy balance, 412
Electrification, of European chemical industry, 348–349, 349f
Energy consumption, 1
Energy Information Administration (EIA), 231
Enzymes, 200–201
Escher Wyss process, 214

Essential oils extraction units, 311, 311f
Ethanol dehydration, reactive distillation (RD), 201–203, 203f
Ethylene glycol (EG), 207–208
Ethylene oxide (EO), 72–73
Eulerian approach, 238
Eulerian-Lagrangian approach, 248–249
European Process Intensification Center (EUROPIC), 327
Exa-Joules (EJ), 406–408
Extractive distillation, 55, 56–57f, 76t

F

Fatty acid alkyl esters (FAAEs), 362
Fensk–Underwood–Gilliland equations, 12
Finite-difference methods, 242
Finite-element methods, 242–243
Finite-volume methods, 243
Fischer-Tropsch synthesized isoparaffinic kerosene (FT-SPK), 119, 122, 132
Fischer-Tropsch synthetic paraffinic kerosene with aromatics (FT-SPK/A), 119
Fixed capital investment (FCI), 298
Fluid flow mathematical modeling, 238–242
  boundary conditions, 240–242
  governing equations, 238–240
Fluorinated ethylene propylene (FEP), 253
Forward osmosis (FPO), 67
Fossil fuel, 231
Four-pillar strategy, in sustainable development of aviation sector, 117–118, 118f
Fuel pellets
  conventional production process, 22, 22f, 168–170
  minimum net calorific power, 22
  process intensification, 174
    case study, 174–175
    conventional pellet production, 175–179
    intensified pellet production, 180–183
Fungi, in bioalcohol production, 52–53
Furfural, 198, 214–217
  conventional purification, 215, 215f
  dividing wall columns, 215–216, 216f
  hybrid ED, 216–217
  intensified purification, 215–216, 216f
  reactive distillation (RD), 214–215
Furnace-boiler system, 177, 178f

G

Gaseous biofuels
  conventional process, 19–21
  process intensification, 28–29
Gasification, 298
GDP. See Generalized disjunctive programming (GDP)
Gene- and genome-editing approaches, 305–307
Generalized disjunctive programming (GDP)
  relaxation of, 372–373
  supply chain modeling and optimization, 369–372
Genetic engineering, 273–274
Genome-wide association (GWA), 305–307
Gibbs free energy, 166–167
Glycerol, 210, 212
Green diesel, conventional production process, 17–18, 17f
Greenhouse gas (GHG) emissions, 74, 87–88, 231, 297–298, 389
Guthrie's method, 145

H

HAU-M1, 164–165
Heat exchanger reactor, 6, 7f, 8
Heat pump-assisted distillation (HPD), 23–24, 203–204, 204f
Hemicellulose, 44, 52
Heterogeneous reactors, 6
Hhydrophobic PV membranes, 63, 65f
High-pressure water stream (HPW), 185
High value-added products, 362–363, 363t
Homogeneous reactors, 6
HPD. See Heat pump-assisted distillation (HPD)
Human development index (HDI), 405–406
Hybrid distillation process, 200
Hybrid extraction-distillation (ED)
  biobutanol, 276
  biorefinery, 204–206, 205f
  furfural, 216–217
  levulinic acid (LA), 222, 222f
Hybrid membrane-distillation (MD), 206f, 207–209
Hydraulic retention time (HRT), 161
Hydrocarbons, 26–27
Hydrodynamic cavitation, biodiesel production, 98–106, 104–105t, 107f
Hydrolysis, 14–15

Hydrophilic PV membranes, 64, 65f
Hydrophobic-hydrophilic pervaporation system, 64, 65f
Hydroprocessed ester and fatty acid synthetic paraffinic kerosene (HEFA-SPK), 119, 132
Hydroprocessed fermented sugars to synthetic isoparaffins (HFS-SIP), 119, 122
Hydroprocessed hydrocarbons, esters and fatty acids synthetic paraffinic kerosene (HHC-SPK), 119
Hydroprocessing
  of *jatropha curcas* oil, 131
  one-step, 6, 26–27

I

IAE. *See* Integral absolute error (IAE)
IEA. *See* International Energy Agency (IEA)
Induced pluripotent stem cells (iPSCs), 305–307
Industrial biorefinery
  current status, 313–316
  evolution and implementation, 309–316
  key bio-based fuels and chemicals for, 319–322, 321t, 323–324t
  natural flavors and fragrances, 310–311
  naval stores and lignocellulosic, 310
  oleochemical, 312
  starting 1900s, 312–313
  sugar and starchy, 311–312
Industrial natural flavors and fragrances, 310–311
Inertial cavitation, 98
Inherent safety, 75
Inoculum, 161–162
Inorganic salts, 207–208
Integral absolute error (IAE), 76, 280, 280f, 282–287, 282t
Integrated pellet systems, 409–410, 409f
Intensification technologies, for biodiesel production, 88–111
  costs, 108–111, 111t
  hydrodynamic cavitation, 98–106, 104–105t, 107f
  microwave, 88–92, 91–92t, 93f
  reactive distillation, 106–108, 109t, 110f
  ultrasonic cavitation, 93–97, 98f, 99–101t
Intensified distillation columns, for biojet fuel production, 126–130, 127–128f

Intensified pellet production process, 180–183
Intensified reactors, for biojet fuel production, 121–126
Intensified single-column extractive divided wall configuration, 56–58, 59f
International Air Transport Association (IATA), 117–118
International Energy Agency (IEA), 1–2, 2f, 42, 157–158
International Monetary Fund, 1–2
International Renewable Energy Agency (IRENA), 117–118, 132
Ionic liquids (ILs), 207–208

J

Jatropha curcas, bio-oil from, 134, 135t

K

Key industrialized biofuels and biopolymers, 321–322, 324t

L

LA. *See* Levulinic acid (LA)
Lagrangian approach, 238
LCA. *See* Life cycle assessment (LCA)
Levulinic acid (LA)
  dividing wall columns (DWC), 198, 223
  hybrid ED, 222, 222f
Life cycle assessment (LCA)
  in biorefineries sector, 394–396
  at early stage of development, 398–401
  environmental impact, 399, 400f
  Gabi, 392
  issues, 395–396
  methodology, 390–391, 391f
  OpenLCA, 392
  regionalization, 392–394
  SimaPro, 392
  social impacts, 396–397, 397t
  steps, 390
  tools, 391–394, 393t
  Umberto, 392
Life cycle costing (LCC), 390
Life cycle impact assessment (LCIA), 390
Life cycle inventory (LCI), 390
Lignin, 44
  elimination in newspaper, 23
  removal from sugarcane bagasse, 23

solubilization, 46
Lignocellulose, 44
Lignocellulosic biomass, 3, 43–44
   and high value-added products, 197–198
   pretreatment technologies, 44–55
      biological, 52–53, 54t
      chemical, 50–52
      combined, 53–55
      physical, 45–50
Lignocellulosic biorefineries (LCB), 310
Lignocellulosic hydrolysates, 162–163
Lignolytic fungi, 52
Liquid biofuels
   conventional process, 13–19
   process intensification, 23–27
Liquid hot water (LHW) pretreatment
   for lignocellulosic biomass, bioalcohol production, 46–47
   in sugarcane bagasse, 46–47, 47t
Liquid–liquid-assisted divided wall intensified configuration, 61, 61f
Liquid–liquid equilibrium (LLE), 200
Liquid–liquid extraction, 24–25, 61, 222, 274–276
Low-pressure water (LPW) stream, 179

M

Macroscale reactors, CFD in, 247–249
Magnetron, 88
Mass conservation equation, 238
Mass intensity (MI), 74, 76
Mass transfer resistance, 93
MEC. *See* Microbial electrolysis cell (MEC)
Mechanical pretreatment, for lignocellulosic biomass, 45–46
Mechanical stirring, 102–103
Membrane-assisted reactive distillation, bioethanol separation by, 72–78, 73f
   controllability analysis, 75–78, 77f
   effective mass yield (EMY), 75
   greenhouse gas emissions, 74
   inherent safety, 75
   mass intensity (MI), 74, 76
Membrane-based processes, for bioalcohol separation, 61–68
   forward osmosis (FPO), 67
   membrane distillation (MD), 66
   microfiltration (MF), 68
   pervaporation (PV), 62–65
   pressure-driven membrane process, 67–68
   ultrafiltration (UF), 68
Membrane bioreactors (MBRs), 62
Membrane distillation (MD), 66
Mesh generation, 236–237, 237f, 243–246, 244f
Mesh geometry, 244
Metastillation, 23–24
Methane, 297–298
Methanol, production from MSW gasification, 297–298, 299f
Methyl terbutyl ether (MTBE) production, via reactive distillation, 324–325
Methyltetrahydrofuran (MTHF), 198
Mexico, optimization of supply chain for production of bioethanol and high value-added products in, 373–386
   levulinic acid and bioethanol, 384, 384t
   problem statement, 374, 375f
   production costs, density, and yields, 374, 374t
   raw material, 384, 385f
Michoacan State
   CDMX in education, comparison with, 419–420, 420f
   CDMX in human health, comparison with, 420, 421f
   CDMX in income, comparison with, 420, 421f
   economic activity, 416
   emissions reduction by pellets use, 420, 421f
   energy satisfication by pellet combustion, 422, 422f
   localization, 416, 417f
   pellets production from agro-residues, 418, 418f
   social impact assessment in designing supply chains for biorefineries, 416–422, 417f
Microalgae oil extraction, 25–26
Microbial consortia, 53
Microbial electrolysis cell (MEC), 164–166, 172–173
Microchannel distillation, for biojet fuel production, 130
Microchannel heat exchangers, 12
Microchannel reactors, 6, 7f, 8
Microdevices, vegetable oil distribution in, 254–255, 256f
Microfiltration (MF), 68
Micromixer with static baffle elements (MSE), 254–255

Micromixer with triangular baffles (MTB), 255, 257f, 265–266
Microorganisms, 200–201
Microreactors, 6, 7f, 8, 123–124
    biodiesel synthesis in, 234
    characteristics, 233–234
Microreactor with static elements (MSE), 251
Microscale reactors, CFD in, 250–253
Microwave
    assisted hydrothermal pretreatment, 49–50
    bioalcohol production, 49–50
    biodiesel production, 88–92, 91–92t, 93f
    CO_2 gasification, 29
    irradiation, 88–89
Minichannel heat exchangers, 12
Momentum conservation equation, 239
MRF. See Multiple reference frame (MRF)
MTB. See Micromixer with triangular baffles (MTB)
Multiphase Eulerian model, 247–248
Multiple reference frame (MRF), 247, 260
Municipal solid waste (MSW), 297–298

N

Nanoferrosonication, 28
National Renewable Energy Laboratory (NREL), 208
Naval stores, 310
Navier-Stokes equation, 239–240
Neumann boundary condition, 241
Newton's second law of motion, 239
Next-generation sequencing (NGS), 305–307
Nodes, 244
Noninertial cavitation, 98

O

Oil mass fraction, in biodiesel synthesis via alkaline ethanolysis, 238, 238f
Oleochemical biorefinery, 312
Oleyl alcohol, 221–222
OpenFOAM package, 249, 253
Open-loop control test, 75
Organic load rate (OLR), 161
Organosolv pretreatment, for lignocellulosic biomass, 51–52
Oscillatory flow reactors, 8
Oxo-synthesis process, 273–274

P

Parastillation, 23–24
Partial differential equations (PDEs), 236–237
Passive saccharification, 164–165
PDEs. See Partial differential equations (PDEs)
Pelletization, 168–170
Petlyuk distillation column, 130–131
Petroleum refinery, 198–199, 199f
Photo fermentation, 164–165
Photovoltaic (PV) panels, 180, 188
Polycrystallin solar module, 183, 184t
Pressure-driven membrane process, 67–68
Pressure-velocity coupling, 247–248
Primus Green Energy, 294–295
Process control, bioalcohol production, 68–72
    control of separation process, 71–72
    control strategies for fermenter, 70–71
    state estimation for fermentation process, 69–70
Process intensification (PI)
    in biofuel production, 5–13, 22–29
        advantages, 6
        conditioning equipment, 10–12, 11f
        defined, 5–6
        design methodologies for intensified equipment, 12–13
        disadvantage, 6
        reaction equipment, 6–8
        separation equipment, 8–10
    biogas, 170–171
    in biojet fuel production, 130–132
        economic assessment, 145–147
        environmental assessment, 147–148
        industrial scale implementation, 132–134
        intensified distillation columns, 126–130, 127–128f
        intensified reactors, 121–126
        one-step hydroprocessing, 122, 123t
        problem statement, 134–136
        process simulation, 139–145
        reactive stage modeling, 136–138, 137–138t
        separation zone modeling, 138–139
    biorefinery, 322–327, 326f, 327–328t
    2,3-butanediol, 219–222, 220–221f
    challenges and future directions, 341–349
    criteria and metrics for industrial industrialization, 327, 328t
    early attempts, 307–309

fuel pellets, 174
   case study, 174–175
      conventional pellet production, 175–179
      intensified pellet production, 180–183
   renewable hydrogen, 172–173
   succinic acid production, 341, 342–347t
   and supply chain, 367–368
   in transformation and production of biobased fuels, chemicals, and polymers, 328, 329–336t

Q

Quadrature method of moments (QMOM), 247–248
Quantitative risk analysis (QRA), 75

R

Rankine steam cycle, 179, 180f, 187t
Reactive distillation (RD), 72
   biodiesel production, 106–108, 109t, 110f
   biojet fuel production, 129–131
   biorefinery, 201–203, 203f
   ethanol dehydration, 201–203, 203f
   furfural, 214–215
   methyl terbutyl ether (MTBE) production via, 324–325
Reactive divided wall column (R-DWC), 212–214, 213f
Reactors
   conventional, 6, 7f
   intensified, 6, 7f, 8
   multifunctional, 6–7
Recuperation unit (RECU), 179
Refuse derived fuels (RDFs), 297–298
Renewable aviation biofuel, 18, 18f
Renewable energy supply, 406–408, 406–407f
Renewable hydrogen
   conventional production process, 20–21, 21f, 163–167, 164f
   process intensification, 172–173
Return on investment (ROI), 298–300
Reynolds number, 252
Rice husk
   particle size distribution (PSD), 175, 176t
   proximate and ultimate analysis for, 175, 176t
Robbin boundary condition, 241
RYIELD model, 177

S

Saccharomyses cerevisiae, 14, 23
Safety and sustainability weighted return on investment metric (SASWROIM), 300–301, 301t
SARS-CoV-2, 118, 305–307
Shear stress transport (SST) turbulence model, 257–258
SIMPLE algorithm, 247–248
Skewness, of cell, 245, 245f
S-LCA. *See* Social life cycle assessment (S-LCA)
Smoothness, in mesh, 245, 246f
Social impact assessment, in designing supply chains for biorefineries
   biorefineries evaluation using optimization tools, 408–409
   case study, 416–422
   conventional fuel replacement by bioenergy sources, 406–408
   mathematical model, 410–416, 411t
   Michoacan State, 416–422, 417f
   problem statement, 410
   supply chain optimization, 409–410
Social life cycle assessment (S-LCA), 390, 396–397
Solar bioreactor, 28, 171
Solid biofuels
   conventional process, 23–27
   process intensification, 29
Sonochemistry, 172–173
Soybean oil, 123t
Spinning disk reactor, 8
Starch, 3
Steam explosion
   effect on sugar degradation and sugar yields from rice straw, 47, 48t
   for lignocellulosic biomass, bioalcohol production, 47–48
Steam reforming, 166
Stirred tank reactor, geometric parameters of, 254, 255f
Succinic acid
   derivatives and applications, 338f
   production, 337–341, 339f
      companies, 337–340, 340t
      petrochemical and biochemical pathways, 337, 339f
      process intensification, 340–341, 342–347t
   properties, 337t

Sugar and starchy biorefinery (SSB), 311–312
Sugars, 3, 14
Supercritical water (SCW), 249
Supercritical water fluidized bed (SCWFB), 249
Supply chain
    of biofuels and high added value products, 368, 369f
    for decision-making, 369–370, 370f
    elements for biorefinery, 366–367
    modeling and optimization, 368–373
        discrete decisions, 369, 369t
        generalized disjunctive programming (GDP), 369–372
        relaxation of GDP, 372–373
    objective functions for optimization, 364–365
    optimization for production of bioethanol and high value-added products in Mexico, 373–386, 374t, 375f, 384t, 385f
    process intensification and, 367–368
Sustainable Recovery Plan, 1–2, 2f
Syngas, 158
    conventional production process, 20, 20f
    purification, 298
Synthetic paraffinic kerosene (SPK). *See* Renewable aviation biofuel
System of algebraic equations (SAE), 237

T

Thermally coupled distillation columns, for biojet fuel production, 127–128, 130
TORBED technology, 174
Transesterification, 25–26, 233
    of *Aegle marmelos* Correa seed oil, 95–96
    of glycerides, 210
    reactive distillation-assisted, 108, 109t
    ultrasonic cavitation-assisted, 97, 99–101t
    of waste cooking oil, 97
    of waste cotton-seed cooking oil, 89–90
Triglyceride biorefinery (TGB), 319
Triglycerides, 16–17
Tubular microreactor (TMR), 251–252
Two-dimensional CFD modeling, 249

U

Ultrafiltration (UF), 68
Ultrasonic cavitation, 93–97, 98f, 99–101t
Ultrasound-assisted alkaline treatment, 23
Ultrasound pretreatment, for lignocellulosic biomass, 48–49, 49t
UNIFAC (UNIQUAC Functional-group Activity Coefficients), 200

V

Vaccines, 1
Vacuum membrane distillation (VMD), 66
Vapor–liquid equilibrium (VLE), 200
Vapor recompression effect, 203–204, 204f
Vegetable oil, 232–233
    distribution in microdevices, 254–255, 256f
    transesterification with parallel saponification reactions, chemical kinetics of, 258–260, 259t
Vinasse, 162–163
Volume-of-fraction (VOF) approach, 247–248

W

Wet stirred media milling (WSMM), 174
Working capital investment (WCI), 298

X

Xylan hydrolysis, 214
Xylose, 214

Printed in the United States
by Baker & Taylor Publisher Services